THE SOUTHEASTERN NATURALIST AND ANTIQU
11 • SOUTHEASTERN UNION OF SCIENTIFIC SOCIETIES

Publisher's Note

The book descriptions we ask booksellers to display prominently warn that this is an historic book with numerous typos or missing text; it is not indexed or illustrated.

The book was created using optical character recognition software. The software is 99 percent accurate if the book is in good condition. However, we do understand that even one percent can be an annoying number of typos! And sometimes all or part of a page may be missing from our copy of the book. Or the paper may be so discolored from age that it is difficult to read. We apologize and gratefully acknowledge Google's assistance.

After we re-typeset and design a book, the page numbers change so the old index and table of contents no longer work. Therefore, we often remove them.

Our books sell so few copies that you would have to pay hundreds of dollars to cover the cost of our proof reading and fixing the typos, missing text and index. Instead we usually let our customers download a free copy of the original typo-free scanned book. Simply enter the barcode number from the back cover of the paperback in the Free Book form at www.general-books.net. You may also qualify for a free trial membership in our book club to download up to four books for free. Simply enter the barcode number from the back cover onto the membership form on our home page. The book club entitles you to select from more than a million books at no additional charge. Simply enter the title or subject onto the search form to find the books.

If you have any questions, could you please be so kind as to consult our Frequently Asked Questions page at www.general-books.net/faqs.cfm? You are also welcome to contact us there.
General Books LLC™, Memphis, USA, 2012.

※ ※ ※ ※ ※ ※ ※ ※

BEING THE TRANSACTIONS OF THE SOUTH EASTERN UNION OF SCIENTIFIC SOCIETIES
ALSO THE
PROCEEDINGS AT THE EIGHTH ANNUAL CONGRESS,
HELD AT
DOVER,
JUNE 11th, 12th, 18tb, 1903.

The objects of the Union are to systematise Scientific Work among the different Societies composing it, to give greater impetus to Scientific research, and, in general, to promote the study and advancement of Science by Co-operation. In view of these objects, School Natural History Societies receive special consideration and are admitted on payment of a nominal fee.

Authors are entirely responsible for the facts and opinions contained in their papers.

Headers of Papers are requested to send to the Editor, J. W. Tctt, Rayleigh Villa, Westcomhe Hill, S.E., a list of any *errata* they may detect in the present volume.

The Congress for 1904 will be held at Maidstone,

On June 2nd, 3rd, and 4th, under the presidency of

F. W. RUDLER, Esq., F.Q.S., C.I.S.O.

The Editor would he glad to exchange Transactions with other Unions and Natural History Societies. All communications relating thereto should he addressed to J. W. Tutt, 'Westcomhe Hill, S.E. CONTENTS.

Places of Meeting and Past-Presidents
   List of Officers for 1903-4
   Dover Congress Local Committee
   Rules of the S.E.U.S.S. (amended)
   Byelaws
   Botanical Research Committee
   List of Affiliated Societies, with names of Secretaries
   Delegates, etc...
   Work done by Members of Affiliated Societies, in—
   Anthropology
   Botany
   Geology...
   Zoology...
   Miscellaneous
   List of Lecturers
   Seventh Annual Report
   Balance Sheet for 1902-8
   S.E.U.S.S. Lantern Slides
   Referees
   Museum Notes—Report of the Museum Committee
   Photographic Survey and Record of Surrey
   Proceedings of the Congress, 1903
   Presidential Address
   The Seedlings of Geophytes
   A Late Keltic Cemetery at Harlyn Bay
   Atmospheric Moisture as a Factor in Distribution
   Diminution and Disappearance of South-Eastern Fauna and Flora
   International Communication
   List of Life-Members, Members, Delegates, etc., for 1903...

And the Officers and Committee of the Dover Sciences Society.

RULES. *As revised at the 4th Annual Congress, held at* Rochester, May 27th, 1899.

i. Objects.—The objects of the Union shall be to systematise work among the various Societies composing it, to give a greater impetus to research, and to promote the interests of the Societies by co-operation. 2. Management.—The affairs of the Union shall be managed by a Council and a General Committee. 3. The Council shall consist of a President, Vice-Presidents, General Secretary, and Treasurer, and seven other persons, three to form a quorum; all to be elected annually, and none except the Vice-Presidents, Secretaries, and Treasurer, to be eligible in the same position for more than two years in immediate succession. The filling of casual vacancies to be at the discretion of the Council itself. 4. The General Committee shall consist of the Council, Past-Presidents, and the Delegates. 5. Affiliation.—All Scientific Societies in Hampshire, Kent, London, Middlesex, Surrey, and Sussex, shall be eligible to join the Union, provided that the Society claiming to join comprises at least 10 members. 6. Congress.—A Congress for the furtherance of the general work of the Union, and for the reading and discussion of papers, shall be held annually in June, at such place and at such date as may be decided on by the General Committee at the preceding Congress, or, failing such decision, by the Council. 7. Delegates. —A minimum Annual Subscription of 5s., *payable in advance at leant a fortnight before the Congress,* shall entitle a Society to affiliation and a voting ticket for one Delegate at the Annual Congress. Societies with more than 50 members, exclusive of honorary members, shall if they so desire, be entitled to voting tickets for additional Delegates in the proportion of one for every additional 50 members, and one for the number (not less than 10) in excess of every multiple of 50, on payment of 5s. for each ticket. 7a.—School Natural History Societies may affiliate for a subscription of 2s. 6d. (rule added June, 1900). 8. Members.—Members of Affiliated Societies shall be admitted to the Congress on payment of 2s. 6d. 9. Associates.—Persons unattached to any Affiliated Society may, at the discretion of the Council, bo admitted to the Congress on payment of 8s. 6d. 9a. Life-Members.—Members, Associates, and other persons at the discretion of the Council, may compound for the Annual Subscription by a single payment of £2 2s. for Life-Membership. (Rule added June, 1902.) 10. General Meetings.—The meetings at the Congress shall be for the reception of reports of work and the reading and discussion of papers. 11. The General Committee shall, at some time during the Annual Congress, receive a Statement of Accounts, appoint an Auditor, elect the Council for the ensuing year (by ballot if demanded), appoint such Sub-Committees as may be required, decide on the next place of meeting, and, when necessary, revise the Rules. 12. Executive.—All other affairs of the Union throughout the year shall be managed by the Council. 13. Transactions.— Such Transactions of the Union as may be published shall be issued free to all Affiliated Societies, Members and Associates. 14. Local Receptions.—Each Society or Town inviting a visit of the Union shall appoint a Local Committee and Local Secretary to assist the General Secretary in drawing up the Programme of the Congress, which shall be arranged at least a month before the said Congress. 15. Expenses.—The expenses of printing and general management shall be paid out of the funds of the Union; those of providing rooms for the meetings of the Congress by the Society or Town issuing the invitation. 16. Changes in the Rules may be proposed and discussed at any meeting of the General Committee, but cannot be passed until the following year, unless they have been submitted to the General Secretary at least three months in advance, so that he may report the proposals to the Affiliated Societies before the Congress.

I.—The Union shall have the right, at its discretion, of printing *in extenso* in its Transactions all papers read at the Annual Meeting. The Copyright of a paper read before any meeting of the Association, and the illustrations of the same which have been provided at his expense, shall remain the property of the author; but he shall not be at liberty to print it or allow it to be printed elsewhere, either *in extenso* or in abstract amounting to as much as one-half of the length of the paper, before the first of November next after the paper is read.

2.—The author of any paper printed in the Transactions shall be entitled to receive 25 separate copies of it gratis, and to have any further number printed at his own expense by private arrangement with the printers to the Association. 3. If proofs of papers to be published in the Transactions be sent to authors for correction, and are retained by them beyond four days for each sheet of proof, to be reckoned from the day marked thereon by the printers, but not including the time needful for transmission by post, such proof shall be assumed to require no further correction. 4. Should the extra charges for small type, and types other than those known as Roman or Italic, and for the author's corrections of the press, in any paper published in the Transactions, amount to a greater sum than in the proportion of ten shillings per sheet, such excess shall be borne by the author himself. 5. A time limit of 25 minutes is prescribed for each paper, with 5 minutes for each speaker, and the discussion of the subject is to be closed at the end of one hour.

March, 1900.

BOTANICAL RESEARCH COMMITTEE.

This will be added to from time to time, so that it may embrace 2 or 8 representatives from each County in the district.

Prof. G. S. Boulger, 11, Onslow Road, Richmond, S.W.

W. H. Beeby, Hildasay, Thames Ditton.

Jas. Groves, 58, Jeffreys Road, Clapham, S.W.

E. Chas. Horrell, 58, Copleston Road, Denmark Hill, S.E.

Thos. Howsk, Glebefield, Guildford, Surrey.

Rev. E. N. Bloomfield, Guestling Rectory, Sussex.

Wm. Mitten, Hurstpierpoint, Sussex.

W. E. Nicholson, Lewes, Sussex.

E. M. Holmes, *Chairman,* Ruthven, Sevenoaks.

OF AFFILIATED SOCIETIES. (Furnished by the Secretaries.)

Balham and District

Aniiquarijin and N.H. Hoc.

Croydon N.H. and Sc. Hoc.

Brighton and Hove N.H. Soc.

West Kent N.H. Soc...

Woolwich and District Antiquarian Soc.

Holmesdale Natural History Club.

ANTHROPOLOGY.

The Archaeological Remains and Early

Thomas W. Shore,
Historical Associations of Streatham, F.G.S.
Tooting, and Balham.
On the recent discoveries (Pit Dwellings) George Clinch, F.G.S.
at Vaddon.
Hollingbury Camp H. S. Toms.
The Roman Wall A. 8. Gover.
Dartford Priory Miss Annie Sharpe.
Harlyn Bay and its prehistoric remains. Rev. R. Ashington (Published by Sonnenschein and Co.) Bullen, F.L.S., F.G.S.
Eoliths from S. and S.E. England, Rev. R. Ashington *Geological Magazine,* March, 1902. Bullen, F.L.S., F.G.S.

BOTANY.
Study of the Structure of Woods Prof. G. S. Boulger,
F.L.K., F.G.S.
The Flora of Hayes Common Dr.H. Franklin Parsons, F.G.8.
The Flora of the Eastern Border of Dr. H. Franklin Parsons,
Dartmoor. F.G.S.
Records of Haslemere Fungi E. W. Swanton (Mem.
Brit. Mycol. Soc.) Observations upon Trees; defective E. W. Swanton (Mem. exfoliation in deciduous Trees. Brit. Mvcol. Soc.)
British Oak Galls E. Connold, F.E.S.
List of Flora of the Bromley, Kent, District
Additions to List of Flora of Northwest Kent.
Verification of Records of Flora Captain McDakin.
Rust Fungi R. R. Hutchinson.
Observations of Seedlings Mrs. T. R. R. Stubbing.
Orders Coinpositu?, etc., Drawing of Mrs. T. R. R. Stebbing.
Seed Vessels.
Photographs of Wild Flowers to illus- D. Johnson.
trate structure.
The Extermination of British Plants.. A. D. Webster, F.R.S.E.
Primroses Dr. WThitney.
Insects and Flowers J. W. Bulman, B. Sc.

GEOLOGY.

The Perched Rocks of Dartmoor
Holocene motlusca from North Cornwall (Proc. Malacological Society, October, 1902).
Holocene mollusca from Cambridge, Dorset, and Kent.
Pleistocene mollusca from Portland Bill
(Proc. Malacological Society, March, 1002).

ZOOLOGY.
City of London Ent. Early stages of Microlepidoptera and N.H. Soc. Life-histoiy of *Staurapus Jagi*
Revision of the Pterophorids
Life-histories of the Hepialid Moths..
Geometrid Moths of the World
Lepidoptera of British Guiana
The importance of certain larval characters as a guide in classification..
South American Lepidoptera
Alf. Sich, F.E.S.
A. W. Mera.
J. W. Tutt, F.E.S.
J. W. Tutt, F.E.S.
L. B. Prout, F.E.S.
W. J. Kaye, F.E.S.
A. Bacot, F.E.S.
A. F. Bayne.
Life-histories of Lepidoptera (British) J. V. Tutt, F.E.S.
Natural History of British Lepidoptera J. W. Tutt, F.E.S.
(Vol. IV. in press).
Migration and Dispersal of Insects J. W. Tutt, F.E.S.
(Published by Elliot Stock, 1903)
Practical Hints for the Field-Lepidop- J. W. Tutt, F.E.S, terist (Part II. Published by Elliot
Stock, 1903».
Alpine Lepidoptera J. W. Tutt, F.E.S.
Classification of Lepidoptera Dr. T. A. Chapman.
Life-histories of Pterophorids Dr T A. Chapman.
Biological Entomology generally Dr. T. A. Chapman.
European Butterflies,. Dr. T. A. Chapman,
Eggs of Lepidoptera (Photographs)... '. F. Noad Clark.
British Odonata W. J. Lucas, B.A.
Orthoptera of the World M. Burr, B. A., F.E.S.

Hemiptera of the World G. W. Kirkaldy, F.E.S.
Life-histories of the Coleophorids H. J. Turner, F.E.S.
On the Foraminifera from the Gault at W. Murton Holmes.
Merstham.
Records of Surrey Mollusca, 2nd C. Pannell, jim., Mem.
series. Conch. S. of Gt. Brit.
Haslemere Mollusca (in Haslemere and C. Pannell, jun., Mem.
Hindhead Guide, 1903). Conch. S. of Gt. Brit. List of the Fauna of the Bromley
District of Kent.
Hydroida, especially family Sertularia.. Rev. H. A. Soames.
Additions to the Land and Freshwater
Mollusca, Lepidoptera, and Birds of North-West Kent.
Verifications of the Birds and Lepidop-8. Webb.
tera.
Plumage Miss E. L. Turner.
"Amphipoda from Costa Rica" in Pro-Rev. T. R. R. Stebbing, ceedings of United States National M.A., F.R.S.
Museum.
"Crustacea" in Victoria History of Rev. T. R. R. Stebbing,
Counties *(continued).* M.A., F.R.S.
Articles, Crustacea, Entomostraca, Rev. T. R. R. Stebbing,
Malacostraca, Thyrostraca, in Ency- M.A., F.R.S.
clopredia Britannica, 10th Edition,
"Pyenogonida " in "Knowledge " con-Rev. T. R. R. Stebbing, *tinned).* M. A., F.R.S.
TheHairof Animals viewed as a Stream. V. Kidd, M.D., F.Z.8.
Contributions to the Entomology of Rev. E. N. Bloomfield,
Sussex and Suffolk in the Victoria M. A., F.E.S.
History of these Counties.
Hastings Hydroid zoophytes P. J. Ruflord, F.G.S.
Dentition W. W. Esam, B.A.
Structure and form of shells Ed. Connold, F.E.S.
Birds Mr. Bedford.

MISCELLANEOUS.
Optics of the Camera and Microscope..

Horace F. Cheshire
 B.Sc.
 History of Greenwich Park A. D. Webster, F.R.S.E.
 The Chemistry of Photography Mr. Pendlebury.
 The Tudor Crowns and other Royal Sydney Webb.
Crowns of Britain and the Continent.
 Woolwich Ships Wm. Norman.
 Woolwich Bibliography W. T. Vincnt.
 Note.—The object of this List is not only to form a record of work done, but also to assist workers to communicate with each other—see also last year's list. Mere lists of Lectures given before Societies are not wanted for insertion, but only particulars of *original* work, whether published or not.
 Where no fee is mentioned it may be assumed that none is expected, but the travelling expenses must be paid and accommodation for the night provided if required. For Lantern Lectures the Society will please provide Lantern and operator. In future issues of this list names of ladies or gentlemen *recommended* by Affiliated Societies will be inserted. It is expected that many of them will be able to give at least the name of one capable Lecturer willing to repeat his or her lecture before other Societies in this Union.
 Lectubers.
 Mr. J. H. Allchin, The Museum, Maidstone, private address, Chillington House.
 Fee, 2 guineas, and travelling expenses.
 Prof. G. S. Boulger, F.L.S., F.G.S., Ed. of *Nature Notes,* 11, Onslow Road, Richmond, S.W.
 Fee on application.
 Mr. A. W. Brackett, F.S.I., 51, Queen's Road, Tunbridge Wells.
 Rev. R. Ashington Bullen, F.L.S., F.G.S., etc., Pjrford Vicarage, Woking, Surrey.
 Mr. Ed. Connold, F.E.S., 7, Magdalen Terrace, St. Leonard's-onSea.
 Fee, 2 guineas, and travelling expenses.
 Mr. Martin Duncan, South Park, Reigate, Surrey.

Mr. F. Enock, F.L.S., F.E.S., 13, Tufnell Park Road, London, N. Fee on application.
 Mr. R. R. Hutchinson, Hon. Sec. Tun. Wells N.H.S., 28, Prince's Street, Tunbridge Wells.
 Mr. E. A. Martin, F.G.S., 23, Campbell Road, Croydon (West).
 Mr. Paul Mathews, M.A., South Avenue, Rochester.
 Miss E. Turner, Upper Blrchetts, Langton, Tunbridge Wells.
 Mr. W. H. Griflin, (i, Rutland Park, Perry Hill, S.E.
 1. A Naturalist's Ramble on the Seashore. 2. Flesh-feeding Plants. 3. Devil Fish and Kraken; some Long armed Monsters of the Deep. 1. Wonders and Romance of Insect Life. 2. British Trap-door Spiders. 3. Insect marvels in a back garden.
 All with specially prepared lantern slides.
 1. Rust Fungi. 2. Mycetozoa. 3. Dispersion of Seeds. 1. The History of Valleys, submerged and exposed. 2. The Physical Future of the Earth. 3. The Coal Problem, geologically and economically considered. 1. Aims of Natural History Societies. 2. The Classification of Animals. 3. Artificial Languages. 1. Plumage. 1. Plant Folklore.

SEVENTH ANNUAL REPORT.

Our 7th Annual Congress met last year in the ancient City of Canterbury and received the valuable support of the Mayor, as well as that of the local Society. The East Kent Science Society was established in 1857, and is therefore one of the oldest in the district. The meeting, both as to the welcome and accommodation provided for us, was all that could be desired.
 The Temporary Museum had a spacious room allotted to it in the Simon Langton Schools, and was most conveniently arranged. The numbers visiting it proved it the most successful of our local Exhibitions. Its attractiveness was in a great measure due to the large number of specimens sent from the Haslemere Educational Museum by our retiring president, Mr. Jonathan Hutchinson.
 The visits to the Cathedral, the late Dean Farrar's house, and Wye College,

are doubtless still very pleasant memories amongst those who attended.
 Your Council has met at the President's house on two occasions since Congress. Arrangements for the management and development of the Congress Museum were considered and Mr. E. W. Swanton, Curator of the Haslemere Museum, was requested to superintend this work. He very kindly agreed to act as Museum Secretary.
 Your sets of Lantern Slides have been used four or five times by the affiliated Societies, but no additional sets have been secured. It is suggested that this branch of our work might be much helped if we offered prizes for photographic work. The Portsmouth Society has tried the plan and found it succeed. The only difficulty seems to be to find the money for the necessary medals.
 Photographic Surveys.—With a little effort it is hoped this Union may undertake in Kent and Sussex the work that has been so well begun by the Surrey Photographic Survey. A meeting to effect this object will shortly be arranged.
 The Corresponding Societies' Committee of the British Association is anxious to get all local Nat. Hist. Societies on its list. Our Union has been on from the first. Beyond sending a delegate to the Brit. Assoc, meetings it is doubtful how far this Union can help. It urges all local societies who publish Transactions to apply at once; they will then receive all papers direct.
 Our Annual Congresses cannot be expected to accomplish all that they should do in promoting co-operation and giving a stimulus to scientific work until the railway companies look upon us with more favour and recognise the value of the work we have undertaken. When they do they will grant us concessions for travelling, which hitherto they have steadfastly denied us. This year they have again refused unless we guarantee to take 200 tickets on the same line. Many Societies in the country and some even in the S.E. district enjoy the privileges which we have asked for in vain.
 The number of Affiliated Societies shows a further increase to 41; the Re-

gent Street Polytechnic Nat. Hist. Soc, Hastings Collegiate School, and the Maidstone Nat. Hist. Soc. have joined our ranks. If we omit Loudon, there is now in our district only one Society that can be reckoned as an outsider.

The list of Life members has also been augmented, amounting now to 17. Local Societies can help very much in getting additions to this number.

The Treasurer's statement shows, for the first time, a satisfactory balance in hand, yet this does not mean that we have an income sufficient for our requirements, for in many ways our work has hitherto been limited by want of funds. A special contribution of £3 8s. from the Rev. R. A. Bullen has enabled your Council to engage an Assistant Secretary. This it is proposed to continue so long as funds permit.

The Yorkshire Union of Nat. Hist. Soc. was invited to send a delegate to this Congress but was unable to accept this year.

BALANCE SHEET, 1902-8. RECEIPTS. £ a. A. £ s. d.
Balance brought forward 0 9 1
Canterbury Congress.—
Proceeds of Tickets sold by Local Society 15 0 0
Dek-gutcs' Subs., 1901 2 6
„ 1902 12 7 6 Delegate's Subscription outstanding Members' Subscriptions Associates' Subscript'ns Donations.-R. Harrison for Plate in Transactions Rev. It. A. Bullen for Assistant Secretary 3 3 0 3 15 6 Kale of Transactions and Reprints 0 13 9 EXPENDITURE.
£ s. d.
Printing, 4c—Dee..070
Saunders 0 2 6
Baldwin 6 4 0
Carriage of Parcels
Printing Transactions,
300 copies, and reprints
Stamps and Postage..
Assistant Secretary
Delegate's Subscription outstanding..
0 12 6
£40 19 10
Balance in Hand 0 2 6 1 O 11 £40 19 10
June 3rd, 1903.

Examined with the Vouchers and found correct.
R. R. HUTCHINSON, *Auditor.*
S.E.U.S.S. LANTERN SLIDES.
The following sets of Lantern Slides are available for use by affiliated Societies on application either to the General or Photographic Secretary: 1.—*Some British Orchids* (50 slides) contributed by Mr. S. Horsley,

M.I.C.E., with explanatory lecture. 2. —*The Gault and Lower Greenland* (about 80 slides), with lecture. 3.—*The Wealden Formation* (about 50 slides), with explanatory notes. 4.—*Ice Flowers and Crystal* (small set), with explanatory notes by G. Abbott.

No charge is made except for carriage both ways. The orchid slides are a new and interesting set, dealing with the general and detailed structure of many British species and their adaptation to insect-fertilisation.

The Society for the Protection of Birds, 826, Holborn, W.C., also lend *to their subscribers* very beautiful Lantern Slides relating to Birds, which are well worth the attention of Secretaries and Lecturers.

The notice of Secretaries is particularly called to the suggestions made in the Photographic Secretary's Report that the Union should solicit loans or gifts of *small* sets of slides (a dozen or so) illustrating *any* particular scientific phenomenon or limited branch of scientific work. Such sets, with full explanatory notes, to occupy about half-an-hour for exhibition, would doubtless be much appreciated for the purpose of soirees or other occasions in which time is necessarily limited, while two sets might furnish material for an ordinary evening meeting. Many members of our Societies possess sets of this character which they have prepared for their own use, and which they would be willing to lend for the use of the affiliated Societies. Secretaries are hereby asked to furnish the Photographic Secretary as soon as possible with the names and addresses of any of their members who, in their opinion, might be induced to cooperate.

Contributions are still solicited towards the following larger sets that are in course of formation: 1.—*Pre-Historic Man in S.E. England.* 2.—*English Wild Flowers,* with special reference to forms of capsules and their dehiscence. 3.—*Photomicrographs.*
4.—*Coast Erosion in S.E. England.*

Contributions of lantern slides may be sent to the Photographic Secretary, Mr. H. E. Turner, B.A., B.Sc, Lindfield Lodge, Folkestone, who will be glad to give any information as to this branch of the work of the Union.

REFEREES. BOTANICAL. (Additions to this list would be welcomed).

The following gentlemen have kindly consented to name a limited number of specimens for our Members and Associates.

(A stamped directed envelope should always be sent, or no replies need be expected).

Cryptogams (not microscopic).— Thomas Howse, Glebefield, Guildford.

Freshwater Alga;.—W. West, 15, Horton Lane, Bradford.

Specimens of species of *Zygnema, Spirogyra,* and *Mougeotia* should be fruiting. They are best sent in small tubes in water. Habitat must be always stated. Permanent reedy ponds and ditches yield the best results, especially those where *Utricularia* occurs. Plants like *t'triculuria,* leaves and peduncles of *Nymphaea, Nuphar,* Ac, might be sent in tin boxes; and Mr. West will examine these for minute forms. Gelatinous or slimy coverings of damp, shady, or trickling rocks should also be sent in small tins.

Marine Alga; (excluding diatoms and desmids).—E. M. Holmes, Ruthven, Sevenoaks.

Fresh algaa should be rolled separately in old muslin or calico, so that one plant does not touch another; then packed so as to be free from pressure in tins or boxes.

Mo.sses.—W. Mitten, Hurstpierpoint, Sussex.

Phanerogams.—A. Bennett, 148, High Street, Croydon, and Rev. E. N. Bloomfield, Guestling, Sussex.

In some orders like *Crueifcrae, Cyperuceae,* and *Umbelliferae,* the fruit is almost a necessity.

Cyperaceae.—A. Bennett, 148, High Street, Croydon.

Mieracii.—Rev. W. R. Lynton, Shirley Vicarage, Derby.

Rubi.—Rev. W. Moyle-rooers, Chetnole, Grosvenor Road, Bournemouth West.

The work of the *Itiilri* referee would be very greatly lightened, and his determinations and suggestions proportionately more satisfactory, if correspondents would send only good *representative* specimens, and place them always *on paper stout enough ami large enough to bear them safely.*

There is, of course, least room for uncertainty when panicles show both flower and fruit, and the stem pieces mature leaves from about the centre of their length; other less satisfactory pieces *in addition* may sometimes help a referee. But if in any case a correspondent can feel justified in sending only such pieces, he should at least press them carefully and supply them in extra quantity.

In explanation of the term "stem" and "panicle," it is to be remembered that all the fruticose *liubi* throw out long leafy shoots (the "stem " or "barren stem") directly from their roots, which, normally, produce no flowers in the first year. From these spring, in the following year, the flowering shoots, and it is to the flowering part of these, including all the branches and branchlets, that the term "panicle " is applied.

No "specimen" can be determined with certainty, unless it consists of both panicle and stem pieces; and it is often of great assistance to the referee when the label accompanying the specimen contains—in addition to the usual memoranda of locality, county or vice-county, date, c.—further notes made from the living plant, of the colour of the flowering organs and the comparative length of styles and stamens, with any other conspicuous character lost in the process of drying.

ZOOLOGICAL.

Diptera.—Rev. E. N. Bloomeield, Guestling, Sussex.

Tenthredinidae.— Ditto.

He will also be glad to hear of any "finds" in either the Fauna or Flora of his part of Sussex.

Coleoptera.—H. St. J. K. Donisthorpe, F.Z.S., F.E.S., 58, Kensington Mansions, South Kensington, S.W.

Lepidoptera.—J. W. Tutt, F.E.S., Rayleigh Villa, Westcombe Hill, S.E.

Orthoptera.—M. Burr, F.Z.S., F.E.S. , 15, Fitzjames Avenue, West Kensington, S.W.

Galls.—E. Connold, Hon. Sec. of Hastings Nat. Hist. Soc, 7, Magdalen Terrace, St. Leonard's.

Hydroida (Calyptoblastea).—Bev. H. A. Soames, Hawthorns, Otford, Sevenoaks.

MUSEUM NOTES.

The temporary museum, which, as quite a new departure, was instituted last year, has again proved a success; indeed, it may be remarked that it has now more than justified its existence. The exhibits were doubly as numerous as those of last year, and (this is extremely gratifying) there was little or no tendency to exhibit the same things two years in succession, an evil which—as everybody knows who has had anything to do with museums of this kind—is a very insidious one. If the museum once becomes stereotyped, then both its interest and value are lost.

We are deeply grateful to all exhibitors for the trouble they have taken.

Special thanks are due to Messrs. Sydney Webb, W. H. Griffin, and B. Lowne, not only for bringing such a magnificent series of exhibits, but also for their very generous help in the management of the same.

The Union would always greatly appreciate help from the museums of the various towns visited, especially towards illustrating papers of a strictly local character. For example, special emphasis and interest would have been given to the paper by Mr. Sydney Webb and Captain McDakin on the "Diminution and Disappearance of S.E. Flora and Fauna within the Memory of Present Observers," had it been illustrated by specimens from the local collection at the Dover Museum. The exhibits were as follows:

British Plants: Mr. B. Lowne, Catford.—A novel feature of the exhibition was a part of Mr. B. L. Lowne's herbarium of British plants, mounted to show various stages in life-history (see Figs. 1 and 2). The educational value of this collection cannot be over-estimated. Mr. Lowne considers a sheet complete when it contains, not merely the mature flowering plant, but (1) the plant (or part) showing a fruit capsule; (2) two or more leaves pressed separately so as to show a perfect outline, and upper and lower surfaces; (8) two or more flowers pressed separately, flat and sideways; (4) an unopened flower; (5) parts of the flower (when practicable; (6) the seed; (7) the cotyledons; and as many stages of the subsequent development as possible. We have pleasure in appending a list of the plants (808 in number) which Mr. Lowne has already mounted in this manner. It is noteworthy that all were raised from seed in the exhibitor's experimental garden at Catford:—*Clematis ritalba, Ranunculus dammula, R. sceleratus, It. auricomu, R. arvensis, (altlta palustris, Paparer somniferum, P. hybrid urn, Chelidonium niajus, Glaucium liiteum, Eumaria officinalis, Corydalis lutea, ('. claviculata, Cheiranthus cheiri, ISarbarea ritlfjari, Nasturtium officinale, X. palustre, Cardamine hirsuta, Hesperis matronalis, Sisymbrium officinale, Alliaria officinalis, Erysimum clieirantluddes, llrassica muralis, 11. niyra, Alyssum maritimum, Draba rerna, Thlaspi arvense, Capsella hnrsapasturis, Lepidium draba, L. rnderale, Senebiera cormwpus, Reseda luteola, 11. lutea, 11. alba, Helianthemum vulyare, Viola odorata, V. canina, V. tricolor, Dianthnt armeria, Saponaria officinalis, S. raccaria, Silene cucubalus, Lychnis vespertina, L. diurna, Sayina procuiubenx, Arenaria serpylli folia. A. trinenis, Cerastium vulyatum, Stcllaria media, S. uliyinosa, S. holostea, Speryularia rubra, S. arvensis, Claytonia perfoliata, luntia fimtana, Hypericum perforatum, 11. Itirsittum, II. quadranyulum, 11. lutmifitsum, 11. pxdehrum, Linum catharticum, Malra rotnndifolia, 31. sylrestris, M. moschata, Geranium dissectum, G. pratense, G. molle, G.*

pyrcnaicum, G. robertianum, G. litcidum, G. pusilliim, O. columbinum, Erodium cicutarium, Inijiatiois fulra, Acer camestris, A. pseudoplatanus, Ilex aquifulium, Euonymus europaens, Ulejc europaeus, Genista anylica, G. scopariis, Ononis arrensis, Medicago falrata, 31. lupulina, 31. satiia, 31elilotus artensis,3I. alba, Trifolium incarnatum, T. arvense, T. pratense, T. frayiferum, T. minus, Lotus corniculatus,L.uliginosus, Anthyllis vulneraria, Omitlwpmperpusillus, Vicia hirsiita, V. tetrasperma, V. cracca, V, satira, V. amjustifolia, Lathyrns aphaca, L. protends, Spiraea ulmaria, S. jilipendula, Geum urban/tut, G. rivale, Rubus fruticosus, Potentilla tormentilla, P. aryentea, Alchemilla arvensi», Poterium sanyuisorba, Ayrimonia eupatoria, Rosa arrensis, R. canina, Pyrus mains, P. anciiparia, Crataegus oxyacantha, EpUobium am)ustifolium, E. hirsutum, E. rose urn, Circaea lutetiana, Lythrum saliearia, Peplis portula, Bryonia dioica, Cotyledon umbilicus, Sedan) acre, Ribes yrossularia, Saxifraya yramdata, S. tridactylites, Hydrocotyle. vulgaris, Sison amonium, Pimpinella saxifraya, Bupleurum tenuissimum, (Enanthephellandrium, JEthma cynapiuiu, Foeniculnm rulyare, Pastinaca sativa, Scandix pecten-reneris, Choerophyllum sylrestre, C. anthriscus, C. temulum, Caucalis nodosa, C. anthriscus, Daucns carota, Smyrninm olusatrum, Hedera helix, Adoxa moschatellina, Sambucus nigra, Lonicera xylosteum, Galium veruin, G. saxatile, G. mollugo, G. aparine, G tricorne, Asperula odarata, Sherardia arrensis, Centranthus ruber, Valeriana dioica, V. officinalis, Valerianella olitoria, V. dentata, Dipsacus sylvestris, D. pilosns, Scabiosa suceisa, S. columbaria, Eupatorium cannabinum, Eriyeron arris, Solidayo viryaurea, Bellis perennis, Filayo yermanica, Gnaphalium sylvaticum, G. uliginosum, Inula conyza,I. dysenterica, Bidens tripartita, Matricaria chamomilla, Anthem is arrensis, A. tinctoria, Achillea millefolium, Artemisia absinthium, Senecio vulgaris, S. sylvaticus, S. aqnaticus, S. jacobaea, S. erucitolius, Carduus nutans, Centaurea scabiosa,

C. calcitrapa, Tragopogon pratensis, Helminthia echioides, Picris hieracioides, Lactua muralis, L. scariola, Sonchus oleraceus, Taraxacum densleonis, Hieracium sabaudum, Lapsana communis, Jasionc montana, Phyteuma orbiculare, Campanula trachelium, C. rotundifolia, C. hybrida, Primula vnlyaris, P. rem, Anayallis arrensis, Samolus ralerandi, Eraxinus excelsior, Ligustrum vulgare, Vinca major, Erythraea centaurium, Gentiana amarella, Chlora perfoliata, Convolvulus septum, Echiiim vulgare, Litlwspermum officinale, Myosotis palustris, M. arrensis, M. collina, M. versicolor, Lycop.iis arrensis, Borayo officinalis, Cynoylossum officinale, Hyocyamus niger, Solanum dulcamara, S. nigrum, Atropa belladonna, Verbascum thapsus, V. lychnitis, Linaria vnlyaris, L. minor, L. cymbalaria, L. spuria, L. elatine, Scrophularia nodosa, S. aquatica, Veronica serpyllifolia, V. officinalis, V. anayallis, V. beccabunya, V. hederaefolia, V. buxbaumii, Bartsia odontites, Rhinanthus crista-galli, Melampyrum pratense, Salvia verbenaca, Lycopus europoeus, Mentha aquatica, Thymus serpyllum, Oriyanum rulyare, Calamintha acinos, ('. officinalis, C. clinopodium, Nepeta cataria, Prunella vnlyaris, Scutellaria yalericulata, S. minor, Marriibinm vulgare, Stachys sylratica, S. palustris, Galeopsis ladanum, G. tetrahit, Ballota nigra, Lamium amplexicaule, L. album, L. maculatnm, Teucrium scorodonia, T. botrys, Verbena officinalis, Plantayo major, P. lanceolata, P coronopus, Scleranthus annuus, Suaeda maritima, Clwnopodiumpolyspermum, C. bonus-henricus, Rumex acetosa, Polygonum aviculare, P. convolvulus, P. hydropiper, P. bistorta, Euphorbia lielioscopia, E. peplus, E. exigua, Mercurialit annua, Vrlica urens, U. diuica, Parietaria officinalis, Hamulus lupulus, limits montana, Abuts glutinosa, Fatjtut sylcatica, Quercus robur, Taxus baccata, Potamoyeton densus, Triylochin maritimum, Btitomus nmbellatus, Alisma plantago, A. ranunculoides, Iris pseudacorits, 'Taunts communis, Asparagus officinalis, Srilla nutans, Litxttla maxima, Scirpus sylvaticus, Panicum crus-galli, Lagurus oratis and Scolopendrium vulgare.

Land And Freshwater Mollusca.— Mr. Lowne also exhibited about 78 species of shells and land and freshwater Mollusca, collected in West Kent, 1901-8, including, amongst others:— Anodonta cygnaea (shell measuring 7 X 8j inches), Pisidium roseum, Xtritina fluviatilis, Yalrata cristata, Planorhis nautilus, P. nitidus, Physa hypnorttm, Umax loevis, Testacella haliotidea, Yelletia lacustris, Vitrea fulva, Helix pom a tia, H.pygmaea, 11. aculeata, Azrca tridcns, Vertigo edentula, V. pygmaea, Balea perversa, Acltatiua acicula.

West Kent And Surrey Plants: Mr. W. H. Griffin.— Mr. Griffin's collection of West Kent and Surrey plants attracted much attention, being mounted with exquisite care. The colours were, in most cases, so well preserved, that, viewed at a slight distance, the sheets presented the appearance of a series of coloured drawings rather than actual herbarium specimens, which are, in only too many cases, calculated to repel rather than attract. We regret that want of space makes it impossible to give a list of the many rare and local plants exhibited by this indefatigable collector. Mr. Griffin also exhibited a series of various Eolithio Implements from the North Downs, Kent, collected by Mr. Benjamin Harrison and himself; one found at West Wickbam in March, 1908, was remarkable for its great size.

Lki'idoptera: Mr. S. Webb, Dover.— The British butterflies, with the exception of the Skippers, 22 cabinet drawers; and the Sphingids, Sesiids, Hepialids, Lachneids, Dimorphids, and Arctiids, 42 drawers in all, the specimens exhibiting marked aberrational forms. Many of the more important, however, were unfortunately without the detailed data, which are now considered so exceedingly valuable. Many of these specimens were formerly in the "Harper," "Gregson," and "Bond" collections. The exhibit at the congress was altogether an extremely rich one, comprising many variations of colour and aberrations of markings in the series of some of the species. So-called hermaphrodites abounded, there being no less than seven *Gonepteryx rltamni*, eleven *Ettchloe cardamines,* twenty *Polyonnuatiis icarns(alexis),* fourteen /'/eftriMxav/ «M,ifcc.,aswellasspeciniensexhibiting similar phenomena in *Pajdlio machaon, ('olias edusa, Melanan/ia galalltea, Hipparchia semele, Polyomvtattts corydon, Amorplta populi, Smt'rintltusocellata, Portitetvia dispar, Dasycltirafascelina,Malacosoma castrensis, Pachygastria trifolii* (? imago with? body), and *Lasiocampa quercii.* Other specimens of biological interest showed males resembling females in coloration, and *vice vena,* without any apparent mixing of the sexes in the body, others only partly doing so, emphasised by example of *H. semele, C. edusa, C. hi/ale,* and the Polyommatids, *Dryas paphia, Aylais urticae* (with larval head but maxillary palpi). The well-known dark, greasy, and partly metallichued *A. urticae,* which is generally associated with the larva having fed upon dry food", was well represented, and the same variety exhibited in *Pyrantels atalanta, Vanessa io,* and *Melampias epiphron.* Increase or disappearance of ocelli in *Hipparchia,* pale centres to the spots in *Argynnis adippe,* and silvery-blue centres in the marginal underside ones of *Polyommatus eorydon* and *P. bellargus,* fittingly connected these species with *Plebeius aegon.* The series of I', *io,* large numbers of aberrations of which have recently been artificially produced by temperature experiments, exhibited specimens with more or less obliterated "eyes," whilst' one example with two ichneumons is said to have emerged from the same chrysalis, and is normal in coloration. There was also a pale-bordered form (of which we see a counterpart in *Aglais urticae)* from the " Bond" collection; this form is known to occur in Madeira, but the example from Mr. Bond's collection was taken long before the Madeiran form, as it is now called, was known to exist; another has apparently attempted to set up a fascia across the wing, and others, again, have a large black spot on the lower wings in place of the usual peacock-hued one. The blue in these ocelli are in some of these specimens most strikingly replaced by emerald green. Mottled specimens of *Apatura iris, Limenitis sybilla,* and *Kpinephele janira* were said to have emerged (in confinement) during a severe thunderstorm; ochreous-tinted *Aglais urticae* upon an exceedinglyhot morning; dull-coloured *Colias edusa,* that emerged after a spell of very cold weather in October, 1887, the dull coloration especially noticeable in the dark patches towards the end of the wings, and a general look of having been dipped in oil, but the orange pigment of the wing scaling was unchanged; pale-coloured *Polyommatus bellaryus,* also connected with emergence in cold weather.

Numismatics: Mr. S. Webb, Dover. —A very choice and wellselected collection of the larger silver pieces of English money, *e.g.,* three pound and three half-pound pieces of Charles I., struck at the mints of Oxford and Shrewsbury; also crowns in silver, from the first issued in 1551 to the close of the late reign, a number of halfcrowns, a few siege pieces of rather smaller dimensions and represen We always associate it with an unhealthy larval condition, the larva; producing such usually having been shut up in an unventilated tin or breedingcage, and having been allowed to " sweat."—Ed. tative values— 280 specimens in all. Amongst the crowns was a very fine Edward VI. of the first coinage; a Portcullis dollar of Elizabeth struck for the West Indian Colonies; an Armada dollar of Philip II. of Spain, in which he styled himself King of England; the finest known Exeter crown; Oxford crown; Cromwell, without flaw in die, most rare; Ormond crown; Anne, made of silver from Wales; Jamos VIII., Old Pretender; George I.; pattern crown of George III., known as Mudie's; Spanish dollars counterstruck with 2d. and Id. heads of George to make them pass current in 1797; pattern of 1817, England, Scotland, and Ireland as three draped female figures symbolical of the union, very rare; Whiteave's pattern of George IV., very large head; Pistrucci's pattern, extremely rare and brilliant; and Victoria Gothic crown, dress plain and undiapered, also rare, and in perfect condition. The half-crowns and pieces of somewhat similar value had amongst them a fine specimen of Edward VI.; portcullis of Elizabeth; and half-dollar Philip II., as already mentioned; an assortment of Charles I. half-crowns of the Tower, Aberystwith, Bristol, Chester, Oxford, Shrewsbury, Salisbury, Weymouth, Worcester, York; and numerous coins classed as uncertain from a doubt as to where they were issued during the troublous times when the mint masters, like Charles himself, had to be constantly upon the move. Of these, perhaps, the most striking, although the most crude, were the "Blacksmith" half-crowns, so called from their coarseness of execution, three distinct types of which were shown; also an unique specimen of rough work with the horse like a buffalo, and, as a conclusion to the Charles series, siege pieces of towns and castles that had been held for him by his adherents, such as Newark, Pontefract, in its three varieties, during his life *(iiiiiii spiro n/iero),* after he had fallen into bis enemies' hands (/wt *mortem jiatrix pro lilio),* and after his death (*'arolns secondnx)* the last three coins were represented by pieces of silver of the value of Is., and the more rare Scarborough Is. 9d., cut out of a silver platter, and 2s. Gd. a piece of silver turned in at the side and end to cause the thin plate from which the coin was made to adhere closely together. An Inchiquin crown; a Carlisle piece of the value of 8s.; a rebel coarsely executed half-crown before the parliamentary forces acquired supreme power; and a specimen of gun money of James the Second of the size of half a crown, but struck from the five shilling die in silver, not in copper, completed this interesting Stuart exhibition. Commonwealth patterns by Raniage and Blondeau; Cromwell, showing the frosting on neck and garment, and thence specimens representing all the reigns to the present day.

A Series Of Original Colourkd Drawings (by Mrs. Procter Hutchinson): Mr. Jonathan Hutchinson, F.R.S., F.R.C.S. ,LL.D., Haslemere.—These were illustrative of botanical subjects: I. *Witches' Brooms.*—(a) Large example from a Scots' pine at Haslemere. Such growths are by no means common on this conifer, though several examples have come under our notice during the past two or three years. Their cause is not properly understood, but is said to be induced by a fungus, (b) Witches' broom on silver fir, caused by *Peridermium elatinum*. The fungus caused cankered swellings at the base of a branch, which was much dwarfed in consequence, and produced abnormal erect branchlets. The needles upon these were dwarfed, yellowish, and deciduous. The cecidia occurred upon the needles only, and not upon the

Fio. 1.—Ivy (hedera Helix). branches, as in other members of the genus *Peridermium*. We shall be pleased to receive at the Haslemere Museum any information respecting " witches' brooms." II. The work of *Nepticula aitvclla* and other leaf-mining insects.

III. Canadian galls. IV. Some oak galls. The rich tints of many of our galls were beautifully reproduced in this series of illustrations.

V. Proliferation or prolification—(a) in the common larch; (/) in a rose; (e) in the spruce fir gall *(Chermea ubietis)*, which simulates *a* cone.

VI. Branch of the common laurel, which had developed roots at a point about 2ft. from the ground, and upon which it had never rested. These adventitious roots had pushed away the bark; viewed superficially, they much resembled a fungus. It is noteworthy that similar roots in the common ivy never break the cortex. VII. A melon found (upon section) to contain germinating seeds. The rind must have been transparent, chlorophyll being present. VIII. /Ecidiospores of *l'ucciiiia caricis* on the common nettle, causing distortion of the stem. It is the *Aeeidium tuticae* of the 4th edition of Cooke's *Micro-funyi*. It is very common.

Flo. 2.—Ml'sk MALLOW (malva Hobcuata). IX. Potato flowers infested by *Phytophthora infitans*. A potato field was attacked by the disease in August, 1902. A week previously it was in full flower and apparently quite healthy, the flowers dropping off normally. The disease spread with amazing rapidity over the whole field. The fungus attacked the petals of the flowers, and extended centripetally, blackening the footstalk. The attacked flowers shrivelled and became brown, but did not drop off.

Educational Exhibits.—Mr. Hutchinson also loaned from his educational museum at Haslemere: (1) A skull of a Babirusa pig from the Celebes, illustrating excessive development of canine teeth. (2) A cuckoo's egg in nest of the Great Shrike. It much resembled, both in size and markings the eggs of the Shrike. (8) Skeleton of a mole, to show modification of structure consequent upon fossorial habits. The palmar surface is increased by the presence of the bone known as the *os falciforme;* the humerus is large and ridged to accommodate the powerful digging muscles. The pelvis is compressed to prevent lateral expansion of the hind legs, which are comparatively weak.

British Lepidoptera: Mr. C. P. Pickett, F.E.S.—Three cabinet drawers contained *Awjerona pnmaria* (over 600 specimens), the result of four years' interbreeding between a dark-speckled ? taken at Eaindene Wood, Folkestone, and a banded female bred from larvro obtained in Epping Forest. Some of the forms were very remarkable. Two have been named (1) *Pickettaria*, a form with bands much broken up in both ? and 5; (2) *Pallidaria*, a very pale-yellow banded form, with bands only just discernible occurring in both ? and J. One drawer of *Lymantria monacha*, showing results of ten years' interbreeding (1892 to 1902), with the idea of getting a dark race. Each year's results are shown, with the carefully-selected parents of each brood. From the fifth year (we quote from Mr. Pickett's notes) they began to show signs of becoming suffused, and have kept so, the tenth year's imagines being very dark (but not the colour of the usual black types), forewings grey-black, hinder slate-black, body of same markings and colour as ordinary type. This strain is still continued. One drawer of *Polyommatus con/don*, taken at Dover during the last eight years, containing many aberrations, also a dwarf race of males and females. Some of the males are exceptionally large, and vary in colour from chalky-blue to a blue approaching that of *Polyommatus bellargm* var. *adonis*. One cabinet drawer of *Mimas tiliae* (mostly bred from dug pupie from many localities), containing many aberrations. We may mention two males with coloration of female; one female with male coloration; a male with usual spots on left side replaced on right side by a deep band; a female with usual spots rust-red instead of green, the same hue being suffused over the whole of the wings.

Electrical Apparatus: Mr. W. H. Holt (Dover Sciences Society).—Model of a fog signal (with bell) made with the works of an old eight-day clock; magnetic motion produced by using the magnetic pole of the earth as a field magnet. An American electric motor producing a power equal to the eighth part of the strength of a man. A motor of same size, as above, but of only one-eighth the power. Excepting the two motors, all the material in connection with the exhibits was of home manufacture, and displayed much ingenuity and skill on the part of the demonstrator. Mr. W. Holt attended during each evening and explained his apparatus.

Horsham Museum Society.—Nine photographs of salt crystals.

Photographs And Specimens Of Magnesian Limestone: Mr. George Abbott, F.G.S., Tunbridge Wells.—These were to illustrate progressive development in inorganic matter, probably without the aid of organisms. As all are doubtless aware, Mr. Abbott's investigations in this direction have been patiently carried on for savaral years past. His well-labelled specimens and large series of photographs are always of the greatest interest.

Diagram Of The Boring At Dover: Captain McDakin, Dover.— Showing in detail the beds passed through. Pebbles of chalk gravel, found in Dover Valley, 140ft. above sea-level. Map to illustrate local geology.

Ego Ball Of Copris Lunaris From Guernsey: Rev. E. W. Bloomfield, M. A., F.E.S.—This species is nearly allied to the sacred *Scarabaeus* of Egypt.

Geological: Mr. Edward G. Martin, F.G.S., Croydon.— Pseudomorphs in flint; fossils from the Folkestone gault; cretaceous fossils from Texas and Greenland. Mementoes of Gilbert White, including photograph of the family bible, and photograph of a window at Crondall, Farnham, showing arms of the White family.

Micro-photographs Of Eggs Of Lepidoptera: Mr. W. H. Hammond, Canterbury.—Showing development of the embryo.

Drawings: Mr. E. W. Swanton (Mem. Brit. Mycological Soc), Haslemere.—A series of original coloured drawings, illustrating decay of leaves, induced by attacks of micro-fungi, etc. Coloured sketches of abnormal fungi, including *Hypnum repandmn*, with hymcnial surface on the top of the pileus as well as in the normal position beneath it. It was found growing in the loose soil (lower greensand) of a bank, and was well sheltered by an overhanging part. Becoming partially detached, it grew pileus downwards. As a consequence of this inverted position, the spires were all turned in a direction opposite to that normally assumed, and a hymehial surface, with well-developed spires, was produced on the (normally) upper surface of the pileus.

86 Photographic Illustrations Of Vegetable Galls, Plant VagaRies, Etc: Edward Connold, F.E.S., Hastings Nat. Hist. Soc.— Galls due to various fungi, some kinds causing dark-brown scabrous pustules, mostly ovoid or circular in shape. Another kind, causing concave-convex fleshy swellings of the blade of the leaf of *Popnlim niijra*, the cavity being lined with a beautiful golden yellow fungus. The fungus, *Aecidium elatinum*, causes galls exceptionally variable in the forms they assume, both while in growth and when mature. It attacks various trees; the specimens illustrated were on branches of *Picea nobilin*. Another form was seen in peculiarly shaped swellings on the roots of *Alnus ylutinosa*, caused by *Schinzia alni*. It had been found in the Tunbridge Wells district several years ago, and quite recently in considerable abundance at Hastings by Mr. Connold. Of the multitudinous galls, caused by Diptera, a few choice and selected examples only were shown, two of the most striking being the larvae of *Diplosis loti*, causing the leaflets of *Vicia sepium* to swell into pods, and the larviB of *Cecidomyia sisymbrii*, causing the flowers of *Barbarea vulgaris* to develop abnormally and the petals to remain unfolded. The *Acarina* or Mites were seen to be responsible for a fine series of galls. The flowers of *Scabiosa columbaria*, of *Origanum vulgare*, the flower pedicels of *Fraxinus excelsior*, and the staminate catkins of *Corylu avellana* being amongst the most noticeable examples. The attacks of various species of *Aphides* yielded some interesting and instructive illustrations. Fasciated stems of several plants were also portrayed in half-plate photographs, as well as leaves rolled by the beetle *Attellabus curculionoides*, and by the larvre of the moth *Cleodora cytisella*.

Photographic Survey Of Surrey.—A selected collection of about 50 prints in platinotype, representing work done by the Photographic Survey and Eecord of Surrey. The work is divided into six sections, each worked by a committee with a chairman, comprising (1) art and literature, (2) anthropology, (8) geology, (4) natural history, (5) architecture, (6) scenery and passing events, and the prints exhibited were representative of each section. The object of the exhibit was to show what had been done in Surrey (over 800 prints in all had been sent in so far), and to stimulate other counties to undertake similarly useful work. This year Kent was chosen as the Congress met in that county, where it will also meet next year, but it is hoped that Middlesex, Sussex, and Hampshire will in due course follow suit. One or two counties, notably Warwickshire, in the Midlands, have already produced useful work, which will, of course, become more valuable as old landmarks and buildings gradually disappear. The collection was brought to Dover by Mr. J. H. Baldock, F.C.S. (one of the delegates from the Croydon Natural History and Scientific Society), who was anxious to say a few words about it, explaining its objects, etc., but unfortunately time did not permit of this.

It may be noted that the committee will be glad to receive information of the existence of objects of interest falling within their province, since a great deal that is most valuable is in private hands, and may otherwise remain unknown to them.

We are indebted to the respective exhibitors for the majority of the notes.

In concluding this brief commentary, we would ask secretaries of our affiliated societies to kindly bear in mind the claims of the museum, and to impress upon their members the necessity of preparing exhibits for the Maidstone meeting. All readers of papers should certainly, whenever it is practicable, bring specimens to illustrate the same. We want the temporary museum to be primarily the focus of original scientific work done by members annually, and, secondly, of *educational* interest and value to the inhabitants of the towns we may visit.

E. W. SWANTON.

PHOTOGRAPHIC SURVEY AND RECORD OF SURREY.

This survey, though not the first in the field, Warwickshire having preceded it, was established in 1902, under the Presidency of Viscount Middleton, the Lord Lieutenant of the county. IU object is to collect photographic records of anything of interest in the county of Surrey. The officers consist of an Hon. General Secretary, an Hon. Survey and Record Secretary, a Treasurer and Curator, acting under a Council and Committee. The work is divided into six sections, each section having an Hon. Sec. and Committee, and comprises the following, *i.e.* :— 1. Architectural. 4. Natural History.

2. Art and Literature. 5. Geological. 8. Anthropological. 6. Scenery and Passing events.

Members having photographs of

houses, streets, etc., which have disappeared, are especially invited to send copies. Objects which are doomed to disappear come next, while objects whose disappearance is not very remote constitute a third class. Inasmuch as the collection is intended to be *permanent,* platinum or carbon prints are most recommended, though well-prepared bromides will not be refused. Silver prints, by reason of their tendency to fade or discolor, are not very suitable for the purpose.

It is hoped that this, which may be called " photography with a purpose," will appeal to a large number of photographers, especially amateurs, as providing some definite object, instead of the promiscuous taking of negatives, which only cumber the house, and, presenting little or no permanent interest, are put aside and forgotten.

The S.E.U.S.S. appeals to all its members, and to its affiliated societies in Kent, Sussex, and Middlesex, to assist in forming a collection for those counties which cannot but be of great value, especially in time to come. The Surrey Society, though so recently established, already has a collection of over 800 prints, and its Hon. Secretary will be pleased to give any information in his power to any society wishing to start a survey in either of the counties referred to.—J. H. Baldock. PROCEEDINGS OF THE CONGRESS OF THE SOUTH-EASTERN UNION OF SCIENTIFIC SOCIETIES, 1903.
Thursday Evening (Captain McDakin in the chair).
June 11th, Thursday Evening.—At 8 p. m. a very well-attended meeting took place in the Town Hall, under the presidency of Captain McDakin, in the unavoidable absence of the President, Mr. J. Hutchinson. Among those present were the past-presidents the Reverend T. R. R. Stebbing, F.R.S., Professor G. S. Boulger, F.L.S., and Mr. W. Whitaker, F.R.S., the President-elect Sir Henry H. Howorth, K.C.I.E., F.R.S., General Newmarch, and many others interested in the work of the Union.

Captain McDakin opened the proceedings by extending a warm welcome on behalf of the Dover Sciences Society, to the members of the South-Eastern Union, which he stated now includes some 41 Societies.

The Rev. T. R. R. Stebbing then delivered a brief address. He said he had to perform a small duty to two men of great distinction —one of them the outgoing President and the other the incoming one. Mr. Stebbing then made complimentary reference to the work of Mr. Hutchinson, and stated that they had in their incoming President a man so well-known in science, literature and society, that he need not waste their time in bestowing compliments on Sir Henry Howorth, whom he now asked to take the chair as their President for the present year, and on whom he then called to read his Presidential address. This address is printed in full pp. 1 *et seq.* A hearty vote of thanks to the President for his stirring address brought the proceedings to a close.

Friday Morning (Sir Henry H. Howorth, the President, in the chair).
June 12th, Friday Hominy.—A Council meeting was held at 10 a.m. and concluded just before 11 a.m.
At 11 a.m. a general meeting of Delegates, Members, and Associates was held, and the Chairman stated that some rearrangement of the programme became necessary owing to the fact that Mr. Hutchinson could not be present until Saturday. Two papers were then read; one entitled "Atmospheric moisture as a factor in distribution," by Mr. A. O. Walker, V.P.L.S., and the second, a paper, particularly interesting locally, by Mr. A. T. Walmisley on "International Communication."

Both papers were exceedingly well-received. The President complimented Mr. Walker on his paper, and expressed the hope that he would prosecute his researches further. They had a great deal to go upon, in the geographical distribution of plants and animals, but they greatly needed an examination of the problem from the side of hygrometrical distribution in these western part? of Europe. The puzzles were tremendous, and led one to think sometimes that it was not possible to draw any quite general law?. He would mention one or two curious facts to illustrate his meaning. Two English trees, which grow splendidly, and were found in almost every English park—the towering feathered elm and the lime—had never been known, in his experience, to ripen their seed in England, but, in Berlin, where the winter climate was more severe than in England, but much drier, and the heat of the summer month? was great, the lime did ripen its seeds with ease. On the other hand, it was found that a tree like the chestnut, introduced into England in the time of Henry VIII., from the southern zone of Europe, ripened its seeds anywhere. Another great puzzle was the distribution of insects. In Iceland there was not a single butterfly, but plenty of moths; whilst in Greenland there was no: a single moth but plenty of butterflies (1). It seemed to him that the only possible explanation was that some particular winter or season must have been so severe as to exterminate a particular class of lepidoptera, and they had never been able again to reach the country. In connection with this question, the President drew attention to the extraordinary collection of lepidoptera of the British Isles, belonging to Mr. S. Webb, and shown in the Hall. He had never seen such a collection in private hands before. It contained many specimens which were the very key-stone to many biological problems. Professor Boulger and the Rev. T. R. R. Stebbing also spoke briefly on the subject, and a hearty vote of thanks was accorded the lecturer.

Sir H. Howorth, in proposing a vote of thanks to Mr. Walmisley for his paper, mentioned that he was a member of the House of Commons at the time of the Channel Tunnel controversy. Hi? belief, strengthened by the experience of the difficulties of ventilating long tunnels both in London and through the Alps, was that it was impossible to successfully work a tunnel of this enormous length under the Channel. He was not so hopeless, however, about a scheme of suspended railways such as was in use at Lyons and several other places. The vote of thanks was heartily accorded, and acknowledged by Mr. Walmisley. The papers are printed in detail, the former at pp. 43 *et seq.,* the latter at pp. 61 *et xeq.* June 12th, Friday Afternoon.—The afternoon meeting was held at 2.80 p.m., when a lengthy paper by Captain McDakin and Mr. Sydney Webb on "The diminution and disappearance of the southeastern flora and fauna within the memory of present observers" was read, and a hearty vote of thanks to the authors passed. The paper was essentially one for discussion and criticism, so many points insisted on by the authors being, in the opinion of other observers, doubtful, *e.g.,* the remarks relating to *Grapholitha caecana (postea* p. 51), *Papilla machaon* (p. 52) etc., and others as wanting in definiteness of evidence, *e.g.,* the inclusion of such species as *Tort Ha; heparana, Dicrorampha petirerdla, Calmetia niyromaculana,* etc. The paper is printed at length pp. 48-60. This was followed by another paper, a most interesting piece of detailed observation, "The seedlings of geophilous plants," by Miss Ethel Sargant. The paper is printed in detail pp. 22 *et set/.*

After Miss Sargant's paper a large section of those present at the meeting took advantage of permission kindly granted by General Bundle, commanding the south-eastern district, to visit the Keep of Dover Castle, and a most enjoyable hour was spent inspecting the many things of interest there.

Friday Evening.

At 8 p.m. a reception was given by the Mayor and Mayoress of Dover—Councillor and Mrs. F. G. Wright—at the Town Hall. The host and hostess received a very large number of delegates, members, associates, and friends of the local scientists, as they entered the Connaught Hall, which had been exceedingly prettily decorated with palms and flowers, and along the sides of which the exhibits forming the temporary Museum made a brave show. When the Hall was becoming comfortably crowded the Mayor, the President, and other shining lights of the Congress, adjourned to the platform, where the Mayor delivered an address, including a most hearty civic welcome, and expressing in no measured terms his personal delight at the choice of Dover for the Congress. Without being scientific himself, he said he had a great appreciation of the value of science, and was proud of the results achieved by workers in its various branches. He detailed the many interesting things that were to be seen in Dover, his native town, and trusted the Congress would again visit the city at no very distant date.

The President thanked the Mayor very warmly for his kind hospitality, and the cordial terms in which he had referred to the Society. He was glad to hear the patriotic way in which the Mayor had spoken of this famous old town of Dover. It seemed to him that, in trying to discover the real cause of the greatness of this Empire, we must not lay too much stress upon what some people were apt to lay all stress upon, the great capacity of our statesmen and the valour of our soldiers and sailors. But this empire had been made—and it preserved the privilege now, against all the communities of the world—by the public spirit of its citizens, as shown in the way in which they governed themselves in these great communities. It was the English towns which won their liberties for England, and it was these English towns which continued to be the real democratic centres of life in Europe. In conclusion, Sir Henry expressed the wish to the Mayor "I hope the sunshine will always rest upon the doorstep of your wife and yourself wherever you may live."

During the earlier part of the evening an enjoyable programme of vocal music was contributed to by the Misses I. Boyton and Ormsby, and Messrs. Eaton and Davies, whilst Mr. II. L. Taylor gave an organ selection. Between the various items of the programme, two most interesting papers (illustrated by the Lantern) were read, in the Maison Dieu Hall. The first one was by Dr. Arthur Rowe, F.G.S. and was entitled "The White Chalk of Dover." He commenced by saying that there was no coastline in chalk which was so well-known as that of Dover. The English people had thoroughly made it their own, and it had become part of their history, literature, and everyday life. The term "the white cliffs of Albion" which might also be very well applied to other parts of the coast, had by common consent been applied to those of Dover alone. The reason was obvious; it was geographical; the Dover cliffs stood alone in their uprightness and snowy whiteness. The lecturer then threw on tho screen a number of striking views of cliff scenery around the English coast, in order to illustrate the "sculptured " effects of these as compared with the straight face of the Dover cliffs. Dover cliffs were the richest section on the coast in regard to fossils and also the most normal in distribution. It was said that all good Frenchmen when they died went to Paris. Varying that, he would say "All good geologists when they die come to Dover." For himself he should like to put in a claim for the bit near Fan Hole. He had spent 8 months working on these cliffs, and had robbed the local scientists of no less than 5000 fossils. It was when he came to Dover that he first really learned what zonal geology and zonal variation meant, and it was there he first learned the meaning of evolution in *Mirraster.* The lecturer then dealt at some length with the question of zonal geology, and contended that the theory could be proved by any man who had eyes and hands and common gumption, on going to a chalk section at any time. A hearty vote of thanks was accorded the lecturer.

This was followed by a very interesting address on "The discovery of an ancient Keltic Cemetery at Harlyn Bay, near Padstow, delivered by the Rev. R. Ashington Bullen, B.A., F.L.S. He stnted that the discovery was made about three years ago, and between 15ft. and 16ft. of sand had to be removed from over some portions of the cemetery. A number of illustrations were shown of the skeletons, etc., discovered. The skeletons were all in what is known as " the crouched position," and, judging from this and the evidence of the ornaments, tools, etc., the cemetery dated back to about the "iron age"—*i.e.,* late

Keltic or pre-Eoman. The lecturer also gave a description of a Keltic hut which was discovered in the same district. It was 13ft. long, 8£ft. wide, and 4ft. 6in. high. It contained a hearth inside and another outside. The hut was excavated on quartz gravel, and was roofed over with slabs of slate. A number of illustrations of the implements of these ancient people, which had been found in the district, were also shown and. explained by the lecturer. On the proposal of the President, a cordial vote of thanks was given Mr. Bullen. The paper (of which the above is a *resume)* is printed at length, pp. 27 *et seq.*

The visitors did not disperse until well after 11 p.m. During the evening a bountiful supply of refreshments was provided by the Mayor and Mayoress.

Saturday Morning (The President in the chair).

June 13th, Saturday Morning.—-At 9. 30 a.m., a fully attended Council meetingdiscussed finally the various matters to be presented to the delegates at their annual meeting, to which an adjournment was made at 10.15 a.m.

Delegates' Meeting.—The meeting at 10.15 a.m. was only moderately well attended. The minutes of the last Delegates' meeting at Canterbury having already been printed in the *Transactions,* it was agreed that they should be considered as read.

The Secretary read the seventh annual report of the Council, including the Treasurer's report, which gave rise to some discussion, and some proposals arising therefrom.

As a result of the recommendations of the Council it was proposed by Mr. Whitaker and seconded by Professor Boulger, the Chairman strongly supporting, that Mr. F. W. Budler, F.G.S., be the President of the Union for 1904-5. This was carried by acclamation.

The Chairman then moved, on the recommendation of the Council, the names of Mr. G. A. Boulenger, F.B.S., and Mr. J. Hutchinson, F.B.S., as Vice-Presidents of the Union. Carried.

To fill the vacancies that had arisen on the Council, Mr. Martin proposed and Mr. Adkin seconded that Mr. F. G. Fleay, M.A. (Balham), Mr. H. N. Gray, P.A.S. (City of London), Mr. L. Green (Maidstone), become members of the Council. Carried.

The Chairman stated that it had been agreed by the Council to recommend to the delegates that Mr. J. W. Tutt, Editor of the *Transactions,* should join the official Staff of the Council as Editor, and that Rule 3 be amended accordingly; he stated that if this were agreed to it would still leave a vacancy to be filled. He therefore moved from the Chair that Mr. Tutt's name be added to the official staff as Editor and that the necessary alteration of Rule 3 be made. Agreed.

The Chairman then asked for a name to fill the vacancy. None being forthcoming, Mr. Whitaker moved and Dr. Treutler seconded that the Council fill up the vacancy at the next Council Meeting. Agreed.

The Rev. T. R. R. Stebbing drew attention to the fact that, in order to relieve the General Secretary of some of the clerical work attached to his office, the Rev. R. Ashington Bullen had given a sum of three guineas to the funds, to be paid to an assistant who should help the General Secretary in the direction indicated. He also considered that it was high time that an annual honorarium should be given to the General Secretary for personal out-of-pocket expenses. In view of the condition of the funds, however, no motion was made on the subject.

The Chairman then moved the adoption of the Secretaries' and Treasurer's reports, which were duly agreed to.

The Rev. T. R. R. Stebbing then proposed as officers for the ensuing year— as Treasurer, The Rev. R. Ashington Bullen; as Hon. Secretaries, Mr. G. Abbott (General), Mr. H. E. Turner (Photographic), and Mr. E. W. Swanton (Museum); as Editor, Mr. J. W. Tutt. The proposition, supported from the chair, was carried by acclamation.

The Rev. R. Ashington Bullen was then appointed the Representative of the South-Eastern Union to the British Association.

A deputation from the Maidstone Natural History Society, consisting of Mr. Hoare and Mr. Allchin, then invited the Union to hold its Congress at Maidstone. Mr. Hoare said that, on behalf of the Mayor of Maidstone, who unfortunately could not be present, he attended to offer a very warm invitation to the Union to hold its next Congress at their town. He could assure them of a hearty welcome from the citizens as well as the members of the Natural History Society.

Mr. Allchin, Curator of the Museum, supported Mr. Hoare in a most interesting speech, and pointed out the attraction that Maidstone offered to naturalists, whatever their special branch of study.

The Chairman remarked on the pleasure that he, the officials, and the delegates had in accepting the warm invitation of the Mayor and members of the Natural History Society of the. ancient town.

The Rev. T. R. R. Stebbing proposed a vote of thanks to the retiring President and officers and members of Council (humorously remarking that it had only just struck him that he was including himself in the vote of thanks), for although vacating his office of Treasurer he still remained a Vice-President, and pointing out how much Mr. Hutchinson had done for the Union. His indomitable energy and scientific attainments were reflected on the Union, and his generous welcome at Haslemere was fresh in the minds of all.

Mr. Hutchinson replied thanking the Congress for the kind treatment he had received, and while regretting he had not more leisure to bestow on the work of the Union, yet promised attendance at the meetings in future years.

Mr. Whitaker then proposed a vote of thanks to the Mayor and Mayoress of Dover for their kind hospitality, to the Dover Sciences Society, for its generous work, and to the hosts of the delegates who had treated them most handsomely. This was seconded by the Rev. R. Ashington Bullen, and Captain McDakin replied, thanked the delegates for coming to Dover and expressed the great pleasure it had been to each and every member of the Dover Society to receive them.

Mr. Adkin then proposed a vote of thanks to the Commanding Officer of the South-Eastern District for permitting the members of Congress to visit the Castle. He also proposed a vote of thanks to the Harbour Officials who had generously offered to take them (with the kind permission of the contractors) over the new harbour works in the afternoon. He particularly wished to state his appreciation of the kindness of the Castle authorities on their visit the previous day. Mr. Connold seconded, and the votes were carried.

Captain McDakin stated that Mr. Dixon, the harbour-master, had placed a steamer at their service for the afternoon. It would be at Granville Dock at 2.15 p.m., and would start without delay, so that the visit could include an actual inspection of the work now being done. A special vote of thanks to Mr. Dixon was passed.

The President then thanked the Exhibitors for the pleasure they had given everybody in the arrangement of the temporary Congress Museum, and particularly drew attention to the botanical and entomological exhibits. Under Mr. Swanton's supervision an excellent show had been made, and this reflected great credit on all concerned. Mr. Swanton briefly replied.

This closed the business of the delegates' general annual meeting.

At 12.0 (noon), the final meeting of the Congress commenced, and there was a good attendance when Professor Boulger read an extensive series of notes "On the preservation of our indigenous fauna and flora." The Rev. T. R. R. Stebbing hoped that Professor Boulger would summarise his remarks so that they might appear in the *Transactions*, and thus reach the societies interested. A long discussion on the advisability of legislation for the purpose of protection of plants from plant-stealers (for the purpose of trade), from the tripper, and from the covetous destroyer, took place, in which it appeared to be generally held that, except in the neighbourhood of certain well-patronised excursion resorts, the tripper did little or no harm compared with the professional plant-stealer (who sold what he stole), and the covetous collector (who destroyed anything not wanted for his own personal collection).

Mr. J.Hutchinson, F.R.S., then gave a most lucid and instructive lecture entitled "Experiences of Leprosy in India. " Mr. Hutchinson, who had placed in front of the platform a number of tables of statistics, " writ large," to show the facts that he had ascertained and verified as to the effect of fish-eating in producing leprosy, spoke from the floor of the hall so as to point to and elucidate his statistics. He stated that he had formed the idea that leprosy now and in past times arose mainly from eating bad fish. He had been told, however, that there was much leprosy in India, amongst the natives, who, owing to their religion, were strict vegetarians, and also amongst people who were so far removed from the sea and other fish-yielding waters that it was impossible that the leprosy amongst them could have arisen from eating fish of any kind. He had visited India, and had travelled extensively in that country to see for himself if that were so. He had visited Ceylon, Madras, Calcutta, and Darjeeling, at the foot of the Himalayas. In all these places he found lepers. In some of the places he found the lepers in leper-houses, and in others they came with other out-patients to the hospitals. The leprosy in the country, however, was not confined to the hospitals and the leper-houses, and there was much in the country that was not isolated. He enquired from the medical men, and others whose information could be relied on, and he found that fish-eating was common amongst those who suffered from leprosy. He went to a station on a sandy plain far inland, and he was told that be would find that the leprosy there was not caused by fish eating, as none could be got there. He found the leper-bouse, containing 500 lepers, under the management of an American doctor, who was a Doctor of Divinity as well as of Medicine. His enquiries, however, disclosed the fact that, even there, fish was an usual article of diet, and every one of the lepers had been accustomed to a fish diet, except one man who said he had never eaten fish. But he had found that leprosy could be contracted by contagion, although not in the ordinary way. For instance, he had no fear of being in the vicinity of lepers, of touching them, or breathing the atmosphere which they breathed, but in eating that which they had handled there was danger, and the man he had referred to might have taken leprosy in that way. At the present time, in India, there was no great increase or decrease in the disease, but, in proportion to the population, there was a great variation in different parts of the country, in some parts the average being as low as 1-8 per 10000, and in others as high as 150-0 per 10000; and it was remarkable to note that the 18 was 'where the fish was scarce, while the 150.0 was where it formed a regular article of diet, and in many cases was badly cured. At Assam, for instance, where there was a large importation of salt fish, leprosy was excessive and on the increase. In the year 1872 the proportion was 6 in 10000, and in 1891 it had risen to 12 in 10000. There was a small island near Bombay which had been the seat of a Roman Catholic Mission. There the consumption of fish was specially encouraged by the Church on two days a week, and there the leprosy was much more prevalent amongst the Roman Catholic Christians than amongst the native population. As a matter of history, it was a fact that leprosy died out in England at the time of the Reformation. He was reminded of this by a classical friend, who called his attention to a passage in the writings of Erasmus to the effect that the Pope in his time proposed to stop the use of fish on fast days in England owing to the spread of leprosy. Unfortunately, that Pope died before he could carry out his intentions, but it was to be hoped that some future Pope would make the change which the Reformation did as far as the Protestants were concerned. In India, Hindoos who were nominally vegetarians, ate fish whenever they could get it. The high class Brahmin did not eat fish, but even they lived as the other people did until eleven years of age, and no doubt the leprosy, which was rare amongst them, was contracted before that age. On the other hand, there was another religious sect called the Jains, who abstained from using as food anything that had life, in the strictest manner possible, and they were almost entirely free from leprosy. There had been a few cases, and these he attributed to some of the less strict members of the sect eating fish when driven to it in times of famine. He came to the conclusion, therefore, that leprosy in India largely arose from the same causes as in otber countries, *viz.,* from the use of bad fish as an article of diet, and, in India, the fish was very badly cured owing to the dearness and the bad quality of salt. He suggested as remedies the abolition of the salt tax, and the alteration of the regulation of the Roman Catholic Church which provided for the eating of fish on fast days. He recommended the latter because leprosy followed the Roman Catholic Church, whereas the Greek Church, which prohibited both flesh and fish on fast days, were free from it.

The Chairman entered heartily into the subject, praised the excellence of the lecture, suggested that the moderate tone and logical presentation of the facts almost disarmed criticism on the one hand, and made difficult facts quite clear to the lay mind on the other. He offered the lecturer his own best thanks (a remark that was almost lost in the cheers of his audience) and wished our grand Ex-President a hearty godspeed in his excellent work, and hoped that, when the time came and his energy did fail, his mantle would fall on a worthy successor. The formal vote of thanks was carried by acclamation.

Saturday Afternoon. *June 13th, Saturday Afternoon.*—In the afternoon there was an excursion around the National Harbour Works, the delegates viewing them from the sea in the Harbour tug, Lady Vita. The visit concluded with a tour of inspection of the Dover-Ostend steamer, Leopold, on which the visitors were given a practical demonstration of wireless telegraphic communication between the ship they were aboard and another vessel of the same line at sea.

Sir H. Howorth, on behalf of the visitors, expressed thanks to Mr. V. Grant, the agent of the Belgian mail boats, and to the other officials, for the courtesy which had been shown them. He referred to the cordial relations which had always existed between England and Belgium, concluding by calling for three cheers for the Belgian nation— which were heartily given.

TRANSACTIONS OF THE SOUTH EASTERN UNION OF SCIENTIFIC SOCIETIES. *1903.*
PEESIDENTIAL ADDRESS.
The Uncertainties Of Science.
By Sir HENEY H. HOWOETH, K.C.I.E. , F.E.S., Ac.

It strikes me very plainly, in coming face to face with such an audience as I see before me, how very incongruous our respective positions are. You a grave, sensible, sane, orthodox body of students engaged in disentangling the mysteries of nature by many patient methods, and I a heretic and a rebel against scientific convention and authority, whose proper position would be as the victim of some scientific *auto dafi'* rather than occupying this chair. The only thing that excuses me in my own eyes is that you put me there, and that it is you rather than myself who are most to blame for it all, and if you go away discontented and sulky, and disagreeable, after contrasting my vagaries and heresies and dulness with the matured wisdom and sense which you have heard from those who went before me, you will remember that it is all your doing.

I confess that, in addition to feeling very humble, I also feel very embarrassed in knowing exactly what I am to talk to you about. Most scientific men have got a special province of their own where they can speak as authorities, and they travel along well macadamised roads. I have always disintegrated myself by being interested (like Martha, in Holy writ) in many things, and have carefully avoided macadamised roads, and preferred to go over hedge and ditch, and, if there is anyone present who, like myself, likes to travel over hedge and ditch, he or she will perhaps lend me an umbrella or a parasol if a heavy shower of disapproval should fall. These being my qualifications and my tendencies, how can I preach to you upon any profitable subject?

It has struck me, as most of you are specialists and believers in the omnipotence of science, that a little tonic would not be amiss, and that I may profitably advocate two things which you may think impertinent; first, the frequent frailty of science, even at its best; and, secondly, the necessity of questioning all forms of dogmatic teaching, positive and negative, which come to you in a scientific garb, and all scientific Popes, however eminent, and if I bid you beware above all things of those Popes who lay it down that science is capable of all things, and that its vast harvests are mostly garnered in impregnable fortresses, where doubt and difficulty have no right of entry because the great Pundits are agreed, and, therefore, small men like you and I must say amen. There is no need in our day to exalt the capacity and to belaud the victories of science. Everybody does that, even the halfpenny papers. What we do need most, as a perpetual tonic, is the reminder that the greater part of science is merely "hypothesis," and consists of ingenious speculations consistent with partial knowledge, but which may be quite inconsistent with fuller and complete knowledge. That the very greatest men have made the most stupendous errors, if the magnitude of errors is to be measured by their far-reaching effects, and that every generation ought most certainly, as a plain duty, to verify all the premises of its knowledge, however supported by the high priests of the time, and to be continually vigilant lest the shadow of great success and the heavy load of dominant schools of thought should bring back to us the darkest of dark ages, when every sheep followed a bell wether, and all read the same scientific catechism.

I propose to put together some elementary thoughts on this question, not with the object of making you mistrust and turn aside from knowledge and science as full of uncertainty, but to encourage those among you, if I dare, who feel timid and frightened sometimes when your commonsense revolts against any dogmatic assertion which claims to condense the results of science, and yet seems to you mere foolishness, and to bid you remember that the hypotheses upon which science mainly stands are every one of them tentative, every one of them relative and provisional. I am not in all this as mad as you may think me. I am not a circle-squarer, a believer in the primitive notion that the earth is flat and is covered with a solid vault, that the lives of men are controlled by the stars, that certain geese grow out of barnacles, &c. It is not upon old-fashioned fables like these that I wish to speak, nor, indeed, about any fables fashioned by old wives, but about fables, quite modern in date, fashioned by some of the keenest of keen intellects, and published with the very highest scientific sanction. Fables I call them, because they profess in some cases to solve by plausible, but quite illegitimate, methods what is insoluble by any methods, and, in other cases, offer solutions which are inconsistent with similar solutions derived from similar data in other sciences, and because, in every case, they ignore either some of the facts or some of the laws composing legitimate knowledge. The stupendous discoveries of the last century have made their authors in many cases adopt a tone of boastful assurance as to the possibilities of science, especially in matters outside their own special study, which is quite unjustified. They have come in many cases to look upon empirical knowledge as the only true knowledge, and to despise what is called philosophy as mere metaphysical word-splitting, instead of being, as it is, the necessary basis of all legitimate inquiry. Philosophy, among its most elementary lessons, teaches us that knowledge is necessarily and inevitably limited, both in fact and method, and that no amount of sophistical juggling or make-believe, either in the closet or the laboratory, will enable men to transcend the limitations of their minds, and by mere analysis and experiment to rob the universe of its innermost secrets.

The fact is, the capacity of the human mind is limited to the acquiring of such truths only as are eventually based on human experience, and the only avenues between us and the world outside us are our senses. We have no other roads available than those which men have possessed since they first began to think, and we are as far off gazing into the real mysteries, the underlying causes of things, as Socrates was. We have, no doubt, enlarged the boundaries of knowledge beyond human expectation. That is true enough, but we have not by any legitimate means been able in doing so to transcend experience any more than the Greek philosophers could, nor found any method of securing truth inaccessible to them. We can and do continually invent new and high sounding names and phrases, and fancy that new names conceal new explanations, and enable us somehow to look further into the inner machinery of things, their ultimate cause, origin, and purpose. Sometimes we invent a new calculus, but we do not thereby multiply our ultimate sources of knowledge, and the Sphinx is as dumb now as she was at the beginning of things to our inquiries. She replies not when we ask her what thought and consciousness are, how thought and consciousness are united with material things, what gravity or force, or chemical affinity, or cohesion, or light, or heat, or electricity, &c, really are. We know certain phenomenal effects produced upon us by them. We construct intricate hypotheses which have a logical sequence, and are consistent with certain asceitained sequences among such effects, and we fancy that with terms like ether, atom, vibration, vortex-motion, &c, we bridge over the gap and get more nearly face to face with things as they are, but we do not, in fact, get one atom nearer to the essential explanation of things, or to understanding what they are, and how they act apart from our sensations, and the different ways they affect them, and all efforts to scale that particular heaven seem to me to to be doomed to everlasting failure.

Nowhere is this more obvious than in the field of mathematics, the most abstract and empyrean of scientific studies whose professors most loudly claim to be above criticism. Mathematics is merely an elaborate form of logic applied to certain properties of numbers and figures, and we all admit it to be a very perfect form of formal logic in which the conclusions are derived from the premises by rigid deduction.

While the logic of mathematics is very perfect, it does not mean that the conclusions of mathematics necessarily represent actual facts of nature. This depends on whether the materials or data underlying our mathematical arguments are true in fact or not, and this is not always immediately obvious, because so much of mathematics is not expressed in language but in symbols, which have to be interpreted into actual thought before we can be certain that we are applying our logic to real facts, and not to mere phantasms of the imagination, and to mere definitions of abstract concepts embodied in formula'. If, among these materials, these data and premises of our reasoning process, we find such things as infinity (either the infinitely great or the infinitely small) in number or magnitude, or attempts to establish a definite numerical ratio between incommensurable quantities, if we find negative roots to equations, and impossible or irrational numbers, if in plausible symbolic clothing we find formula? dealing with space of more than three dimensions, &c, we may be sure that our logical chain cannot be stronger than these links, and we must conclude that such data can only lie consistent with purely imaginary, or irrational, or illegitimate conclusions, conclusions inapplicable to the world as we know it. These " *Humeri absurd i infra nil*" or "*numeri jicti,*" Ac, as the early mathematicians called them, can only lead to sophistical conclusions if we treat the process as applied to the facts of nature and not as a mere logical game with symbols. Yet, and it is an astounding fact, some of the very greatest mathematicians of modern times, Gauss, Riemann, and Helmholz, Salmon and Cayley, and Clifford and Tate, and others have, at all events in the realm of geometry, claimed not merely to frame transcendental conclusions with transcendental data (which would be legitimate enough), but to be framing conclusions of real value in analysing the properties of things. Sylvester, second to none of the great men above quoted, quite humbly justifies his adherence to their view, and says of these very men, that if they claim to have an inner assurance of the reality of transcendental space, that he strives to bring his faculties of mental vision into accordance with theirs, and accepts it on the authority of the illumination of privileged intellects. The infallibility of scientific Popery can take us no further than this. Let me explain more clearly what I mean, for the position is a very interesting one. Space as we know it is of three dimensions, and has length, breadth, and thickness. The acute mathematicians above cited claim that there is no reason why there should not be other forms of space with less or more dimensions than three in other parts of the universe than ours, and proceed to work out its qualities. Gauss used to say he had put aside several questions which he proposed to attack in the next world, when his conceptions of space should have become amplified and extended, for he says, "As we can conceive beings (like infinitely attenuated bookworms in an infinitely thin sheet of paper), which possess only the notion of space of two dimensions, so we may imagine beings capable of realising space of four or a greater number of dimensions."

"Our Cayley," says Sylvester, "the central luminary, the Darwin of the English school of mathematicians, started and elaborated at an early age, and with happy consequences, the same bold hypothesis." It may interest you to hear an example or two of the results of this analysis of another space than ours. Professor Newcomb showed that in space of four dimensions a closed material surface or shell like a football could be turued inside out by simple flexure, without either stretching or tearing. Felix Klein showed that in such space no knots could exist; while Professor Koellner accouuted for some of the feats of the American medium, Slade, in which be produced knots on a rope, the ends of which were sealed together and held in Koellner's hands, as evidence of such space existing. Professor Tait, on the basis of similar arguments, raised a question as to whether space had the same properties throughout the universe, and argued that other kinds of space may exist in those regions to which the solar system is hurrying.

I have not the slightest doubt that all this is mere moonshine, and that Professor Donkin was quite right when he treated the notion of generalised space as a mere disguised form of algebraical formulation, a mere treating of symbols, not as symbols of imaginative ideas, but as representing real things. Mr. Stallo, a most acute critic, seems to me to have applied a most destructive criticism to the whole notion, except as an exercise in mathematical logic, and he ventures to conclude with the homily that even mathematics, the exactest of all the sciences, whose methods are said to be as infallible as its foundations are supposed to be permanent, and which, ever since the dawn of human intelligence, has pursued the even tenor of its way amidst all the vicissitudes of speculation, is thus not exempt from the prepossession of ontological realism, that is, of the frailty apparently attaching to all science, of pursuing at times false and delusive metaphysical Jack-o'-Lanterns outside of human experience.

Let us now turn from mathematics to physics, which stands next to it in its claim to deal with rigid methods of argument and proof. Let us consider some of its best established postulates. Take, for instance, the theory of gravitation. I suppose no one could be found to dispute the theory of gravitation as propounded by Newton. It is only a theory or hypothesis however; we all know that he and its discoverer only claimed it to be such, but it accounts so completely for all the facts involved in the mechanics of the universe, that it is a common axiom of all science. Granted; yet it involves certain considerations which are absolutely staggering to the physicist who, having framed very ingenious and remarkable theories to explain, as he thinks, the intricacies of light, heat, electricity, magnetism, &c, finds himself compelled to acknowledge that in gravitation he has a crux to meet in which none of these theories will help him, and which seems to set at naught the most elementary principles underlying his science.

In his original promulgation of his theory, Newton confessed that he did not know how gravitation acted or what caused it. Subsequently he published an attempt at an explanation of it in the Queries in the second edition of his "Optics." In this he suggests that there is an elastic medium pervading all space, and increasing in elasticity as we proceed from dense bodies outwards, and that this "causes the gravity of such dense bodies to each other. Every body endeavouring to go from the denser part of the medium to the rarer." "Of this hypothesis," says Whewell, "we may venture to say, that it is in the first place quite gratuitous; we cannot trace in any other phenomena a medium possessing these properties; and in the next place, the hypothesis contains several suppositions which are more complex than the fact to be explained, and none which are less so. Can we, on Newton's principles, conceive an elastic medium otherwise than as a collection of particles repelling each other? And is the repulsion of such particles a simpler fact than the attraction of those which gravitate? And when we suppose that the medium becomes more elastic as we proceed from each attracting body, what cause can we conceive capable of keeping it in such a condition, except a repulsive force emanating from the body itself, a supposition at least as much requiring to be accounted for, as the attraction of the body" (Whewell, *Hist, of the Inductive Scienies,* i., 224). No one now accepts Newton's attempted explanation of gravitation, nor did anyone apparently ever accept it but himself.

Others have suggested that gravity

may be due to waves in ether running in the direction of the wave translation and not athwart it, as in the case of light and heat, which waves would enormously exceed in velocity the transmission of the wave motions due to thwart-vibrations, somewhat in the same way that the transmission of sound waves in water exceeds in velocity the transmission of waves along the water's surface; but, says Proctor, we have no other evidence of any such ether waves, and we should have to imagine qualities in the ether to account for the supposed action of such waves, which, though not perhaps more inexplicable than some of the qualities already attributed *ex necessitate* to the ether, are utterly outside all that is known respecting matter. In particular, we should have to imagine such etherwaves acting directly, not on matter as we know it, not even on the molecules or particles of such matter, but on the ultimate individual atoms, since otherwise the property of gravity, in being proportional to the masses, would not be explained" (Proctor, *Astronomy,* 323 note).

Proctor, in opposing this explanation, speaks of it as involving assumptions outside the range of physical experience, and as *a way of escape from the absolutely inexplicable.*

Euler, the great German analyst, would have it, that the action of gravity must be due either to the intervention of a spirit, or to that of some subtle material medium escaping the perception of our senses; and he insists that the latter was the only admissible alternative. Professor Challis, who spent many years of a great mathematician's life in trying to unravel the mystery of gravity, concludes similarly, that all physical force being pressure, there must be a medium by which the pressure of gravity is exerted, and had recourse at last for its explanation to some other kind of ether than that demanded, for the explanation of light, heat, *&c,* having the same relation to it that it has to air, and that gravity is "due to the attractive action of a molecule of a higher order as to magnitude than the molecule of molecular attraction "; that is, he Hies away further and further into the recesses of a transcendental world unknown to experience.

Apart from such difficulties, the fundamental objection to all these and similar theories, involving processes of conveyance of energy or force by waves through a medium, and which seems quite unanswerable, whatever gravity may be, is that it seems to require no time to traverse space, but acts instantaneously at all distances.

Arago, an astronomer and physicist of the first rank, says of it: "If attraction is the result of the impulsion of a fluid, its action must employ a finite time in traversing the immense span which separates the celestial bodies, whereas there is now no longer any reason to doubt that the action of gravity is instantaneous. If it were otherwise, if gravity, like light and electricity, were projected with a measurable velocity, it must necessarily accelerate the rate of the motion of the planets in their orbits, and the apparent line of attraction would be directed to a point in advance of the real place of the sun, just as the sun's apparent position is displaced in the direction of the earth's orbital motion by the aberration of light. Such an effect, if it had any existence, would have been detected long ago."

Laplace, more than a hundred years ago, laid it down that if the action of gravity were propagated in time, its velocity must be at least fifty million times greater than that of light. No wonder, then, that Stewart and Tait affirm so positively that no undulations of any supposed ether can be made available for a physical theory of gravitation, and that every attempt to connect it with the luminiferous ether or the medium required to explain electrical and magnetic action has completely failed.

Apart from this, it seems to me that any such explanation is unthinkable. How waves proceeding or emanating from any portion of matter, whether longitudinal or transverse, can cause another piece of matter to be pulled towards it, seems to me unthinkable.

In view of this infirmity of a radiating force acting through a medium, Professor Oliver Lodge has had recourse to Lord Kelvin's theory of vortex motions (by which he resolves all material things into motions of ether), for an explanation both of cohesion and of gravity. The medium in which such vortices are supposed to arise, according to Professor Oliver Lodge, is a perfectly homogeneous, incompressible, continuous body, incapable of being resolved into simplerelements or atoms; it is, in fact, continuous, not molecular.,

These vortex rings, the substitutes for matter, which were first postulated by our Father Anchises, Lord Kelvin, are claimed to be rotational movements in such a medium, and it is claimed further for them that they are permanent, of invariable volume, due to an invariable quantity of motion, though susceptible of a great variety of form, and incapable of interpenetration or coalescence, and that their mutual approach results in rebounds similar to the resistance of perfectly elastic bodies. Clerk Maxwell, in his article on atoms in the *Encyclopaedia lIritannica,* subjected the notion of such vortex rings as substitutes for an explanation of matter to most destructive criticism. *Inter alia,* he says, " such vortex rings would have no inertia; they would consist not in the substance of the omnipresent fluid, but of motions in it merely. Of these motions, the persistence both of mass and energy would have to be predicated, and from them the concretions of mass, together with all the phenomena exhibited by sensible matter, would have to be derived. From its very nature, motion cannot be the bearer of motion.nor be the generator of momentum, which is the product of two antagonistic factors, which would be utterly extinguished by the suppression of either. Upon the basis of the mechanical theory, the fundamental antithesis between mass and motion, inertia and energy, cannot be destioytd without obliterating all the distinctions which constitute the elements of our conceptions respecting the nature'of physical action. " Mr. Stallo, a most acute person, further says: "It seems to be evident that motion in a perfectly homogeneous, incompressible, and therefore continuous fluid, is not sensible motion. All partition of such a fluid is purely ideal; in spite of any displacement of any portion of it by any other, a given space would at any moment present the same quantity of substance absolutely indistinguishable from that present there a moment before. There would be no phenomenal difference or change. A fluid both destitute and incapable of difference is as impossible as a vehicle of real motion as pure space; it is as useless for the purpose of accounting for the phenomena of material action as the quasi-material medium without inertia, of which Eoger Cotes said that it was not to be distinguished from a vacuum" (Stallo, *Concepts of Modern Physics,* 48 and 44).

It is clear, therefore, that vortices are as helpless in explaining gravity as waves in a medium are? and with Stewart and Tate we are compelled to turn to some form of the impact theory as the only possible one. This, however,-also completely fails us.

The only impact theory seriously discussed by modern physicists and astronomers as explaining gravitation was published by Le Sage in his *Liicrece Newtonien,* and further illustrated by M. Prevost. According to their theory, "all space is occupied by currents of matter moving perpetually in straight lines in all directions, with a vast velocity, and penetrating all bodies. When two bodies are near each other, they intercept the current which would flow in the intermediate space if they were not there, and thus receive a tendency towards each other from the pressure of the currents on their farther sides.", "This theory, again, is perfectly gratuitous, except as a means of explaining the phenomena. If it were proved," says Whewell, "it would still remain to be shown what necessity has caused the existence of these *two hinds* of matter; the first kind being that which is commonly called matter, and which alone affects our senses, while it is inert as to any tendency to motion; the second kind being something imperceptible to our senses, except by the effects it produces on matter of the former kind; yet exerting an impulse on every material body, permeating every portion of common matter, flowing with inconceivable velocity, in inexhaustible abundance, from every part of the abyss of infinity on our side, to the opposite part of the same abyss; and so constituted that through all eternity it can never bend its path or return, or tarry in its course" (Whewell, *id.).*

Again, as Stallo says of the theory, it utterly ignores the necessity of accounting for the origin of the enormous energy constantly expended by the supposed streams of ultra mundane corpuscles; both the agency postulated and the mode of its action are unknown to experience, and it is doubtful whether its assumptions, if they could be granted, would serve as an explanation of all or any of the features of gravitation, in the presence of which every hydrodynamic theory is doomed to fail. The futility of Le Sage's theory, however, is most strikingly exhibited by Clerk Maxwell, who tests it by the principle of the conservation of energy. (See *Enc. Brit.,* Art. " Atom "; also Stallo, *op. cit.* 64 and 65.)

We have thus exhausted the various methods of accounting for or explaining gravity, and they leave us in a stupendous quandary. If it acts regardless of time and without the intervention of an intermediate medium through which stress or strain, or pull or push can b'e exercised in the only ways known to experience, there remains only the conclusion so fiercely protested against by physicists of all ages and schools from Newton himself downwards, that matter can act upon matter at a distance, and across a void. If this be unthinkable, so is every substituted theory hitherto suggested to explain it. After a review of the whole position, Stallo says, surely unanswerably, "Once more then science is in irreconcilable conflict with one of the fundamental postulates of the mechanical theory. Action at a distance, the impossibility of which the theory is constrained to assert, proves an ultimate fact inexplicable on the principles of impact and pressure of bodies in immediate contact, and this fact is the foundation of the most magnificent theoretical structure which science has ever constructed, a foundation deepening with every new reach of our telescopic vision, and broadening with every further stretch of mathematical analysis." The fact remains, therefore, that either matter can act where it is not, or the theory of gravitation must go.

The mysteries and difficulties involved in gravitation are not exhausted by the two which I have discussed. Its action is wholly unsusceptible of interference by intervening obstacles, or, as Jevons says, "all bodies are, as it were, transparent to it; it is not subject to reflection or refraction, unlike the forces of cohesion, capillarity, chemical affinity, and electric or magnetic attraction, it is incapable of exhaustion or rather saturation. Every body attracting every other body in proportion to its mass, it is wholly independent of the nature, volume, or structure of the bodies between which it occurs, and its energy is unchangeable, incessant, and inexhaustible." It stands apart, therefore, like thought does, among the wonderful mysteries at our very elbow, to which we have no key, of which we have no explanation, a perpetual reminder of the impotence of the human mind, which has failed in understanding it, as of its strength in postulating it.

I do not propose, as I might, to take you through the similar difficulties attending a final and complete, or even an incomplete, explanation of light, heat, electricity, and magnetism, in regard to which, also, the theories that have done service for so long are crumbling away. I would merely refer to one or two. That what we call light, heat, electricity, magnetism, &c, are different forms of energy or force, and that this force can be ultimately analysed into modes of motion only, has been an almost universally accepted axiom of physics for a long time, to deny which was to raise issues about the questioner's scientific sanity. The latest analysis of the phenomena of radiant matter goes a long way to make all this very doubtful, and to show that some of these phenomena may have a really substantial corpuscular existence apart altogether from force, and the older philosophers may have been right after all in treating these phenomena as material. I do not say this latter conclusion has been established, but merely that it is no longer possible to treat a man who holds it as demented. What I further mean, is that this change of view has shown that either theory is equally an hypothesis. It ought to make us exceedingly modest when we find such fundamental truths, as those referred to have been supposed to be, resolved into mere ingenious theories invented by men to co-ordinate a certain number of facts. It is further consoling to think that if we have to go back to a corpuscular and material theory of light, we shall only be returning to what sufficed as a theory to Newton, who made his wonderful discoveries about light when treating it all the time as a very attenuated material substance.

Again, the kinetic theory of gases, which insists that their ultimate atoms are perfectly elastic bodies in constant motion and constant collision, is directly at issue with the general mechanical theory at the basis of nearly all physics that the ultimate atoms of matter are unalterable, hard, and incompressible.

Gases, says Kroenig, consist of atoms which behave like solids, *perfectly elastic* spheres, moving with definite velocities in empty space. Clausius and Clerk Maxwell say the same, while Lord Kelvin says: "By the modern theory of the conservation of energy, we are forbidden to assume inelasticity or anything short of perfect elasticity of the ultimate molecules, whether of ultra mundane or mundane matter."

On the other hand, speaking from the side of molecular physics, as taught in the highest seats of learning, Professor Williver declares that the concept of elastic atoms is a contradiction in terms, because elasticity presupposes parts, the distance between which can be increased and diminished. Father Secchi, another great authority, says: "While it is possible to admit the existence of elasticity in a compound molecule, the same thing cannot be done in the case of elementary atoms. Indeed, elasticity in the received sense presupposes void spaces in the interior of the molecule whose form is changed by compression so as to return afterwards to its original figure. Now, we regard the atoms as unpenetrable, and not as groups of solid particles, hence they cannot include void spaces which permit their dilation and contraction." Here, then, we have another huge impasse and mutual contradiction between the fundamental theories of closely allied sciences.

Again, it seems impossible to reconcile theindivisibleand ultimate atoms of the chemist with the newly postulated and infinitesimally smaller atoms by which electric discharges are supposed to be explained, and which form the substantial feature of radiant matter according to the accepted theory of its votaries. If the latter be true, then it would seem that the whole of the foundations of modern chemistry must go under, and we must reconstruct the science on other foundations than those laid by Dalton, Gay-Lussac, and Faraday.

Let us now pass on to astronomy. The most potent and influential theory which was ever proposed to explain the facts of astronomy was, I suppose, the so-called Nebular theory, the credit of which has to be divided between Kant and Laplace. Until recently it has dominated the whole position, and he would have been a bold man who dared to raise any serious questions about it. The Nebular theory postulates that the material now found largely aggregated together in the form of the stars and planets, including the earth, &c, was once dispersed throughout space, and was uniformly diffused there in a very attenuated form. Presently, by the action of some general attractive force, it came to be gathered together into large nebulous masses in the form of spheres, which began slowly to rotate, being impelled to do so from causes not quite clear, such as their forcible diversion or from their internal differences in density and irregularity in force. As these spheres parted with their heat they contracted, and as they contracted they revolved more rapidly round their axes in accordance with a wellknown law of mechanics. This increase of velocity in rotation led to the increase of centrifugal force in the equatorial region of the rotating masses, until it balanced and eventually exceeded the tendency to gravitate to its centre. This led first to the conversion of these spheres of rotating matter into spheres more and more flattened at the poles, and until they were shaped like discs or telescopic eye-pieces, and eventually to the successive detachment of equatorial rings or zones, which at first circulated round the residual mass in the direction of its original rotation; but by reason of their instability, due to the least departure from absolute regularity of form or constitution, they eventually broke up into parts forming one or more minor spheres or spheroids. These continued to revolve round their suns with a speed almost equal to that at which their materials were rotating at the time of their detachment. In most cases the whole material of such rings coalesced into a rough mass forming a planet, or a group of similar masses like our asteroids. Each of these planets also began to rotate on its own axis, all fn the direction of their motion of revolution. It again threw off rings, as in the parent system, which either retained the form of rings as in Saturn, or formed moons.

In favour of this magnificent scheme a great number of arguments were accumulated, *e.g.,* the existence in stellar space of nebulous masses in various stages of condensation, the evidence of increase of temperature in our planet as we approach the centre; the approximate coincidence of the motion of planets round their orbits, both in direction and plane; and its further coincidence with the direction of the sun's rotation, and of the rotations of the satellites with those of their parent planets; the spheroidal shape of the earth and of the other planets: the fact that as theory requires the number of detached rings forming planets increases as it should increase with the ever increasing rapidity of the contraction of the whole mass. All these facts and others seemed to have planted the Nebular theory in a position of impregnability.

Alas, however, our knowledge is always outgrowing our theories, and men have continually to confess that the hypotheses and schemes of the wisest among them have to be renovated or replaced periodically as our facts increase. If true, the Nebular theory must conform to certain conditions which are subject to precise mathematical treatment. The first serious blow it had to maintain was the discovery that all the planets do not move round their axes in the same direction as was thought. Uranus was shown by its discoverer, Herschell, to rotate in a direction opposite to that of the other planets, and its moons to do the same. Babinet, a very competent mathematician, in an article on the cosmogony of Laplace, next showed that the velocities of rotation of the several planets are really in excess of what they should be if our planetary system were the result of condensation from a diffused nebulous mass in the way he postulated. Starting with the fact that the sun moves round its axis in twenty-five and three-tenth days, he shows that instead of the earth revolving round the sun in a year, it ought to take 8181 years to do so, while Neptune ought to take 27000 centuries.

He further concludes from this that, if the entire mass of the sun had been expanded to the limits of the planetary system, it must have had a motion of rotation far too feeble to enable the centrifugal force to balance the force of gravity so as to lead to the separation of an equatorial ring from the total mass." These calculations of Mr. Babinet seem conclusive against the Nebular hypotheses.

The discovery of the fact that the satellites of Uranus do not revolve approximately in the same plane with the parent planet, but at right angles to it, and that one of the satellites of Mars revolves round its parent planet in one-third of the time required for the planet's motion show further how dangerous it is to draw general conclusions from limited facts. The Nebular theory has largely lost its hold, and is still doing so upon the more thoughtful students of cosmology, for every effort hitherto made to explain away or mitigate the difficulties, except by other difficulties, has failed, and we thus find ourselves compelled to say good-bye to another induction of the highest quality which, having done its duty in explaining the temporary knowledge of our fathers, must give place to some other induction equally tentative, which shall enable us or our children to explain all the accessible facts.

Let us proceed again. We have hitherto dealt with what men call the exact sciences, since they are based on mathematical induction and the rigid application of the laws of physics. Let us now turn to other sciences, which do not make the same claim, and which consequently lead their votaries into much greater and wilder excesses. Take the science of geology, for instance, upon which I have ventured to write a good deal, chiefly as a critic of its methods and its premises.

No science known to me is so dominated by metaphysical and scholastic premises. "The doctrine of uniformity," as taught and understood by the scholars of Lyell, upon which modern geology is based, is just as much a scholastic dictum as that Nature abhors a vacuum, and as little founded upon true induction. That the primary laws of Nature are immutable, that all we now see is subordinate to those immutable laws, and that we can only judge of effects which are past by the effects we behold in progress "are truisms of science. When this sound doctrine is perverted, however, to teaching men that the physical operations now going on are not only the type but the measure of intensity of the physical powers acting on the earth at all anterior periods, is to assume an unwarrantable hypothesis with no *a priori* probability." These are the words of a great geologist, and a distinguished observer and generaliser; they condense the geological philosophy I have unceasingly ventured to preach. Conybeare translated the argument he laughed at into a moreconcrete form. "Because, he says, a child grows two inches every year, therefore, that is the normal growth of a human being during the three score years and ten, which are his allotted pilgrimage." This is hardly a parody of the logic of some unilormitarians.

I have no more doubt that the geological record proves the former existence of catastrophes and periods of rapid and widespread destruction, unmatched in our day, than I have of my own existence. It was the creed of my masters, Sedgwick and Murchison, and Conybeare and Hopkins.

It has been well formulated by a great mathematician and physicist, Babbage. "No one appreciates more highly than I do," he says, "the labours of these older geologists" (that is Lyell and his school) " who first taught an earlier generation to estimate more truly the power and efficiency of certain forces, which act very slowly but continuously during long periods of time. The error in scientific speculation, against which they fought, was the error of laying exclusive or exaggerated stress upon forces of a particular kind—forces, the existence of which they did not deny, but which had worked only at comparatively distant intervals, and in alliance always with other forces, the operation of which is ceaseless. It is precisely the same error in principle, though exhibited in a different form, which is now exhibited by those geologists who attribute almost everything to running water and scraping ice. They are simply catastrophists in a new dress. They attribute extravagant power and stupendous effects to one form of force instead of to another. There is, indeed, one difference, and it is a difference in favour of the older school of catastrophists rather than of the younger, whereas there never could be any doubt of the adequacy of subterranean force to produce the effects ascribed to it, there is the greatest doubt of the adequacy of rain and ice to effect in any time, however long, the stupendous changes ascribed to them by Mr. Geikie. On the other hand, there is really nothing stupendous about these effects when they are regarded in connection with the known and visible effects of subterranean force. The highest ranges of mountains we have are, relatively to the circumference of the earth's crust, infinitely smaller than the puckers on an orange skin." Again he says:— "Magnitude is all relation. The store of time and force may be regarded as both unlimited, but it does not follow that in accounting for any given effect we are enabled to draw to an unlimited extent upon the one or upon the other. Extravagant demands may be easily made upon the one as upon the other."

To myself the moon is a perpetual object-lesson in these matters. So quiescent is its surface, so absolutely peaceful its life, that the keenest search with very powerful telescopes has hitherto failed to find traces of movement in it; yet its surface is so terribly rent and torn, so full of great gaping craters, each one big enough to hold all the world's volcanoes, of vast cliffs and scarps, which we can attribute to nothing save violent action at one time. I would as soon conclude from an examination of the moon's face that all this ruin was the result of the actual causes now acting on it, and acting with the same intensity that they do now, as I would that a man whose face is pitted with smallpox never passed through any catastrophe in the shape of that terrible disease, but that his appearance was the product of the ordinary wear and tear of life.

In pursuit of this mediaeval bogey of Uniformity, the current school of geology has perpetrated every kind of absurdity as it seems to me, for I deem it scientific absurdity when a man absolutely refuses to verify his premises, absolutely refuses to test the capacity of the forces to which he appeals for his effects, and, in questions where physics and mechanics must have the last word, pours contempt upon the methods of both, and simply contents himself with perpetually crying out "Great is our Goddess," not Diana of the Ephesians, but Uniformity.

This school continues to appeal to a fluid nucleus to the earth long after mathematicians of the analytical gifts of Lord Kelvin and Professor G. Darwin have proved that the earth must be as rigid as steel, or it would deform under the operations of the forces acting upon it and become a heap of ruins.

In order to avoid the stupendous difficulty of accounting for the distribution of the surface beds without the intervention of a widespread catastrophe, they have appealed to effects of ice action quite inconsistent with every known quality of ice. They have conceived it possible to pile ice up in mounds many miles high, oblivious that under such pressure it must inevitably crush, and thus dissipate its energy. They have treated as possible the movement of ice for many hundreds of miles across level plains, and up and down great depressions, unaware that no thrust can be applied to ice of more than a certain amount without also crushing it.

When asked to explain how this ice could be produced, or whence it came, they have appealed to astronomical and mundane causes of an ice age, everyone of which has been rigidly shown to be impossible and to involve absurdities.

At every turn these champions of an Ice Age are false to the creed which it was introduced to sustain, namely, the creed of Uniformity. As Babbage again says, "The inventions and imaginations to which the extreme glacialists resort are beyond all comparison, more violent than those which were common with the old convulsionists. Whole continents are built up upon the top of the existing mountains, which there is no proof whatever ever existed; and then these continents are all ground down by ice or washed away by ordinary surf, and yet so that not a fragment shall be left behind." This criticism seems to me the truest wisdom. After the most patient work of 80 years I myself abide by the lessons I learnt from my old masters, who taught me that no plainer witness is to be found of any physical fact than that Nature has at times worked with enormous energy and rapidity, and at others much more evenly and quietly, and that the rocky strata teem with evidence of violent and sudden dislocations on a great scale. These catastrophes were neither aimless nor lawless. On the contrary, they were the result of a law whose entire tendency we have not yet ascertained. It may never be ascertained, but it seems most certain to me that if ascertainable, its discovery has been postponed for a long time in consequence of the great mass of geologists having been in pursuit of the fantastic shadows of metaphysical reasoning.

I was still very young when I rebelled against the doctrines of Uniformity in the extravagant form it was being taught by Lyell's scholars. It was, however, rather on biological grounds than purely geological ones. You have all heard of the Mammoth, that great extinct hairy elephant whose carcases have been found buried in the frozen earth of Siberia, almost from one end of it to the other, and whose skeletons have occurred intact in so many parts of the Old and New World. The finding of these carcases with their flesh preserved so that the wolves can eat it, seemed to me, as a boy, to involve a great unsolved problem, and to be quite inconsistent with any doctrine of Uniformity known to me. It is clear these animals could not bury themselves after they were dead. It is clear that being buried in hard frozen gravel their flesh would have been pounded to jelly if it was attempted to thrust it violently into the ground, nor was the quite undisturbed condition of the ground consistent with such forcible burial. The ground must have been soft when the carcases were buried and must have become frozen afterwards. Unless it became frozen very quickly, the carcases would all have decayed away in the course of one or two of the very hot Siberian summers. This, again, is supported by other facts, such as that the Mammoths actually lived where their remains are now found, in an area now void of trees and of any possible food they could eat, that their stomachs are distended with the remains of trees that could not grow within many degrees of where they are found. That the young and the old are found together, and that they are found mixed with all kinds of incongruous beasts, horses and bison, and rhinoceroses and reindeer, and with great branches of trees, the former all intact, all buried directly after they died under gravel. Not deposited by any rivers, for they chiefly occur in the high ground forming the watershed of the so-called tundras; and the Siberian rivers have been shown to deposit hardly any wharp. This seemed to me, as a boy, as it seems to me now, and as it seemed to Cuvier, a puzzle and a paradox incapable of solution without the intervention of a catastrophe which could kill the animals in great masses in full health, could immediately cover them over with unbroken beds of gravel stretching for many hundreds of miles, and could so affect the climate that was previously a moderately temperate one that it became one in which the ground was perpetually frozen. The story is a very romantic one, and everything I could find to illustrate it I put in a big book published many years ago,'not a single argument of which has been as yet answered, and in which the lessons of Siberia were applied to many other parts of the world.

So much for geology. I must now say a few words, only a few, on the subject of biology. Here I feel rather afraid of my audience, for I am disposed to think that while you will tolerate heresies on other subjects, you will not on this one. Nevertheless, I must speak frankly to you. Recently a discussion—a not very edifying discussion —has taken place in the *Tiuwa* between two distinguished physicists (Lord Kelvin and Professor Lodge) and two distinguished naturalists (Professors Ray Lankester and Dyer)—all friends of mine—on the subject of whether the universe had an architect or not, and whether a tree or an animal affords greater evidence of this than a crystal. The problem as stated seemed to me a barren one, because, if I may presume to say so, these great men were all discussing, not a scientific issue, but a philosophical one, or, perhaps, I ought to say were discussing what looked like a scientific issue with metaphysical arguments. I felt what an advantage it would have been to them—all so much greater than myself in their own special subjects—if they had taken the pains to read a little up-to-date philosophy before entangling themselves in such obvious knots.

The drift of the argument on the part of the biologists was that animals and plants are mere automatic machines, and that the key to the machinery which constitutes them and makes them live and grow was discovered by Darwin, who explained everything by the ordinary laws of physics, and left no insoluble secrets, no mysteries, no room for unseen and unknown handsbehind any screens pulling the panorama of life along and arranging its scenes. It would not be opportune to discuss here the last of these clauses, but I would say something of the others. We are told, then, that there are no mysteries that cannot be solved by such mechanical or physical processes as are known to experience. May I humbly ask then: Has Darwin or anybody else helped us in any way to understand what thought, or consciousness, or memory, or imagination is, or how they work at all? Have they shown us how thought and consciousness are united with the bodies of men and animals? Would anyone dare now to repeat the experiment of the doctor of the 18th century, who said he had looked inside the skull of a living man and found no separate mind there, and argue that the consequence of this premise was that mind was a mere function of matter to be equated with physical forces with which it has not one other single factor in common, because one of its concomitant phenomena is progressive transmission through nerves?

Let us pass on, however, from thought to life. Are we one single step nearer the tremendous problem of explaining the difference between living and dead matter than Aristotle was? We no doubt have accumulated vast quantities of facts— phenomenal facts—about life, but is there any analysis which will enable a zoologist to distinguish a living egg from a dead one, or a living seed from a dead one, and to tell us what constitutes the difference, except the mere blind answer to the blind that the one is living because it lives, and the other is dead because it is dead? It does not enable us to march one single step to put together a number of odd and picturesque names by which the principle of life has been called by the simpler philosophers of an old time, and to laugh at them all in the *Times,* unless the substantial underlying puzzle of it all can in some way be met other than by a change from homely to learned sounding names, which mean just as much and just as little. The underlying puzzle, we may at once confess it, is entirely beyond our ken, and we do not seem to have any analytical method of reaching and solving it. We have no chemical or physical test by which we can distinguish a man immediately before from a man immediately after he is dead, except that the particular force (call it vital force or what you will), which made his heart beat and his brain think, has ceased to be. To equate that force as some would, with electricity or magnetism, or heat, or motion, which have no known factor in common, seems to me as preposterous as to attempt to square a circle or to equate any incommensurable quantities. The ideas are incommensurable.

To proceed, however. Darwin is naturally a great name to conjure with. In my view he stands with Newton, on the highest platform of British science, and was the greatest experimental naturalist who ever lived. It is, perhaps, a presumption in me to say so. All this, however, does not mean that Darwin's hypotheses and Darwin's explanations of the variety of Nature are beyond criticism, and as my friend Professor Dyer seemed to argue, ought to be accepted as Pope Leo's pronouncements are accepted by his devotees. This was not Darwin's own view. He was continually altering and modifying his hypotheses to meet the objections of his critics, and he was not one-tenth part such a Darwinian as his scholars are. He found many and striking analogies in the methods employed in producing and altering domesticated animals and plants, and applied the lessons to wild animals and plants with great success; and so long as the laws he laid down are treated as limited laws and are limited to the cases and classes of cases where they have been verified, all is well, but to treat them as laws of universal application seems to me now, as they seemed to me when I dared to say so in print soon after I was at school, quite illegitimate and preposterous. The result has been the crowding of zoological and botanical literature with a good many scholastic dicta which have no empirical basis at all, or very little.

I tfaink myself, notwithstanding what has been said on the subject, that to compare selection, where the human will is involved, in trying to produce a result, with selection as the result of the survival of the fittest, is a mistake of language, and opens a wide door to misconception, and that the term " natural selection " was itself an unfortunate one. Selection ought surely to designate something, involving the exercise of the will. Again, the " survival of the fittest" is a term susceptible of every kind of misapplication, and very ambiguous. If it means merely that the fittest things to survive survive, that is an identical impression like saying red is red and green is green.

When we translate the phrase into more concrete language it will be seen how difficult it is to admit it, as used by Darwin, for any but limited classes of facts. It is easy to speak of the struggle for existence in crowded areas where creatures elbow each other, and to argue that those competitors which are the strongest and best equipped must survive, and to test it by the struggle of grass on a lawn, trees in a forest, and deer in a park. But as I ventured to urge long ago, and I was never answered, this induction must be completely qualified when we find that the well-fed, the hearty and physically strong, as has been shown by great classes of cases, are much less fertile than the weak and the sickly, and tend continually to extinction in consequence; that animals and plants stricken by mortal disease or punished to the verge of death are much more fertile than animals and plants in vigorous health. The individuals may elbow each other so that the fat kine may make the position of the lean kine a very uncomfortable one, but the latter have their revenge in producing much larger families, as the Israelites in Egypt had, according to Bishop South, who said it was because the Egyptians insisted on starving them with lentils.

Again, it seems to me that the main contention of Darwin, that the great variety of life is due ultimately to the accumulation of small differences which have proved useful in the struggle for existence, has been very greatly, if not completely, answered by those who have urged that until differences have become appreciable they are useless. The same applies to the theory of mimicry or protective colouring. The initial stages of change in all such cases are useless, and the mechanical law of changes being dependent on survival of the most useful, that is, of the fittest, cannot work until the change is sufficient to handicap its subject.

These are only suggestions, some of them made long ago, and never really answered. They might be greatly multiplied, and are only meant as examples. My purpose in quoting them is to again press upon you the necessity of watchfulness, lest you should be tempted to treat great men's guesses and hypotheses as anything but guesses and hypotheses; to bid you not to be frightened by the production of bogeys, in the shape of great names, from pressing home enquiry wherever it may lead you, and to always adopt a critical attitude when you find a man particularly dogmatic and assertive, however eminent he may be, and, lastly, to remember that, whatever the secrets of Nature may be, whether they are to be eventually wrung from her by human skill and perseverance or not, our ignorance is, and must always remain, so stupendously greater than our knowledge, that we should be at least modest, reverent, sympathetic, and humble. Lastly, to me, and here I am trenching on forbidden ground, I feel constrained to say that after manifold studies in many fields I daily and hourly feel more and more the impossibility of realising how the whole thing arose and how it continues as it is, unless it lies in the grasp of something or somebody which I am not permitted to see or hear or feel, except through the analogies of mine own mind, and thought, and will, and initiative, a controlling and directing influence not unlike in quality, if stupendously unlike in degree and power, to my own mind, and which has in its hands the myriad strings that make the puppets in the show move on. I will conclude with some favourite lines of mine, written by a fine old philosopher and doctor, the author of the *lielvjio Medici,* sometimes described as a Free Thinker.

Search while thou wilt; and let thy reason go
To ransom truth, e'en to the abyss below,
Bally the scattered causes, and that line
Which Nature twists be able to untwine.
Give thou my reason that instructive flight
Whose weary wings may on thy hands still light.
Teach me to soar aloft yet ever so,
When near the sun to stoop again below.
Thus shall my humble feathers safely hover,
And though near earth more than the heavens discover,
And then, at last, when homeward I shall drive,
Rich with the spoils of Nature, to my hive,
Then will I sit like that industrious fly
Humming thy praises, which shall never die,
Till death abrupts them, and succeeding glory
Bids me go on in a more lasting story.

The Seedlings Of Geophytes.
By ETHEL SA1lGANT.

Plants are said to be geophilous when their green parts live above ground for a part of the year only. Strictly speaking, all perennial herbs, with the exception of a few evergreens, are geophilous in the colder parts of the temperate zones, for their green aerial shoots are killed by the frost early in the winter, and the plant has then nothing but its underground organs left. These will throw up fresh green shoots in the spring after the annual hibernation.

In such a climate as ours, however, the aerial parts of perennials are not killed until late in the year, and they reappear early in the spring. In mild winters, indeed, we must all have noticed how many of our common wayside weeds struggle through to the next spring without completely losing their foliage. Even some annuals survive to enjoy a second season. Such are chick weed, groundsel, and the purple deadnettle, which may all be found in flower about Christmas time in the South of England, provided there has been no prolonged frost. On the summits of Scotch mountains, however, the growing season is very much shorter, and here a few plants of truly Alpine habit are to be found. They have squat massive underground stems, with a wonderfully well-developed root-system, while through the short summer they show above ground only a compact rosette of leaves and a few flowers, often large and brightly coloured. These plants take refuge beneath the soil for a good half of the year. Their geophilous habit is pronounced.

Geophilous plants or geophytes are characteristic of all climates where a short growing season is followed by a long period in which the conditions are unfavourable to plant life. Besides alpine situations all over the world, such conditions prevail in the Arctic and Antarctic regions, with their short hot summer and long frostbound winter. Or, as in many parts of the temperate and subtropical zones, vegetation may be checked by recurrent periods of drought.

Such climatic conditions as these, when well marked, end by forming races of plants adapted to them. Geophytes withstand frost and drought by getting rid of their tender aerial shoots during the bad season, and seeking the protection of the soil for their tougher members. This is one device, but there are others. The winter's cold of a Scotch mountain is resisted as well by the wiry leafless twigs of the heather as by the hidden root-stock of the lady's mantle. The thorny and heath-like shrubs of Cape Colony are as characteristic of the veldt as the bulbous vegetation which we know so well, because Cape bulbs are largely imported to keep our greenhouses gay in winter..

Although the mild and uniform climate of these islands does not encourage the geophilous type, yet a number of species with this habit are included in our flora. Some of these, to be sure, are alpine—found only in our highest mountains—but others are common over lowland districts. A large proportion of them are woodland plants. They send up leaves and flowers early in the spring before the trees are full of leaf, and for the rest of the year remain underground, with the additional protection of fallen leaves throughout the winter. Such are the primrose, blue-bell, woodsorrel, snowdrop, lily of the valley, and many others. Some are hedgerow plants. The wild arum, with its massive tuber, is one of the first to send up shoots in the spring. Its great clump of leaves soon overtops neighbouring plants. But search for it in July, and you find nothing but a few pods of scarlet berries, overgrown and half hidden by a tangle of grass and hedgerow weeds.

The short season enjoyed by woodland and hedgerow plants depends on the want of light in such situations during the greater part of the year. This cannot explain the very limited period during which the orchids of our chalk downs show above ground. The flower-head lasts for about three weeks, I suppose, but I do not know in the case of any one species how long the leaves remain above ground, and I should be very glad of observations on this head from field naturalists.

The seeds of woodland and hedgerow plants must germinate readily enough in England, and with a fair chance of survival. They fall into a rich light soil, and are covered in the autumn by fallen leaves or the *debris* of the summer hedgerow flora. They are not exposed to extremes of cold or drought during the winter, and they have a long summer in which to establish themselves after germination. Orchis seedlings probably have a much harder time. I should much like to know whether the orchids of our chalk downs are frequently reproduced from seed. Some species ripen seed freely, but there must be immense odds against any single seed finding a convenient cranny in which to lodge, and suitable conditions for germination in the following season. The scanty covering of coarse grass on the downs can give little shelter to the seedlings.

Two years ago, at the Haslemere meeting of the South-Eastern Union, I read a little paper which described the difficulties besetting the seedling of a bulbous plant which has germinated in its native haunts. Bulbs are perhaps the most specialised form of geophytes, and similar difficulties are encountered by the seedlings of all plants in which the geophilous habit is well-marked. Since that time I have had occasion to consider the question more carefully, and have studied the literature of the subject. Two points commonly observed in the seedlings of geophilous seedlings, the meaning of which had then escaped me, now appear as clear adaptations to the circumstances under which such seeds germinate.

These characters are, the slowness with which the seeds of geophytes germinate, and the economy in the aerial shoot of the seedling during the first season in which it appears.

The seeds of geophilous plants commonly lie dormant in the ground for a long time after they are sown. In the Haslemere paper I attributed this habit—probably with justice—to a reminiscence of the long rest forced on the ancestors of these plants through the long dead season which followed the shedding of the seed. Further light is thrown on this subject by the work of Ilegelmaier, Sterckx, Schmid, and others, on the internal structure and development of such seeds.

Every seed contains within itself the germ of the future plant and a store of nourishment with which to start it in life. At the time when the seed is shed the germ may be more or less developed, and the food may be packed away either in some part of it or within a distinct body called the endosperm or albumen. The presence or absence of an albumen in the ripe seed is a character much used in classification, and it is commonly easy to determine from systematic books whether an albumen is found in the seed of any particular species or not. In this way I have ascertained that the seeds of a great many geophilous species have a minute germ enclosed in a massive albumen.

Now, in four geophilous species at least, the structure of such a seed has been followed throughout the long period of maturation, that is, after the seed has been ripened on the plant and shed from it, and before actual germination. In every case the germ is found to grow considerably during this interval. It begins as a spherical or egg-shaped mass of tissue, but by degrees the rudiments of the seed-leaf or leaves appear at one end and the rudimentary root is formed at the other. M. Sterckx, who followed this process with great care in the Lesser Celandine *(Ranunculus jicaria),* found it lasted from May, 1896, to the spring of 1898, in almost all the seedlings which he raised. I am told on good authority that in England the seed sown in the summer of one year comes up in the spring of the next, and this was also Dr. Schmid's experience at Tubingen in the south of Germany. The growing season in Belgium is shorter, if hotter, than ours, and this may account for the tprdy maturation of the germ described by M. Sterckx.

The seeds of *Ranunculus Jicaria, Eranihh hiemalis,* and other geophilous species, which ripen early in the season here, and germinate in the following spring, would not probably ripen seed until the end of the short summer in Alpine situations. Maturation would then in all probability occupy the whole of the succeeding season, and germination be postponed until the third.

The key to the slow germination of seeds belonging to geophilous species lies, I believe, in the necessity for rapid seed production, which is forced on such species in their native habitat by the shortness of the growing period. The germ is left undeveloped while the endosperm is stored with food, and by the time this is accomplished the dead season has arrived. This means that the following season must be devoted to the maturation of the seed: that is, to the development of the germ within it. In milder climates, as we have seen, the process is shortened.

The reduction of the green parts of the seedling during its first season distinguishes a number of well-marked geophytes, when compared with related species in which the geophilous habit is less developed. To understand the significance of this character, we must consider the conditions under which the seed of a geophilous plant, growing in its native station, germinates. The most critical period of a plant's life then lies before it.

Seedlings are peculiarly sensitive, both to frost and drought, and one or other of these foes must be faced in a very short time. The first care of the geophilous seedling is to transfer the food stored within the seed to an underground receptacle. Some seedlings never appear above ground during their first season of growth. They prefer to use up a certain part of the food at their command, in order to place the rest in safety; and they do not attempt to repair this waste by sending up a green shoot to procure fresh supplies. This, however, is an extreme course. Most seedlings form at least one green leaf during the first season of growth, but it is characteristic of geophilous seedlings to reduce the green surface in size.

Some instances of this may be given. The most specialised geophytes are found among Monocotyledons, and Monocotyledons have only one seed-leaf, which rarely possesses a blade of any size. Many plants belonging to this class send up no other leaf during their first season. Others keep their seed-leaf underground, and within the seed, using it as a sucking organ only. The first foliage leaf is sent up in its place. Some produce more than one green leaf in the first year.

Among Dicotyledons a considerable number of species possess two seed-leaves joined together for a great part of their length, the blades expanding at the top. In this way, as several observers have remarked, economy of material is gained. The two blades are kept in the erect position by the combined leaf-stalks at less cost to the plant than if they were separate. All these species are of more or less geophilous habit.

A few Dicotyledons possess one seed-leaf only: these are all well-marked geophytes. The converse, however, is not true: there are many geophytes with two distinct seed-leaves.

Sometimes the economy of green surface is obtained in another way. The seed-leaves of *Anemone nemorosa,* for example, are subterranean, and serve as storehouses for food. A single green foliage leaf is sent up between them in the first season.

In conclusion, I must again express my conviction that there is a great field for research in the correlation of the structure of seedlings with the conditions under which they thrive in Nature. I have seen here for the first time a herbarium in which the immature plants are preserved side by side with the mature form. Such a collection, accompanied by notes as to the conditions under which the seedlings were growing, would be of untold value to the philosophical naturalist.

A Late Keltic Cemetery At Harlyn Bay.

By E. ASHINGTON BULLEN, B.A. Lend., F.L.S., F.G.S., etc.

"Man is the Tale of narrative old Time."—Young.

I.—The Ancient People Of Harlyn, Their Characteristics, Culture, And Mode Of Burial.

Harlyn Bay is about *1* miles W. of Padstow on the North Coast of Cornwall. Its antiquarian interest was unknown until 1866, when some fine gold lunettes were discovered in digging a dewpond for sheep close to the "fish-cellars" near Cataclew. Thence onward till 1900 nothing transpired, in the August of which year, however, in commencing building operations, and searching for water, a slate cist was discovered about 15 feet below the surface of the sand. Mr Mallett, the owner, communicated with the Royal Institution of Cornwall, and that Society undertook the excavation of the blown sand there: something like 2000 tons were removed in 1900, and more has been removed since, with the result of laying bare about 180 slate cists. The Members of the Royal Institution of Cornwall took away for examination all the bones, skulls, and metal objects found during their excavation. Two skulls are in the Plymouth Athenaeum, others are at Truro. Those now at Harlyn Museum the proprietor has unearthed since.

Slate cists containing human skeletons had been discovered in Cornwall previously, notably at Trevone, about a mile distant, and also at Harlyn Bay, at the time when the road from St. Merryn was excavated to the shore, but the present discovery is far more extensive than any previously made; and is the only "find" that has led to systematic examination of the evidence on the spot.

The interments are of what are known as the " contracted " or "crouched" type. In Neolithic times (and afterwards, during the Bronze and Iron ages, among the descendants of Neolithic men) when a person died the legs were placed, before stiffening, in the attitude they occupied when sitting and sleeping. Probably the legs may have been tied up with withes or ropes, as we know the Peruvians used to do at the time of, and anterior to, the Spanish Conquest, as did also the prehistoric Neolithic race of Egypt, and as the Eskimo of the Lower Yukon still do. This custom of "crouched " burial has been well nigh universal, and is not See Schenk, *Lee Sepultures de Chamblandet*, p. 173. Bulletin Vaudoise, etc., 4e Serie, No. 144 the characteristic of a single period. In addition to those regions just indicated, it is known to have extended over France, Belgium, England, Italy. Germany, Switzerland, Hungary, Austria, Poland, Russia, Algeria, India; also in New Caledonia, and among the Mincopies of the Andaman Islands, the Maories of New Zealand, the Central Africans, the Patagonians, the Araucanians, the Puelches, and Charuas of South America, and the Australian natives. Thomas (in 1851) found that the ancient Babylonians practised "Crouched" burial. It is mentioned by Diodorus Siculus as occurring among the Troglodytes, a pastoral people of Ethiopia, and by Herodotus (7th century B.c.) among the Nasamones of Libya. The latter people did not allow their sick to die on their backs, but in a sitting posture. +

Many are the reasons which have been given for this posture, but most probably the two which are nearest to truth are, first of all the ease with which a body so treated can be carried: and secondly the fact that the posture is that of sitting and of sleep among races of a low stage of culture, and that when the dead was put into his last house, with food and implements (his "funeral furniture ") around him, the posture considered most comfortable when living would be accorded to him when dead. At Harlyn Bay the persons so buried were laid in the crouched position, on the right or left side, in this case the position in a sitting posture being impossible owing to the small height of the cists.

The same type of neolithic burial is also found in the next age, that of Bronze. In some cave-and barrow-explorations made by Salt, near Buxton, Derbyshire, in one instance at least a "crouched" burial was accompanied by a bronze dagger: Sir John Lubbock (now Lord Avebury), mentions fifteen "contracted" burials in tombs of the " Bronze " age and two in tombs of the " Iron " age (on the authority of Bateman), and four in tombs of the " Bronze" age in Wilts (on the authority of Sir R. Colt Hoare).

Dr Schenk, of Switzerland, produces similar evidence from Sion, Verchiez, Derriere la Roche, Chardonne, Belvedere (near Lausanne), from Bardonette, and from the wood of Sembres.

Since objects of bronze and iron have occurred among the Harlyn Bay interments, and cinerary urns of undoubted bronze age have occurred close by on the Harlyn cliffs, it seems simplest to suppose that here we have *Neolithic customs* continuing into the Bronze and the Early Iron ages.

As the skeletons at Harlyn Bay are found lying sometimes on the right, sometimes on the left side, it is possible that they were disposed in the contracted position in the cist in a limp condition after the *riyor mortis* had ceased. t Schenk, *op. cit.* p. 182.
A

Rev. D. Gath Whitley" contends that the whole cemetery is of Neolithic age, with later intrusions, and he gives some cogent illustrations of Roman and even later relics occurring in cave-earth at Eozarnia (Poland) with the mammoth and rhinoceros; in another cave, with the cave lion and hyaena; he also states that in dolmens Roman coins have often been found, but in all these cases the later objects have of course been intruded into deposits or structures of Paleolithic and Neolithic age.

The type of skull, however, is the strongest point in all such evidence. Neolithic man is generally considered to have possessed a long oval head, finely moulded, and curving from a somewhat narrow brow to a full round occiput. The "cephalic! index " in Neolithic skulls is about 70. The cephalic index of Bronze age man, a brachycephalic race, is about 81. If then we find that the cephalic index is between 70 and 81, and especially when it approximates to a mean between these figures, we must assume that there has been an admixture of races. That is just what we find at Harlyn Bay, and leads Dr Beddoe, F.R.S., to put the date of the people who inhabited the spot as living about the age just preceding the Roman Invasion. The lowest cephalic index at Harlyn Bay is 70, the highest is 82 22, the mean of 11 skulls being 75-19.

Neolithic man, found as far North as the Orkney Islands and in Ireland (at a later period), is generally considered to have been related to the old inhabitants of Spain (the Iberians or Iverni ans). Their descendants are now known as Basques, and they occur in Prance as well as in Spain. Since the Basques call their country Euskara, the Neolithic men are sometimes referred to as *Euskarianv.*

General Pitt-Rivers, in excavating at Cranbourne Chase, in a Romano-British Settlement, found the skeletons measured 5ft. 2£in. for the men, and 4ft. 4in.

for the women. These measurements correspond fairly well with those of the Basques and the Berbers of Africa.

The men of Harlyn, however, were 5ft. 4iin. and the women 5ft. lin. in height, and this indicates that some admixture of race had taken place to differentiate the height from the Neolithic type of skeleton.

Dr. Beddoe says *pp. cit.),* "judging by the eye, I should say that the bronze type is even now not uncommon in Cornwall, whereas in Wiltshire it is certainly rare." He compares the Harlyn Bay skulls with those from the Romano-British burials at Rotherley, and remarks on their general agreement.

The Harlyn Burial in the Light of Recent Archaeological I)ieorerie» in Europe, Journal of the Boyal Institution of Cornwall, no. xlviii., 1902. t *I.e.,* length-breadth index. See Dr Beddoe's masterly and laborious analysis of the Harlyn Bay calvaria, etc., in *Journal Royal Institution of Cornwall,* no. xlviii.

The next race, after the Euskarian, that entered Cornwall and subdued Neolithic man was the Goidelic or Gadhelic race, a Keltic people from the Continent. They conquered, but did not exterminate, the Neolithic race. Probably the urns' found on Harlyn cliffs were the last resting-places of Chiefs of either this race, or the next, the Brythonic wave of Kelts. These men of the "Round Barrow" race are thought to have introduced a higher culture. Among the bones from Harlyn Bay and Constantine Bay are those of a small ox (which Mr. E. T. Newton, F.R.S., thinks may possibly be the long-faced Keltic ox), sheep, perhaps goat, horse, pig, and goose. The horn cores of the ox are too fragmentary to be certain as to the variety to which they belong. The " Round Barrow " men lived by tilling the soil and rearing cattle, they could weave, they manufactured pottery, and made or traded in articles of bronze.

Whether they were really more cultured or only followed the business of war and conquest, enriching themselves by the work and industry of others, their bronze weapons made them more than a match for the men armed only with weapons of stone and wood, and so everywhere this Goidelic invasion pressed the earlier Euskarian population to the westward.

The Goidels or Gaels were followed by a powerful horde of Kelts from N. E. Picardy—these were the Brythons whose name has been generally given to our whole island.

The characteristics of these fair brachycepbals (or shortheads), were jaws macrognathous, strong and massive; and prognathous, *i.e.,* with prominent teeth, but with a good chin; cheek bones prominent; bones large; stature about 5ft. 9in. These men became everywhere the leaders and chiefs of the conquered population, *i.e.,* of the intermixture of Neolithic Euskarians and Goidelic Kelts. The Keltic Chiefs are described "as men with blonde hair, beetling brows, aquiline nose, and resolute mouth, commanding figures among their swarthier tribesmen."

It is safe to say broadly, though it is not well to dogmatize, that at Harlyn Bay, at the Late Keltic (or Early Iron) epoch, the interments of Keltic race were cremated, those of Neolithic descent, and those of mixed blood, were buried in the Neolithic manner.

Professor Windlet suggests that the Goidels buried, the Brythons cremated, their dead.

We can reconstruct slightly the life of these Harlyn people. They were intelligent, to judge by their skulls. They practised the arts of the potter and the weaver (as is evidenced by the fragments of pottery and the spindle-whorls found); they were perhaps fisher Windle, *Early Man in Britain,* pp. G9 and 70.

t Windle: *Life in Early Britain,* p. 112. men since certain shell implements! resemble line-sinkers; we have no trace of their boats, which would be probably of basket work covered with skins; they kept domestic animals, including the goose, which, however, if Caesar is to be trusted, the old Britons did not eat; perhaps they practised agriculture of a sort, Mrs. Mallett having found at Constantine cliffs a hoard of seed, under the ruined wall of a primitive building; this seed has been identified by the British Museum authorities as *Lolium perenne,* common Ray vGrass. Whether or no the Harlyn Bay race grew and ate such inferior grain there is no evidence on the spot to show. The occurrence of large quantities of broken shells of *Purpura lapillus* in the kitchenmidden, close to the cemetery, together with rounded pebbles of quartz large enough to break them, warrants us in the conclusion that the dwellers near Harlyn practised the dyer's art; a very large quantity of *Purpurae* would be required, since each mollusc would only yield a small drop of dye,

Fig. 1.—Broken Shells Of Pabpura Lapillu.s.

These people seem to have had some belief in a future state as edible landsnails and marine mollusca are found in the cists J No fish bones have, I believe, occurred in the cists or kitchenmiddens. But as fishing would be quite feasible from several places on the cliffs, boats may not have been needed. f Bullen: *Harlyn Bay,* 2nd edition, p. 14, and plate 5. *Helix nemoralis,* L; *H. hortentU,* L; *Pomatias elegans,* Miller; *Helcicn pellucidum,* L; *Patella vulgata,* L, similar to those found in their kitchenmiddens; also materials for procuring fire, with personal ornaments in some instances, such as bronze brooches of La Tene type, iron armlets, and slate and shell tools. In one case a *Heleion pellucidum* found in a cist was adherent to a human lower jaw, suggesting that such "shellfish" were put in alive for food. The cist is too far from the habitat of this mollusc for it to have reached the cist across sand by its own exertions. The occurrence of numerous human teeth, not belonging to the bodies interred, suggests a belief in charms, still a characteristic of the Cornish peninsula. From lumps of ruddle found in these cists we may perhaps infer that they painted their bodies, similar discoveries having been made in France, Germany, Russia, Moravia, and elsewhere.

Some of their customs were certainly cruel; two flattened skeletons, for instance, found under a V-shaped wall, suggest the practice of " foundation-sacrifice." These skeletons were covered by a very heavy slab of slate. Baring-Gould in his "Book of Brittany " says that such sacrifice was a result of primitive ideas of land-tenure, property in land not belonging to a tribe until the dead had taken possession of it.t

In another cist, in which the skull is broken and the bridge of the nose also severed by a straight cut, we have a decided suggestion that the dying were sometimes put into a cist before death, and deliberately "put out of their misery," a custom known to be practised among some North American Indians and by the Kamskatdales of N.E. Asia, and concerning this custom Mr. George Bonsor has brought forward indubitable proof of its having occurred among the inhabitants of pre-Roman agricultural colonies of the Guadalquivir, in Spain. J II.—The Implements Of Flint, Slate, &c.

We will now pass to their implements of slate, flint, and shell. The flint implements are of the ordinary Neolithic type —rather small, and for the most part roughly flaked. Their small size is due to the fact that their makers depended upon beach pebbles for their material. They were unable to procure flint locally by mining like the Neolithic men of Cissbury (Sussex) or Grimes' Graves, Brandon, Norfolk, of Aveyron in France, and Spiennes, in Belgium. The source of the flint here is a puzzle, and some geologists think that the chalk-with-flints once covered Cornwall and lands farther West, and that it has been removed by Sehenk, *op. cit.,* d. 179. See also Kvuns: *Ancient Stone Implement of Great Britain,* 2nd edition, pp. 202-4. And *Reliquiae Aqnitanicat,* pp. 22, 251, 297; and explanation ot plates, p. 94.

I See also Haddon " The Study of Man ", 1898, p. 347.

J Bonsor: *Let colonies agricole prt-Iiomaines de la valUeda Bitii.* Revue Archeologique, tome xxxv., 1889. Section: "Les lapidi's d'Acebuchal." denudation. The large deposit of chalk-flint on the summit of Haldon Hill, Devon, confirms this view and suggests that a large area of chalk has disappeared from the West of England, for, according to Jukes-Browne, F.G.S., the Haldon gravels represent the riddlings of 500ft. of chalk. The coarse gravel which occurs at Haldon Hill is believed to be of Eocene age, and to have travelled from West to East. He quotes Clement Reid, F.R.S., in support of his view. Dr. G. J. Hinde, F.R.S., has kindly examined the flint flakes, and reports that they are undoubtedly chalk-flints.

The slate implements are not quite easy to explain; they were found both in and above the cists. Slate is not a substance that suggests general suitability for the manufacture of tools; the Harlyn Bay implements, however, cannot have been the production of natural forces. Like the Eoliths, they can be arranged in classes. They are nearly all of the pointed type; tbey are neither to be found in the fields with the ordinary slate-rubble, nor on the seabeach, as some, without due examination or appreciation of the evidence, have suggested.

The slate tools from the Harlyn Bay cists are made of a slate that has a fine, even texture, and the local workmen do not

Fig. 2.—Slate Tools From Harlyn Bay. recognise the material as being from the immediate neighbourhood. It is quite hard enough for the purpose of boring, provided the material to be pierced is not too obdurate. Deal and soft woods are easily pierced with them. Slate tools have been found in Wilts, A. Jukes-Browne, *The occurrence of Marsupitts in Flintx on the Hal/Ion JIUU,* Geol. Mug., Decade iv., vol. xi., 1902, pages 449, 450.

Worcestershire, Inverness, and Skye, in interments of the Bronze Age. I have collected evidence of the use of slate for various implements, in my little book on Harlyn Bay, 2nd edition, pp. 17-21.

Mr. Reddie Mallett deserves to be congratulated on the assiduity with which he has examined every piece of slate and upon the beautifully made bodkins and other piercing instruments he has collected. He has lately (June, 1903) unearthed three slate tools of such beautiful workmanship, one pierced *for suspension,* as to substantiate the fact of their being artefacts (see Fig. 2).

It is possible that some of these tools were manufactured (like some of the pottery" otherwise useless) for "spiritual" purposes—as part of the " grave furniture" to accompany the spirit in the other world.

Some of the more finely pointed boring tools may have been used for the purpose of tattooing!; others would serve as awls

Fio. 8.—Tool For Bevelling Edges. for sewing skins, in the making of garments and sandals, or the skin-coverings of wicker coracles.

The shell implements are not so easily explained, but Mr. Santer Kennard, who has seen them, considers that they resemble the T. Bupert Jones, in a review of Greenwell's *British Barrow.* t Mr. A. Pott, F.S.A., in suggesting this, quotes Darwin, *Descent of Man,* p. 574, as stating that the ancient Britons tattooed themselves.

line-sinkers used by the race of fishermen who frequented the caves near Hastings. This purpose was suggested by Mr. Lewis Abbott, in 1895.

In addition to these a slate implement, for bevelling edges,! taken from a cist and a piece of slate with a bevelled edge from the Potter's Hut at Constantine are interesting.

Fio. i.—Slate Tool From Potter's Hut. III.—The Bbhains Of Terrestrial And Marine Mollusca Found In The Burials And In The Kitchen Middens.

We now come to the third branch of our subject—the mollusca found at Harlyn Bay. It will be well to add to these, those from the kitchen-middens: (i) At Constantine Bay, and (ii) surrounding the ruins of Constantine Church.

The bulk of the mollusca call for no more remark than has already been devoted to them in my " Malacological paper." But there are three or four points of interest to which I would allude.

1. There is the curious occurrence of a cuttle-bone *(sepiostaire)* in one cist. The question arises whether the *Sepia, officinalit* was put in the cist as focd, or whether the cuttle-bone alone was inserted as an ornamental offering. 1 incline to the first alternative. *Primeval*

Refuse Heaps at Hastings, Nat. Sci., vol. xi., pi. vi. and p. 94. + Bevelled edges are characteristics of the Harlyn Bay slate implements. J *Notes on Holocene Mollusca from North Cornwall,* Proc. Malac. Soc. vol. v., 1902, pp. 185-8. 2. Many of the "Mytilus" shells from the Constantine kitchen-midden contain charcoal. Evidences of hearths occurred in several places and at various levels in the kitchen-midden at Constantine Cliff. 3. The occurrence in great abundance also of the garden snail *(Helix atperta)* is interesting, its horizon at Harlyn and Constantine constituting evidence of its pre-Roman date.

All my time and labour spent at Harlyn Bay has been amplyrepaid by the discovery of *Hygromia montivaga,* Westerlund. This is

Fig. 5.—Hygrosiia Montivaga, Westerlund. a Lusitanian shell, and an entirely new record for England.! Our connection with Portugal dates to a time far anterior to the reign of Charles II! and the origin of the fauna and flora of England will always be a fascinating subject, the new shell being a fresh link in the chain of evidence. Mr. J. P. Johnson has recently discovered, near St. Michael's Mount, that curious underground "snail-slug," *Testacella maugei*—another Lusitanian species. This, however, is not the first time of its recorded appearance in Britain, but is an extension of its *habitat* to the southwest of England. In the south-west of Ireland, near Kenmare, occurs a beautiful slug— *Geomalacus rnaculosus,* also a Lusitanian species. Edward Forbes Mr. A. Pott suggests that the people of that day opened the shells by inserting them in the tire.

+ I have expressed my thanks elsewhere to Messrs. Edgar A. Smith, F.Z.S., and B. B. Woodward, F.L.S., etc., and Dr. Bottger, for their kindness in identifying this puzzling shell. Proc. Malac. Soc, vol. v., p. 188, *Gcohgical Magazine,* Jan. 11103, p. 28. explained "the origin of the plants and animals of the British Isles on the hypothesis that they were all diffused from a common

Fiu. 6,—The Kerry Slug. centre, and that consequently they must have been disseminated when these islands were continuous with those countries where the identical species were found. He brought forth geological evidence to support his assertions, and even went so far as to point out the fact that at one time, and that recently, dry ground existed between the S.W. portions of the British Isles and America."

In the short space that can be allotted to me, I can only summarise the evidence for the peninsular position of our islands before the commencement of the Pleistocene period.

According to Dr. Scharff, Edward Forbes "held that the Lusitaniant element of the British flora was of Miocene age, and that it survived the glacial period in this country. Among mollusca common to Britain and southwest Europe, he instances *Vyramidula rotunda* as occurring in Miocene freshwater deposits near Bordeaux. This hardy little snail has since flourished throughout Britain and Europe and it also occurs in the Azores.

Oneindication of the antiquity of a species is its" discontinuous distribution." Applying this principle to several of our British terrestrial mollusca Dr Scharff has reduced Forbes' ten molluscan provinces to two:— 1. England arid Wales (except southwest).

2. Southwest England and Wales and the whole of Scotland and Ireland.

In this second division six species occur which are wholly absent from the first. These six are all members of the "Lusitanian " fauna in the broad sense of the word.

They are:—*Genmalaeus maeulostu,* Allman, *Textacella maugei,* Ferussac, *Helix pisana,* Miiller, *.liygromia revelata,* Michaud, E. Forbes, *Memoirt Veol. Surrey (it. Britain, &e.,* vol. i., 184G, pages 33G *et teq.,* and Scharff, *History of the European Fauna,* 1899, passim, (in the Contemporary Science Series, Walter Scott Publishing Co.).

t The term "Lusitanian" indicates southwest Europe and northwest Africa. From this centre, and probably from now sunken land to the west of it, issued a fauna and flora of which we have abundant evidence in our own islands, especially in Ireland. *Scharff,* p. 308. *Helicella acuta,* Miiller, and *Pupa rinyens,* Jeffreys *(= anylica,* Ferussac). The importance of the discovery of a new locality of a known Lusitanian species nearly allied to *Hyyromia rerelata* is at once apparent. Moreover, it occurs in the lowest stratum at Harlyn Bay, namely the clay and rubble close to the underlying Devonian

Fio 7.—Map or British Islands When The Earlikr Members or The Southern Mioration Reached England. slate rock, and from its occurrence in that position it is probably to fce reckoned as of Pleistocene age. The following table gives the extension in space of the above six recent species as far as I have been able to ascertain it.

The table is compiled from the Distribution List of the ConcholoK'cal Society contained in Lionel Adams' *liritih I.tind mid F.W. Shell,* 1896; Lovell Keeve's *llritith Land and t II'. Mollmkx,* 1863; and Ur. B. B. Woodward's authority, the last for *Helicella acuta* only.

In discussing the subject of faunal emigration, Dr. Scharff considers that wood-lice, spiders, landsnails, and slugs require a land-bridge to convey themselves to an outlying island. He quotes Darwin's experiments and those of Baron Aucapitaine as showing this necessity.

Darwin immersed *Helix jiumatia* in sea water for 7 days, and again put it in for 20 days, and each time it perfectly recovered. During that time it might have been carried with an average ocean current, say on timber, G60 geographical miles, about 700 English miles. Darwin then removed the calcareous hibernaculum and, when it had formed a membranous one, immersed it another 14 days, and it again recovered and crawled away.

Aucapitaine's experiments were with 100 shells of 10 species, placed in a box pierced with holes, and immersed in sea water for 14 days: 27 recovered. Among these out of twelve *Powatias eleyam,* 11 revived. *Powatias eleyans* was better protected than the other varieties, because that species possesses an operculum.

Scharff, *op. cit.,* pp. 14-17-The introduction especially treats of the migration of animals from mainlands to islands.

Here, then, there is positive evidence that *H. pomatia* and *P. eleyans* might be transported to an island from the mainland.

Yet neither of these species inhabits the Canary Islands, Madeira, or Ireland. On all sides of Ireland dead specimens of *P. eleyans* have been picked up on the shore; but the species during all the centuries this has happened has not established itself in Ireland. If so well protected an animal as *P. eleyans* has failed to be propagated by sea, what chance would slugs or small nonoperculated shells have?

It is a well-known fact that slugs and their eggs are killed by a short immersion in sea-water, and even a light artificial spray of sea-water is too trying to their tender skins.

Dr. Scharff accepts Edward Forbes' view that the Lusitanian molluscan fauna is the oldest in the British Isles, and, from the occurrence of the same littoral forms of marine and semi-marine animals on the west coa3t of France and on the southwest coasts of England and Ireland, he concludes that the land-bridge was between the points which are now separated by a deep sea and submerged river valleys.

Among these forms there is one Echinoderm—the purple, rockboring, sea-urchin *(Stronyylocentrotm liridus);* two semi-marine beetles *(Ochthebius lejolitii and Aepophilus bonnairei):* five crustaceans *(Achaeus cranchii, Inachtts leptochirus, Gonoplax anyulata, Thia asrid.ua, Callianassa snbterranea);* two fishes *(lilenniits yalerita, Lepadoyaster decandnllii);* three molluscs *(Otina utix, Donax pnlitns, Amphidesma castaneuni).*

Scharff's reasons for believing this land-bridge to have existed seem unassailable:— 1. Common shore forms migrate along the coast just as land animals do.

2. Their eggs are carefully attached to fixed objects. 3. The young remain and grow old in some particular rock-pool, rarely venturing but a few yards from the spot where they first saw the light of day.

The case is strengthened by the fact that a millipede *(Polydesmns yallicus)* exists in Ireland, France, and the Azores; two earthworms *(Allolobophora veneta* and *A.yeoryii)* occur in Ireland, Spain, and the Mediterranean region; a weevil *(Otiorrhynchus auropnnctatus)* in Ireland and France, south of the Auvergne Mountains (see also Chapter VII. of Scharff, *op. cit.).*

Reasons for believing that a land-bridge stretched across the Atlantic come from a somewhat unexpected quarter. Dr. Alcock instances one of the weevers, known as *Bembrops caudimacula,* Stdnr., as occurring in Japanese waters, from which locality it was named by Steindachner in 1877; it was discovered in 1880 in the Gulf of Mexico, not recognized and renamed by Goode, *Hypsicumetes yobioides.* Some years after it was discovered off the Madras coast and again received a name. "Seeing that it is a ground fish its Alcock, *A Naturalist in Indian Heat,* 1902, p. 120.

occurrence in three regions so utterly remote.... requires a special explanation. My own opinion, he says, is that the curious range of this fish, which is by no means an isolated instance, can only be explained by the assumption of a continuous sea and shore connecting those points at some former geological period."

If a ground-fish require a continuous coast line at a former geological period to account for its present habitats, the case is

Flo. 8.—Map Of British Islands When The Sea Had Probably Invaded The Country From The East In Or Before The Pleistocene Period.

much stronger for littoral invertebrata, and stronger still for terrestrial mollusca.

The map, Fig. 7, explains how the Lusitanian species immigrated into our islands, and the map, Fig. 8, helps to account for their present "discontinuous distribution." Ibidem.

My sincere thanks are due to the Walter Scott Publishing Co. and to Dr. Scharff for permission to reproduce Fig. 6 and the maps from p. 60 and p. 126 of ScharfTs *Hhtory of the European Fauna,* and also to Professor T. Rupert Jones, F.R.S., &c, for criticism of MS. and reading proofs.

In concluding this all too brief notice of Harlyn Bay and its interesting problems, we can only regret that we are left, as in all such problems, to comparison, conjecture, and hypothesis; and must feel how true it is, of these men of the grey dawn of civilisation, that

"Man dwells apart, though not alone
He walks among his peers unread,
The best of thoughts which he hath known
For want of listeners are not said."

List Of Figuiies In Text.

Fig. 1. Broken shells of *Purpura lopillus;* scale J; from kitcheu-midden near cemetery. Fig. 2. Slate tools, Harlyn Bay; scale J linear. Drawn by Mrs. B. Ashington

Bullen. Fig. 3. Tool for bevelling edges, from x-a cist, Harlyn Bay; greatest length, 3in. (87mm.); greatest width, ljjin. (44-5mm.). Fig. 4. Slate tool from Potter's Hut, Constantino Island; length, *4J*in. (102-5mm.). Fig. 5. *Hyi/romia inoutivoga,* Westerlund (slightly enlarged). Fig. 6. The " Kerry slug ": *Geomalacm macuUu,* Allman. Fig. 7. Map of the British Islands and surrounding area at a time when the earlier members of the southern migration reached England. Only some of the rivers have been indicated: shaded parts, water; light, land. (Schaiff, *op. cit.,* p. 60.) Fig. 8. Map of the British Islands, showing approximately in what manner the sea may have invaded the country from the east during, or shortly after, the Forest-Bed period. Daik shading = water; lightly shaded and white = land. This map, of course, shows one stage only, the sea covered more giound a little later. (Scharff, *op. cit.,* p. 126-.)

Atmospheric Moisture As A Factor In Distribution.

By ALFRED O. WALKER, F.L.S.

Having lived in, or on the borders of

North Wales from 1852 to 1899, since I came to live in Kent, I have been struck by certain differences in the fauna and flora of the two regions, and I venture to bring before this Union of Scientific Societies a subject which has long occupied my attention. I do so in the hope of inspiring some of the members of the associated societies to assist by their observations in applying meteorology to the elucidation of biological phenomena.

As far as I know, the question of the effect of comparatively trifling differences of climate on the distribution of animals and plants has not received the attention it deserves. Of course, when the climatic differences between two regions is great, it is generally understood that the fauna and flora will also differ largely. Nobody in their senses would attempt to cultivate cocoa-nut palms in Kent or to acclimatise birds of paradise even in a Devonshire park! But it must have sometimes occurred to many who are here present to wonder why the nightingale, to whom, after its long migratory flight from Central Africa (where, according to Seebohm, it winters), an additional 50 miles or so would be nothing, should practically never be heard "north of the Trent," as the saying is, nor, I believe, in Cornwall or Devon, and again, the nuthatch, which remains with us through the winter, and whose various notes are so distinct and so insistent that it is impossible to mistake or overlook them in the southern or midland counties, is, so far as my observations go, unknown in North Wales.

As to plants, it is sufficient to point to what must be familiar to everybody who has travelled in North Wales, viz., the abundance of various species of fern in every hedgebank there, while here, in Kent, it is rarely that one sees anything but bracken *(Pteris aquilina).* In short, anybody who is accustomed to compare the plants and animals of the districts in which he has resided or travelled will admit that there is a marked difference in the fauna and flora of the East and West of England. The question before us is, how to account for it.

Taking the case of the nightingale, a bird that, as everybody knows, visits us only in summer, it is clear that the winter climate can have no effect upon it. It is also a purely insectivorous bird, and unless it can be shown that it feeds only upon some peculiar species of insects that are found in the eastern and not in the A most interesting paper on this subject by Dr. T. A. Chapman, is to bo found in *Entomologist's Record and Journal of Variation,* xi., pp. 29, 60. —Ed. western counties (in which case the question would be transferred from the bird to the insect), its distribution can hardly be determined by its food, for, as regards the quantity of common insects, the parts of North Wales with which I am most familiar are at least as abundantly supplied as Kent. Indeed, nothing has struck me mo.re than the freedom of the Kentish lanes from that plague of pedestrians and horses, the common house-fly, as compared with those of North Wales, and some of the nightingale's nearest allies, ri;., the chiff-chaff, willow-warbler and wood-warbler, all purely insectivorous birds, appear to me to be more abundant in those parts of North Wales than here.

If, then, the distribution of the nightingale is not affected either by the winter climate or the food supply in summer, we are compelled to seek some other cause. This can hardly be other than the summer climate; and we will now consider the differences between that of the eastern and that of the western portions of England. The following condensation of some remarks on "The Climatic Causes affecting the Distribution of Lepidoptera in Great Britain," in a paper read by me at Liverpool, before the Lancashire and Cheshire Entomological Society, in 1888, and which is not likely to have been seen by any of those present, may throw some light on the subject.

The total number of species of butterflies given in Stainton's *Manual* (1857), exclusive of the extinct *Chrysophanus dispar,* and what I may call the "casuals"—m., *Pontia (Pieris) daplidicje, Euvanessa (Vanessa) antiopa,* and *Issoria (Anjynnis) lalhonia,* is 62. Of these the southeastern counties! have 67 specimens; or over 90 per cent., and the eastern counties 56. On the other hand, the southwestern counties, though the most southerly, and the warmest in winter, have only 52; and the district of the Chester Society of Natural Science, comprising West Cheshire, Flintshire, and Denbighshire, which have also a comparatively mild winter climate, has only 87, or 58-7 per cent.

We find, therefore, that the number of species decreases as we go from east to west, and the same law holds good on the Continent of Europe, for, in Silesia, there are 124 species, against 94 in Belgium. The cause of the decrease can hardly be other than climatic. It is well known that the rainfall is, as a rule, greater, and that the climate is moister and warmer in winter on the west of the British Isles than on the east, but it is not so well known that the summers are cooler. The cause of this lies in the capacity of aqueous vapour in air to absorb radiant heat, in consequence of which less of the sun's heat reaches the earth in the west than in the east. This is shown by the following five years' averages of the *Proceedings of the Chester Society of Natural Science,* Part iii. , 1884, p. 62. + I use the divisions of Svmon's " British Rainfall." mean maximum temperature for the three summer months, June, July, and August, for 1878-82—

S.-E. Blackheath 69-5
S.-W. Plymouth 66-4
N.-E.Leeds 68-1
N.-W. Liverpool 64-2

Blackheath and Leeds, therefore, are hotter in summer than Plymouth, though nearly 100 miles further north in the one ease, and 250 miles in the other, and, to show what this signifies, I may say that of the above five years the difference between the mean maximum of the same months of the warmest summer (1878), and the extraordinarily cold summer of 1879 was only 4-2 while the *average* difference between Blackheath and Liverpool is 6-3. As regards the diurnal lepidoptera, there can be little doubt that it is the direct action of solar heat on the perfect insect that is necessary to their welfare, and the relative amount of which principally determines their abundance or otherwise. But with the nocturnal species it is not easy to see how this can be the case, and as we find that the proportion of those species whose larvte feed on the leaves of trees and shrubs, compared with those that feed on herbaceous plants is relatively smaller in the west than in the east, it is probable that this is due to the imperfect ripening of the wood of such trees, and the consequent injurious excess of sap in the leaves—a condition which is known to be prejudicial to the health of lepidopterous larvas.

It must be admitted that the difficulties in the investigation of the effect of climate on organisms are great. First, there are the meteorological difficulties. In spite of the enormous number of meteorological observations that have been recorded, the investigator will often find himself at fault. To find the amount of aqueous vapour in the atmosphere at any given meteorological station, he will find himself restricted to the results from the wet and dry bulb thermometers. Yet these are most unsatisfactory. As long ago as June, 1878, at a meeting of the Meteorological Society, Mr. Dines (a high authority on the subject) said that whatever tables may be used, the dry and wet bulbs could never be depended upon as giving more than an approximation to the dew point, and in this he was supported by the late Mr. J. G. Syuions, F.R.S. The conclusion to which I have come after some 20 years' experience in meteorological observations is that the range of temperature is the best practical guide to the relative dryness of the air in any locality. If we take Cambridge and Truro as representing the climates of the east and west, we shall find that their respective annual ranges on an average of five years, taken at random (1895 to 1899 inclusive), are Cambridge 51-2, and Truro 44-1. It will hardly be disputed that the climate of Truro is more likely to be a moist one than that of Cambridge, and there is no doubt that this is the cause of the difference in the ranges. But even the higher range is small, compared with what it is in some countries, such as the interior of Australia, where it is not uncommon to have a daily range of 40 to 50 degrees in the same 24 hours! It is obvious that this must have an effect primarily on plants (and probably on many animals), and secondarily, on animals that feed on plants, including man. And here I cannot help referring to the effect of the low temperature of last summer on the apple crop of this season. The fruit growers of this country have to complain, not of injury to the blossoms from frost, but of the scarcity of blossom. And I shall be surprised if the crop, even on trees that have had a fair show of blossom, is not in a majority of cases below the average, both in quantity and quality. The moisture of the atmosphere is, therefore, a matter of great importance from an economic, as well as from a purely scientific, point of view. It is true that we cannot alter climatic conditions, but we can adopt our modes of cultivation to them, and that this has been done to a great extent is shown by Kent having become "the Garden of England," for in no county are these conditions more favourable for fruit growing, and, should anyone be tempted, by the comparative immunity of our west coast from spring frosts, to plant orchards there, he will find that the lower temperature in summer is a far greater disadvantage. At Colwyn Bay, on the Denbighshire coast, in an exceptionally warm and sheltered situation, I found it impossible to grow plums, damsons, or cherries on standards.

Another meteorological difficulty is the deceptive character of mean temperatures. For instance, let us suppose a locality with a mean maximum temperature of 70 and a mean minimum of 60—the mean temperature will be 65. But another place has a mean maximum of 80 and a mean minimum of 50, again, the mean temperature is 65. Yet the two climates are entirely different, and so will be their fauna and flora. So, again, in comparing the temperature of, let us say, July, in one year, with the same month in another, even if the mean of the maxima and minima be compared, it does not tell you all you want to know, for, if you have in one year a tolerably constant range of moderate temperature, and in the other a fortnight of very hot weather, followed by a fortnight of very cold, bringing the mean of the two months in question to about the same value, the effect upon vegetation will be very different, and there will be a much better ripening of the wood in the latter case than in the former.

At the last meeting of the British Association, Professor A. Schuster, F.R.S., Chairman of the sub-section of Astronomy and Cosmical Physics, used the following words:—" What shall we say about meteorology? That science is bred on routine, and drudgery is often its highest ambition. The heavens may fall in, but the wet bulb must be read. Observations are essential, but though you may never be able to observe enough, I think you can observe too much."

He goes on to suggest that all observations should be stopped for five years, and the energy of the observers concentrated on a discussion of the results obtained. While I fail to see the advantage of stopping the observations, yet I feel very strongly the importance of making some practical application of the results obtained, both from the point of view of pure science, and of agricultural and horticultural economy. Every county should have at least one station at which meteorological and biological observations should be carried on together, with a view to the improvement of methods of cultivation and the general advancement of knowledge. It is in the hope that they may have some influence in this direction on the members of those societies whom I see before me, that I have ventured to submit my views to you to-day.

One word before I conclude. Whatever observations you may make, be sure of them before you record them. Of all the many hindrances with which science has had to contend, I do not believe there is one that can compare, for powers of mischief, with the inaccurate observer!

The Diminution And Disappearance Of The South-eastern Fauna And Flora

Within The Memory Of Present Observers.

By SYDNEY WEBB; assisted by CAPTAIN McDAKIN and GEOBGE GBAY.

Published life-histories, so much inculcated by the British Association for the advancement of Science, have, in the past, been often defective and even unreliable; for their details were too frequently arrived at by conjecture rather than observation, and we are all now agreed that, for the last fifty years, there has been far too much copying by authors from the works of preceding writers, without verification of the statements asserted as facts; and, this being so, every ascertained plant, mammal, reptile, or insect, and those groups of life still lower in the scale, assured to us as no longer obtainable, must be an irreparable loss to the more accurate and systematic recorders of the present day, who, better trained than those who preceded them, desire to unlock the storehouses of knowledge appertaining to animals and plants, with a view of reconciling and establishing points hitherto too readily assumed concerning their life-histories, by an investigation of their present condition, and so placing our knowledge upon a solid and lasting basis. The flora of our country is continually being added to by accidental circumstances, but the fauna only occasionally. It is no object of this paper to deal with either of these. The chief sources of destruction are due to man's agency, and nature's surface changes. The species lost or disappearing by man's agency are suffering from, among other causes, the eradication of rare plants and birds by too eager collectors, the urban and suburban builder, the country surveyor, the railway engineer, and those employed in drainage works or the recovery of waste lands. Those disappearing by the hand of nature herself, suffer from landslips, changes of surface soil, drought, and (its opposite) floods, alterations of sea and river levels, and starving out by overgrowth of other plant life, whilst the use of cattle to browse down and aid in the recovery of the soil of woods, and sloping fields, may be said to belong to both these classes.

It is not our intention in this paper to enter into any argument of tho causes mentioned. These have we believe been dealt with before, but it is admitted that they do exist, and in a greater or less degree influence the decadence of species both of the animal and vegetable kingdoms. This you all know. The main object of this paper is to place upon record, by the means of tabulated lists, the plants and animals, otherwise than very common, that occurred generally in East Kent from 20 to 30 years ago, and the position that they occupy now, June, 1903. Consequently this is only a tentative list, and no doubt it might be much increased at the present moment. We hope also that, improved and added to (or reduced) in the future by other workers, it may, if considered worthy of publication by our Committee, become a work of reference and usefulness in the county. But tabulated lists *(per se)* are dull reading, even to those interested in them, so we have inserted some notes under the different headings, not intended to be exhaustive of each list, but only explanatory of some of the items in them.

Of the larger animals we have little to say. Neither badgers nor foxes are less common in the district than thirty years ago, although the latter are not seen near the towns as formerly. They often breed in the cliffs, and no doubt their number is attributable to the votaries of the chase. If foxes were not hunted they would assuredly not be preserved, nor would living ones be imported to be turned loose for breeding purposes; remove these elements and reynard would quickly become one of our rare creatures. The polecat, on the contrary, has no friends, and it has become almost extinct in East Kent, not one having been recorded by a local taxidermist for over thirty years. Although rare in this neighbourhood it would appear that otters are rather upon the increase, if we may trust the report that a hunting kennel of otter hounds has this spring been established at Robertsbridge in the adjoining county. The small colony of seals at Beachy Head precariously holds its own; almost annually one or two of the number detach themselves from the main body, but they generally meet with death for their temerity. The old English black rat continues about the railway stations and wharves at the lower end of the town of Dover, but is dying or crossing out, not one having been recorded now for five years."-The water rat, or vole, is far from common even in the country districts, and is entirely absent from the neighbourhood of towns, the common brown rat taking its place.

May we be excused for saying that we are pleased to note a great decrease in our ophidians! Snakes, be they venomous or harmless, are not usually held in esteem, and we do not regret that they are much less numerous than formerly. They vary much in coloration and length; as regards the latter, it may be generally said that the common green snake is not so long, and the viper considerably longer, than the usual dimensions in other counties. No doubt the habit of the green snake and viper in associating in numbers for the purpose of hibernation, and the different changes of the surface soil during the winter months, will in some measure account for this dying out in the undercliff known as the Warren lying between Dover and Folkestone, as well as other similar localities; Since this was written, one has been taken in a trap here.— S. W.

but the diminution is not confined to these situations only. Dry inland banks and woods where there is little change from year to year tell us the same tale, and a walk through especially selected localities will now-a-days yield but one or two examples where twenty years ago ten might be noticed without fail. In East Kent vipers vary in colour from nearly black to pale tawny, and in many cases the latter are unieolorous, the ornamental diamond pattern down the back being entirely wanting. Among the working-classes the black examples are reputed to be more venomous in their bite, bnt this is probably a mistake, for the dark specimens are very scarce, and a comparison between the two cannot possibly be made in the open.

Turning to the birds, we note our first loss in the raven, which formerly nested in the chalk cliffs. Thirty years ago its familiar croak might commonly be heard, and,, turning the head upwards, a soaring black speck would be discerned in the sky almost beyond one's vision. It was the first bird sound the writer heard when he came to Dover, but for several years now it has had only a passing significance. All the falcon tribe are scarce, with the exception of the windhover, but one of the rarer of the harriers, seems loth to leave our marshes, and almost yearly rears a brood in one or other of them, but the bird we are most proud of is the peregrine falcon, which still frequents the cliffs and builds there. Five or six pairs of this noble bird nested both east and west of the town of Dover in 1897, but they do not seem to like the smoke from the colliery works on the west, and resent the Admiralty ones upon the east, yet one pair still continues in its old haunts. From the causes named, the gun, and the taking of the eggs, the species will probably not long remain with us. Equally with the hawk tribe, the gamekeepers are constantly reducing the list of our owls and shrikes, from common to scarce, and soon *rare* will have to be placed against their names; the long and short-eared owls and the great grey shrike are conspicuous among these, and, of course, many of the water-, wading-, and shore-birds which, unprotected by game laws, fall victims to the itinerant sportsman at all seasons of the year, whilst the neglectof prohibitive statutes on theothersideof thechannel has certainly contributed largely to the decrease of our smaller migratory birds of late years. Of these, the goldfinch, now quite rare in East Kent, wheatear, garden warbler, and the chaffinch, may be cited with others; and we have elsewhere noted the decrease in numbers of both the housemartin and swallow, the latter of which we seldom now see near the coast. The swift, on the other hand, holds its own, and the sand martin is increasing in some localities. Of the family of the tits it is difficult to speak, like the birds themselves, their numbers seem to come and go, sometimes seeming about to be lost to us and again increasing, but the longtailed tit is certainly not nearly so plentiful as formerly. Nor does diminution in the numbers of our visitors apply only to those coming over sea to us from warmer climes; the same is noticeable among our boreal and winter migrants, of these, the redwing and fieldfare at once may be mentioned, as well as several of the geese, ducks, and gulls. The list of our disappearing birds comprises 88 species, of which 25 are, or should be, with us all the year round, 11 are occasional or partial migrants, 18 summer visitors only, and 84 only show themselves here in winter.

They are as follows:—Peregrine falcon *(Falco pereyrinus)*; Hobby-hawk *(Falco subbuteo)*; Merlin *(Falco aesalon)*; Buzzard *(Buteo rnlyaris)*; Honey buzzard *(Perms apicorwi)*; Rough-legged buzzard *(Arehibitteo layopus)*; Kite *(Milvus ictinus)*. Thjs is almost too rare to be included in this list, nowhere in England is it now otherwise, but 80 years ago it might occasionally be recognised in this part of the country. Marsh harrier *(Circus aeruyinosus)*; Hen harrier *(Circus cyaneus)*; Montagu's harrier *(Circuit cineraccim)*; Short-eared owl *(Asio accipitrinus)*, a winter visitor only; Long-eared owl *(Asio otus)*, a winter visitor only; Waxwing *(Ampelis yarrulus)*, an occasional visitor; Hoopoe *(Upupa epops)*, an occasional visitor; Great grey shrike *(Lanius excubitor)*, an occasional visitor; Ring ouzel *(Merula torquatus)*, only seen in the district when passing through; Hedge sparrow *(Accentor modularis)*, greatly decreased in numbers latterly; Redstart *(Fttttticella phoeniciirus)*, now to be considered scarce; Black redstart *(R. tityus)*, in early spring, scarce; Wheatear *(Sa.cicola aenanthe)*, summer migrant; Sedge warbler *(Acrocephalusphraymitis)*, summer migrant; Redwing *(Tardus iliacus)*, winter migrant; Fieldfare *(T. pilaris)*, winter migrant; Pied flycatcher *(Muscicapa atricapilla)*, getting very rare; Garden warbler *(Sylvia lwrtensis)*, summer visitor; Grasshopper warbler *(Locustdla naecia)*, summer visitor; Blue-headed wagtail *(Motacilla fava)*, scarce to rare; Rock pipit *(Anthns obscurim)*, winter months only; Shore lark *(Otocorys alpestiis)*, migrant; Wood lark *(Alaula arborea)*; Chaffinch *(FrimjUla coelebs)*; Tree sparrow *(Passer wontanus)*; Brambling *(Frintjilla uwntifrinyella)*, a winter resident; Corn bunting *(Kmberiza miliaria)*, formerly very common, but now somewhat scarce; Snow bunting *(Plectrophanes nivalis)*, rare of late years; Goldfinch *(Carduelis eleyans)*, very seldom seen; Hawfinch *(Coccothrastes vulyaris)*, autumn and spring visitor; Mealy redpole *(Linota linaria)*, winter; Siskin *(Carduelis spinas)*; Twite *(Linota jiavirostris)*, winter resident; Wren *(Troylodytes pamilus)*, much more scarce than formerly; Spotted crake *(Cre.v maruetta)*; House martin *(Hirundo urbica)*; Swallow *(H. rnstica)*; Jay *(Gamdus f/laudarius)*, although not usually considered a migrant, a large number arrived from abroad and alighted at St.

In East Kent young thrushes are frequently called by this name.—S. W.

Margaret's about 20 years ago, and smaller parties have since been observed; Carrion crow *(Corvus corone)*; Raven *(Corvus corax)*; Kentish plover *(Aeyialitix cantiana)*, rapidly getting scarcer; Grey plover *(Syuatarola helvetica)*; Oyster catcher *(Haematopus ostraleyix)*, a summer visitor only; Turnstone *(Strepsilax interpret)*, winter visitor only; Dotterel *(Fndromias morineUus)*, only seen occasionally; Stone curlew *(Aedktremux xcolopax)*, already somewhat more than scarce, must become before many years one of our rarities; Curlew sandpiper *(Trinya subarquata)*, winter visitor; Green sandpiper *(Totanus ochropus)*, winter visitor; Purple sandpiper *(Trinya maritima)*, winter visitor; Grey phalarope *(Phalaropus ftdieariits)*, winter visitor; Red-necked phalarope (F. *hyperborean)*, winter visitor; Whimbrel *(Numenius phaeopus)* ; winter visitor; Knot *(Trinya canutu)*, formerly summer and winter, now winter only; Crossbill *(Lo.via curvirostra)*, summer visitor; Bittern *(Bntaurits stellarix)*, scarcely more than an occasional visitor now; Avocet *(Hecurvirostra aeocetta),* threatens before long to become one of our rarest British birds; Greenshank *(Totanus canescens),* goes north to breed. As winter visitors, the Bar-tailed godwit *(Limosa lapponica);* Black-tailed godwit *(L. belyica);* Bewick's swan *(Cyynus beicicki);* Hooper swan *((,'. miixicus);* Grey-lag goose *(Anser cinereitx);* White-fronted goose *(A. alhifrons);* Golden eye *(Clanyida ylaucion);* Tufted duck *(Fuliyula cristata);* Sheld duck *(Tadorna coniuta);* Long-tailed duck *(Harelda ylacialis);* Widgeon *(Mareca penelope);* Scoter *(Aedemia nigra);* Surf scoter *(A. perspieillata);* Velvet scoter *(Aedemia fusca);* Bed-necked grebe *(Podicepx yriseiyena);* Sclavonian grebe *(P. auritus);* Goosander *(Meryus nieryanser);* and the Bed-breasted merganser *(M. serrator);* Arctic tern *(Sterna macrura),* now very rare; Glaucous gull *(Larux glaucus).*

For reasons that are obvious no list of fishes is included in this paper, a short digest of them was published in the Handbook of the British Association, 1899.

The consideration of our losses in the insect kingdom would alone occupy a session; but as the lepidoptera are the most popular amongst young collectors, and best known to us, we devote our attention to this branch alone. The annual ebb and now of insect life is governed by extraneous circumstances of which we have at present no precise knowledge. A species is in abundance one year and almost wanting in the next, as, for instance, the tortoiseshell butterfly *(Aylais urtieae),* which lays its eggs upon a plant abundant everywhere. The larva; of this insect almost devastated the beds of stinging-nettles throughout East Kent two years ago, and by their numbers attracted universal attention, whilst the butterflies which succeeded them swarmed in the autumn, as well as in the following spring after hibernation, and therefore promised an equal if not greater number last year, but the conditions had somehow altered, scarcely a nest of young larvie was observed in a circuit of several miles around Dover, and but very few in the surrounding country. Any collector could multiply this case by many that he has observed, *but* the fact remains unexplained, although we think not hitherto generally recognised, that prior to a decadence of a species, there is, usually, if not invariably, one of these great outbursts of vigour, and that the number of aberrations or varietal changes from the ordinary type, is at that time greatly increased, as though the waning physical powers of the insects were for a time abnormally stimulated. After a series of years in which there is a declension in numbers, a decline so great that it would appear that the species would never again occur, another turn of the wheel will present it to our view once more, and an almost doubted British insect is recognised and speedily becomes a common one. *Myeria chrysidiformis* (scarcely obtainable outside the Dover radius) is an example of this, for its name was struck out of our lists and no British example was known in any cabinet for forty years, until it again forced itself before our notice. This ebb and flow of insect life naturally renders the subject of its decadence a difficult one with which to deal, for no sooner has it been observed that a species is lessening in numbers, and the fact been recorded in the magazines, than the next year, perhaps, seems to contradict the statement, and too often, a conscientious, youthful, and accurate observer may in this way be *deterred from aijain communicating with an editor.* Thus records, and often valuable records, are lost, and it is only to persons residing within a radius where certain species are known to occur, that we can look for exact knowledge when the loss of a species has taken place, whilst frequently it is lost to us by natural causes without any. one being able to say in what particular year it was last seen.

This part of our island, being so near the continent, is naturally more susceptible than any other to the incursions of foreign insects which for a time establish themselves and then die out. Several of our reputed rarities among the butterflies and hawkmoths do this, and one quite recent addition to our microlepidoptera, *Orapliolitha caecana,* for some years locally abundant on our cliffs, has thus proved itself of foreign origin. With these few preliminary remarks we let the list speak for itself, but give first a few names of species never common here, but which persistently occur to the number of one or more each year, and therefore appear to hold their own as securely as they did some 80 years ago. These are: *Euyonia polychloros, Polyyonia c-qlbum, Zephyrns betulae, Hippotion celerio, Hemaris bombyliformis,. Hyena bembeciformis, Oenistis quadra, Etitricha t/uercifulia,* ('*ymatopliora fluctuosa, Ayrotis dnerea, Taeniocampa leucoyrapha, Xylina semibrunnea, Ayrophila sulphuralis, Acontia luctuosa, Pericallia xyringaria, Eurymene dolobraria, Loboplwra hexapterata, Scotosia retulata,* and many among the Micros. Our list of decreasing species is as follows: *Papilio machaon,* from 20 to 80 years ago taken in small numbers every year in this neighbourhood; *F. ncldo'e cardai.iines: Melanaryia yalathea*—the practice of setting fire to the last year's grass stems on the slopes of the downs is greatly decreasing this pretty butterfly in places where it formerly was exceedingly abundant; *Pararye aegeria,* now reduced to a single vernal brood each year, and then not common; *P. meyaera,* not nearly so plentiful as formerly; *Epineplwle tithonm* and *E. hyperanthiis* are also much lessened in their numbers, and *Vanessa io,* rapidly becoming one of our rarer insects. We believe this will ultimately prove the case with all our gregarious and social lepidopterous insects; *Melitaea aurinia* and *M. athalia,* the marsh and wood species, are alike very scarce excepting in a few favoured localities; *Chrysophannx phlaeas* is still common, but not nearly so abundant as formerly; *Polyommatiu bellaryus (adonis)* is absent from many slopes and localities where it was quite common twenty years ago; *Thymelicus thaumas (tinea)* is reported to be one of our lessening species; we cannot, however, so regard it in this part of the country, but *Xisnniades tayes* is not so common. Of the *knthroceriis, Adscita ylobulariae* has become very rare in East Kent, as has its companion forester, *A. yeryon;* *Anthrocera trifolii* seems to have quite died out in numerous places where it was abundant 20 years ago. The Sphingids, generally, retain their numbers, *F.yeria chrysidif'ornris* and.*/',' ivlineumoniformis,* are certainly very much scarcer, as are both the chocolate-tips, *Pyyaera curtitla* and *l'. reelusa.* The Lithosiids are also retrogressing in all the genera. The incursions of the sea and systematic search for the lame of *Hypereompa doiiinida* has greatly reduced the numbers of this handsome moth; whilst *Arctia rillica* and *A. raja*, also, are not nearly so plentiful as formerly, nor is *Xemeophila plantayinis: Spiloxoma itrtuae (papyratia),* once so abundant in the Deal marshes, is very scarce indeed; whilst among the Lachneids, *Lachneislanestris&nAMalacosoma neustria*, like other gregarious species, are annually becoming scarcer; *Dimorpha vrrsieolora* is almost extinct now in the Kentish woods. Of the Psychids, *Epichnopteryx pulla (radidla)* and *Fumea casta (nitidella)* are much more rare than formerly. Among the Noctuids, the.following (88 in number) all seem scarcer, viz.:— *Cyiiiatophiira diiplaris,* (*'. diluta, V. or, Asphalia riaricornis, iiryophila perla, Craniophora (Acronycta) liyustri, Pharetra atrricoma,* 80 years ago abundant, how very rare indeed; *Leuva'nia littoralis, I., straminca, LK,phrayuitidis, Chortodes tnorrisii (bondii),* precariously lives on in its old locality in Folkestone, but many nights must be passed if a series be desired, instead of its flying by hundreds as formerly; *Xonayria fulva, X. despecta,* and.*V. sparyanii* all scarcer, the latter owing to reed birds, the bane and banned of Fen collectors; *Hydroecia micacea* (why this species should become rarer we are at a loss to know, but the fact remains!); *Mamextra abjecta, Charaeas yraminis, Ayrotis corticea,* and *A. cinerea; Triphaena interjecta, Xortna c-nigrum, X. baja, TracJtaea piniperda, Ceraxtix erythrocephala, Cosmia trapezina,* and *C. dijfinix; Eucosmia ochroleuca, IHanthoecia albi-*

maciila, Ciiculiia axterix, C. chamomillae, Heliothix armigera, for some years very abundant, probably the outcome of immigration; *H. peltiyera, H. maryinata, Heliudex arbuti, Acontia luctuoxa* (where formerly common), *Erattria fuxcula, Ilydrelia uncana,* and *Phytometra aenea,* and we think this list might be extended even further. Of the Geometrids, there are 50 species, viz.:—*Epione advenaria: Iiinnia crataeyata, Anyenma prunaria, Ellopia faxciaria, Eurymene didobraria, Pericallia syrinyaria, Ennomox J'uxcantaria, E. eroxaria, E. aiujidaria, Tephroxia bixtortata (crepuxctdaria), T. pnnetidata, Pseudoterpna cytixaria, Iodix rernaria, Hemithea thymiaria, Ephyra IHirata, E. pimctaria, E. pendularia, Axthena litteata, Eupixteria heparata, Acidalia triyeminata,* has become quite rare; *A. promntata, A.immutata,A.striyilata,eiyscarce; Cnryciapiinctata,Macarianotata, Minoa enphorbiata, Asjnlatex citraria, A. yilvaria, Abraxax ulmata,* formerly common, but now extinct for miles around Dover; *Lomaspilix marijinata, Larentia miaria, Kmmelexia albulata, E. deeolorata, Enpithecia pumilata, E. pimpinellata, E. exiyuata, t'ollix xparxata, Lubopliijia hexapterata, L. polyconniiata, Melanthia albicillata, Melanippe tiixlata, M. rivata,* much less common; *M. yaliata, Ant idea derirata, Camptoyramma bilineata, Pliibalapteryx vitalbata, Sruttixia itndtdata,* now quite scarce; *Cidaria fulrata, Eitbolia paluwbaria,* and *E. bipunctaria.* Of the Pyrales, 22 species:— *Odontia dentalix, Pyralix coxtalix, Endotrieha Jiamwealis, Stenia piinvtalix, Hotys lanceaiix, li. hyalinalix, Ebulea croeealix, E. verbaxcalix, E. xtachydalix, Pionea waryaritalix,* formerly annually at Folkestone on the *Cliortodex worrixii (bondii)* ground, but now extinct; *Spilodexpalealix, S. cinctalix, Lemiodex pulreralis,* now extinctsincethe changes in the Warren ten years ago; *Xola crixttdalix, X. ventimalix, X. xtriyida, Scoparia lineola, Melixxoblaptex anella,* not been seen for -some years; *Homemoma xinuella, Gyninancyla canella, Pemjielia camella,* and *P. ornatella.* Of the Tortricids

there are 58 species:— *Eidia minixtrana, Tortrix corylana, T. heparana, Lozotaenia xyloxteana, L. costana, L. xemialbana, Litlioyraphia campnliliana, L. penkleriana, Pldaeodex tetraquetrana, L'atoptria axpidixcana, C. iiiirroyraiiiinana, Halonota inopiana, //. foenella, Dicivraniplia lietirerella, D. xeiptana, 1. politana, Hedya ncellana, H. dealbaua, Anrhylopera diininntana, A. lundana, Bactra lanceulana, II. fuifurana, (1.1-ijjirapha literana, Peronea crixtana, Paramesia ferriujana, Semaxia -janthinana, S. rnjillana, Ephippiphora aryyrana, Stiyinonota nitidana, S. cuiupositana, S. leplaxtriana, Carpocapsa xplendana, (Jrapholitha citratia, Sericnrix fuliyana, S. cexpitana, S. euphorbiana,* very scarce now, and quite lost from the Folkestone Warren; *Eriopxela quadratio, Aryyrolepia stubborn))anniana, A. dubrixana, Calosetia niijrnnmciilana, Enpoecilia macubixana, E. atricapitana, E. cardnana, E. anyuxtana, E. notulana, E. mwtsehliana, E. rupicola, E. flaviciliana, E. rnficiliana, E. heydeniana, Lozopera xmeathmanniana, L. yiyantana,* and *Aryyridia dipoltana.* Of the Tineina there are 84 species:—*Dwtytoma salicella, Uhimabacche fayella, Tinea iniella, T. yranella, T. lapella, T. xemifulrella,* now quite rare; *Teichobia rerliiii'llella,* 20 years ago quite common, but since the food (hartstongue) fern has been so much sought after, probably extinct, certainly we know not where to meet with it; *Xemaphora meta.rella, Ailela deyeerella, Xemotois minimelUu, Micropteryx aruncella, M. tluinberyeUa, Erincrania.snbpitrpitrella, Sirammerdamia aidceUa, Hyponomenta riyintipunctatws, Phibaloccra quercana, Depre$xaria';nanatella, D. atmnella, D. propinquella, D. purpurea, D. pimpinella, D. pastinacella, Gelechia rufexcenx, G. rilella, G. domextica, G. luculella, G. maculea, (jr. tricolorella, G. moiiffetella, G. atrella, G. lucidella, G. naeriferella, G. inopeUa, Paraxia metzneriella, P. neuropterella, llarpella ijeofl'rella, llypercaUia clirixtiemana, llutalix senexccnx, Glyphipteryx fuscoriridella, Tinayma xericiellum, Cedextix yyxselinella, Gracilaria sired-*

erella, G, xtit/Hiatella, G. trinyipennella, G. onnnidix, Coleophora lutipennella, ('. viminetella, ('. yryp/iipcnnella, I', paripennella, ('. fuxcocuprella, ('. anatipennella, ('. rulneruriae, ('. albicuxta, ('. lixella, ('. onoxwella, C. conyzae, ('. lineidea, ('. aryentnla, Cosiiinptery.r orichalcella, Clianliodes UliyereUnx, Earerna cunturbatella, Chryxoclixta Jiaricaput, Cliryxitcoryx fextaliella, Elachixta apicipunctelta, E. atricniiiella, E. obxcurella, E. rufucinerea, E. ochreeUa, Tischeria dodonaea, T. waryinea, lAthocolletix cranierella, L. lautclla, L. fayinella, /,. lantanella, L. quinqueyuttella, L. trixtriyella, L. trifaxeiella, Bucculatrix ulwella, Xepticnla rujicapiteUa, X. xeptembrella, X. arcuata, X. mierothtriella, X. puterii, Trifurcula iniiiiundella: or a grand total of 288 species which are decidedly scarcer than formerly.

The loss of the larger animals, birds, and a few of our insects may be renewed from time to time from outside the radius, but with the exception of the Orchidacere and a few sporadic species amongst plants, the absence of a plant for several years in succession generally means a botanical extinction. Latterly, but too late for many of our scarce and beautiful species, the withholding of exact localities has been adopted. A true botanist and lover of flowers does not resent this, but only the selfish tin-filler, who will spoil a colony for the ideal hope of adding to his own herbarium by the means of an exchange of specimens. It is not sufficient, we hold, The substitution of tiled or slated farm buildings instead of thatch is probably the cause of apparent decrease.

to tell a collecting friend of a favourite locality, we should know the man very thoroughly before we introduce him to the plants he requires, for the more inveterate the tin-filler the more he will try to impose upon you, even to coming down to the spot again behind your back. For these reasons we do not give localiies in the following list, but all the finds or losses have been verified by at least two gentlemen with a knowledge of botany.

Following the plan adopted by us in this paper we prefix to our list of disappearing or threatened plants a few remarks upon some of the most interesting:—*L'altha paluttris.*—The king cup of Shakespeare, called also the marsh marigold, simply from the brilliant colour of the calyx, has but recently attracted the attention of the vagrant gatherers for sale, but their ravages during the last ten years, aided by the more important farm improvements of drainage of marshy lands, have already had great effect, and the plant is fast becoming more scarce. *Crambe niaritiwa.*—Seakale, although formerly occurring in the Isleof Thanet, and even 20 years ago sparingly near Dover, is now, we believe, nearly, if not absolutely extinct in Kent, Marshall and Hanbury's notes being all old ones. *Bramca oleracea*, L., is generally received as the classical name of the wild cabbage, although doubtless old Turner's, *Bramcum dobricum*, 1551-65, should have the preference; this plant has greatly decreased in all come-at-able situations, the sprouts being removed by coastguards and others for their rabbits and pigs. *Geranium pratenxe.—* Introduced to our notice in 1871; it then occurred in a small patch of several plants occupying barely a yard in length, at the edge of a bramble clump about 40 feet from the railings at the foot of the railway embankment, by the Rakemere pond, in Folkestone Warren. It continued to spread in a similar way, until it had, in 1890, covered an area estimated at 20 yards deep from the original spot, forming an edging to some half dozen clumps of bramble and hawthorn, and even invaded a portion of the railway bank itself, at this time some of the plants showed pale blossoms, but the flowers were constantly plucked by visitors who made its neighbourhood a favourite picnic station, so that probably no seed had a chance of ripening, since that date numerous earth slips and small changes have taken place in the soil, the plants have been much circumscribed in area, and now but a few specimens close to the railway bank remain; extinct in Thanet. *Vomarum paluxtre* has been recorded by many living persons from Minster marshes, but it appears now to have disappeared from there, and is confined to the neighbourhood of Dungeness. *Eryrujium maritimum* is carried away in large bundles for sale in adjacent towns, it is becoming very scarce and stunted in the majority of its former stations, and will soon be quite rare; its showy companion, *E. catiegtre,* has been exterminated from near New Romney. The different species of Orchis are all becoming scarcer, as are the Orobanches, for, with the exception of the ferns, the orchids, especially the more showy ones, fare more badly at the hands of the itinerant vendor of wild flowers than any other of our wild plants (the primrose only excepted), but few remarks need therefore be made upon them individually, drainage of lowlying ground has decreased the numbers of some species, this is especially the case with *Malaria jialudosa, Kpipactis palustris,* and *Orchis latifolia,* whilst within the last few years (. *purpurea (fusca)* has been rooted up by hundreds, and even the common field species, *O. mono* ani *().* maculata, are beginning to suffer in a less degree. All the species are scarcer than they were twenty years ago. The chalk of the Downs is not favourable for ferns but in some of the green lanes, as at Guston, the following have been recorded:—*Polypodium vulyare, Lastrea Jilix-mas, Polystichum aculealum,* and *Scolopendrium vulyare,* the latter, both in the broad-and narrow-leaved varieties, are more general than others, but fast disappearing. On some of the old buildings, as on St. Radigund's Abbey, *Rata muraria* and *CeUrach officinarum* may be found, with *Axplenium trichvmanex. Pteris ai/ uilina* is fortunately still common in its own localities. In some of the swamps, as at Ham ponds and Romney Marsh, *Lastraea thelypteris* is abundant, but the whole tribe has been greatly depleted in the last twenty years, and the number of specimens and species almost annually decrease. A little piece of sea-marsh below the cliffs, near Lydden spout, about 80 years ago, yielded some very interesting specimens of. plants, such as *Lactuca rirosa, Frankenia Uteris,* and

Euphorbia paralias, but this locality has been for years totally destroyed by sea encroachments, although published lists still record its flora without comment.

Our decreasing list is as follows:— *Helleborus viridis, Paparer hybrid urn, Glaitcium luteum,* much lessened by sea action; *Uochlearia anylica,* probably lost now from Thanet; *Erysimum cheirantlwides; Brassica oleracea* has been already referred to; *Diplotaxis tenuifolia, Lepidium ruderale L. draba,* an introduced species is rapidly becoming a noxious weed in East Kent; *Crambe maritima,* previously mentioned; *Cakile maiitima,* has disappeared from many of its recorded localities; *Frankenia laeris,* quite lost west of Dover, also at New Romney; *Silene nutans,* becoming annually rarer; *Silene maritima,* continues common in many places, but the area is more circumscribed and the plants are lessened in number; *Silene conica,* much rarer, and almost extinct near Deal; *Arenaria peploides, Sayina nodosa, SperyuLaria marina,* of this species Marshall and Hanbury say, the station Dover cliffs, seems an unlikely one for this species, indeed its station is a very unusual one, close to the Clock Tower on the Dover Parade; *Geranium pratense* we have already dealt with; *Krodium cicutarium,* much decreased by sea inroads; *lihamnus franyula, Genista tinctoria; Genista spinosa* seems to be getting scarce in Thanet; *Vicia sylvatica,* this pretty species was common in all the southeastern woods in Kent until twenty years ago, since which it has entirely disappeared in many where it was previously abundant; *Lathryus maritimus, Eryngium maritimnm,* our previous notes have mentioned this and the next species *E. campestre. Bupleurum tenuissimum,* disappears occasionally for a few years; *Foeniculnm vulyare, (Fnanthe fistulosa, (E. pimpinelloid.es,* not nearly so numerous owing to the removal of ballast shingle at Stonar near Sandwich; *Peucedanum officinale,* decreasing, but holds its own at Whitstable; *Galium verum,* amongst grass common, but on dry banks, &c, not nearly so plentiful as formerly; *G. cruciatum,* local, and as it generally grows in the open by hedgerows or roadsides is gradually being killed out or destroyed by plough; *Eriyeron acre, Matricaria salina,* killed out by the shifting of the shingle west of Dover; *Onopordon acanthium,* holding its own precariously in its own localities; *Carduus marianus,* not so equally distributed as formerly; *t'entaurea calcitrapa, C. scabiosa,* common but not so plentiful near the coast; *Lactuca virosa,* exterminated or destroyed by sea encroachments in places where formerly common; *L. saliyna* seems to be an anomaly, it has increased west of Whitstable, but is decidedly scarcer upon the east; *Statice armiyera* has decreased almost to total loss upon the cliffs of Folkstone and Dover, but is still very abundant on common land approaching sea level; *Statice limonium* is only becoming scarcer below the cliffs; *S. auriculaefolia: Samolus ralerandi* has but passing notice bestowed upon it in the flora of Kent. Although not otherwise recorded, Captain McDakin found it between Folkestone and Sandgate, but since its recognition in 1882 it has apparently become scarcer; *Convolvulus soldanella, Solatium dulcamara* var. *marinum, Hyoscyamus niyer,* sporadic in its appearance, perhaps slightly losing ground; *Linaria dantina, F. spuria, Mentha puleyium, Nepeta cataria, Beta maritima, Chenopodium vulvaria,* the loss of this plant with its peculiar odour of decaying fish will probably not be missed or regretted excepting by botanists, *Salicornia herbacea, Salsola kali,* is fast disappearing through sea inroads; *Atriplev pedu'nculata,* has greatly decreased, as also many of the exclusively maritime plants in certain localities; *Hippophae rhamnoides,* by falls of cliff and alterations by slipping of the surface soil several stations for this shrub have been lost; *Euphorbia amyydaloides,* dying out in the warren and many other places, but still common in and about woods; *Euphorbia cyparissias,* this plant is so conspicuous and peculiar in its growth that its large circular patches incur the "enmity of farmers, who are doing their best to eradicate it from their upland pastures in the few situations where it grows near Dover; *E. paralias,* below the west cliffs, probably nearly extinct if not entirely so; *F. exiyua. Asparayus officinalis,* our latest Kentish authority says of this plant, " Native (formerly) or an escape;" he admits to have seen *one* plant about two miles from Deal, upon which he evidently formed his opinion, we could take him to a spot where he could at his leisure gather fifty, but extraneous circumstances threaten the locality, and the removal of the surface soil to get more readily at the shingle, bids fair before long to make this a rare Kentish plant, although it has been met with in other localities. *Allium urunuin,* the broad-leaved garlic, is decidedly decreasing. *Juncus* species affecting damp situations are, of course, becoming fewer in numbers through farm drainage, and the same remark applies to *Scirpus* and *Carex,* especially *Scirpus maritimus. Trit/lochin maritiwum* appears to be holding its own in many stations, but it has been destroyed by sea encroachments in numerous others where it was once plentiful.

No doubt this list could have been extended further, but we have aimed at accuracy rather than an enumeration of what might be but transient fallings back among the species, and we have in no case included a specimen in the lists upon a note made in any one year, these are the results of several, and in numerous instances, many, years' observations, and as such we present them to the Congress members assembled here to day. They represent the decadence, not to speak of especial losses, of a grand total of no less than 500 species, which we think sufficient of itself to point a moral, although it naturally cannot adorn the tale.

International Communication. By A. T. WALMISLEY, M.I.C.E.
Engineer to the Dover Harbour Board.

This is a subject which is not only intimately connected with the growth of trade and progress, but is also associated with the restoration of health. Europe being no longer regarded as a theatre for war, the answer to the universal question in table talk—" when are you going away?"—is deemed to be more or less incomplete unless it is contemplated to cross the ocean, and the means for this sea-passage, coupled with the provision of well-sheltered harbours and landing stage accommodation upon both ends of the sea journey, have involved the consideration of some of the largest schemes that have been projected by engineers of any country.

From the days when Julius Caesar crossed the channel by galleys propelled by oars and sails, the Straits of Dover have always formed a favourite route to and from the Continent, rendering the county of Kent, the gate, as well as the garden, of England. The Society of Antiquaries' *Transactions* for 1862 contain interesting correspondence on the question of the exact spot where Caesar landed in Britain, B.c. 55, in connection with the views expressed by the Admiralty on the effect of the tidal stream off Dover, and a very clear.and concise historical account of the route by Dover and Calais by Mr. James D. Paterson, Assoc. Mem. Inst. C.E., was published in the year 1894, a copy of which is here open for inspection. More recently we have a popular illustrated guide to coast resorts accessible by the South Eastern and Chatham Railway, which has been ably prepared by Mr. W. T. Perkins, and is published by Messrs. McCorquodale and Co., of London, in which he reviews the development of sailing, paddle, and screw vessels, adding an account of the new turbine steamer "The Queen," the latest addition to the service, upon which the author will have more to say at the conclusion of this Paper.

Tunnels, bridges, and submerged tubes have been proposed to obviate the rolling and pitching of a floating means of transit. From an engineering point of view, no scheme is impracticable, but the great outlay attending their construction would prevent them becoming a commercial success.

As regards tunnels driven beneath the bed of the sea, geological inquiry reveals no great fault or break of continuity in the strata of the lines selected.

The route from near the South Foreland, crossing to three or four miles west of Calais, passes through the lower or grey chalk, which is less permeable to water than the upper or white chalk, and being freer from cracks or fissures, suggested desirable strata to work in. Trials made upon the English coast to a depth of 470ft. below high water level, gave 175ft. of upper or white chalk, and 295ft. of lower or grey chalk. Trials on the French coast to a depth of 750ft., showed 270ft. of upper or white chalk, and 480ft. of lower or grey chalk.

As an alternative route, Dungeness to Cape Grisnez has also been suggested so as to avoid any chalk fissures, the work on this line of route being in the Wealden formation, consisting of very strong clay beds of freestone and freshwater limestone. Tunnelling through water-bearing strata is no longer a difficulty, with the use of the Greathead shield and compressed air, as evidenced in the construction of the Blackwall tunnel under the river Thames, and several parts of the modern tube railways throughout London. Submerged tube railways, some with ventilating shafts at intervals, have also been proposed.

From a point in close proximity to Dover upon the English coast, to a point in close proximity to Cape Grisnez upon the French coast, the sea bed is stated to be comparatively free from hard rocks and to consist of coarse sand, gravel, and clay. Average depth of water 110ft., maximum about 200ft. It was argued that as the current alternated, a bank would be formed against each side of the tube by silting up. This effect at the stated depths is, however, questionable. Some schemeshave proposed a submerged tube built inside a horizontal cylindrical chamber, which is constantly pushed forward as the construction of the permanent tube proceeds.

As regards *bridges,* interference with the navigation of the Straits has to be considered—a bridge was suggested 50 years ago, M. Thome de Gurnard, Engineer of the Department of the Straits of Calais, first promulgating the idea. Another French engineer, M. Verard de Sainte Anne, later on suggested an iron bridge supported by 840 piers.

There would in the case of a bridge be no trouble in ventilation, and no risk of inundation as in its rival, the tunnel. The scheme is perfectly feasible, but subject to exposure of storm and tempest. From Shakespeare's Cliff to Cape Blanc Nez, a short distance from Calais also from the South Foreland to Cape Blanc Nez, were the sites suggested.

In *The Engineer* of January 27th, 1893, is given a design for a proposed channel bridge in connection with a suggested Channel Bridge and Railway Co. Cape Grisnez to Folkestone, 24 miles, with 120 piers, was another proposition. The late Sir John Fowler and Sir Benjamin Baker advised thereon, and confirmed the opinions expressed upon the solidity and stability of foundations, and with the assistance of M. Renaud, a French engineer, recommended a straight line from Cape Blanc Nez to the South Foreland, with 78 piers, 147ft. long by 65ft. wide, and spans arranged alternately in distances of 1812ft. and 1640ft. The examples afforded by the Brooklyn Bridge, with piers of masonry 275ft. high, supporting spans 1640ft. between bearings, and, later on, our wellknown Forth Bridge, encouraged the projectors. For the channel bridge it was proposed to place the platform 200ft. above the level of low water, and the deepest pier would be 167ft. below low water. It was intended to adopt masonry piers to a height of 46ft. above high water, surmounted by steel columns, upon which would rest the girders carrying the platform of the bridge providing for a double track, at a total cost of £82,740,000.

Dealing with the history of the development of steam vessels, the author may note: That in 1808, Symington built on the Clyde the first practical stern wheel steamer, and that it was run on the Forth and Clyde Canal. The first steamship to perform a sea voyage was also built on the River Clyde in 1818, and was 79ft. long, beam 16ft. Views of Dover and Calais early passenger boats, "King George" and the " Ondine," are given in Perkin's *Popular Coast Guide,* pages 146-147. The Dover and Calais,

"Spitfire" was built in 1828; 88ft. long, 80ft. 8in. over paddles, and took three hours to cross. The "Firefly" was built in 1880, and the "Onyx," built in 1845, 137ft. long, 21ft. beam. The "Onyx" carried the mails for the Admiralty. The Admiralty ceased to convey mails in 1854. The "John Penn" was running from 1854 to 1862; 180ft. long, 20ft. beam. (See view in Perkin's *Popular Coast Guide*, p. 148.)

Between 1820-1880, in the good old times of sailing packets, the number of passengers between England and the Continent did not exceed 80000 per annum. To and from Calais, 1899-1902, average 286000; to and from Ostend, 1899-1902, average 122801; total (say) 409000 annually. It is not the distance, 21£ knots (about 24f statute miles), from Dover to Calais, but the nature of that distance, which produces the horrors of the channel passage, to alleviate which, improvements are constantly made with the view of obviating what has been not inaptly styled "one of the greatest evils that can affect humanity." Captain Morgan was the first Marine Superintendent under the L.C. & D.R. Co. He was 84£ years in office, and was succeeded by Captain Dixon. During 1846-1861 the South-Eastern Railway Co. ran a boat to Calais and back daily. The "Onyx" was the first vessel to land passengers at the Admiralty Pier, January, 1851 (the year of the Great Exhibition). The line between Calais and Paris was completed in 1848. The London, Chatham, and Dover Railway was opened in 1861.

The Belgian Mail Packet Service was established in 1846, being at first jointly carried on by the British Admiralty and the Belgian Government. In 1868, the Belgian Government took the entire service into their own hands. The steamships known as "La Flandre" and "Ville de Douvres" are exactly similar to the "Prince Albert," while the "Princess Henriette" is exactly like the "Princess Josephine." "Leopold II" and "Princess Clementine" are also exactly like the "Princess Henriette." There is an installation of wireless telegraphy on all the Belgian Government's mail steamers.

Ferry steamers to convey the train over have been proposed. The Danish Railway Ferry steamers are a success, aided by a lifting bridge to adjust to tidal level with the shore. The length of the platforms required in Dover and the rise and fall of the tide there, are against the application of the system. Passengers need shelter in rough weather, and to be able to walk on deck in fair weather.

The Bessemer Channel steamer, 1875, was an independent venture. It was 850ft. long, 40ft. wide along deck beam, double ended, 65ft. wide over paddle boxes, drew 7ft. 6in. of water, speed 20 knots, and was designed by Sir E. J. Reed, the late chief constructor of the navy. Two pairs of paddle shafts 106 feet apart, between which the Bessemer saloon, 70ft. long by 35ft. wide by 20ft. high, was swung on a longitudinal axis at its fore and aft ends, having a central enclosure, in which, by means of a hydraulic apparatus actuating pistons, a skilled operator sought, with the aid of spirit levels and a small, heavy disc wheel, to counteract the roll of the waves. The vessel had a low freeboard of about 50ft. long at the bow and stern ends, which was deemed to contribute freedom from pitching, and give longitudinal steadiness, while the side paddles offered resistance to rolling. The effect of the broken water of the forward paddle wheels was found to be slight in its action upon the after pair of wheels, so that the difference between the revolution of the two wheels was never greater than one or two revolutions. The "Bessemer" was externally an ugly looking vessel, but internally quite palatial, and as the occupants of the saloon were those who paid, the owners studied their customers more than the criticisms of those who did not cross the sea. The saloon rested on massive beds of indiarubber, but, as a sea-going vessel, proved unmanageable, and the saloon had subsequently to be made a fixture.

The "Castalia," or twin ship, was suggested to the inventor, Captain Dicey, by the outrigger boats of the east, and was 290ft. long. Two halves of a longitudinally divided hulk, having their inner sides flat and vertical, to obviate any obstruction to passage through the water, were placed 26ft. apart, connected by braced framework carrying a raised deck enclosing cabin space. The twin vessel had a beam of 60ft., and ran during the summer of 1876-77, and was also an extra to the mail service. The first Calais-Douvres ran in 1877, and was 800ft. long, 61ft. beam. It consisted of two complete vessels placed 26ft. apart, and strongly bound together by a system of girders, and attained speed 18·2 knots. The boilers and funnels soon became worn out. Though purchased by the L.C. and D.E., the vessel soon became a white elephant. The "Calais-Douvres" of 1889 was 824ft. 6ins. long, 68ft. 6in. over paddles. The Wellington dock entrance was widened to 70ft. to admit her. "The Empress," "Victoria," and "Invicta" are other vessels, the latter fitted with two rudders so as to travel in and out of Calais Harbour (where there was no room to turn). The friction of the sides of these experimental vessels was great. So important an element is the friction of the immersed surface that a captain and his engineer can always tell the effect of an extra few inches of immersed depth. An efficient draught, however, prevents the vessel acting like a tea-tray. Long vessels are considered to steer more steadily in rough weather than short ones, though when the bow of a long vessel gets under shelter of a pier at a harbour entrance, a heavy sea striking her on the quarterdeck might drive her against the other pier if the entrance is not wide enough for clearance.

The new turbine steamer which was christened "The Queen" when launched at the works of Messrs. Denny and Brothers, Dumbarton, is an enlarged example of the vessels which have proved satisfactory for river purposes on the Clyde, with the use of steam turbines. The new vessel is 810ft. long, with a moulded breadth of 40ft. and a depth of 85ft., and has a complete awning deck. In large dock houses on this deck, there are a number of special cabins provided for the convenience of passengers.

There are five propellers fitted by the Parsons' Marine Steam Turbine Co., Lmtd., of Wallsend-on-Tyne, one centre propeller which causes the vessel to travel forward, and is fixed in the stern of the vessel, and two propellers upon each side at suitable distances apart, which serve for manoeuvring. These can be caused to move ahead or astern as required. The machinery actuating these propellers consists of three of Parsons' compound turbines, *viz.,* one high pressure in the centre of the ship, and two low pressure, one on either side thereof. Each turbine drives a separate shaft. Vanes are fitted on the shaft which revolves within a fixed cylinder, having vanes fixed on its inner surface in such a way as to direct the passage of the steam from one vane surface to the other, thus forming the turbine which turns the shaft and actuates the propeller. Inside the exhaust casing of each of the low pressure turbines are placed the astern turbines, which are in one with the low pressure turbines, and operate by reversing the direction of rotation of the low pressure turbines which drive the outer shafts.

In ordinary travelling forward, the steam from the boilers is admitted to the high pressure turbine, and from thence to each of the low pressure turbines to the condensers, and the whole three shafts are at work. The comparative estimated revolutions of the centre shaft are 500, and with each of the wing shafts 700. When coming alongside a jetty, or for manoeuvring in and out of harbour, the outside shafts only are used, and the steam is admitted by suitable valves directly into the low pressure turbines, or into the reversing turbines, as required. By this arrangement great manoeuvring power is attained, and the port or starboard engines are capable of being worked ahead or astern independently of each other, and of the high pressure turbine. The high pressure turbine will, in the meanwhile, revolve idly, its steam admission valve under these circumstances being closed, and its connection with the low pressure turbines being also closed by means of nonreturn valves.

Turbines are capable now-a-days of being made to run at any speed between zero and a maximum. The " Queen Alexandria," a vessel at present running on the River Clyde, experiences no difficulty in being easily manoeuvred when coming alongside a pier, and there seems no reason why the S.E. & C.R. Co.'s new vessel, after attaining the average speed at sea of 21 knots, should not be equally easily handled at a pier or landing stage, and give as great satisfaction as an ordinary twin screw, while in the going astern there will be none of that objectionable vibration which is felt even in the most modern twin screw balanced arrangement.

The author visited the works at Dumbarton while the new turbine vessel was in progress, when the following description of the steamer was given to him, for this lecture, by the builders. Amidships in the vessel is the smoking-room, framed in wax oak in a free classical design, and upholstered in leather. The seats are arranged in bays with small tables, and the floor is laid with ornamental tiles. Above this first-class accommodation, there is a long promenade deck extending out to the ship's side, which shelters the awning deck in wet or rough weather, and provides a large promenade for passengers in fine weather. The ladies' accommodation is in a large apartment on the main deck, framed in oak and enamelled pine with broad frieze of anaglypta. The doors are leaded glass, richly coloured. A feature of this apartment is the fireplace, with brass interior and tiled panels of special design, the overmantel having a beautiful dolphin clock. The settees of the apartment are upholstered in leather and arranged with spring seats to form berths for use on the night passage. Immediately below this is a gentleman's sleeping saloon framed in panelled oak with shell pattern frieze and upholstered in rich leather. Abaft this is the restaurant, also panelled in oak, with anaglypta frieze representing ancient war galleys. Large dining tables are provided, with revolving chairs capable of accommodating about 60 passengers. The galley, which is in connection with a pantry above by means of a hoist, is fitted with a very complete appointment of steam cooking appliances. All the first class vestibules are framed in light oak waxed, and the interior doorways have leaded glass panels. The second class accommodation is situated aft, the ladies' cabin being in the large deck-house on awning deck, the roof of which extends out to ship's side, forming a sheltered promenade in wet weather. The gentlemen's accommodation is on the lower deck, and consists of comfortably upholstered sofa beds for the night service. A large open space is also provided on the main deck as a shelter during the day service. The crew are accommodated on the lower deck right aft, and the officers and engineers on the main deck amidships.

The *Journal of the Society of Arts* for July 3rd, 1903, p. 688, also reported further details of this vessel, given by the author.

The appliances for working the vessel are exceptionally powerful, and consist of a large windlass and capstan by Messrs. Harfield at the fore end of the vessel, and a warping capstan aft by Messrs. Paul. The rudder, which is large and of the single plate description, is worked by Messrs. Brown Bros.' steam tiller, controlled from the flying bridge by means of the maker's patent telemotor. For convenience in canting and backing out of harbour, the vessel is fitted with a large bow rudder worked by steam-stearing gear of Messrs. Bow and McLachlan's manufacture controlled by a wheel on the flying bridge. As very heavy mails have to be frequently handled on this service, the vessel is provided with derricks, and two large winches for dealing promptly with this matter. The vessel is heated throughout the passenger accommodation by means of a system of steam piping, and is fitted throughout with electric light and electric bell installation. The sidelights, which are large and numerous, are of the Denny-Porterfield patent pattern. Arrangements are made to carry crates conveying passengers luggage in bulk, and mails which are transferred at Dover and Calais respectively by electric cranes, to and from railway tracks. Messrs. Preece and Cardew have advised upon the electrical gear for the cranes at the Admiralty Pier, and these cranes are now in working order. They were built by Messrs. Grafton and Co. , of Bedford, to the author's design, Messrs. Siemens supplying the electrical parts.

LIFE MEMBERS. *(Instituted at tits Canterbury Congress, 1903.)*
Adkin, It., F.E.S. 4, Lingards Road, Lewisham, S.E.
Adkin, Mrs. H. E. ,,,,
 Bennett, F. J., F.G.S. West Mailing, Kent.
 Bullen, Kev. R. Ashington, F.L.S., F. G.S. Pyrford Vicarage, Woking, Surrey.
 Bullen, Mrs. Ashington, Pyrford Vicarage, Woking, Surrey.
 Coomaraswamy, Ananda K., F.G.S. Walden, Worplesdon, Guildford.
 Foran, C. Elm Grove, Southsea.
 Meeson, F. 98, Sutherland Avenue, Maida Vale, W.
 Mrrificld, Frederic, F.E.S. 24, Vernon Terrace, Brighton.
 Neate, P. J., J.P. Watts' Avenue, Rochester.
 Rudler, F. W., F.G.S., vc. 18, St. George's Road, Kilburn, N.W.
 Stebbing, Rev. T. R. R., M.A., F.R.S. Ephraim Lodge, The Common, Tunbridge Wells. Stebbing, Mrs. T. R. R. Ephraim Lodge, The Common, Tunbridge Wells. Stebbing, Miss Grace, Eastbourne. Stirling, Sir James, Bart., F.R.S. Finchcocks, Goudhurst, Kent.
 Turner, Miss E. L. Langton Green, Tunbridge Wells.
 Walmisley, A. T., M.Inst.C.E. Atherstone, Castle Avenue, Dover.
 Whitaker, W., F.R.S., F.G.S. 3, Campden Road, Croydon.
 Vardon, Rev. S. A., M.A. Langton Green, Tunbridge Wells.

MEMBERS, ASSOCIATES AND DELEGATES FOR 1908.
Abbott, Miss. 1, West Parade, Whitstable (M).
 Abbott, George, F.G.S. 33, Upper Grosvenor Road, Tunbridge Wells (M).
 Abbott, Mrs. G. 33, Upper Grosvenor Road, Tunbridge Wells (M).

Adeney, Mrs. E. L. Mt. Sion, Tunbridge Wells (M).
Adkin, Mrs. R. 4, Lingards Road, Lewisham (M).
Adkin, B. W. Brandon House. Morden Hill, Lewisham, S.E. (A).
Adkin, Mrs. B. W. Brandon House, Morden Hill, Lewisham, S.E. (A).
Adkin, R. A. 4, Lingards Road, Lewisham, S.E.
Adkin, Miss Jessie. 4, Lingards Road, Lewisham, S.E.
Allcliin, J. H. The Museum, Maidstone (D).
Baker, A. 36, Upper Grosvenor Road, Tunbridge Wells (M).
Baldock, J., F.C.S. St. Leonards Road, Croydon (D).
Bannerman, W. Bruce, F.S.A., F.G.S. , F.L.S. The Lindens, Sydenham Road, Croydon (M).
Beach, Mrs. 11, Parkhill Road, Huverstock Hill, N.W. (M).
Beer, Rudolf, F.L.S. Westwood, 5, Edridge Road, Bickley, Kent.
Bennett, A. 143, High Street, Croydon (M).
Bird, C, B.A. Mathematical School, Rochester (M).
Bird, Mrs. C. 38, Frindsbury Hill, Rochester (M).
Bishop, A. 3, Earl's Road, Tunbridge Wells (M)
Bloomfield, Rev. E. N., M.A. Guestling Rectory, Sussex (V.P).
Boulenger, G. A., F.R.S. 8, Courtfield Rd., South Kensington, S.W. (Ex-Pres).
Boulenger, C. R. 8, Courtfield Road, South Kensington, S.W. (M).
Boulger, Prof. G.S., F.L.S., F.G.S. 11, Onslow Road, Richmond, Surrey (V.P). Box, J. W. 25, Henry Street, Chatham (D). Brackett, A. W. 51, Queen's Road, Tunbridge Wells (M). Brackett, Mrs. A. W. 51, Queen's Road, Tunbridge Wells (M). Britton, C. E. 35, Dugdale Street, Camberwell, S.E. (A).
Burr, Malcolm, B.A., F.Z.S., F.E.S. 12, Fitzjames Avenue, West Kensington (Referee).
Chapman, T. A., M.D., F.E.S. Betula, Reigate (M).
Charles, F. Bedford Lodge, South Parade, Oxford.

Cole, Miss. 53, London Road, Canterbury.
Connold, E., F.E.S. 7, Magdalen Terrace, St. Leonard's-on-Sea (D).
Crafer, Mrs. 6, Dyke Road, Brighton (M).
Dodd, C. Tattershall. Grosvenor Lodge, Tunbridge Wells (M).
Donisthorpe, Horace St. J. K., F.Z.S., F.E.S. 58, Kensington Mansions, South Kensington, S.W. (Referee).
Dowson, Miss. Dover.
Earle, Dr. W. G.
Edwards, Stanley, F.Z.S.,F.L.S. 15, St. German's Place, Blackheath, S.E. (D)
Elvery, Mrs. The Cedars, Maison Dieu Road, Dover (M).
Enock, F., F.L.S., F.E.S. 13, Tuffnell Park Road, London, N. (M).
Ewell, Mr. Townhull Street, Dover (M).
Fitzgerald, Rev. H.P., F.L.S. Wellington College, Berks (M).
Fleay, F. G., M.A. 27, Dafforne Road, Upper Tooting, S.W. (D).
Gray, H. Norman. Newlyn House, 131, Enrlham Grove, Forest Gate, E. (M).
Griffin, W.H. 3, Rutland Park, Perry Hill, S.E. (D).
Groves, Jas. 55, Jeffreys Road, Clapham Rise, S.W. (Referee).
Gruner, Miss Joan F. Oakhill, Hindhead (M).
Habgood, H., M.D. Stafford House, Upperton Road, Eastbourne (M).
Hammond, Miss. 48, Upper Grosvenor Road, Tunbridge Wells (M).
Harris, Dr. P. 63, Lower Addiscombe Road, Croydon (M). (Loc. Sec, 1898).
Hatchard, S. Glendore, Camden Park, Tunbridge Wells (M).
Hatchard, Mrs, S. Glendore, Camden Park, Tunbridge Wells.
Holmes, E. M., F.L.S. Ruthven, Sevcnoaks (V.P).
Horsnaill, A.E. Swaithmore, Barton Fields, Canterbury (M).
Howes, Prof. G. B., LL.D., F.R.S. Royal College of Science, South Kensington (Ex-Prcs).
Hurst, Miss. Fairborne House, Dane John, Canterbury.
Hutchinson Jonathan, F.R.S. The Li-

brary, Inval, Haslemere (Ex-Pres).
Hutchinson, R. R. 28, Princes Street, Tunbridge Wells (Sec. N.H.S.).

Inge, E. G. Haslemere (D).
7200511

Jackson, R. J. 40, Lee Street, Woolwich (Sec. W. and D.A. Soc.).

Jenner, J. H. A., F.E.S. 209, School Hill, Lewes (M).

Kensett, Miss. 15, Upper New Street, Horsham (Sec. Museum Soc.) (M).

Lobley, Prof. Logan. 28, Palace Road, Buckingham Gate, S.W. (M).

Lowne, B. T. Bromley Road, Catford, S.E. (D).

Mann, W.P., B.A. Simon Langton Schools, Canterbury (D).

Martin, E. A., F.G.S. Campbell Road, Croydon (West) (D).

Mathews, Paul, M.A. South Avenue, Rochester (D).

Mathews, Mrs. P. South Avenue, Rochester (D).

McDakin, Capt. G. 15, Esplanade, Dover (D).

Mrs. G. McDakin, 15, Esplanade, Dover (M).

Mitton, W. Hurstpierpoint, Sussex (Referee).

Morgan, J. 18, The Steyne, Worthing (M).

Morris, J. 17, Throgmorton Avenue, E.C.

Munro, W. 138, Britton Street, New Brompton (Sec. N.B.N. Soc).

Newman, T. P. Hazelhurst, Haslemere (M).

Newman, Mrs. T.P. Hazelhurst, Haslemere (M).

Newmarch, Major-Gen., R. E. 6, Norfolk Terrace, Brighton (M).

Nicholson, C, F.E.S. 22, Crouch Hill Road, Crouch End, N. (M).

Nicholson, Mrs. C. 22, Crouch Hill Road, Crouch End, N. (M).

Nicholson, W. E., F.L.S. Lewes (M).

Nottidge, A. J. Dry Hill Park, Tonbridge (M).

Otter, J. L. 10, Vernon Terrace, Brighton (D).

Pankhurst, E. A. 3, Clifton Road, Brighton (Lac. Sec. 1900 Congress).

Pannel, C. Haslemere (M).

Pannel, Mrs. C. Haslemere (M).

Payne, Mrs. Linden Gardens, Tunbridge Wells (M).

Payne, Miss. Linden Gardens, Tunbridge Wells (M).

Payne, E. S. Hampstead (D).

Peirson, H. 19, Ware Road, Hertford, Herts (A).

Pickett, C. P., F.E.S. Leyton (D).

Pollock, Sir Frederick, Bart., LL.D 48, Great Cumberland Place, Hyde Park, W., and Hindhead Copse, Shottermill (V.P.).

Poore, A. S. 47, GriBin Road. Plumstead (Sec. N.K.N.H.S.).

Potter, Dr. G. W., M.D. 19, Molyneux Park, Tunbridge Wells.

Proudfoot, Miss. Simon Langton Schools, Canterbury.

Roberts, C. J., M.A. Folkestone (D).

Robertson, Miss Agnes (A).

Rogers, Rev. W. Moyle. Grosvenor Road, Bournemouth (Referee).

Roods, Alfred. Croydon (D).

Russell, The Hon. Rollo, M.R.Met.S. Dunrozel, Haslemere (M).

Russell, Mrs. Dunrozel, Haslemere (M).

Sargant, Miss E. Quarry Hill, Reigate (M).

Saunders, Sibert. Bank House, Whitstable (M).

Shaw, Miss. The Museum Library, Haslemere (M).

Soaines, Rev. H.A. Hawthorns, Otford, Sevenoaks (M).

Stainer, Miss,. Folkestone (A).

Stallworthy, Rev. G. B. The Manse, Hindhead (D), (Loc. Sec. 1901 Congress).

Stallworthy, Mrs. G. B. The Manse, Hindhead (M).

Starling, Dr. E. A. Chillingworth House, Tunbridge Wells (M).

Stephenson, Miss. High Street, Guildford (A).

Storr, Baynor. Hindhead (M).

Starr, Mrs. R. Hindhead (M).

Swanton, E. W. College Hill, Haslemere (M).

Swanton, Mrs. E. W. College Hill, Haslemere (M).

Tremayne, L. J. 29, Cockspur Street, London, S.W. (M).

Treutler, Dr. 8, Goldstone Villas, Hove (D).

Treutler, Mrs. 8, Goldstone Villas, Hove (M).

Trollope, Mrs. W. T. Hawthorndene, Camden Park, Tunbridge Wells (D).

Turner, H. E., B.A., B.Sc. Lindfield Lodge, Folkestone (M).

Tutt, J. W., F.E.S. Rayleigh Villa, Westcombe Hill, S.E. (M).

Walker, Mrs. Ulcombe Place, Maidstone (M).

Webb, Sydney. 22, Waterloo Crescent, Dover (Sec. Dover Sc. Soc).

West, W. 15, Horton Lane, Bradford (Referee).

Wilson, Henry, M.A., F.S.A. Bromley, Kent (D).

Window, Miss. "Howberry," Haslemere (Loc. Sec. 1901 Congress).

Young, W. P. 251, Lavender Hill, Clapham Junction, S.W. (D).

Young, Mrs. W. P. 251, Lavender Hill, Clapham Junction, S.W. (D).

Books written b; TUHAL HISTORY 01 Vols-I. II. and III. c. £1 e&cb now In Press, Subscription Price oof or publication, the price will be raised to £'.

I

The objects of the Union are to systematise Scientific Work among the different Societies composing it, to give greater impetus to Scientific research, and, in general, to promote the study and advancement of Science by Cooperation. In view of these objects, School Natural History Societies receive special consideration, and are admitted on payment of a nominal fee.

Authors are entirely responsible for the facts and opinions contained in their papers.

Readers of Papers are requested to send to the Kditor, J. W. Tutt. Rayleigh Villas, Westeombe Hill, S.E., a list of any *errata* they may detect in the present volume.

The Congress for 1905 will be held at Reigate, on

June 8th, 9th, and 10th, under the presidency of

Professor WILLIAM MATTHEW FLINDERS PETR1E,

D.C.L., LL.D., F.R.S.

The Kditor would be glad to exchange Transactions with other Unions and Natural History Societies. All communications relating thereto should be ad-

dressed to J. YV. TuTT, *Westcombe Hill, &.E.*
April, 1890
May, 1897
.lime, 1898
May, 1899
June, 1900
June, 1901
June, 1902
June, 1903

PLACES OF MEETINGS AND NAMES OF PRESIDENTS.

Tunbridge Wells Rev. T. R. R. Stkbbtno, M.A., F.R.S.
Tunbridge Wei
Croydon
Rochester-
Brighton-
Haslemere
Canterbury
Dover
June, 1904 Maidstone PAGE.
Places of Meeting and Past-Presidents ii
List of Officers for 1904-5... iv
Maidstone Congress Local Committee......... v
Rules of the S.E.U.S.8. (amended) vi
Byelaws........................ vi ii
Botanical Research Committee.......... viii
Work done by Members of Affiliated Societies, in—
  Anthropology................. ix
  Botany..................... ix
  Geography................. ix
  Geology..................... ix
  Meteorology................. ix
  Microscopy................ ix
  Photography................. x
  Physics..................... x
  Physiology................. x
  Zoology x
  Miscellaneous................. xi
List of Affiliated Societies with names of Secretaries, Delegates, fcc................. xii
List of Lecturers................. xiv
Eighth Annual Report............. xvi
Balance Sheet for 1908-4................ xvii
S.E.U.S.S. Lantern Slides............... xviii
  Referees.................. xix
  Museum Notes—Report by the Honorary Curator...... xx
  Exhibition of Photographs............... xxv
  Delegate's Report of the British Association...... xxvii
  Proceedings of the Congress, 1904.... xxix
  Presidential Address.................. 1
  Allington Castle 22
  The Friars, Aylesford............... 25
  On the Meridional Position of Megaliths in Kent compared with those of Wilts, and also with those of Earthworks and Churches................. 29
  The Use of 25in. Ordnance Maps for Estate and Agricultural Purposes..................... 87
  The Practicability of an Artificial International Language 40
  Ice Streams and Ice Caves............... 45
  Notes on the Lepidoptera of Mid-Kent......... 47
  Note upon some interesting Neolithic Weapons recently found on Blackdown, near Haslemere......... 55
  Memoranda respecting some late Keltic Pottery found at Haslemere, Surrey, in November, 1908......... 56
  The Teaching of Nature-Study 58
  The Abbey and St. Leonard's Tower at West Mailing... 69 List of Localities, not recorded in recent Floras of Kent and Surrey, for some comparatively rare plants...... 78
  A few Notes on the Corporation Museum, Maidstone... 78
  List of Life-Members, Members, Delegates, etc., for 1904... 84

His Worship the Major of Maidstone (AM. Moiling, J.P.).
The Rt. Hon. Baron Avebury, P.C., D.C.L., LL.D., F.R.S., etc.
Sir Francis Evans, Bart., K.C.M.G., I M.P.
Aid. Day,.J.P., Deputy Mayor.
„ Brownscombe, J.P., F.H.G.S.
„ Dr. Oliver, J.P. I
.Coun. G. F. Baker, J.P.
„ W. Day, Jun. „ E. Vaughan.
Lieut.-Colonel. H. K. Allport.
Mr. J. Arkooll.
Mr. J. E. Austin.
Mr. F..1. Bennett, F.G.S. I
Dr. Boyce, J.P.
Mr. B. P. Boorman.
Rev. Canon H. Collis, M.A.
Mr. G. Foster Clark.
Mr. R. P. Evans.
Mr. D. C. Falcke.
Rev. C. H. Fielding, M.A.
Mr. R. J. Fremlin.
Mr. Walter Fremlin.
Lady Members of the Committee.
Mrs. Morling, Mayoress. Mrs. Silcock.
Mrs. Day, Deputy Mayoress. Miss C. Smith.

The following are the President, Vice-Presidents, and Council of the Maidstone Natural History Society—
Rev. Gardner-Waterman, M.A.
Mr. J. Hepworth, Rochester N. Society.
Mr. Herbert Green.
Mr. E. Goodwin.
Mr.S.Harvey, F.I.C., F.C.S., President of East Kent Scientific Society.
Hon. H. Hannen.
Mr. Edward Hills, J.P.
Mr. R. Hoar.
Mr. H. Lamb.
Rev. G. M. Livett, B.A.
Mr. P. Mathews, M.A., President of New Brompton Natural History Society.
Mr. Randall Mercer.
Mr. Herbert Monckton, Town Clerk.
Mr. G. Pavne, F.S.A.
Capt. S. G. Reid, R.E., F.Z.S., etc.
Mr. L. Stansell, F.I.C.
Mr. F. S. Stenning, M.A.
Mr. A. O. Walker, F.L.S.
Mr. H. Snowden Ward.
Rev. C. E. Woodruff, M.A., Editor of *Arclurnlogia Cantiana.*
Major W. Haynes, J.P., President.
Vice-Presidents—
Mr R J Balston, F.Z.S.
Mr. W. H. Bensted.
Rev. C. G. Duffield, M.A.
Mr. Laurence Green, F.C.S.
Dr. C. E. Hoar, J.P.
Mr. Frederic Laurence. Members of Council—
Mr. J. H. Bridge.
Mr. G. T. Cook, Jun.
Hon. Local Secretaries—
J. H. Allchin, The Museum, Maidstone

W. H. Day, 12, Earl Street, Maidstone
Mr. R. P. Grant.
Mr. J. B. Groom.
Dr. C. Pye Oliver, M.D., State Medicine, Lond.
Rev. W. A. H. Legg, M.A.
Rev. S. Richards, M.A.
Mr. G. H. J. Rogers, F.R.M.S.
Mr. F. W. Ruck.
Mr. P. H. Silcock, B.A.
Mr. A. Barton, Treasurer.
Mr. W. H. Day, Secretary.
Photographic Section—
H. J. Elgar, The Museum, Maidstone.
H. W. Witcombe, 85, Holland Road, Maidstone.

RULES.

As revised at the 4th Annual Congress, held at Rochester, May 27th, 1899, with subsequent amendments. i. Objects.—The objects of the Union shall be to systematise work among the various Societies composing it, to give a greater impetus to research, and to promote the interests of the Societies by co-operation. 2. Management.—The affairs of the Union shall be managed by a Council and a General Committee. 3. The Council shall consist of a President, Vice-Presidents, General Secretary, Treasurer, Editor of the Transactions, and seven other persons, three to form a quorum; all to be elected annually, and none except the Vice-Presidents, Secretaries, Treasurer, and Editor, to be eligible in the same position for more than two years in immediate succession. The filling of casual vacancies to be at the discretion of the Council itself. 4. The General Committee shall consist of the Council, Past-Presidents, and the Delegates. 5. Affiliation.—All Scientific Societies in Hampshire, Kent, London, Middlesex, Surrey, and Sussex, shall be eligible to join the Union, provided that the Society claiming to join comprises at least 10 members. 6. Congress.—A Congress for the furtherance of the general work of the Union, and for the reading and discussion of papers, shall be held annually in June, at such place and at such date as may be decided on by the General Committee at the preceding Congress, or, failing such decision, by the Council. 7. Delegates.—A minimum Annual Subscription of 5s.. *payable in advance at least a fortniijlit before the 'ant/iess,* shall entitle a Society to affiliation and a voting ticket for one Delegate at the Annual Congress. Societies with more than 50 members, exclusive of honorary members, shall if they so desire, be entitled to voting tickets for additional Delegates in the proportion of one for every additional 50 members, and one for the number (not less than 10) in excess of every multiple of 50, on payment of 5s. for each ticket. 7a.—School Natural History Societies may affiliate for a subscription of 2s. fid. (rule added June, 1900). 8. Members.—Members of Affiliated Societies shall be admitted to the Congress on payment of 2s. 6d. 9. Associates.—Persons unattached to any Affiliated Society may, at the discretion of the Council, be admitted to the Congress on payment of 8s. 6d. 9a. Life-Members.—Members, Associates, and other persons at the discretion of the Council, may compound for the Annual Subscription by a single payment of £2 2s. for Life-Membership. (Rule added June, 1902.) 10. General Meetings.—The meetings at the Congress shall be for the reception of reports of work and the reading and discussion of papers. 11. The General Committee shall, at some time during the Annual Congress, receive a Statement of Accounts, appoint an Auditor, elect the Council for the ensuing year (by ballot if demanded), appoint such Sub-Committees as may be required, decide on the next place of meeting, and, when necessary, revise the Rules. 12. Executive.—All other affairs of the Union throughout the year shall be managed by the Council. 13. Transactions.—Such Transactions of the Union as may be published shall be issued free to all Affiliated Societies, Members and Associates. 14. Local Receptions.—Each Society or Town inviting a visit of the Union shall appoint a Local Committee and Local Secretary to assist the General Secretary in drawing up the Programme of the Congress, which shall be arranged at least a month before the said Congress. 15. Expenses.—The expenses of printing and general management shall be paid out of the funds of the Union; those of providing rooms for the meetings of the Congress by the Society or Town issuing the invitation. 16. Changes in the Rules may be proposed and discussed at any meeting of the General Committee, but cannot be passed until the following year, unless they have been submitted to the General Secretary at least three months in advance so that he may report the proposals to the Affiliated Societies before the Congress. i.—The Union shall have the right, at its discretion, of printing *in extenm* in its Transactions all papers read at the Annual Meeting. The Copyright of a paper read before any meeting of the Association, and the illustrations of the same which have been provided at his expense, shall remain the property of the author; but he shall not be at liberty to print it or allow it to be printed elsewhere, either *in extenm* or in abstract amounting to as much as one-half of the length of the paper, before the first of November next after the paper is read. *i.*—The author of any paper printed in the Transactions shall be entitled to receive 25 separate copies of it gratis, and to have any further number printed at his own expense by private arrangement with the printers to the Association. 3.—If proofs of papers to be published in the Transactions be sent to authors for correction, and are retained by them beyond four days for each sheet of proof, to be reckoned from the day marked thereon by the printers, but not including the time needful for transmission by post, such proof shall be assumed to require no further correction. 4. Should the extra charges for small type, and types other than those known as Roman or Italic, and for the author's corrections of the press, in any paper published in the Transactions, amount to a greater sum than in the proportion of ten shillings per sheet, such excess shall be borne by the author himself. 5. A time limit of 25 minutes is prescribed for each paper, with 5 minutes for each speaker, and the discussion of the subject is to be closed at the end of one hour.

March, 1900.

BOTANICAL RESEARCH COMMITTEE.
This will be added to from time to time, so that it may embrace 2 or 8 representatives from each County in the district.

I'rof. G. S. Boilcier, 11, Onslow Road, Richmond, S.W.

W. H. Bebby, Hildasay, Thames Ditton.

.tas. Groves, 08, Jeffreys Road, Clapham, S.W.

E. Chas. Horrell, 58, Copleston Road, Denmark Hill, S.E.

Thos. Howse, Glebefield, Guildford, Surrey.

Rev. E. N. Bloomfield, Guestling Rectory, Sussex.

Wm. Mitten, Hurstpierpoint, Sussex.

W. E. Nicholson, Lewes, Sussex.

E. M. Holmes, *Chairman,* Ruthven, Sevenoaks.

ORIGINAL WORK DONE OR BEING DONE BY MEMBERS OF AFFILIATED SOCIETIES. (Furnished by the Secretaries.) ANTHROPOLOGY.

Croydon N.H. and Sc. Flints found at Waddon Marsh H. II. Sower.
Soc.

Battersea F. Club. Neolithic Man in Surrev. (Published Walter Johnson, and by E. Stock.) William Wright.

Hastings and St. Leon-On prehistoric remains of man dis-W.('.J. Ruskin Butter ards. covered at Fairlight. field, M.B. O.U.

BOTANY.

Haslemere Micro, and Notes on the Forestry Section of the K. W. Swanlon (Mem.

N.H. Soc. Haslemere Museum (in Haslemere Brit. Mycol. Soc.) Herald).

A preliminary list of Haslemere Fungi. E. W. Swanton (Mem.
Brit. Mycol. Soc.)

East Kent N.H. Soc... Botany Notes from Thnnet F. Hewett.

Maidstone and Mid-List of Flora of the Maidstone district. H. Lamb.

Kent N.H. Soc. List of rare and local plants of the H. Lamb.
Maidstone district.

Catford N.H. Soc Additions to previous records of the Messrs. Griffin and
Flora of N.W. Kent. Lowne.

Tunhridge Wells N.H. The primrose, the oxlip, and the cow-Dr. E. G. Gilbert.

"Soc. Blip (Journal of Botany).

Notes on Hybrids (Journal of Botany).. Dr. E. G. Gilbert.

South London Cactuses H.J. Turner.

City of London Sc. Soc. The preservation of Wood with Sugar.. Prof. G. S. Boulger,

F.L.S., F.G.8. Rochester Nat. Club... Flora of 10 miles radius GEOGRAPHY.

Hastings and St. Leon-The River Rother from the Source to A. M. Apel. ards N.H. Soc. the Sea. Eastbourne N.H. Soc.. The Farne Islands E. J. Bedford.
GEOLOGY.

Rochester Naturalists' Geology of District
Club

Eastbourne N.H. Soc.. Some Aspects of Geology G. W. Bulman, M.A.,

B.Sc. Croydon N.H. and Sc. List of Fossils from the Chalk at W. Murton Holmes. Soc. Whyteleafe, Coulsdon, and Purley.

On a Section of Clay with Flints, near N. F. Robarts, F.G.S.
Woldingham.

On the Plateau Gravel, Upper Norwood. N. F. Robarts, F.G.S.

East Kent N.H. Soc... On Fissure Flows of Lava t'apt. McDakin.

Tunhridge Wells N.H. "Rambles in Search of Geological G. Abbott, M.R.C. 8., Soc. Specimens" in "Nature Study F. G.S.

Readers," Book 111.
METEOROLOGY.

East Kent N.H. Soc... Full Meteorological Records and New A.Lander.
Self-recording Apparatus.
MICROSCOPY.

Eastbourne N. H. Soc. Life in Ponds and Ditches R. Watts White,

M.R.C.S. Maidstone N.H. Soc... Photo-micrographic Study of Bacteria. E. Clements.

East Kent N.H. Soc.
Maidstone N.H. Soc.

South London Entoni. and N.H. Soc.
West Kent N.H. Soc...

Hastings and St. Leonards N.H. Soc. Eastbourne N.H. Soc.
PHOTOGRAPHY.

Features of the River Stour, with 70 C. Buckingham.
original photos.

Birds' Nests and other Natural Objects. F. C. Snell.

Photographs of Kentish Plants H. Elgar.

Photographic Enlargements of Insects1 ,, ,,

Wing, to illustrate their Neuration..

Insects and their Habits F. Enock, F. L.S.

Eggs of Lepidoptera F. Noad Clark.
PHYSICS.

Recent Developments of X-Rays W. Webster, F.C.S.

Waves: Their Nature and Place H. F. Cheshire, B.Sc.,
F.I.C.

Goethe's Theory of Colours R. C. C'ann Lippincott.
PHYSIOLOGY.

Variability and correlation of the Heart W. Greenwood.

and Spleen (methods of recording variation.

The relation between muscular act- W. Kidd. M.D., F.Z.S.

ivity, and beauty of form of animals.
ZOOLOGY.

West Kent N.H. Soc...

Brighton and Hove N.H.
Soc.

South London Entomological and Natural
History Society

North London N.H.
Soc.

East Kent Sci. Soc
Maidstone N.H. Soc...

Cat ford N.H. Soc.

Haslemere M. and N.H.
Soc.

Tunbridge Wells N.H.
Soc.

Birds of " the Cedars," Lee J. F. Green.

Protective Coloration J. F. Green.

Birds of Brighton and Neighbourhood.. E. Robinson.

Life-histories of Lepidoptera J. W. Tutt, F.E.S.

Alpine Lepidoptera J. W. Tutt, F.E.S.

Revision of Pterophorids Dr. T. A. Chapman, and
J. W. Tutt, F.E.S.

Lepidoptera of Dawlish G. B. Browne.

Maternal Solicitude of Female Insects G. W. Kirkaldy.
for their young.
The Norwegian Lobster E. Step.
Recent researches in Protective Re- Prof. E. B. Poulton, .semblance, Warning Colours, and F.R.S.
Mimicry in Insects.
Marine Fishes E. Kiev.
Pearly Nautilus W. Munger.
Biology of *Eupitheeia eriguata* R. Adkin, F.E.S.
Eggs of Lepidoptera F. Noad Clark.
European Lepidoptera Dr. T. A. Chapman, and
J. W. Tutt.
Lepidoptera of West Indies W. J. Kaye.
Orthoptera of the World M. Burr.
Life-histories of the Coleophorids H. J. Turner.
Spiders of the Family Linyphiidae F. P. Smith.
Coleoptera in the Canterbury District. . B. F. Maudson.
Lepidoptera in the Canterbury District F. A. Small.
Lepidoptera in the Ashford District.. W. R. Jeffery.
List of the Anthophila of Maidstone H. Elgar.
and neighbourhood.
Preliminary list of the Syrphida? of H. Elgar.
Maidstone and neighbourhood.
Additions to the previous records of the W. H. Griffin.
Fauna of N.W. Kent
The Land and Freshwater Mollusca of C. Pannell, M.C.S.,
Haslemere. M.M.S.
Observations on the Homing Instincts C. Pannell, M.C.S., of Snails and Slugs (in Haslemere M.M.S.
Herald).
"Pycnogonida " in " Knowledge" *(con-*Rev. T. R. R. Stebbing, *eluded)*. M.A., F.R.S., Sec.
L.8.
"Crustaceans" in Victoria History of Rev. T. R. R. Stebbing, the Counties of England *(continu/d)*. M.A., F.R.8., Sec. L.S.
Holmesdale N.H. Club.
City of London En torn, and N.H. Soc.
Rochester Nat. Club
Gregarious Crustacea from Ceylon. fi Rev. T. R. R. Stebbing, plates in Willev's Spolia Zeylanica.j M.A., F.R. S., Sec.
L.S.
Isopoda of the Maldive and Laccadive Rev, T. R. R. Stebbing, Islands. Stanley Gardiner Expedi-M. A., F.R.S., Sec.
tion, 5 plates. L.S.
The Head of a Common Wasp E. Connold, F.E.S.
Birds and their Eggs T. Parkin, J.P.r M.B.O.U.
New Species of Non-marine Shells from Rev. Ashington Bullen, Java, and a New Species of Corbicula F.L.S., F.G.S.
from New South Wales (1 plate Proc. Malac, Hoc).
Natural History of the British Lepidop-J. W. Tutt, F.E.S.
tera, Vol. IV., published 1904.
Migration and Dispersal of Insects J. W. Tutt, F.E.S.
(Elliot Stock).
Life-histories of Pterophorids i J.W. Tutt, F.E.S.,and
Life-histories of Hepialids t" A. W. Bacot, F.E.S.
Spilosoma fuliginosa: Habits, Distribu-J. W. Tutt, F.E.S.
tion, and Variation.
Spilosoma fuliginosa: Larva, Pupa, and A. W. Bacot, F.E.S.
Larval Habits.
Variation in Gnophos ohscnrata L. B. Prout, F.E.S.
Geometrides of the World L. B. Prout, F.E.S.
West Indian Lepidoptera W..J. Kaye, F.E.S.
Phorodesma pustulata Rev. C. R. N. Burrows.
Insects and Birds of the District M. G. Palmer.
Eastbourne N.H. Soc..
Hastings and St. Leonards N.H. Soc.
Tunbridge Wells N.H. Soc.
Woolwich Antiq. Soc...
Sidcup Lit. and Sc. Soc.
Brigh ton and Hove N.H. Soc.
East Kent Sci. Soc
Maidstone N.H. Soc...
Northfleet and District Sc. Soc.

MISCELLANEOUS.
Pictures and Portraits of the Coinage H. D.Gaidner,F.R.G.S., of Ancient Times. F.R.Met.S.
History of Local Nat. Sci E. Connold, F.E.S.
First Ideas of Nat. Hist H. King, A.C. P.
Skins A.Edmunds, M.B.,B.Sc.
Woolwich and the Dutch Wars of the R.-J. Jackson.
Commonwealth and Charles II.
Woolwich Bibliography *continued)* .. .. W. T. Vincent.
Chapters on Paper-making (Vol. II Glayton Beadle.
The Relation between Poetry and Rev. Felix Asher.
Science.
Nature Notes: Principally Botanical.. W. H. Hammond.
The Rambles of a Naturalist (a series J. B. Groom.
of articles in the Local Press).
Ancient Roads and Defences of the M. H. Heys.
locality.
Notk.—The object of this List is not only to form a record of work done, but also to assist workers to communicate with each other—see also last year's list. Mere lists of Lectures given before Societies are not wanted for insertion, but only particulars of *original* work, whether published or not.
1902 19
18 Hi
IW3 1M17 1HUH
1900
1H9C, LONDON.
Battersea Field Club
Crity of London Entomological and Natural History Society
City of London College Science Society
North London Natural History Society
I ! Polytechnic Natural History Society.
-South London Entomological and

Natural History Society
Selborne Society
Hampsteari Scientific Society Fulham Field Club and Literary and Scientific Society
SURREY.
Balham Antiquarian and Natural History Society
Croydon Natural History and Scientific Society
Croydon Camera Club
Holmesdale Natural History Club...
Haslemere Microscopic and Natural History Society
Photographic Survey and Record of Surrey
Tiffin's Boy's School Natural History Society
Wokiiw Field Club
Secretary's Name and Address.
Delegates (19041.
Established.
No. of1
Mem-Subscriptions.
Publications.
Transactions.
Journal (monthly), Is.
Proceedings.
"Nature Notes" (2d. monthly.
Annual Reports and Proceedings.
Annual Transactions.
Proceedings.
5C S303 isb » 3 a = S £ Ad'S5 to m 5-I
LIST OF LECTURERS.

Where no fee is mentioned it may be assumed that none is expected, but the travelling expenses must be paid and accommodation for the night provided if required. For Lantern Lectures the Society will please provide Lantern and operator. In future issues of this list names of ladies or gentlemen *recommended* by Affiliated Societies will be inserted. It is expected that many of them will be able to give at least the name of one capable Lecturer willing to repeat his or her lecture before other societies in this Union. Lectuken.
Mr. J. H. Allchin, The Museum, Maidstone, private address, Chillington House.
Fee, 2 guineas, and travelling expenses.

Prof. G. S. Boulger, F.L.S., F.G.S., Ed. of *Nature Notes,* 11, Onslow Road, Richmond, S.W.
Fee on application.
Mr. A. W. Bracket, F.S.I., 61, Queen's Road, Tunbridge Wells.
Mr. Ed. Connold, F.E.S., 7. Magdalen Terrace, St. Leonard's-on-Sea.
Fee, 2 guineas, and travelling expenses.
Dr. Vaughan Cornish, F.G.S., F.C.S., F.R.G.S., 72, Prince's Sq., London, W.
Fee on application.
Mr. Martin Duncan, South Park, Reigate, Surrey.
Mr. F. Enock, F.L.S., F.E.S., 13, Tufnell Park Road, London, N.
Fee on application.
Mr. R. 11. Hutchinson, Hon. Sec. Tun. Wells N.H.S., 28, Prince's Street, Tunbridge Wells.
Mr. A. B. Harding, F.l'hys.Soc. Lond., Belmont, Catford, S.E. Fee on application.
Mr. Richard Kearton, F.Z.S., Ardingly, Caterham Valley.
Fee and titles of other lectures on application.
1. Haunts and Habits of British Birds.
2. Peeps into Nature's Secrets. 3. Wild Life at Home: how to study and photograph it.
Each lecture is illustrated by photographs taken direct from nature by the lecturer.
Mr. F. Lambert, 213, Putney, S.W. Fee on application.
Rev. W. A. H. Legg. M.A., Bower Mount Road, Maidstone.
Fee on application.
Mr. E. A. Martin, F.G.S., 23, Campbell Road, Croydon (West).
Mr. Paul Mathews, M.A., South Avenue, Rochester.
Miss E. Turner, Upper Birchetts, Langton, Tunbridge Wells.
Mr. W. H. Griffin, (j, Rutland Park, Perry Hill, S.E.
Borneo St. 1. Marvels of the Subterranean World
(Jenolan Caves, N.S.W.).
2. A Tour through the Crystal Caves of New South Wales.
3. Through the Mammoth Caves of Kentucky in search of Eyeless Fish and other Blind Fauna.
The dazzling nature of the Stalactites is conveyed by means of a patent crystalline screen.
1. Buried Cities or Northern England, 2'JOO.years ago. 1. The History of Valleys, submerged and exposed. 2. The Physical Future of the Earth. 3. The Coal Problem, geologically and economically considered. 1. Aims of Natural History Societies. 2. The Classification of Animals. 3. Artificial Languages. 1. Plumage. 1. Plant Folklore.
EIGHTH ANNUAL REPORT.

We have to report at this,-the Ninth Congress, that the S.E.U.S.S. is making satisfactory progress in numbers, finance and usefulness; although the number of affiliated societies cannot show much increase.

Our Meeting at Dover (the Eighth Congress) had a smaller attendance than usual, owing doubtless to the unsettled weather. In every other way it was marked by the kindness, attention and enthusiasm of the Members of the local Society, as well as the Mayor, Town Clerk, and other officials of the town. The accommodation provided for the Meetings and the Museum was exceedingly good. Mr. S. Webb's lepidoptera, and the plants by Messrs. Lowne and Griffin, with other contributions, made an exceedingly fine display in the Town Hall, The Union is indebted to the local Committee and Secretary for their co-operation and liberality.

In the *South-Eastern Naturalist* of last year a start was made with a List of Lecturers, which should prove useful to Secretaries, who, it is hoped, will help to make the list complete and reliable.

It will have been noticed that this year's programme is marked by many changes. There are already signs that the alteration has led to a greater appreciation of the value of our Meetings, and an increased attendance of Members and Associates. In this way we hope to benefit our funds also.

The Council has had under consideration, at its two Meetings, the proposal made by Mr. H. Norman Gray, that one winter meeting should be held in London. It is thought that this would help many Members to become better acquainted with the London Museums and Societies. No definite plan has yet been decided on, but the Council has approved the principle and appointed a Committee to arrange details. Then we shall trust to Secretaries and Delegates making the suggestion as widely known as possible, for if not well-attended the first meeting is not likely to be repeated.

We have now *22* Life Members, and the usefulness of this alteration of the Rules cannot be doubted.

Photographic Surveys of Kent and Sussex were mentioned in our last report. The proposed joint meeting in London could not be arranged, but a meeting held in Maidstone on April 16th, has succeeded in starting an association for Kent, and we are glad to know that Sussex has made a similar start.

Mr. H. E. Turner has for some time wished to be relieved from the Photographic Secretaryship. Having succeeded in finding him a successor in Mr. E. A. Martin, we cordially thank him for his many years' service. During the past session our slides have been used to some extent, but we should like to see still more interest taken in this part of our work. More slides should belong to the Union for all the members to use, and in this we may obtain valuable assistance from our three County Photographic Surveys.

The Commons Preservation Society, in common with the Rambling Clubs in London, are making an effort, in which the Council is co-operating, to induce the Railway Companies to give us some concession as to fares, which we have so long tried for.

Our roll of Affiliated Societies is about the some as last year—42. We welcome a newly established Nat. Hist. Society at NorthHeet, the Kent Photographic Survey, and the Rochester Philosophical Society. Two School Societies, which were included last year, but about which we were uncertain, have withdrawn, and one other will perhaps ultimately retire.

Copies of the *South-Eastern Naturalist* have been sent to the following Societies, etc.:—British Association, British Museum (Natural History), Cambridge University Library, Linnean Society, Nottingham Naturalist's Society, Owen's College Museum Manchester, North of England Institute of Mining, etc., Newcastle-onTyne.

The Treasurer's statement shows an unusual balance this year, due, however, to his own generosity. Your Council desire to acknowledge the gift with many thanks.

BALANCE SHEET, 1903-4. RECEIPTS.
Balance brought forward
  Dover Congress.— Proceeds of
Tickets sold by Local Society
  Contributed by Local Society..
  Delegates' Subscription 1901..
  Delegates' Subscriptions outstanding
1903
  Members' Subscriptions 1903..
  Associates' Subscriptions 190.1..
  Donation
  Paid for Advt. in S.E. Naturalist
  Paid for Reprints-
Rev. R. A. Bullen
  Interest on Current Account..
  Interest on Deposit Account EXPENDITURE.
  Printing—Mr.Snunders
  Mr. Dee
  Mr. Pelton..
  Mr. Baldwin
  Printing.'100 copies S.E.
Naturalist
Postage ditto.
Carriage of Parcels..
Postage of Letters
Expenses *re* M useum,
Dover
  Delegates' Subscriptions outstanding
Lantern Slide Box for
travelling
Balance in Hand
  Sale of S.E. Naturalist..
7 in
£58 17 9
  Examined and found collect as per Vouchers.
  W. T. TROLLOPE, *Auditor.*
S.E.U.S.S. LANTERN SLIDES.

The following sets of Lantern Slides are available for use by affiliated Societies on application either to the General or Photographic Secretary: 1.—*Some Itritish Orehidt* (50 slides) contributed by Mr. S. Horsley,

M.I.C.E.. with explanatory lecture. 2. —'*The (iiiidt ami Lower (Ireensantl* (about 80 slides), with lecture. 3.—*The Wealden Formation* (about 50 slides), with explanatory notes. *i.—lee Flower and Crystal* (small set), with explanatory notes by G.

Abbott.

No charge is made except for carriage both ways. The orchid slides are a new and interesting set, dealing with the general and detailed structure of many British species and their adaptation to insect-fertilisation.

The Society for the Protection of Birds, 826, Holborn, W.C., also lend *to their subscribers* very beautiful Lantern Slides relating to Birds, which are well worth the attention of Secretaries and Lecturers.

The notice of Secretaries is particularly called to the suggestions made in the Photographic Secretary's Report that the Union should solicit loans or gifts of *small* sets of slides (a dozen or so) illustrating *amj* particular scientific phenomenon or limited branch of scientific work. Such sets, with full explanatory notes, to occupy about half-an-hour for exhibition, would doubtless be much appreciated for the purpose of soirees or other occasions in which time is necessarily limited, while two sets might furnish material for an ordinary evening meeting. Many members of our Societies possess sets of this character which they have prepared for their own use, and which they would be willing to lend for the use of the affiliated Societies. Secretaries are hereby asked to furnish the Photographic Secretary as soon as possible with the names and addresses of any of their members who, in their opinion, might be induced to cooperate.

Contributions are still solicited towards the following larger sets that are in course of formation: 1.—*Pre-IIUtoric Man in S.F. England.* 2.—*English Wild Flower,* with special reference to forms of capsules and their dehiscence.

3.—*Photomicrograph.*

4. — *Coat Erosion in S.E. England.*

Contributions of lantern slides may be sent to the Photographic Secretary, Mr. E. A. Martin, F.Ci.S., 28, Campbell Road, Croydon (West), who will be glad to give any information as to this branch of the work of the Union.

REFEREES. BOTANICAL. (Additions to this list would be welcomed).

The following gentlemen have kindly consented to name a limited number of specimens for our Members and Associates.

(A stamped directed envelope should always be sent, or no replies need be expected).

Cryptogams (not microscopic).—Thomas Howse, Glebefield, Guildford.

Freshwater Algae.—W. West, 15, Horton Lane, Bradford.

Specimens of species of *Zygnema, Spirogyra,* and *Mougeotia* should be fruiting. They are best sent in small tubes in water. Habitat must be always stated. Permanent reedy ponds and ditches yield the best results, especially those where *I'tHcularia* occurs. Plants like *I'tricularia,* leaves and peduncles of *Nymphaea, Nuphar, Ac,* might be sent in tin boxes; and Mr. West will examine these for minute forms. Gelatinous or slimy coverings of damp, shady, or trickling rocks should also be sent in small tins.

Marine Algae (excluding diatoms and desmids).—E. M. Holmes, Ruthven, Sevenoaks.

Fresh algre should be rolled separately in old muslin or calico, so that one plant does not touch another; then packed so as to be free from pressure in tins or boxes.

Mosses.—W. Mitten, Hurstpierpoint, Sussex.

Phanerogams.—A. Bennett, 1-13, High Street, Croydon, and Rev. E. X. Bloomfield, Guestling, Sussex.

In some orders like *Cruciferae, Cyperaceae,* and *Umbelliferae,* the fruit is almost a necessity.

Cyperaceae.—A. Bennett, 143, High Street, Croydon.

Hieracii.—Rev. W. R. Lynton, Shirley Vicarage, Derby.

Rubi.—Rev. W. Moyi.e-Rogers, Chetnole, Grosvenor Road, Bournemouth West.

The work of the *Rubi* referee would be very greatly lightened, and his determinations and suggestions proportionately more satisfactory, if correspondents would send only good *representative* specimens, and place them always on *paper stout enough ami large enough to bear them safely.*

There is, of course, least room for uncertainty when panicles show both flower and fruit, and the stem pieces mature leaves from about the centre of their length; other less satisfactory pieces *in addition* may sometimes help a referee. But if in any case a correspondent can feel justified in sending only such pieces, he should at least press them carefully and supply them in extra quantity.

In explanation of the term "stem" and "panicle," it is to be remembered that all the fruticose *liubi* throw out long leafy shoots (the "stem" or "barren stem") directly from their roots, which, normally, produce no flowers in the first year. From these spring, in the following year, the flowering shoots, and it is to the flowering part of these, including all the branches and branehlets, that the term "panicle" is applied.

No "specimen" can be determined with certainty, unless it consists of both panicle and stem pieces; and it is often of great assistance to the referee when the label accompanying the specimen contains—in addition to the usual memoranda of locality, county or vice-county, date, Ac.—further notes made from the living plant, of the colour of the flowering organs and the comparative length of styles and stamens, with any other conspicuous character lost in the process of drying.

ZOOLOGICAL.

Diptera.—Rev. E. N. Bloomeield, (Jnestling, Sussex.

Tenthredinidae.— Ditto.

He will also be glad to hear of any "finds" in either the Fauna or Flora of his part of Sussex.

Coleoptera. H. St. J. K. Domsthorpe, F.Z.S., F.E.S., 58, Kensington Mansions, South Kensington, S.W.

Lepidoptera.—J. W. Tutt, F.E.S., Rayleigh Villa, Westcoinbe Hill, S.E.

Orthoptera.—M. Burr, F.Z.S., F.E.S., Royal Societies' Club, S.W.

Galls.—E. Connoi.d, Hon. See. of Hastings Nat. Hist. Soc, 7, Magdalen Terrace, St. Leonard's.

Hydroida (Calyptoblastea). Rev. H. A. Soames, Hawthorns. Otford, S?venoaks.

MUSEUM NOTES.

By The HONORARY CVRATOK.

We expect that our temporary museum this year appeared very insignificant to many, it being completely overshadowed by the magnificent collections in the Maidstone Museum, but an exhibition almost entirely limited to original work done by members of the Union must naturally be a small one, and it is to be sincerely hoped that next year's exhibition will not suffer in consequence. Part of the room containing the county collection was placed at our disposal, and Messrs. Allehin and Elgar spared no trouble, not only in preparing it for us, but also by kindly assisting in many ways. We thank them heartily for their generous help. Before proceeding to tabulate the members' exhibits in alphabetical order we would like to express our appreciation of the excellent arrangement of the Kent countv exhibits, and, indeed, of the whole museum. Mr. Elgar's fine collection of bees caused many of us very frequently to break the tenth commandment. Mr. Elgar has very kindly sent us, at request, the following note upon local Anthophila in the Maidstone Museum.

Hymenoptera Of Maidstone District. —"The cabinet of Hymenoptera contains 142 species, or about two-thirds of the wild bees indigenous to the British Islands, these have all been taken in the neighbourhood of Maidstone, and, with few exceptions, within a seven miles' radius of the town. The number of species obtained is no doubt due to the abundant and varied flora of the district, as well as to the well-drained Chalk and Folkestone Sand, on which formations nearly all the collecting has been done. The species to which attention may be drawn by reason of their being rare or local in the British Islands, are as follows:—*I'roxopis cornuta, I', dilatata, Sphccodes rubicund us,* this species has been only taken in three other localities; *S. spinulosis,* an inquiline on *Halictux xantlmpus, H. quadricinctux, H. maeidatux,* a male taken August, 1900, only three females had previously been recorded for Britain. *//. xanthopus, H. laeciijatus, Andrena apicata,* Smith, *A. lappmiica,* Zett., specimens taken May, 1895, and the first recorded for the British Islands. Smith, in the *Catalogue of British Hyiiienoptera,* Part I., confused *A. lappmiica,* Zett., with what now stands as his *A. apicata. A. fasciata, A. hattorfiaiia, A. dcnticnlata, A. chri/ xoscelex, A. jiolita,* a female taken July, 1901, was the first recorded capture in Britain since 1855, when specimens were taken at Northfieet by the late F. Smith. *A. Iiiiniilix, A. proximo, ('ilisxa Itaemorrlioidalix, C.melanura, Daxypoda hirtipex, Xomada fucata, X. jacobaeae, X. bifida, X. borealix, X. J'crrwjinata, F.peiilnx prodiictux, Ceratina cyanea, Coelioxys quad rid entata, Meyacliile liyniscca, M. versicolor, Oxmia bicolor,). pilicornis, O. leucomelana, Stelix phaenptera, $. octoinacnlata,&n* inquiline on *Oxmia leucomelana, Antliophora /areata,* and *A. quadrimaculata."*

Photographs Of Honeycomb Weathering: Mr. George Abbott, M.R.C.S., F.G.S., Tunbridge Wells.—This special form of decay leaves characteristic marks and patterns both in sandstone, limestone, and marble, whether we examine rock-faces or the same materials used as building-stones. It appears also to affect brickwork in much the same way. The photographs shown were to illustrate the effects in the fine grained stone of Wealden sandstone at Tunbridge Wells, and the coarser sandstone of the Trias at Budleigh Salterton, and Dawlish.

"Honeycomb" Concretions In Magnesian Limestone (permian) Sunderland.—"They are believed to be entirely inorganic, but shew a remarkable resemblance to corals and other organisms. It is surmised that the carbonate of lime molecules, probably when amorphous, had an inherent molecular directive force, which produced the four classes, each with a distinctive pattern. Many specimens appear to have passed through four (or less) stages of 'growth,' and to have undergone rearrangement of the particles while in the solid condition."

Mr. W. H. Griffin, Catford:—(1) Fifty sheets from the Herbarium of A. 0. Hume, illustrating the batrachian *Ranunculi, Yinlarieae* and *UmluiliJ'erae.* (2) Two very large eoliths found at West Wickham and Cbelsfield, West Kent. (8) Specimen (in alcohol) of a carnivorous slug, *Trstacrlla halintiilea,* found (with others) in a garden at Catford. (8) Examples of the fungus, *Dalilinia cimeentriea*—a pyrenomycete—gathered from an ash-tree at Down House, Kent. It is not uncommon, we have seen it in many localities in the South of England.

Mr. Benjamin Harrison, I;htham.—Series of eoliths with sketches and sectional drawings of the pit sunk near Terry's Lodge for eolithic evidence, in October, 1902. The implements occurred amongst a compact mass of oehreous flints in a clayey matrix, eighteen inches thick, resting upon a bed of mottled clay five feet thick. The mottled clay contained large flints but no eoliths. We quote the following from Mr. Harrison's label, which accompanied his exhibits:—

"The large series of eoliths exhibited at the Congress of the Union of Scientific Societies held at Rochester, in May, 1899, consisted principally of implements found on the surface gravels of the Chalk Plateau at a height of about 500 feet 0.1). Since that time the isolated high-level gravels at and above the 700 feet level have been under continuous observation. The exposure by workmen of the Mid-Kent Water Company of gravel on the crest of the North Downs near Terry's Lodge, led to an excavation being made by permission of the owner, Mr. E. B. Evelyn, and, in carrying out this work, I was most ably assisted by my friend Mr. E. J. Bennett, E.G.S. We also opened a pit near the site of a former excavation at Parsonage Farm, Ash, Mr. Edward Pink, the owner, readily permitting us to do so. The specimens displayed have been obtained from these two excavations, and therefore represents finds in gravel *in xitit.* In the high-level gravel at 760 feet O.D. only eoliths were found, at the Parsonage Farm pit, in addition to eoliths, one implement was discovered, which was described by Professor T. Rupert Jones, E.R.S., as 'showing a passage from the eolithic towards the paheolithic' age."

Mr. Jonathan Hutchinson, E.R.S., LL. D.—A series of frames to illustrate the method of pictorial teaching in his educational museums. The contents of the frames are changed frequently, in this way local interest is kept up. Some of the frames are made on the portfolio principle, and thus are picture frames and portfolios combined. The frame is fitted with a false back, and is deep enough to contain a large number of engravings or drawings mounted on card the size of the frame. The illustration displayed gives a clue to the contents of each portfolio frame. For example, four plates from Sowerby's *Ferns of Great Britain* were shown in a frame, the other forty-five plates were placed within. At the Haslemere Museum this plan is being extended to other groups, zoological as well as botanical. The series of frames on exhibition contained (1) portraits of Greek philosophers, (2) Roman Emperors, (3) natives of northern India, (4) Australian Aborigines, (5) an interesting collection of original drawings of the Megalithic monuments of Wilts, made by Mr. Henry Browne, of Amesbury, in the year 1.S26, (6) photographs of Indian coolies carrying heavy loads, and an Aino woman carrying a child. In all the weight is supported (on the back) by a strap passing over the top of the head. With these photographs was placed an illustration, taken from Dr. Thurnam's *Crania Uritannica,* of the skull of an ancient Briton showing a peculiar depression, which might have arisen through carrying heavy weights with the aid of a head strap.

Mr. B. T. Lowne, Catfobd.—(1) A

further series of plants from his unique herbarium, prepared on the lines indicated in the last issue of the *Transactions* (p. xviii), (2) A series of seedlings mounted to show the various stages of growth and the development of the epicotyl. (8) Land and Freshwater Mollusca recently collected in West Kent.

E. W. SwAnton, Brockton, Haslemere.—A series of photographs of " witches besoms" on various trees, including:—(1) Birch, the common form, caused by the fungus, *F.jcaascus tunjiihix*. (2) Larch, agency unknown. Not commonly seen, and of slow growth, we have had one example under observation for fifteen years. (8) Scots Pine, according to Hoffmann it is caused by a species of *('latlosporimn*. (4) Elm, agency unknown. Has been attributed by some writers to the bite of an insect. (5) Whitethorn, agency unknown. Uncommon. (() Beech, from photo negative of ' phytoptus" beech kindly presented by Mr. Sydney Webb, F.E.S. Dense masses of twigs envelop the upper part of the trunk. Has the malformity been directly traced to insect causation? (7) Hazel, we have only seen one example. It is now in the foresting section Haslemere Museum. The branch is abruptly swollen immediately below the point where the fasciation begins, and it looked very much as if fungus mycelium was the cause.

Photograph Of A Norway Spruce Whose Trunk Had Been SplinTered By Lightning.—It showed the hollowing out of the centre of the trunk caused by the mycelium of the fungus *Fames annosns*. Previous to injury, the tree, to all external appearances, was quite healthy.

Pendulous Tumours From Branches Of An Old Beech.—Caused The writer is specially interested in "witches besoms," tumours, and other abnormal growths of trees, and would welcome information upon the subject.

by the fungus *Xectria litisxiina*, which is also responsible for the cankerous growths frequently seen on plum and cherry trees. A series of specimens illustrating morbid conditions, occasionally seen in trees known as "knars," "knobs," and "burrs." They are probably not in any way due to foreign agency, changes in nutrition being the chief etiological factor in the formation of these abnormalities. The peculiar abortive ends, "knars," frequently seen in the bark of beech, holly, and other trees, are occasionally carried by old men in the villages around Hasleinere as an antidote for cramp! and are known as " cramp balls."

Neolithic Implements. Obtained by Mr. Allen Chandler on Blackdown, near Hasleinere (see p. 55), photographs of Hint arrowheads, knives, and scrapers from the same locality. Photographs of cinerary urns and other pottery belonging to the late Keltic period, discovered at Ilasleinere in November 1903 (sec p. 50).

Nature Study Exhibits.—The Nature-Study exhibits, which were shown in the Shell Room of the Museum, were intended to illustrate the paper read by Mr. Wilfred Mark Webb. It was considered fitting that the schools invited to exhibit should be chosen from those situated in Kent, but one or two examples which were not local were brought by Mr. Webb himself. We may begin by considering the secondary schools. The Grammar School, Maidstone, was represented by the results of what have been called leisure time pursuits. These included a number of photographs of insects and of bird's nests illustrating how, in Nature-Study, pictures may be collected instead of specimens. Records of continued observations in the shape of insects with their chrysalides and caterpillars reared and preserved by the exhibitor were also on view. The whole exhibit was contributed by Master Alban DufHeld. A fine series of drawings, sent by The Girls' Grammar School, Rochester, by the kindness of Miss Measkam, the Science Mistress, were more properly part of the actual work of the school. One was of a local nature, and consisted of water colour drawings of Kentish orchids, a second referred to the methods of scattering seeds and fruits adopted by various plants. The fascinating consideration of such plants as live together and form associations was exemplified by a third series of drawings, which depicted the special flowers of the hedgerow, of woods, and of the meadows, together with examples of such forms as grow together on particular soils. It will be noticed that these pictures, many of which were exceedingly well done, were not made primarily for artistic purposes. In Nature-Study drawings and pictures, the dominant idea must be the recording of observations and not merely the making of an artistic production. The methods of keeping creatures in captivity, in vivaria and in aquaria (where a proper balance is maintained between the vegetable and animal occupants of the water) were exemplified among the elementary school exhibits by that from Orleston.

It is to be doubted whether the snake or the blind-worms shown by Mr. E. A. Thomas would have been in such fine condition if they had remained in the state of Nature. In further illustration of the work on pond life, which is a feature of the Nature-Study of Orleston School, the contents of two cases may be mentioned. These consisted of various aquatic animals which had been observed from time to time, together with the skins which they have been seen to cast, and of other objects of interest.

The exhibits from Chislehurst Road, Orpington, which Mr. Boult has made famous for its brushwork, included some excellent studies of the germination of the Spanish chestnut. There were also some of the note-books kept by the scholars with special reference to insects. By means of a special series of exercises the teaching of brushwork in co-relation with Nature-Study throughout the whole school was clearly shown.

The use of natural models in drawing was also exemplified by the exhibit sent by Mr. J. B. Groom, of St. Paul's Schools, Maidstone. He also furnished a description of one of the rambles which he has successfully conducted.

The exhibits brought by Mr. Webb were original drawings illustrating continued observations on the growth of feathers in the young of the robin and on the development of the silkworm. He also showed the lifehistory of the lackey moth, illustrated by specimens, and a label intended as a leaflet for children. The remainder of his exhibits consisted of Nature Notebooks, kept and illustrated by one of the older students at the House of Education, Ambleside (Miss B. M. Goode), and by a young member (Miss Annie Primrose Farren) of the Parents' National Educational Union in Suffolk (this latter was in the special "Eton Nature-Study Notebook," with detachable leaves, designed by Mr. Webb). There were also the drawings and observations upon a germinating mustard seed made by Miss Winifred Allen, at Heidelberg College, Ealing, and natural history observations taken down in the children's own words at the Coombe Hill School, East Grinstead, which has recently been moved to Westerham, in Kent.

A SERIES OF FIFTEEN VERY FINK PHOTO-MICROGRAPHS: Mr. W. P.

Young, E.H.M.S.—Including tongue of blowfly, transverse section of spine of *F.ehinux*, sections of plant stems, sheep tick *(Mdopliai/uit oriiius),* diatoms and foraminifera.

EXHIBITION OF PHOTOGRAPHS.

The exhibition of photographic prints contributed by members of the Photographic Survey and Record Associations of Kent and Surrey, showed conclusively the value of work of that nature, when it is done with a definite object; especially in its application to old examples of domestic architecture, which in this age of change, reconstruction, or total abolishment, are in constant danger of being materially altered or altogether demolished. It is in this connection that photography presents its most useful side as a handmaid to the various arts and sciences, and several instances were noticeable in the Congress Exhibition; *i.e.,* two views of Whitgift's Hospital, Croydon, which is, at the present time, threatened with extinction: five prints of the exterior of Great Auckland, Maidstone, a fine old mansion of the 17th century, which was pulled down about 1894; a series of views of the Museum, Maidstone, showing the various alterations that have been effected in its structure from 1867 to the present day; and a block of old houses in Mill Street, Maidstone, which were removed in 1903 in order to effect a much needed widening of that thoroughfare, one of the main entrances into the town from the Weald of Kent.

The importance of obtaining photographs, from various points of view when possible, of old buildings, or buildings which are interesting through their association with notable personages, cannot be expressed too strongly, and the same may be said of other objects of photographic research, especially those that are not of an enduring character, such as many natural history phenomena, temporary views of geological formations, etc. The Congress Exhibition, arranged in the Bentlif Picture Gallery of the Museum, although not a very extensive one, included many prints of great interest. The principal contributors from the Surrey Survey were Mr. H. (iower, who exhibited a large number of prints of natural history specimens, and Mr. J. H. Baldock, who sent both architectural and landscape views. Prints were alto contributed by the following members:—Messrs. J. H. Anderson. Bunce Brothers, Bender and Lewes, H. Brown, H. C. Collyer, G. Clinch, G. C. Druce (who sent photographs of some very interesting architectural details), R. H. Edgar, C. H. Goodman, J. M. Hobson, M. S. Johnson. M. S. Jenkins, H. Maclean, J. Noaks, J. H. Stanley, and H. Ward.

In the Kent section the exhibits were contributed chiefly by members of the Maidstone Camera Club, the Survey and Record of the county not having been established long enough for any collection of prints to be brought together. Captain G. McBakin, of Dover, sent three very interesting prints of the Roman Pharos at Dover, and the Baptistry, Canterbury Cathedral, from collodioalbumen negatives taken by himself 86 years ago, with ten minutes and twenty minutes' exposure respectively. Mr. J. H. Baldock, of the Surrey Survey, contributed a series of photographs of the Roman Villa at Darenth, discovered in 1895; Mr. W. J. Nichols, of Chislehurst, sent four views of Chislehurst Caves, supposed to be of prehistoric origin; and Mr. J. Russell Larkby, of Bromley, supplied five views, exterior and interior, of St. Mary's Church, Horton Kirby. The local exhibitors were Mr. W. H. Witcombe, Secretary of the Maidstone Institute and Camera Club; the Rev. Gardner Waterman, Messrs. F. Argles, E. Clements (who aent a very fine series of prints of the interior of Canterbury Cathedral) A. F. Corfe, H. J. Elgar (Sub-Curator of the Museum), J. J. G. Greenhill, J. C. Harris, W. Wilson, and Messrs. De'Ath and Dunk.

For several years past Mr. Elgar has been working on a photographic survey of the Maidstone district, and has acquired for the Museum a valuable collection of negatives, which include objects of interest in the following departments: — Geology, Archieology. and the various divisions of natural history, /.-., trees, plants, flowers, and birds' nests, *in situ,* in addition to a very large number of botanical and entomological specimens photographed in the Museum, several of which were included in the Congress Exhibition.

DELEGATE'S REPORT OF THE BRITISH ASSOCIATION MEETING AT SOUTHPORT.

As your delegate at the Southport Meeting of the British Association, it is my duty to present a brief review of the Delegates' meetings held on September 10th and 15th, 1908, under the presidency of Mr. W. Whitaker, F.R.S., both of which your delegate attended. Fifty-five societies were represented.

At the first Conference the President, Sir Norman Lockyer, explained his views as to the necessity of forming a Guild of Science. In the course of his speech he said, "Unless we can control votes in the House of Commons and in the Councils throughout the country, Science will not be any better looked after. If we can control votes, Science will be benefited, and scientific bodies from one end of the country to the other, working with a goal in view, would be a most important factor in our future national life." Many valuable hints were made by practical workers in the course of the debate which followed, in which the Chairman, Principal Griffiths, Messrs. J. 0. Bevan, W. Johnson, W. F.

Stanley, Garstang, and Hopkinson, and Professors Watts and Carr took part.

Mr. W. M. Rankin, B.Sc, read a paper on "The Methods and Results of a Botanical Survey of Counties." He showed that 4000 square miles have been surveyed, including Yorkshire (West Hiding), Edinburgh district, North Perthshire, Forfar, and Fife, which have been completed, and Westmorland and Somerset, which are in an advanced state towards publication. Messrs. Hopkinson, Woodhead, Ewing, and Professor Kendall joined in the subsequent discussion.

Mr. T. V. Holmes sent a note on Maps of the Ordnance Survey, deprecating the omission of some arclueological details from newer maps, which were inserted in earlier issues—a point which the Chairman emphasized.

At the Second Conference (September 15th) the Chairman opened the meeting by urging the need of a speedy organization of the Guild of Science.

Mr. William Cole, F.L.S., read an important paper entitled " A Suggestion with respect to Exploration and registration work for County Local Societies." His view was that many of the recommendations of the British Association to local societies were impracticable, but that many things which he enumerated might be well taken up by them—that the patient work of registering plants, animals, and fossils, the examination of minor earthworks, camps, deneholes, redhills, etc., should be the duty and pleasure of local enthusiasts He thought that the County Councils might allocate a small annual sum from the "Whisky money" for these purposes, according to a scheme which he formulated. Dr. W. R. Scott thought that local societies might add to their investigations *record of old industries and rallini/s* which are generally passed over; Mr. W. P. Stanley suggested *Photography,* but doubted whether Councils could spare the postulated grant. The Hon. Itollo Russell suggested *Meteoroloi/i/,* and the importance of *experiments in plant life* in relation to soils, weather, and various other conditions; and other suggestions were made by Professor Weiss, Messrs. Herbertson, Ackroyd, and Hopkinson.

The practical question of return railway fares at a single fare, or a fare and a quarter, was brought forward in acommunication from Mr. Herbert Stone, of Birmingham. The privilege would only be asked when accredited members of scientific societies were actually travelling for natural history purposes. In the discussion that followed it transpired that what is denied in England (except partially in Yorkshire) *is allowed in Ireland* to those proceeding on field-club business. The discussion was carried on by Messrs. Hopkinson, Parkin, Lamplugh, Scott, and Capt. Phillips, R.A.

Reports from Section C—Geology (Mr. Lamplugh); D—Zoology (Rev. T. R. K. Stebbing); K—Botany (Professor Weiss); E— Geography (Mr. E. Heawood) were presented. I am sure we all hope that Mr. Stebbing received some well-shrimps (we will not ask him what he did with the whisky they came in), and that Miss Sargant learnt some of the facts in relation to plants with underground growth, for which they respectively asked.

"The American Handbook of Scientific Societies" was put forward by Mr. J. David Thompson, of Washington, U. S.A.

At a meeting of the Corresponding Societies' Committee, held on November 9th, 1908, a resolution by Mr. E. P. Knubley was carried "that the members of the Corresponding Societies be requested to give as much help as they can to teachers in those Elementary and Secondary Schools, which are taking up the subject of naturestudy."

Also, in re-considering Mr. Cole's paper before mentioned, it was resolved " that the Corresponding Societies be recommended to enter upon the six-inch ordnance maps any unrecorded natural features and archaeological remains."

I may add that the important "Report of the Corresponding Societies' Committee" may be purchased for one shilling, at the offices of the British Association at Burlington House.—(rev.) R.

ASHINGTOX Bl'i.len, B.A. PROCEEDINGS OF THE CONGRESS OF THE SOUTH-EASTERN UNION OF SCIENTIFIC SOCIETIES, 1904.

Thursday Afternoon. *June 9th, Thursday Afternoon.*—The proceedings of the ninth congress commenced on the afternoon of June 9th, when, after a few words of welcome by the President, Sir Henry H. Howorth, to the early arrivals, an excursion was made by train to West Mailing, for the purpose of visiting St. Mary's Abbey and the St. Leonard's Tower, under the guidance of the Rev. G. M. Livett, B.A., F.S.A., Vicar of Wateringbury. A good muster of delegates assembled on the East Maidstone platform for the 8.12 p.m. train, and the party reached its destination shortly after 8. 80 p.m. The visit to the Abbey was by the kind permission of the Mother Superior, who allowed the excursionists to be shown over the building, and herself accompanied them, with the President of the Union, Sir H. H. Howorth. The members assembled together at the west front of the ruined abbey-church where they were received by the Mother Superior. Sir Henry Howorth opened the proceedings with a few felicitous remarks, and Mr. Livett then gave a brief account of the history of the Abbey, the remaining buildings of which he subsequently described in detail (see page 69). The Rev. C. H. Fielding, M.A., and Mr. F. J. Bennett, F.G.S., made some observations. The latter gentleman, referring to the tufa used in the construction of the early churches, explained why, in his opinion, it was at first employed and afterwards discarded. This stone, as already pointed out by the Rev. G. M. Livett, was practically confined to the quoins and was the only squared stone used. The other stones were naturally shaped, and perhaps in some cases roughly dressed with the hammer. The workmen in those days were unskilled and the tools rude, so that they apparently could not square the hard ragstone. The tufa, he said, was a soft and easily dressed stone readily lending itself to squaring with the saw, a very early tool, hence its use no doubt in early buildings. But the tufa was a rough cavernous stone and, though weathering hard, would not be desirable for the fine ashlar work of later times where more skilled men and better tools were available, and hence its disuse later on. The Rev. G. M. Livett also stated that tufa seemed abundant now at Wateringbury and other places where the Kentish ragstone occurred, the tufa being due to calcareous springs from this limestone.

Before leaving the Abbey Sir Henry Ho worth thanked the Mother Superior for her kind reception of the company, and the members then made their way to the so-called St. Leonard's Tower, where Mr. Livett again acted as cicerone. In the brief time at his disposal Mr. Livett gave a short account of this building also (see page 69). At the close of Mr. Livett's description, Mr. Bennett called the President's attention to the remains of a small arch in the lower part of the solid south wall of the Tower. He pointed out that the arch occurred just where there was an outcrop of the soft hassock between the ragstone, and that the arch seemed to have been made to take the weight off this soft stone, the bearings of the arch being on the rag. At the conclusion of the visit Mr. Livett was cordially thanked by Mr. Laurence Green on behalf of the whole party of visitors.

Professor Boulger records that a Spanish plant, *Linaria oriijanifolia* was found on the walls of St. Mary's Abbey, the plant being also naturalised on the walls of Wells Cathedral.

Thursday Evening (Sir Henry Howorth, K.C.I.E., F.R.S., F.S.A., in the Chair).

June 9tlt, Thursday h'.reniny.—At 8 pm. the opening meeting took place in the Town Hall, when a large and influential company were present, including the President-Elect, Mr. F. W. Hudler, I.S.O., F.G.S., the Ex-Presidents—Professor G. S. Boulger, F.G.S., the Rev. T. R. R. Stebbing, M.A., F.R.S., and W. W. Whitaker, F.R.S., F.G.S.—Vice-President, Mr. F. Merrifield, F.E.S., also the Rev. C. G. Dnffield, the Rev. S. Richards, Miss Dobie, Miss Semark, Miss L. Hills, Alderman Day, Councillors Elmore, and C. Styles, Messrs. Laurence Green, F.G.S., Herbert Green, Dudley Falcke, G. H. J. Rogers, F.R.M.S., etc., the Hon. General Secretary, Dr. G. Abbott: the Hon. Local Secretaries, Messrs. J. H. Allchin and W. 11. Day; the Photographic Secretary, H. J. Elgar, and many others.

Sir Henry H. Howorth, in his opening remarks, said that men felt miserable when they descended from office to the position of ordinary men, and this misery was shared by their wives (laughter). He knew no one more miserable than the wife of a Bishop or Mayor who no longer held those offices (laughter). He was in that position that night. With regard to his successor, who was an old friend of his, he was a scientific man in the 20th century who was modest (laughter). Did they know another? He was afraid he did not. He was one who always effaced himself, and was courteous to all. Everyone spoke well of him. He must therefore be in a parlous position, as Holy Writ said was the position of those of whom everyone spoke well. In that Mr. Rudler was different from him, for his father had gout, and he had suffered from it from his cradle, and found that the only cure for it was to pour out adjectives and adverbs in controversy. He would now resign the presidential chair to Mr. Rudler.

Mr. Rudler then read his Presidential address which is printed at length (pages 1-21). A vote of thanks to Mr. Rudler for his excellent address was proposed by Mr. W. Whitaker and seconded by Mr. F. Merrifield, to which Mr. Rudler briefly replied.

The Rev. T. R. R. Stebbing, being called on by Mr. Rudler to move a resolution said, " Mr. President, Ladies and Gentlemen, we owe it to the cause of humanity and our own self-respect to do what we can towards raising a fellow-creature out of that depth of gloom and depression into which the retiring president has professed himself to have fallen on relinquishing his distinguished official position among us. It might carry an invidious implication, were I to propose a vote of thanks to Sir Henry Howorth for giving up the chair as though he had come prepared to hold the fort against his successor, but, finding Mr. Rudler equipped with torpedoes and submarines, or apparatus of that kind, had concluded to make a virtue of necessity and to beat as graceful a retreat as circumstances would admit. In fact, under our constitution the resignation of the chair is automatic, but it is not every outgoing chairman who is ready to come with kindly and encouraging words to greet the new occupant. The resolution which I am asking you to carry is a vote of thanks to Sir Henry for brilliant service to the SouthEastern Union throughout his year of office, from the first day of it to this the last. He is too well known in literature, science, politics, and society for his versatile accomplishments to need any praises of mine. But those who attended our Congress at Dover will not have forgotten how he there kept us all delighted by his geniality, his readiness on every subject, his pungent criticism and unfailing wit. Our experience of him this evening has been no less agreeable. That he may attend many future congresses of this Scientific Union is beyond doubt the general wish of the members. The reception you have given him this evening is an earnest of the welcome he will have in the future. If became here with the feeling that being out of office was being out of regard, he will carry home the assurance that the more you know him the better you like him."

Professor Boulger seconded the vote of thanks. In replying, Sir Henry, in allusion to Mr. Rudler's remarks on modern athleticism, said that, notwithstanding all that had been said, it was an extraordinary fact that, if they took the last five years at Oxford and Cambridge, the men who had taken the highest degrees were men who had been in the boats, the cricket eleven, or the "fifteen." It was not sport *per se* that they protested against, but he himself would protest against the athleticism of the spectator class. When he was a boy they were all athletes, now few were athletes, whilst the mass of people were merely spectators. In the north this side of sport so-called had become extremely pernicious, it really consisted of nothing but the watching of, and betting on, games. The remedy, lie said, did not lie in a general tirade against the evil, but the establishment of counter attractions in the way of experimental natural history societies, etc. The meeting then terminated.

Friday Morning (Mr. F. W. Rudler in the chair).

June 10th, Friday Mornimj.—In opening the proceedings the President welcomed the delegates and expressed a hope that their deliberations might be of substantial and real value to their respective Societies.

Delegates' Meeting.—The delegates' meeting was opened by Dr. Abbott, who read the Annual Report of the Council to the delegates for the year 1903-4. This is printed in detail *(atitflt* pp. xvi *et sn/.).* The adoption of the Report was moved from the chair and carried.

The Treasurer, the Rev. R. Ashington Bullen, then presented the financial statement for the past year. This is printed in detail *antc('i* p. xvii. The Balance sheet was also adopted.

The Rev. It. A. Bullen then read a letter from the Chairman and Secretary of the Holniesdale N.H. Club, Reigate, inviting the Union to hold its congress at Reigate next year.

On the motion of the Rev. T. R. R. Stebbing, seconded by Mr. T. Hill, the invitation was accepted.

The Rev. R. A. Rullen, B.A., F.L.S., F.G.S., then presented a report as Delegate at the Southport meeting of the British Association. This is printed in detail, *antca* p. xxvii. The hearty thanks of the meeting were given to Mr. Bullen for his report.

Mr. N. F. Robarts, of Croydon, drew attention to an archieological matter of great interest at Croydon. The local authority there intended pulling down a set of very old buildings, of undoubted Elizabethan age, known as Wycliffe's Hospital, for the purpose of street widening. He should like the congress to pass a resolution of protest. There would shortly be an influential deputation waiting on the Town Council, lie proposed a resolution of protest. Mr. Wbitaker (Croydon) seconded. Mr. Neate suggested that, before passing such a resolution, they should ascertain if there were a practicable alternative scheme. It was quixotic to try and stop the development of modern towns. Mr. Robarts said there were several alternatives. He accepted an amendment to the wording of the resolution, and the motion in its amended form was then carried.

Mr. Laurence Green in the course of the discussion, referred to the old gatehouse at Maidstone which was purchased and saved.

The question of the Photographic Survey of Kent was then discussed, when Mr. Snowden Ward read the report of the Provisional Committee, which stated that the objects of the record and survey were: *(a)* To obtain permanent photographic pictures of objects of literary, historical, archwological, and spientine interest; of current customs, costumes, and events; and of prominent men and women within the county of Kent; and to deposit such pictures, with explanatory notes, in the County Museum at Maidstone, and in other places where they may be suitably preserved and readily accessible to the public under proper regulations, *(b)* To facilitate intercourse between photographic societies and archaeological or kindred societies, and to make arrangements for promptly photographing objects which may be only temporarily accessible, *(c)* To assist in the exchange of prints and in their acquisition by scientific societies wishing to make collections dealing with their own subjects. The provisional committee included Mrs. Golding (Sevenoaks), theEev. E. W. Banks (Longfield), Rev. Gardner Waterman (Loose), and Rev. C. E. Woodruff (Faversham), Dr. Abbott (Tunbridge Wells), Captain McDakin (Dover), and Messrs. J. H. Allchin (Maidstone), E. W. Andrew (Bromley), Hubert Bensted (Bearsted), Joseph Chamberlain (Tunbridge Wells), J. C. Dunk (Maidstone), H. J. Elgar (Maidstone), J. Hepwortb (Rochester), Chas. Igglesden (Ashford), Jas. Roach (Sidcup), H. E. Turner, B.Sc. (Folkestone), Ben. G. Wade (Sheerness), Harry F. Wingent (Rochester), H. Witcombe (Maidstone), and Norman Wolters (Catford). A subcommittee of five members had been appointed to deal with the rules. The Secretary pro. tern, had carried on considerable correspondence with gentlemen who were thought likely to undertake the Secretaryship, the first object being to secure, if possible, a Secretary who resided in Maidstone, or was closely identified with that town. Mr. Harry Wingent, of Rochester, who had had much experience as Secretary of the Rochester Photographic Society, and who is now Vice-President of the Rochester Naturalists' Club, had been good enough to undertake the Secretaryship of the Survey. Although numerous promises of support have been received, some of the writers have not definitely taken up membership. Only those who have actually paid their subscriptions have as yet been placed on the list of members, and they consist of: four at 10s 6d each; six at 2s 6d; one at 5s. The subscriptions amounted to *MS* 2s., and there was £1 8s. Id. due to the Secretary.

Mr. Baldock, of the Surrey Photographic Survey, tendered some excellent advice as to the work which such a Society should undertake.

Mr. Griffin (Catford) said he thought it a great pity that Natural History Societies did not work more in conjunction with Camera Clubs. He spoke of the value which would attach to the results of such an amalgamation.

Professor Boulger concurred, and urged the advisability of including the photographing of Natural History Objects, such as the rarer plants of the county, growing *in situ,* the nest of the Kentish plover, etc. It was decided to advise Natural History Societies and Camera Clubs accordingly.

Mr. Snowden Ward then submitted a few names as suitable for inclusion on the Committee.

It was decided to ask either Lord Stanhope or Lord Avebury to be President, and the. following Vice-Presidents were elected: Mr. F. G. Smart (Tunbridge Wells), Capt. McDakin (Dover), Mrs. Golding (Sevenoaks), Dr. Abbott

(Tunbridge Wells), Mr. F. Abel (Ashford), and the Poet Laureate.

The delegates' meeting was then adjourned until Saturday morning at 10.15 a.m., and the meeting was resolved into a general meeting, when Mr. F. J. Bennett read a paper on " The meridional position of Kentish Megaliths compared with those of Wiltshire, and also with those of earthworks and churches" (see page 29).

The time both for the paper, the discussion, and the reply was very limited, but several members spoke, *vis.,* the President (Mr. Rudler), Sir H. Howorth, Mr. Whitaker, Professor Boulger, the Rev. G. M. Livett of Wateringburv, Mr. H. Wilson, the Rev. R. Ashington Bullen, and the Rev. T. R. R. Stebbing.

Sir Henry Howorth seemed to think that Mr. Bennett had included along his mentioned lines so many structures of so many different ages, that this constituted rather a difficulty, but he thought that there might be something in their tendency to assume north and south lines. He considered that to put megaliths, feudal castles, and Christian churches together, was a little difficult and confusing. He thought Mr. Bennett would be doing good woik if he would collect evidence of the sites of broken megaliths. The great problem to solve is how does it come about that megaliths should be so peculiarly distributed, none being found in N.E. England or Scotland whilst they abound in Holland, S. Sweden, and generally in maritime localities.

Mr. Whitaker, referring to the generally accepted statement that the foreign stones at Stonehenge had been transported some vast distance, referred the speaker to the report of the Society of Antiquaries on the recent discoveries at Stonehenge. There is the geological report by Prof. Judd, who considered these older rocks to have been probably transported by Glacial Drift to the immediate neighbourhood of the spot where they are now seen.

The Rev. G. M. Livett seemed rather to demur to connecting churches with such ancient structures as megaliths, though he admitted that the worship of these may have lasted till the Christian Era, as the early Christian bishops had to warn their people against these idolatrous practices. He also thought that many of the lines were as much E. and W. as N. and S.

Professor Boulger expressed a doubt whether a definite prevalence of a north and south arrangement, rather than any other, had been demonstrated, he seemed to think that all the churches had not been marked but had been selected as favouring the speaker's views. He also questioned the mile distance of the churches, asking if it could be an exact one.

The Rev. T. R. R. Stebbing admired the ingenuity with which Mr. Bennett had argued his point, but, looking at the elaborately dotted maps, showing churches, castles, and megalithic remains sprinkled over an extensive district, he ventured to ask Mr. Bennett whether, by an equal exercise of ingenuity, he might not have traced lines running east and west instead of north and south. Even near the north and south line, on which Mr. Bennett's hypothesis seemed to be principally founded, there was another line diverging from it at a considerable angle that would need explanation.

In reply, the speaker said that he had marked all the churches, and had, of course, seen, as was obvious, that some other lines, especially those in the Wealden Forest area, tended rather to the east and west than to the north and south, but when the composite lines, *viz.* , those that included megaliths, tumuli, and camps, were considered, these certainly trended northwest and the megalith lines also. Much further investigation was required; it might then be found that the churches along north and south lines were much older than those apparently almost east and west ones. There certainly seemed to have been a tendency to favour a meridional position, at one time at least, and the Rev. R. Ashington Bullen's evidence was all in favour of this. So that the north and south cult appeared to be the older, and that afterwards this was replaced by an east and west one. Many of the churches were exactly one mile apart.

Friday Afternoon. *June 10th, Friday Afternoon.*—On Friday afternoon, in delightful weather, two excursions were made; one, botanical and entomological, to the North Downs, and the other, geological and archaeological, to the Aylesford gravel pits. The journeys were undertaken in brakes, and those who joined the first-named excursion went on the Chatham Road as far as the Lower Bell, and then rambled along the North Downs as far as Boxley, Professor G. S. Boulger and Mr. Lamb leading the botanists, and Captain S. G. Reid, Mr. H. Elgar, and Mr. E. Goodwin being in charge of the entomologists. The second under the leadership of Mr. W. Whitaker and Mr. W. H. Bensted, went to Aylesford. The two parties subsequently met at Foley House, where they were hospitably entertained by Mr. and Mrs. John Arkcoll.

The Geological and Archaeological party, by the kind permission of Mrs. Hunter, first visited the Friars, Aylesford, an interesting monastic building of the 18th century, when a short paper (printed in detail *poxtea* p. 2") was read by the Rev. C. H. Fielding. The company was quite a hundred strong, amongst the visitors being, the President (Mr. F. W. Rudler), Sir Henry Howorth, Mr. Whitaker, Rev. G. M. Livett, Mr. Bennett, Mr. Bensted, Mr. Bird, *cum muitis aliis.* The paper was read at the gate of the Friars, and then the buildings were inspected under the direction of the Rev. C. H. Fielding, all appearing to enjoy the visit. A vote of thanks to the Rev. C. H. Fielding for his paper and to Mrs. Hunter for her kind permission to view the building was proposed by Sir Henry Howorth, and carried with acclamation. After this, the party inspected the fossils and implements found by Mr. Wagon in the Aylesford gravel-pit near his house, and from thence they proceeded to the gravel-pit, of which Mr. Whitaker gave an excellent description.

Mr. Bennett supplemented Mr. Whitaker's address by stating that, as the results of a supplemental drift survey of his own that he was making, he had found that the southern gravel in the pit there was capped by a chalky gravel drift, only a few feet in thickness, in the pit, and looking like a calcareous wash, as stated by Mr. Whitaker. He had found that this chalky gravel, for which he proposed the name of " scarp drift," thickened to over eighteen feet to the north, so that here, instead of only one gravel, the southern drift, you had two gravels, the one from the south, and another representing the waste of the chalk escarpment of a northern origin. The recent sewer excavations at Aylesford, Burham, etc., had exhibited very good sections of this gravel where bare gault was shown on the drift maps of the Geological survey.

The excursion to the North Downs was attended by about 40 members and associates. The party met at the Town Hall and, under the superintendence of Mr. A. Barton, Hon. Treasurer of the Maidstone and Mid-Kent Natural History Society, journeyed in three brakes as far as the Lower Bell, on the Chatham Road; there they alighted and commenced a two hours' ramble over the Downs to a point where they had been preceded by the brakes. The return drive was along the Pilgrims' Road, through the village of Boxley, and over Penenden Heath to Foley House, where they were entertained as above mentioned.

The Botanical section under Professor Boulger and Mr. H. Lamb was successful in finding and noting the following plants:— *Aquiletjia rulijaris, Reseda Intra, ffelianthemum vuli/are, Volyijala calcarea, Linmn catharticum. Geranium robertianum* with white (lowers, *Kuimymus euro/melts, llliamnus catharticu, Anthyilix rulneraria, Lotus eornicnlatux, hlippoerepis comoxa, Poteriunt sanyuisorba, Rosa pimpinellifolia, Rosa rnbiyinosa, Pyrnx aria, t'onium maciilatitm, Cmtius sa/tyiiinea, 'ibintnim lantana, lUackstonia perfoliata, 'ynoylossum officinale, F. chiiim rubjare, I,ami urn ijaleobdolon, Iris foetidissima, Listera ocata, Orchis purpurea. Orchis maculata, Aceras anthropophora, Ophrys muscifera, Herminium monorchis, Habenaria conopsea* and *Habenaria chloroleuca*. The abundance of *Pyrits aria*, fine specimens of *Herminium monorchis*, smelling like honey-in-the-eomb, and of *Orchis purpurea*, " the old woman orchis" of these parts, were particularly noteworthy, as was also a remarkable profusion of the white-flowered variety of *Geranium robertianum*.

The Entomological section leaders, Captain Savile G. Reid, R.E., F.E.S., Messrs. E. Goodwin and H. J. Elgar, report that the scarcity of insects was most noticeable, the common species of Lepidoptera and Aculeate Hymenoptera being almost entirely absent. The following species were taken or seen— Lepidoptera:— *Nisoniades tatjes, Syrichthux malcae, Pamphila sylvanus, C'upido minima, Polyommatiix icarus, P. axtrarchc, P. btilarijus, Cailophry rubi, Chrysophanus phlaeas, Kiiehloe cardaminex, PyrameU cardui* (worn), *Sesia stellatarum, Hemaris fuciformis, Euclidia mi, E. glyphica, Phytometra aenea, Melanippe montanata, Venilia maculata, Acidalia ornata, Eupithecia sitccenturiata* (teste Captain Reid), *Pyraitxta purpuralis, Scojiula oliralix, Ennyehia octomaeulata, Cram6ms chrysonuchellus* and *llythia semirubella*.

Hymenoptera Tenthredinid/e:—*Tenthredo bicinc.ta, Allantus scrophulariae, Macrophya. neglecta*.

Hymenoptera Anthophila:—*Andrena proximo, Osmia bicolor, Osmia piliconiis, Osmia spbullosa, Bombus pratorum, Bombus terrestru, Psithyrus cestalis*.

Diptera Syrphids:—*Enstalis intricarius, Helophilus trivittatus, Criorrhina asilica, Eristalis pertinax, Enstalis nemorum, Rhinyia eampestris, Syrphiis vitnpennis, Syrphus bifasciatux, Syrphus nitens, Leucoxona lucorum, Pipizella rirens, Chri/xochlamyx cuprea, Vhilosa variabilis, Chilosa fraterna, Vhilosa illuxtrata*.

Mr. and Mrs. John Arkcoll most hospitably received the members of both excursion parties at Foley House, Penenden Heath, on their return. Refreshments were provided in a marquee on the lawn, and a band discoursed sweet music, while the party wandered through the moss-carpeted wood of rhododendrons and inspected the fine old Judas-tree *(Cercis siliquastrum)*, and the pergola covered with the white Banksia rose. Time did not permit a long stay; but before the visitors left, the President, Mr. F. W. Rudler, briefly expressed their indebtedness to their host and hostess.

Friday Evening. *June 10th, Friday Erening.*—The Mayor's Rkception.— On Friday evening the Mayor and Mayoress (Alderman and Mrs. Morling) gave a reception at the Museum, when, amongst their guests, were the President, Vice-Presidents, members and associates of the SouthEastern Union of Scientific Societies. Shortly after 7.80. p.m. the guests began to arrive, and were received by the Mayor and Mayoress in the fine old oak-panelled Great Hall. By H p.m. some 400 or 500 guests were present. Those present included Mr. and Mrs. R. Adkin, Mr. F. Ashdown, Br. and Mrs. Abbott, Mr. and Mrs. Allchin, Messrs. Appleyard, Anstey, H. Ardon, Lieut.-Colonel and Mrs. Allport, Mr. and Mrs. J. Arkcoll, Mr. and Mrs. S. Britt, Mr. and Mrs. A. Barton, Alderman and Mrs. W. Brownseombe, Mr. and Mrs. A. Brownscombe, Mr. AV. H. Bensted and Miss Bensted, Mr. G. Barnes, Mr. and Mrs. H. J. Bracher, the Rev. W. Bovs Roberts, Professor G. S. Boulger, Mrs. Blake, Mr. and Mrs. A. Blake, Mr. and Mrs. W. Bonney, the Rev. R. Ashington Bullen, Mrs. and Miss Ashington Bullen, Mr. and Airs. B. P. Boornian, Dr. and Mrs. Boyce, the Misses Bun yard, Mr. F. G. Bennett, Mr. and Airs. Beaufoy, Mr. and Airs. Bunting, Miss Cochrane, Air. R. Cook, Air. and Airs. A. Clenietson, Air. E. J. Clifford, Air. G. and Aliss H. Cornell, Mr. Cruttenden, jun, Mr. G. T. Cook, jun., and Aliss Cook, Air. W. B. Clark, the Rev. and Airs. Carre, the Rev. W. H. Denovan, Air. Dunk, Air. W. P. Dickinson, Mr. and Airs. AY. H. Day, Air. and Airs. H. Day, Aid. and Airs. V. Day, Air. Y. Day, jun., and Airs. Day, the Rev. and Airs. 0. R. Dawson, the Rev. C. G. Duffield, Air. Stanley Edwards, Air. E. Elliot, Air. H. J. Elgar, Air. and Airs. T. Elmore, Air. G. Farmer, the Rev. and

Airs. Fielding, Air. S. Kremlin, Air. A. M. Flint, Mr. and Mrs. AV. T. Frost, Dr. Gibbs, Air. T. and Airs. Goodwin, Air. Griffin (Catford), Dr. Gilbert (Tunbridge Wells), Air. and AIis. R. P. Grant, Air. and Airs. II. Green, the Rev. E. H. Hardcastle, Alajor Harpur, Dr. Ilallowes, Alajor and Airs. Haynes, Kir H. H. Howorth, Air. and Airs. H. Hoar, Air. J. and Aliss Hobbs, Air. F. and the Alisses Hobbs, Aliss Hailey, Air. F. Hill, Air. and Airs. Halliday, Air. and Airs. Innes, Dr. R. A. H. Johnston, Dr. J. Johnston, Air. and Airs. AV Jenkinson, Rev. and Airs. R. Jeffcoat, Mr. YV. Jackling, Dr. Jones, Air. G. H. and Airs. King, Air. F. Keeley, the Rev. and Airs. AV. A. H. Legg, Air. C. G. Long, Air. B. T. Lowne, Alderman and Airs. AlcVitie, Air. and Airs. AV. Alaskell, Air. S. L. and Airs. Alonekton, Air. R. Alercer, Alderman Dr. and Airs. Oliver, Air. and Airs. F. J. Oliver, Air. and Airs. C. P. Oliver, Air. and Airs. J. Potter, Air. and Mrs. Ewart C. Potter, Air. J. Parks, Airs. Rapley, Air. and Airs. W. Ruck, Air. J. and Aliss Reeves, Mr. and Airs. C. C. Streatfield, the Rev. and Airs. T. R. and Miss Stebbing, Alr.and Airs. Stevens, Air. and Airs. Styles, Air. and Aliss Stansell, Dr. Shaw, Air. and Airs. P. H. Silcock, Miss Semark, Air. and Airs. A. AV. Smith, Mr. E. W. Swanton, Air. J. Tomlin, Mr. J. AV. Tutt, Air. and Airs. Trollope, Air. and Airs. A. B. Urmston, Air. E. and Airs. Vaughan, Air. and Airs. G. Ward, Air. H. S. Ward, Air. AV. R. Ward, Air. and Airs. Wahnsley, Alajor Wright, Air. and Airs. AV. G. Wallond, Air. and Airs. W. Wallis, Air. Silas Wagon, Air. 1. AVilliams, Air. and Airs. AV. B. Young, Air. G. and Aliss Youngman. The reception proper being over, the Alayor and a large part of the company passed into the Brenchley Room, where, from a raised platform, the Mayor and Mayoress welcomed the President, Vice-Presidents, and members of the Union. His Worship said:—

Mr. President, Ladies and Gentlemen—Last summer, when I proposed that the South-Eastern Union of Scientific Societies should be invited to hold the Ninth Annual Congress in Maidstone, I little thought to be occupying so prominent a position as I am doing this evening, or possibly I might have hesitated before making that proposition, because I do not consider myself a scientific man. Thirty years ago it was different. I could with pride point to Queen's Prizes and Certificates from South Kensington, gained for Physiology, Magnetism and Electricity, and Acoustics, and to a Cambridge Extension Lecture Certificate for Geology; but now I find I know nothing, except how little it is possible for man to know, but that little adds very much to the enjoyment of life. It is my duty and pleasure to offer to you, Mr. President, the Vice-Presidents, and all the members of this Union, a very hearty and sincere welcome to the good old Borough of Maidstone. It is a town we are very proud of, it stands very high among the towns in the south of England for health; we pride ourselves also on our streets and on their sanitation, but most of all upon our Institutions; the old Parish Church, the old Palace of the Archbishops, the Municipal Technical Schools, and this splendid Museum with its priceless collections; in fact, the whole neighbourhood is rich in attractions, and cannot fail to instruct and interest. I trust the arrangements that have been made for your visit have been found agreeable and welcome, and that you will leave here with very pleasant recollections of your visit to Maidstone. I noticed a few days ago an article in a paper, headed "Scientists at Play," and, on reading, found that Lord Rosebery had given a reception such as this to the International Scientific Congress. If it was play, it was play of the right sort, to meet and discuss various problems and discoveries, and to have excursions to various places of interest, and, putting aside the personal enjoyment of these, the advantage of friendships formed by intellectual and refining association with men and women of similar tastes and pursuits, must be beneficial. I can only add that I thank you for coming to Maidstone on the occasion of your Ninth Annual Congress.

In reply to the Mayor's speech welcoming the members, the President spoke as follows:—Mr. Mayor, Ladies and Gentlemen— It is with extreme gratification and much warmth of feeling that, on behalf of the South-Eastern Union of Scientific Societies, I beg to express our grateful appreciation of the magnificent reception accorded to us on this occasion by your Worship and the Mayoress. Personally, I was much touched—and I feel sure that those whom I represent share my sentiment— by the cordial words of welcome, and the expressions of friendly sympathy, which have fallen so generously from your Worship's lips. Since we came into Maidstone yesterday, favours have been lavishly showered upon us from all quarters, and these tokens of good feeling have culminated in this evening's function. Although we come as representatives of various scientific societies, assembled for serious discussion, I can assure your Worship that we are by no means insensible of tbe social amenities of the meeting. It is always a privilege and pleasure to visit the excellent museum in which we are now assembled, and after acquaintance with it for many years, I must acknowledge that I never come without learning much and acquiring useful hints in connection with museum work. Nothing, in my opinion, can be more appropriate than the association in one centre of a group of institutions such as are located here—the Art. Gallery, the Museum of Natural History and Antiquities, the Free Library, and the Technical Schools. May this happy union of institutions that have so much in common, all tending to the elevation of the people, long remain unbroken! Has not Tennyson said—

"Beauty, Good and Knowledge are three sisters,
That doat upon each other, friends to man;
Living together under the same roof,
And never can be sundered without tears."

Mr. Mayor, when our proceedings come to a close to-morrow, I feel sure that we shall carry away from Maidstone the most pleasant recollections of our visit, whilst we shall leave behind a legacy of profound gratitude, and, I hope, a stimulus to the local Natural History Society. But among the recollections of the Maidstone meeting there will be none more deeply graven in our memory, none that will call up feelings of more lively gratitude, than that of the brilliant reception which we owe this evening to the generosity of your Worship and the Mayoress.

Sir H. H. Howorth, the outgoing President, being called on by the Mayor to say a few words, remarked that spontaneous utterance was not a gift confided to the male sex (laughter). Referring to the visit of the Union, he said the Maidstone Museum was a sort of Mecca which he had never before visited. Sir Henry then went on to pay a glowing tribute to the unpaid patriotism by means of which the local government of the country was effected, and after a strong advocacy of the value of athletics, he concluded with a humorous, but hearty, eulogy of the Mayor of Maidstone.

The party then divided and such was the wealth of entertainment and space at disposal, that, throughout the evening, the whole of the visitors were not collected together again. During the reception, and at intervals throughout the evening, selections were played by the Corporation string band. In the art room of the Municipal Technical School, a first class musical entertainment was given. This was divided into two parts—Parti. at 8.80 p.m., and Part II. at 9.30 p.m.—Madame Adey Brunei, Miss Adele Lancaster, and Mr. William Llewellyn being the principal artistes. Simultaneously with this, in the Bentlif Picture Gallery, at 8.45 p.m., a lecture was delivered by Mr. Alfred 15. Harding, with lantern illustration, etc., on "Ice Streams and Ice Caves," a summary of which appears *posted,* p. 45, and a vote of thanks to the lecturer was moved by the President and carried by acclamation. Almost directly on the close of this lecture, Mr. Paul Mathews, M.A., read a paper on " The Practicability of an Artificial Language," which is printed in detail, *postea,* p. 40. The President again moved a vote of thanks to the lecturer, which was very heartily received.

Light refreshments were served throughout the evening in the Reference Library and the Victoria Lending Library.

In the Museum and Bentlif Art Gallery the following collections were on view:—(1) *In the Great Hall.*—The Corporation Maces, Plate, Charters, and Ancient Seals. (2) *In the County Room.* —The Congress Museum of Natural History Specimens (arranged by Mr. E. W. Swanton); the Harrison collection of Flint Implements.
(3) *In the Shell Room.*—Nature-Study Exhibition to illustrate Mr. W. Mark Webb's paper on "The Teaching of Nature-Study." (4) *In the Bentlif Winy (upper floor).*—An Exhibition of Photographs, under the direction of the Committees of the Photographic Surveys of Kent and Surrey. (5) *In the Kent Archaeological Society's Museum.*— Collection of Kentish Antiquities of the Romano-British and Anglo-Saxon periods. (6) *On the Ground Floor of Bentlif Wing.*—An excellent display of microscopic objects. Among the exhibitors in this branch were—Dr. C. Pye Oliver, Medical Officer of Health for the Borough; Mr. L. Stansell, F.C.S., Mr. G. H. Rogers, F.R.M.S. (Pond-life), Mr. E. Clement (Bacteria), Mr. J. B. Groom, Mr. J. H. Bridge, F.R.M.S., Mr. F. W. Hembry, F.R.M.S. (Delegate from the Sidcup Literary and Scientific Society), Mr. J. W. Eele (Delegate from the Haslemere Microscopic and Natural History Society).

Special mention should be made of the beautiful programmebooklets, which, ably compiled by Mr. J. H. Allchin, the Curator of the Museum, at the suggestion of the Mayor, contained admirable photographs of the museum buildings as they were in 1857, and as they are today, together with photographs of the interior, a summary of the principal collections in the museum, and a list of the special exhibits. The whole booklet forms a delightful souvenir of the occasion. At no period of the proceedings was the huge work, in the interests of the Congress, carried out by the Local Secretaries, Messrs. J. H. Allchin and W. H. Day, so amply demonstrated as during the whole of the evening's entertainment, at which the richly deserved success of their meritorious efforts culminated.

Saturday Morning (Mr. F. W. Rudler in the Chair).

June 11th, Saturday Morning.—A Council Meeting was held at 9.80 a.m. The adjourned Delegates' meeting *(antea,* p. xxxiv) was opened at 10.15 a.m. , when about 50 delegates and members were present.

Dr. Abbott notified that a new society, the Fulham Field Club, had joined the union, bringing up the total number of affiliated societies to 50.

The Rev. T. R. R. Ktebbing, in proposing that Professor William Mathew Flinders Petrie, D.C.L., LL.D., F.R.S., should be elected to succeed to the presidential chair at next year's congress, said that by this time Mr. Rudler would probably have found the position of president to no small extent arduous as well as honourable. That being its character, it must be a great satisfaction to the Committee, that, if we left out of sight the infancy of the Union when everything was excusable, we had been able to secure a succession of distinguished men to fill the chair, and that, while each and all had filled it to the advantage of the Union, there had been the greatest possible variety among them of character and attainments. Dr. Flinders Petrie, the Egyptologist, had a more than European reputation. Probably in every part of the world, where there were students of antiquity, his name was known and valued. His researches carried us back far beyond what ordinary men counted as ancient history, to periods which were strange, not merely for their chronological remoteness, but from the fact that, in those distant eras, hitherto unsuspected civilisations were flourishing and men of renown making their mark. The Committee might rely upon it that Dr. Petrie would make a good steward of their interests. There was a notorious steward who was reported to have said "I cannot dig, to beg I am ashamed." But Dr. Flinders Petrie would have no occasion for begging, and as for digging, in that he was a most accomplished master, continually unearthing inestimable treasures.

This was seconded by Mr. Paul Mathews, who hailed with pleasure the fact that they had obtained so excellent a prospective successor to Mr. F. W. Rudler.

In putting the motion to the meeting, the President said it was highly satisfactory to have obtained the services of such an excellent scientist, and a man of world-wide reputation. The motion was carried by acclamation.

Mr. Fleay then proposed that the list of Vice-Presidents for the year should be that of last year with the addition of Sir Henry H. Howorth. He observed that it was superfluous to say anything of the ability, tact, and skill of their ex-President, who had obtained the goodwill of every member of Congress. This was seconded by Captain McDakin, and carried by acclamation.

Mr. Tutt proposed that Miss Sargant and Mr. Wilfred Mark Webb should be members of the Council in place of Captain McDakin and Mr. Foran, who retired automatically. He stated that, sorry as the Council would be to lose the services of the two retiring members, they were bringing into the Council a lady of tried ability, who had previously served for some years on the Council, and a new member, whose energy should do much to aid the Union to further the objects it had in view. Mr. Merrifield seconded, also bearing a tribute to Miss Sargant's excellent working qualities, and the motion was then put by the President and carried.

An announcement from the President, regretting that Dr. Abbott had, acting on medical advice, asked to be relieved of his office of Hon. General Secretary, was received with the greatest sympathy by the delegates present. The Eev. K. Ashington Bullen had consented to act as Secretary, and he was the only man who could console them for the loss of Dr. Abbott.

This was formally put by the President and agreed to.

Professor Boulger moved, and Mr. Whitaker seconded, a motion that Dr. Abbott should succeed the Rev. R. Ashington Bullen in the post of Treasurer. Professor Boulger pointed out that the Union largely owed its existence to Dr. Abbott, whilst Mr. Whitaker hoped that Dr. Abbott's position would be as great a sinecure as the post of Treasurer had been during the time that he (Mr. Whitaker) was Treasurer, when, as a matter of fact, the Secretary did not only his own work, but also that of the Treasurer.

The motion was passed with acclamation, the President observing that, as Dr. Abbott was unfortunately compelled to retire from the Secretaryship, he thought the change which was now suggested an excellent one.

Dr. Abbott replied, and said he was pleased to accept the office, as he would have been sorry to drop out of the work. He pointed out that the work devolving on the Secretary at the time of the Congress was more than he could manage. He alluded to the good work done by the Congress in raising the standard of the work carried on by some of the affiliated Natural History Societies, and its value in introducing workers in the same branch of study to one another, such introductions possibly leading to lifelong friendships.

Mr. Whitaker proposed a vote of thanks to the Mayor and Corporation, the Library and Museum Committee, especially to Messrs. Allchin and W. H. Day, to the Maidstone Natural Science Society, and to their hosts and hostesses. He detailed the kindness that he himself had received, and had no doubt that his own case was merely typical of the hospitality lavished on the other delegates.

The Rev. R. Ashington Bullen seconded, and described the hospitality of the Mayor on Friday night as magnificent. He referred to the late Mr. J. L. Brenchley, one of the earliest, and one of the largest, benefactors of the museum, and who was in his early life a curate at Shoreham, a parish of which he himself had been more recently vicar.

Dr. Abbott said, as Secretary, he could judge of the excellence of the work of the Local Secretaries of the local society. There had been no slip from the beginning. Mr. Allchin had worked night and day to get the museum in trim for them, while Mr. Day had done excellent work. Their splendid work had been followed by equally magnificent results.

The President referred to the hearty manner in which the authorities had helped them, and so their gratitude should be unbounded to Mr. V. H. Day and Mr. Allchin. Both had worked hard, and Mr. Elgar bad also ably assisted. With regard to the local society, there were yet only 100 members, so that there was plenty of room for increase.

The resolution was passed with acclamation.

Mr. Paul Mathews then spoke on the advisability of arranging combined excursions between the societies of adjacent districts, *e.g.*, Rochester, Maidstone, etc. Dr. Abbott thought the idea an excellent one, and said such arrangements had been carried out with success in some districts.

Dr. Abbott then drew atttention to a "Photographic Book" belonging to Mr. Box, one of the New Brompton delegates. It was most interesting, and contained a large number of photographs taken at the various congresses that had been held during the last seven years.

Professor Boulger reminded the delegates that he had, at the Dover Congrees, requested them to obtain from their societies an expression of opinion as to the desirability of taking some steps, legislative or other, for the preservation of wild plants in danger of extermination. He mentioned that he was to lay the subject before the Royal Horticultural Society, on August 28rd, next, and he again asked that some mandate should be given to show that the Natural History Societies of the southeast of England were not indifferent in the matter. The Hon. Secretary (Rev. R. Ashington Bullen) had kindly offered to send out a circular on the subject, and this he, Professor Boulger, had already prepared and submitted to Mr. G. C. Druce, of Oxford.

This concluded the meeting of the delegates, and the delegates and members present then resolved themselves into a general meeting. By this time some 500 or 600 teachers, pupil-teachers, and elder scholars from the surrounding district, had filed into the hall to hear Mr. Wilfred Mark Webb's paper on " The Teaching of NatureStudy." The paper is printed in full *postea,* p. 58 *et seq.,* and was followed by a discussion that occupied almost as long a time as the paper itself, and which elicited a variety of views from different members of the audience.

The President in opening the discussion stated that it appeared to be very generally accepted that some form of nature-study in schools was desirable. He congratulated the local society in having attracted so large an audience to hear Mr. Webb's paper, and trusted that the paper, and the discussion that he hoped would follow from the many experts present, would do much in making clear the lines on which the study of Nature might satisfactorily proceed in schools. He rather demurred to Nature-Study being considered a new subject. He even had an idea that Solomon knew something of the subject, but at any rate he would, preferably to speaking further on the matter, ask those present to discuss the paper.

Sir Henry H. Howorth treated the matter in a humorous style, which highly interested the younger part of the audience. He considered that nature-study would possibly be a most fascinating and popular subject among children, especially if properly taught. He touched on the question of animals and birds in confinement, and gave a number of anecdotes tending to illustrate the hypothesis that animals rather liked to be kept in confinement if they were treated well. Among other anecdotes, he related that of the famous rattlesnake, which, after being saved from a dreadful death and taken home, was heard making a hullabaloo one night, whereupon its preserver went downstairs and found that the snake had caught a burglar, had opened the window, and was rattling its tail to call a policeman. He illustrated the advantages of hopefulness and pluck by the story of two frogs which fell into a milktub, one of them being despondently drowned, while the other, by paddling energetically round and round the tub, found itself at last comfortably seated on a pat of butter of its own churning, etc. He added seriously that it was a subject, to the practical application of teaching which, he had paid no real attention.

Professor Boulger, among other remarks, thought it a fortunate circumstance that the promoters of nature-study were not all too much agreed as to the details of the method, because that difference of opinion guaranteed that teachers would have no cast-iron system of uniformity forced upon them. In the matter of detail, for instance, he would himself strongly recommend it as having the highest educational value. Thus Babington's *Manual,* though more difficult to work with than Bentham's *Handbook,* was partly, for that very reason, a better book for the botanical tyro. We wanted to encourage not only the knowledge but the love of nature; and, therefore, especially with young pupils—in the kindergarten stage—sentiment and poetry, which had been called "caterwauling about nature," were not to be despised as educational means. Children might well draw or paint or embroider daffodils, recite or sing poetry about daffodils, and study daffodils in every way for a whole day. He considered that collections were not wanted. To make it attractive the work must not be burdened. He thought that the reasons of modification observed were better left until a later stage, and that " how " rather than " why " is the most important factor of such work.

Mr. Tutt was much interested in the discussion as far as it had gone; at present, he was not clear whether the nature-study ideas of Sir Henry Howorth or Mr. Webb were those most favoured by the audience, but he had no doubt that, although those of Sir Henry would be hugely popular with the scholars, the point might arise in the minds of mere teachers, like himself, whether such tales had any educational value. So far as Mr. Tutt had followed this and other similar discussions, there appeared to him to be three groups of people interested in the matter, (1) People who talked and lectured a great deal about the subject, but who knew little or nothing either of the subject educationally or of the schools. (2) The teachers, into whose schools it was to be introduced. (8) The scholars, to whom the subject was to be taught (?) whether or not their temperament was suited to take advantage of this form of teaching.

He wished briefly to make three assertions as a professional teacher and lecturer on natural-history subjects:— (1) That the time had gone by when the teachers (particularly elementary teachers) wanted simply stirring up, as some of the speakers seemed to think, and that stirring up would not supply the knowledge and enthusiasm that would make the teaching of such a subject moderately successful. (2) That most of the lecturers on the subject of teaching nature-study in schools, were hindering the matter rather than pushing it forward, by the complete absence of constructive criticism in their lectures. (8) That what the teachers wanted was to discuss the matter with those who were both naturalists and teachers, who understood the limitations and possibilities of the school, and could advise in particular cases how to make such teaching practical.

With regard to these points, his criticisms had already, it would appear from the statements of previous speakers, been voiced by others. One of these speakers had noted with grief, that some of these lecturers had been dubbed by the press " outrageous apostles of nature-study," but he was inclined to think, from some of the lectures he had heard, some of the statements made, and some of the pretensions put forward, that the criticism was fully deserved. Nature-study was a most useful branch of work; it was capable, iu right hands, of producing excellent educational results; it was, however, not a substitute for all other, nor for a large part of other, branches of educational training, and it must only occupy its proper share of time and not oust other important subjects. "The outrageous apostle of nature-study," like all other "outrageous apostles" of anything, had no sense of proportion, and was wanting in a proper perspective of the educational field. When one had listened to one of these lecturers and ventured a query as to extent of subject, time, opportunity, and the thousand-and-one points that arise at once in the mind of a naturalist who is also a professional schoolmaster, he is met with the platform declaration, which is, however, not at all convincing: "I do not want to tell teachers what to do, nor how to do it, they will know better than I, nor do I want to tie them down to a cast-iron code, each one must make his own scheme and do it in his own way. I do not profess to teach the teachers how to teach." This is almost word for word what I have heard from many lecturers, and the lecturer sits down, whilst the would-be critic feels that with such an opponent criticism is useless. In considering the subject now the naturalist-teacher can be left out; he is possibly already doing good work and wants no whip to spur him on, nor aid to produce good results educationally, but the non-naturalist teacher, who sees the educational value of the subject, and is eager to do it, does require such help, and as most teachers belong to this class it is just such help that is wanted. He does not want of course to be tied down to a cast-iron scheme, but he does want a naturalist-teacher to show him or her how the new subject can best be taken, how it can be best fitted into the work of the school with the least dislocation of the other work, and even perhaps a suggestion as to how the work of the staff can be utilised to get useful educational results. In this direction none of our lecturers appear to know how to help the teacher. There is also another point of view. Haphazard attempts to teach such a subject may lead to failure, and there is then a waste of public money, and, still more serious, the child's school-life, already short, is being wasted. One must not suppose, the speaker urged, that he was not keenly alive to all the value to be derived from nature-study, no one was more wholly interested in the matter. For more than twenty-five years he had attempted work of this kind in a London school, for some years he had lectured for the late London School Hoard, to their teachers, on natural history subjects and the method of teaching tbem, and he had become unwittingly a sort of referee for all sorts of details relating to the matter, submitted by teachers all over the country, and it was because of this and because he felt so strongly that the present methods of dealing with the subject were not reaching the desired end educationally, that he asked those of bis fellow delegates who were interested in the subject to bear with him if be did not take exactly the same sanguine view as themselves, and wanted to put the matter rather on a workable and business-like footing. He would now like to criticise some of the points touched upon in Mr. W. Mark Webb's paper.

Mr. Webb had suggested that the consideration of Nature-Study as a school subject was of quite recent growth, *i.e.,* of the last two or three years, the President had suggested that he thought Solomon knew something of it, and he, the speaker, had an idea tbat Adam was credited with being a naturalist, in fact, that he named all the created animals. The fact was that some teachers, under the title of natural history, had done the work for years. It was now some six or seven years since Mr. Graham Wallas first arranged for the speaker to lecture to the teachers of the London Board Schools at the Medical Hall of Science, and that Dr. Kimmins, who was then present, brought the matter forward the following year at the Teachers' Conference, held at South Kensington and Shoreditch Museums, when the speaker read another paper, at which Mr. Webb was present. This paper was printed in detail by Mr. C. Jackson, the present senior inspector of the Board of Education, in the official educational year-book of New Zealand, and distributed to the teachers of that country. Most advanced educationalists were now agreed that natural history work, or nature-study, or whatever it might be called, had, under certain conditions, an excellent educational value. Mr. Webb asserted that nature-study was not science teaching. This was a matter for definition, and the speaker insisted that if the work done as nature-study added to the knowledge of the worker accurate facts from which he could deduce sound conclusions, then such work was to be accounted science, however elementary and simple it might be, and, if the results were not to this degree scientific, the study undertaken was not that of nature, and possibly had no educational value. Mr. Webb said nature-study was not a subject, but a method of teaching. The speaker did not understand what this meant, he thought that nature-study was the study of nature, and he had always had an idea that nature was a subject as vast as the universe, and the rest of Mr. Webb's paper showed clearly that his view of naturestudy was synonymous with the study of natural history, and it appeared to Mr. Tutt that, unless there was something definitely done in these lessons, the time was wasted. Mr. Webb's somewhat offhand suggestion that some successful teachers of natural history might know less than the students, might be met by the retort that the teaching of ignorance is not difficult, and the sweeping definition of a teacher as a "disseminator of dry facts which he bad learned in similar fashion," was an assertion that wanted substantiation. The speaker urged that what was wanted in the teachers, as well as accurate knowledge, was the right temperament and enthusiasm, and hence the right teachers wanted selecting for this work. Mr. Webb, apparently to make the subject appear attractive and easy to teachers, stated that nature-study "considered things just as they occurred," "it did not consider matters in any logical sequence or inter-connection," but the speaker ventured to urge that isolated scraps of information or knowledge, even in large quantity, did not constitute education. It had always been laid down to him, and by him, as an educational law, that it was rather the ability to logically correlate facts, than the amassing of unconnected facts, that bespoke the truly educated man, and he did not think that Mr. Webb's statement would convince his hearers that this was not so. There was surely a possible correlation of simple, as well as difficult, facts and observations. Mr. Webb had rightly said that the true place of nature-study was in the fields, but the great mass of our children had to be educated in towns, and Mr. Webb had little or no real suggestion to offer as to how naturestudy was to be brought within reach of such children; he had offered no truly practical alternative, although, to a practical teacher, such alternatives were self-evident, and would, in skilful hands, produce the same educational results; nor had Mr. Webb shown how classes of 50 or 60, or even 30 or 40 children could, apart from a pleasant outing, get any really very useful nature-study work out-of-doors under the guidance of a single teacher; the size of many existent classes must be reduced before such work could be successfully undertaken even in the fields. Enquiry into successful out-of-door classes usually resulted in the discovery that the classes were comparatively small, and this must necessarily be so, but yet the making of such small classes must not be allowed to dislocate the rest of the school work. One of the speakers had, as he (Mr. Tutt) had already noted, referred to those advanced (?) pioneers who saw in nature-study a cure for all educational ills, and an efficient means of converting stupidity into intelligence, and whom the educational press had dubbed " the outrageous apostles of nature-study." A remark made by the same or another of the speakers led him (Mr. Tutt), as he had already said, to incline towards agreeing with the criticism, and Mr. Webb's statement that they hoped to monopolise most of the school time, suggested somewhat similar lines of thought. They had been told by one of the preceding speakers of a lady enthusiast who had had what was called a "daffodil " day; the reading was daffodils, the drawing was daffodils, the songs were to be of daffodils, the grammar was to be about daffodils, and everything possible that day was to be of daffodils. On another day primroses were to be the *yiece de resistance,* and so on; and that same or another speaker had referred to a well-known scientist who did not think anyone could be a botanist till he knew 10000 species of plants, and he (Mr. Tutt) suggested that if seven flowers per week (including Sundays) were done for a year, i.e., if 865 flowers were handled this way in a year, and that if we divided the 10000 flowers required to make a botanist, by 865 flowers thoroughly done per year, it would take a matter of 27 years to make a botanist. Were botanists made this way? And surely nature-study was not, as most speakers assumed, restricted to flowers and botany? He ventured to think that the other branches of natural history were of at least equal importance. There was also another factor. He was a great believer in correlation in education, but not correlation in which reading, singing, grammar, etc., were made subordinate to a knowledge of the daffodil, primrose, etc. Concentration was one of the main factors of success in education, and there was little real concentration in this. And when a child had breakfasted on "daffodils," dined on daffodils, supped on daffodils, and slept on daffodils one day, had breakfasted, dined, supped, and slept on primroses the next day, what was the result on the child? (A voice—Mental stagnation.) Was such a method as this putting an interesting subject (*pace* Mr. Webb) on an educational basis? He, and the teachers at the back of the room, evidently with him, doubted it. It could not be done in this fashion, nor had it any connection with the naturestudy of the lanes and fields, the nature-study of living things, the nature-study of moving animate existence in which children delighted—it was words, all words. It was stated by one speaker, that children were born observers, that work done in school would be continued by the carrying on of observations out of school— he (Mr. Tutt) denied the assertion. It was recognised by all naturalists, who were not merely *dilettanti,* that to observe accurately is one of the most difficult things that faces the trained naturalist. How often could the editor (and reader) of a natural history magazine accept absolutely the observations of a contributor? If children can, as asserted, observe accurately and usefully, naturally, and without aid or training, why, from the educational standpoint, is nature-study wanted at all to train them to do the very thing which, those who argue thus, urge they can already do? Does not then the whole theory of naturestudy as an efficient educational instrument in this direction, fall, as being of none effect? The whole point of success in training by studying nature must be an efficient and sympathetic master or mistress, in complete touch with children who have the temperament to understand what is carefully brought under their notice. With the master or mistress who will give the children with the right temperament no chance of benefiting by nature-study training, he (Mr. Tutt) had no sympathy. There was, in almost every school, at least one teacher with the right spirit and right knowledge. This teacher should be allowed to make all he or she could out of the subject, within the possibilities at his or her disposal. Mr. Tutt added that there were many other points that he had jotted down as material for discussion, one related to the extract read by Mr. Webb from a book stated to have been written by an eight-year-old infant and exhibited by him, that Mr. Webb had " annexed" from a nature-study exhibition. It was, no doubt, the unanimous opinion of every teacher present, that the teacher really thought the child had written his or her own, and not the teacher's, observations, but it was self-evident that children of eight did not, even unconsciously, quote Shakespeare. These showbooks had little at present to do with real nature-study. Another point that wants consideration is Mr. Webb's off-hand remark that "if you have to satisfy County Councils, etc., you must do.... " something not quite within the four walls of what the instructions of the County Councils say shall be done. He, Mr. Tutt, urged that such advice cut both ways, and the forcing of teachers against their will could be met by a passive resistance that might upset the nicest calculations. Then there were some illogicalities in Mr. Webb's paper, too, which there was not time to discuss, *e.g.*, his advice to teachers to do the work without reference to books, and his later advice to read natural history books. His assumption, too, that nature-study supplies a great need in education in leading people to act independently, is cheering—for one supposes that many men, besides naturalists, have learned to act independently when need has arisen. There was also a statement made by Professor Boulger, and worthy of great consideration, that he, Mr. Tutt, supported very strongly, as against the opinion of other speakers, *viz.,* that it was more important for children, in nature-study work, to follow up "how" rather than "why." The children should be led to observe facts rather than speculate too largely on possible reasons. "What is the go of the thing?" is still the most important item in training children to observe, and then when this is learned, " What is the particular go of it?" He, Mr. Tutt, wanted to see nature-study take its right place in the curriculum of every school, but he wanted those who urged its introduction into every school to recognise its limitations as well as its possibilities, he wanted them to see that if the study was sound as an educational instrument, it must take its place with other efficient instruments, and that this will never be done by assuming that it should replace other subjects such as mathematics and physics, and other proved efficient educational means, but that it must be used side by side with them. Among young children particularly, and in a very elementary but carefully considered, and not haphazard, form, its power for good is especially great, but, here, constructive criticism must come in, and something more than advice was needed. He had been listening to advice all his life—good, bad, and indifferent—and it was little that was offered on educational matters by non-educational men that was of real practical value. It was the advice of the expert educationalist who could construct as well as destroy, that was valuable. If those lecturers who were pushing forward the value of nature-study would add some constructive criticism to their wealth of advice, the teacher who was anxious to give the subject a fair trial and was not specially well-equipped might learn how to set about it. But there were many educational problems underlying successful nature-study work, of which the large size of classes, the special training of teachers in special subjects, the organisation of schools so that teachers might teach subjects rather than classes, and above all the choice of the right teachers, are not the least. He thanked Mr. Webb for his paper. If they did not agree as to detail they both desired the same goal, and, if there were two sides to a question, it was better that both should be discussed, and it was his unfortunate temperament that he always preferred to take the other side so as to see what was the logical mean between the two views that could be presented. It, therefore, worked out that he appeared, perhaps, almost as an opponent of Mr. Webb, when, on broad principles bearing on the utility of nature-study as an efficient educational instrument, there was no real difference between them.

Mr. Groom, whose nature-study class is well-known in Maidstone, made some remarks. He feared that the remarks of the last speaker might do some harm to the movement of nature-study. He believed that nature-study work should lie haphazard. Everything that he and his class came across in a walk was considered. He pointed his remarks with one or two illustrations: (1) Called attention to branching of oak and elm, compared them, difference told between the trees even in winter; (2) Speedwell on a sunny bank; blue petals on ground; petals fallen because insects had fertilised flowers and were of no further use, etc.

Mr. Neate was convinced of the value of nature-study work, and, as chairman of an educational committee considering this subject, he should push it forward. He thought the best means was to train young teachers especially to give instruction in the subject, and, in a short time, it would happen in the district that teachers possessing a qualification for the work would be preferred for ordinary appointments. He thought that teachers ought more frequently to join the outings of natural history societies.

Mr. Beecham, as a teacher, thought that on the whole he should agree with what Mr. Tutt had said, although some points raised might, perhaps, be open to question, but he had an especial duty now to perforin, *viz.,* to thank most heartily the South-Eastern Union of Scientific Societies for inviting the teachers to hear the paper and discussion.

Mr. Webb briefly replied. Among other remarks, he thought that Mr. Tutt, whose best known work had been somewhat technical, had taken too advanced a view of the subject; he thought further that kindness to animals rather than the reverse might be readily inculcated in nature-study lessons; he further believed that "Why" as well a;'-"How" could be considered by even the youngest children, because the reasons for most peculiarities were so self-evident, *e.g.,* the hooks of wild roses developed to enable the wild rose to climb; the divided leaves of buttercups to let the grass grow through, etc. He had wished to make suggestions to the teachers present, he did not pretend to show them how to teach the subject, the details of which must be left to each individual teacher. He thanked the audience for the great interest they had taken in his paper.

Saturday Afternoon. *June 11th, Saturday Afternoon:*—A large party of rather more than sixty visitors and friends started at 2.45 p.m. from the Brenchley Gardens on a stroll by the side of the Medway to Allington Castle, accompanied by the Mayor of Maidstone. At some distance down the river, boats were provided, and a crossing was made to Allington Castle, where the company were most hospitably received by Mr. and Mrs. Dudley C. Falcke. The history of the castle was briefly reviewed by Mr. Falcke *(poxtea* p. 22), and he and the Rev. C. H. Fielding then conducted the party over the building. At the end of the visit the party was photographed by Mr. Roods and the photo has been reproduced as a frontispiece to this volume.

After about an hour spent at the Castle, the company visited a Ragstone quarry in the Lower Greensand.

Mr. Whi taker gave a short account of thequarry, and was followed by Mr. Bensted, who, in the course of his remarks concerning the formation of the " Pipes" which the quarry exhibited, said that the appearance of the rock, forming their sides, did not in nil cases accord with the theory that the chasm was due to dissolving action, as the beds were not smooth and rounded, but angular and broken, leading to the impression that fracture was the cause of the separation of the beds, although, in the majority of instances, their appearance, no doubt, pointed to their having been subjected to a dissolving action.

Mr. Whitaker agreed that probably they were due to both causes.

Mr. Bensted also mentioned the occurrence of mammalian remains in the brick-earth filling, especially those of *Elephax primiijeniiix, Hliinocerox tielwrhinux,* and *Eqnnx;* also of the following shells, *Helix, I'n/ia,* and *Surrinea,*

ADDENDUM.

Page iv. As member of C'ouiicil--5lr. V. II. Griffin (Catford and District N. Hist. Soc.l.

ERRATUM.

Page 88, Line 5 from bottom for "Keltic Pottery found at Haslemere— opposite page 7H " read " opposite page 50."

Addendum To Presidential Address.

With reference to the remark as to the waning popularity of the British Association (p. 9), it is pleasing to note that the recent meeting at Cambridge, held since the address was delivered, has witnessed a revival of interest quite exceptional in its character. This was clearly due to two causes. In the first place the meeting was held in a venerable centre of intellectual culture, which had not been visited by the Association for more than forty years; and secondly, the Presidential Chair was occupied by so distinguished a person as the Prime Minister.

The statement in the Address as to the general decline of interest in the Association during recent years is substantiated by the following comparative figures:

Southport in 1903 attendance 1754.
„ 1888 „ 2714.
Belfast „ 1902,, 1620.
„ 1874 „ 1951.
Glasgow „ 1901,, 1912.
„ 187G „ 2774.

TRANSACTIONS OF THE SOUTH EASTERN UNION OF SCIENTIFIC SOCIETIES. *1904.*
PRESIDENTIAL ADDRESS.
Hv F. W. BUDLER, I.S.O., F.G.S.

In exercising the privilege of addressing you from the Presidential Chair I find myself faced at the outset by a double duty—a duty on the one hand to you, on the other to myself. In the first place it is clearly my duty, as it is most assuredly my pleasure, to express, in the warmest terms I can command, my grateful acknowledgment of your confidence in having conferred upon me the very honourable distinction of the Presidency of this Union. And in the next place it is a duty which I think I owe to myself to explain that circumstances have unfortunately prevented me from having had the advantage of attending any of your previous assemblies, so that, ignorant of your procedure, I must needs solicit your kind forbearance for those shortcomings which will be all too conspicuous during my tenure of office. When it was hinted to me last year that my name might possibly be submitted for election, the suggestion came upon me so unexpectedly, that I gave, rather rashly, a kind of qualified assent, without fully realising the responsibilities of such a position— especially the grave responsibility of having to follow so brilliant a predecessor as Sir Henry Howorth. To switch off suddenly the electric light, and have to be content with the glimmer of a mere taper, is enough to try your temper to the utmost! Put any remarks on my own dulness must be so utterly uninteresting that I dare not indulge in apology.

Reference to the programme of our meeting shows that upwards of forty societies are now banded together in this Union; and of these by far the greater number are Natural History Societies. We are, therefore, essentially a gathering of people devoted to the study of Nature. Ky virtue of that subtle influence that draws like to like, we assemble together animated by a common spirit— working for a common purpose—as lovers of nature, as students of science, as earnest seekers after truth. We might be called a gathering of " naturalists" were it not that the word has been so sorely abused. To one mail a " naturalist " may be a priest at the altar of science, humbly seeking to unveil the mysteries of the surrounding world: to another man a " naturalist" may le merely a bird-stuffing barber.

Fortunately we are not likely to be labelled with the exceedingly ugly word "physicist"—a word which has been denounced as unlovely, irregular, and ambiguous". It is true that "physical.science" and "natural science" mean, so far as etymology goes, precisely the same thing. The distinction is purely a matter of convention, and the dividing line so far from being sharp is often shady. We may agree to limit the term "physics" to that department of natural science which is not simply observational but also experimental—to what we sometimes call "experimental philosophy;" yet the naturalist often becomes an experimentalist and so transgresses the artificial boundary. We are indeed all "natural philosophers," in the true sense of that term—lovers of natural knowledge, lovers of that wisdom which springs from the study of Nature.

Nature stands beckoning to all. Some ignore her invitation. Others are attracted merely by her recreative aspect, whilst others, again, give themselves up to the poetical contemplation of nature. No one will be disposed to say a word against either of the latter courses; but still, as true naturalists, we approach her in a different sense. To turn to Nature for recreation is undoubtedly a healthy sign, and the young man or woman who, in place of devotion to "In its sound. *lizzisto,* it is unlovely, and in its formation it is irregular and ambiguous. ... To express by tlie use of the suffix -*to* a student or professor of physics we should make the word *physics-ist*. But that being intolerable in sound we have in its stead *physic-to,* which really, according to its formation, means a professor or student of the art of physic—quite a different meaning from that of which we are seeking the expression; and we pronounce it, instead of *lizzikist,.tizzisitt,* thus not really improving much on */izziksist,* if indeed the latter, by the interruption by a *k* of the continued hissing, is not the pleasanter word, or rather the less offensive. We thus obtain only an incorrect formation, an etymologically ambiguous meaning, and a succession of hisses, which our performance well deserves." *Kvery-day English,* by Richard Grant White. Boston: 1880, p. 470.

artificial pleasures, looks nature-wards, eager to exchange the fever of the city for the freshness of the field, is in a fair way of becoming a naturalist.

As naturalists, however, we turn to Nature not simply for healthy refreshment, not even for poetical inspiration, but we approach her for the serious purpose of scientific study. Natural science is nothing more than the systematised study of the concrete objects and the phenomena of nature, as distinguished from such science as is mathematical, or mental, or moral. It is this Natural Science that our members seek to study, and to which our respective societies are mainly devoted.

The 42 societies incorporated in the South-Eastern Union include, according to the schedule in our last report, more than 5000 members. With a constituency of this imposing magnitude we may justly regard our Union as a corporation of some importance. But though the aggregate number of members in the affiliated societies seems large, it must be borne in mind that it represents a ridiculously minute fraction of the population living within the sphere of their activities. Why should this be so?

It might not unnaturally be supposed that every healthily constituted mind would find a source of exquisite pleasure in the acquisition of a knowledge of the material creation, in seeking to unfold its mysteries, and to comprehend something of the laws by which natural phenomena are governed. We might expect that every educated man and woman would join the local natural history society. But this is unfortunately not the case. What then is the cause of this neglect'? How comes it that our scientific societies are not better supported"? And is there any way by which our unpopularity may be overcome'?

Now that we are met here in annual conference, assembled I presume for the express purpose of taking counsel together, we might do worse than discuss this subject. At any rate it is a subject in which we all have a common interest.

In every Society we must remember that we have to deal with two distinct types of member—the member who does the work, and the member who likes to hear what work is being done. The requirements of each class must be duly met, if the Society is to prosper. The real researchers usually form so small a minority that they could rarely run a local society by themselves. The very existence of the society generally depends on the support of those intellectual people who take a general interest in the progress of science without being in a position to aid it directly themselves. The wants of the majority must be courteously respected.

In these days of excessive specialization it becomes increasingly difficult for the average man to follow the technical work of a specialist. Really important communications are therefore in most cases more or less unintelligible to him; and hence the proceedings of a Society may become intolerably dull to the majority. In some cases, I fear, the proceedings may become almost farcical.

One of the members reads a paper which most of the others don't understand, and then perhaps he who understands it least jumps up at the conclusion and is profuse in gratitude for so instructive a communication! And the pity is, we often feel ashamed to admit that the whole thing has been far above our heads. Many people don't in the least mind being ignorant, but they are desperately frightened lest other people should think them so.

Probably the most satisfactory way of reconciling the wants of both types of member—the general and the special—is by adopting the excellent practice, now becoming common, of dividing the society into sections, and holding sectional meetings for specialists alternately with general meetings for the whole body of members. At a sectional meeting a member reading a paper can be as technical as he pleases, or as the subject demands, and, indeed, the more technical the better in some respects. True, he cannot expect to draw a large gathering around him, but he has the satisfaction of feeling that those to whom he speaks, though few in number, will understand his words and appreciate his work. If he had read a paper on the same subject at a general meeting, he would have known that the majority of his hearers would be unable to follow the details, and he would consequently have taken infinite pains to explain himself in terms divested as far as possible of technicalities; in fact, he would have been under the necessity of removing his subject to an altogether lower plane. There is hardly any task more difficult than that of popularizing a profound scientificsubject, and the result after all is rarely quite satisfactory.

Every year natural science is becoming more highly specialised. Hence there is a growing reason why special meetings should be held by the separate sections in which discussions can maintain a high standard. All members of a Natural History Society may be regarded as belonging to the genus *Naturalist,* hut they need to be sorted out into species—such as *ijeohujixts, botanists,* and soon. The General meeting is literally a meeting of the genus; the Special meeting, an assembly of the species.

When a young person joins a society there is great temptation for the enthusiasm of youth to lead him to desultory study. It is a delightful thing to gad round the circle of knowledge and just sample the sciences one after another. The knowledge of natural history is so fascinating that we feel anxious to follow the advice which Gargantua gives in a letter to his son Pantagruel: "Now in the matter of the knowledge of the works of nature," says Rabelais, "I would have thee to study that exactly:

so that there be no sea, river, nor fountain, of which thou dost not know the fishes; all the fowls of the air: all the several kinds of shrubs and trees, whether in forests or orchards, all the sorts of herbs and flowers that grow on the ground; all the various metals that are hid within the bowels of the earth, together with all the diversity of precious stones that are to be seen in the orient and south parts of the world. Let nothing of all these be hidden from thee. ""

It may not be amiss to add that the letter which gives this comprehensive advice for the study of Natural History was dated from Utopia. If it was Utopian when written, three centuries and a half ago, be assured it is ultra-Utopian to-day.

Seeing then that a two-legged encyclopedia of natural history is an impossibility in our narrow world, let us warn the young member against aspiring to such an ideal. The man who takes an ardent interest in all departments of nature, whose sympathies go forth to every branch of study brought before his society, is no doubt a man of exceptional intelligence, but he is a man in danger of having his reputation buried under his versatility. Many of us do so little because we know so much. It is the specialist who confines his study of natural history to a single group, however small, that wins scientific success. The mensuration of a man's knowledge is concerned with depth, rather than breadth. Profundity is respected, superficiality decried. The prize is to that man who knows one thing better than other people know it. He who can do a single thing which others cannot is respected; he who can do a dozen different things which others do equally well, or better, may pass through the world unheeded, or be labelled a scientific Jack-of-all-trades.

Let us then advise any young member to look around until he finds some department in which he feels that he can take special interest, and then let him settle himself there. Seated contentedly in some little corner of Nature, he may quietly focus his attention on his limited surroundings, and the time will come when verily he shall have his reward. Advise him to keep his eyes steadfastly fixed on one thing until he pierces it to its very centre, till its inmost core is laid bare, and then at last, seeing it through and through, he may feel that he knows something about it worthy of being put into a paper. He will find that it is this thoroughness of knowledge, not mere magnitude, that counts in scientific circles.

I have been induced to insist on the necessity of specialization in the study of Natural History, because it is only by specialists that work of an original character is likely now-a-days to be accomplished, whilst it is by such work—and by that alone—that the scientific position of any given society is determined. It is desirable that every provincial scientific society should seek affiliation to the British Association; and the necessary and Quoted by Mr. Oscar Browning in his *Introduction to the History of Educational Theories.* London 1881, p. 77.

sufficient condition for such relationship is that the Society should publish original scientific observations.

It is true that in many cases such observation is reduced to a minimum, limited perhaps to the record of local rainfall; but it is obviously desirable that the interpretation of the conditions should be broad and generous. Each Society should aspire to add its little brick to the growing fabric of natural knowledge— and the bigger the brick the better.

No doubt original work of a high order will always find its way to the great Central societies. Nor is this, perhaps, to be deplored. He who has had to work up the bibliography of any scientific subject knows too well the difficulty of searching through the numerous publications of provincial societies; and it is to assist in this quest, that the British Association publishes in each Annual Report a selected list of papers from the proceedings of the Societies with which it is in correspondence.

There is, however, a vast amount of quiet local observation that finds nowhere more appropriate record than in the proceedings of the local society. Floristic and faunistic catalogues, descriptions of new species detected in the neighbourhood, observations of freshly exposed geological sections, maps illustrating the distribution of vegetation or of special forms of animal life, local meteorological observations—such are some of the types of work likely to fall within the compass of a Society having on its roll of members a fair proportion of specialists, and such is the kind of work which a local Society should endeavour to publish.

At the same time, in order to avoid the undue multiplication of minor publications, it might be worth while to consider whether it would not be advisable for some central journal, like the organ of this Union, to undertakethe publication of selected papers and reports from the smaller societies, so as to relieve these societies of the responsibility of independent publication. Economy and uniformity might be effected by a central organisation, whilst the constituent societies would still retain the credit of their respective papers.

It must be a matter of profound satisfaction to any society to feel that by its publications, whether independent or federated, it has done something to advance the bounds of natural knowledge— that it has been able to lay down another paving stone, however small, in the path of scientific progress.

From the development of scientific education in this country in recent years, it might be concluded, not unreasonably, that a knowledge of elementary natural history must be more generally diffused than was formerly the case, so that young people who enter our societies to-day are in a position of great advantage compared with the majority of the older members, who at the time they entered had few facilities for learning. Formerly there was but little opportunity in this country for anyone to obtain a knowledge of biology unless he happened to be a student of medicine, and most of us had to pick up our crumbs of knowledge as best we could. To-day there are evening classes for science, including certain branches of biology, in almost every district. In the larger centres of population the science-teacher stands at the corner of every street; scientific literature crowds the shelves of our libraries; scientific journals are scattered over their tables; and as to popular lectures, they have become so common as almost to defeat themselves by their very frequency.

As a consequence of this spread of knowledge, new members are likely to come to our societies prepared with a knowledge of at least the rudiments of natural history—an advantage which enables them to relish papers of a strictly scientific character. Formerly it was generally necessary in reading a paper to assume ignorance of the first principles of natural science; to-day such an assumption ought to be quite unnecessary. The whole character of the work of our societies should therefore be raised to a higher plane than it formerly occupied, and it may be hoped thatuis knowledge continues to grow the elevation may be yet more marked. The special papers at our societies ought, in fact, to become increasingly specialised.

But after all, it must be acknowledged that the proportion of matured students, trained specialists and active scientific members is never likely in any local society to be large. The bulk of the members will necessarily consist of intellectual people who feel interested in the world of nature around them, and have joined the society with the very laudable object of learning something about its marvels and mysteries, yet without the opportunity of devoting themselves to severe and systematic study. To many, indeed, the Natural History Society offers a rational means of recreation.

Such members are to be greatly respected, and their wants and wishes should be met as fully and freely as possible. To put it on the lowest level, their subscriptions form an element of strength which we can ill afford to ignore. Members who have joined a society for the sake of intellectual pastime, or philosophic recreation, often preponderate to such an extent that their withdrawal would spell ruin to the society.

But we will discuss the subject from a rather more elevated stand-point than that of the treasurer's cash-box. I am not ashamed to hold the view that one great function of a local Natural History Society is to shew us how we can secure rational recreation. Everybody needs occasional relief from his daily occupation, whatever that may be, because, as Milton neatly phrased it, " the spirit of man cannot demean itself lively in this body without some recreating intermission of labour and serious things." Different people, however, have marvellously different notions as to how the spirit can be best made to " demean itself lively,"' and he who turns for recreating intermission of labour to the Natural HistorySociety is exceptionally wise in his choice.

If our local societies did nothing more than enable us profitably to employ our leisure they would be doing a grand thing—especially for the younger folk. It was Dr. Creighton, the late Bishop of London, who once said that "The end of all education is not to enable a man to get on, but to enable him to use the time when he is not employed in getting on." The utilisation of spare time is a very serious business—a matter with which no one can afford to trifle." The manner in which a man is likely to spend his leisure is usually determined by the time he is twenty years of age. Having got by this time into a groove it is not easy to get him out of it. Scientific tastes are usually formed in early life, while the intellectual faculties are in process of expansion, while the mind is in a plastic condition, capable of receiving and retaining new ideas with facility.

Most people find it difficult to take up the study of science late in life. It is true that if they already possess some scientific knowledge they may expand it at almost any age; but one who has been an utter stranger to science in his youth is not likely to find a late start very successful. It is a matter then of great moment that people should be induced in early life, while their habits are in course of formation, to cultivate a taste for scientific pursuits. Parents, and especially ladies, who themselves may have, perhaps, no special leaning towards science, would do well to join our natural history societies if only as an example to the young, and for the sake of introducing them into a healthy intellectual atmosphere. In membership of such societies the young people find a mental stimulant of a refined character; their intellectual energies are quickened by "the sympathy of numbers;" they breathe a scientific atmosphere; they are likely to be attracted by the personal magnetism of those enthusiastic naturalists who usually rule in the society, and are apt, unconsciously, to catch their spirit. If the young members become only collectors, it is something; if they become serious students, it is much more.

Even as a mere hobby Natural History is invaluable. Those who are responsible for guiding the tastes of the young dare not speak lightly of the natural history society. It is in the meeting-room of such a body, it is in the excursions which the society organises, that we educate the eye to observe, the mind to understand, the heart to appreciate the marvels of our material environment.

Now that Nature Study is successfully edging its way into our schools, there ought to be a tine crop of young naturalists in the making. Moreover our Colleges and Polytechnics must be busy manufacturing material for the future support of our societies. From such sources the societies may expect their ranks to be See the Bishop of Stepney (Dr. Lang) on *The use and iilmse of Leisure.* in the *University Extension Journal,* December 1903, p. 37.
substantially recruited a few years hence. But the question at once springs to the lip, need the societies wait? And need the youthful naturalist himself wait? Why should a lad, bitten with a love of collecting natural objects, be shut out from the advantage of relationship with a society specially established for the promotion of the study of natural history?

Some of our societies have a practice which, it seems to me, might be advantageously followed by others—the practice of having a class of " Junior Members." Schoolboys with a taste for natural history can join by payment of only a trifling subscription, and in due course may be admitted to full membership. It is pleasing to note with reference to our Union, that School Natural History Societies are specially invited in our programme to become affiliated. Once attached to a society, and having experienced the pleasure of *a,* natural history ramble in company with enthusiasts, under the guidance of matured conductors, the youth is not likely to detach himself from such an association. The early capture of a member is likely to be a permanent capture, a recruit who will not readily desert his regiment.

Notwithstanding the hopeful views which we may be tempted to take as to the future of our scientific societies, we can hardly shut our eyes to the fact that most of them have at present grave difficulties to encounter. I fear it must be admitted that at this time the position of many of the societies is far from satisfactory. With the increase of the population, with the diffusion of knowledge, with the progress of science, an increased membership ought to follow as a matter of course. And yet many societies find that so far from the number of members increasing it is actually diminishing, and the attendance at the meetings is unfortunately smaller now-a-days than formerly. Certain societies and fieldclubs, it is true, have increased their roll of members, but I fear that they are rather exceptional. It is pleasing, however, to find that among these honourable exceptions we may include Maidstone; but then Maidstone is a place of exceptional enlightenment and culture.

As an instance of diminished interest in scientific meetings, take the case of the British Association. This august body, under the highest scientific auspicen, with the most eminent men of the day as attractions at the meetings, has suffered in late years a lamentable falling-off in membership. The great provincial centres which it visits have grown enormously in population between successive meetings, separated as the meetings are by long intervals, yet the British Association is not supported as well to-day as it was some years ago!

What is the cause of this decline in popularity? It is sometimes said with regard to ordinary societies that they have multiplied to such an extent in recent years that the effect of competition becomes felt. No doubt this statement has some truth in it, yet I cannot admit that it is more than a very partial explanation, and it will hardly apply to the British Association—a body which occupies a position absolutely unique. As there is only one British Association it is not here a question of competition.

No! I fear the answer is to be sought rather in the tastes of the present generation, especially in the inordinate devotion of many young people to sport. This, I suppose, is a very unpopular suggestion, but I raise the question in all seriousness. Pray let me not be misunderstood. J am not so foolish as to say a word against athletic sports in their proper place; what I do decry, and do so most emphatically, is the absurd devotion to such sports which characterizes so many young people now-a-days, and seems to be in a fair way of leading them to intellectual ruin. What we want here, as in most other things, is moderation.

Athletic sports, as we all know, develop certain mental qualities of much value; but I fear that love of scientific study is not usually one of them. According to my own experience the boy most given to sport is often the boy least given to study. The cant alxut "a strong mind in a strong body" is heard, I have noticed, not generally from those who are wise enough to keep the two in fine equipoise, but rather from those who are careful to fortify the body whilst reducing the mind to the verge of intellectual starvation. It is a convenient thing for indiscretion to find shelter behind an epigram.

No one of course disputes for a moment the wisdom of the original saying of Juvenal: *Mens nana in corpore xano:* but physical health, or soundness, is a very different thing from the muscularity to which so many people just now are aspiring. Muscle, to be sure, is useful enough in its way, though physical prowess has lost much of its former importance; only we must be careful not to bow down and worship it. After all it is difficult to see how the acquisition of a few extra ounces of muscular tissue can bring us any nearer to the angels; ill-natured people might say it brought us down a trifle nearer to the brutes. Few at any rate will dare to maintain that muscular vigour and mental vigour invariably go together. If it comes to a question between biceps and brains—whether we should cultivate one or the other—then, in the interest of our societies, I hold up both hands on the side of bruins."

The modern passion for sports finds its outcome in the sorry spectacle of huge crowds gathered together, not for their own "This keeping alive of mental activity is one of the uses for which such Societies us ours exist; and although athletics are at the present day much in evidence, still we must always remember that the object of the *corpux minum* is the *menu nana,* and that the higher part of man's nature must not be entirely neglected in order to cultivate physical prowess." Dr. Stolterforth, M.A., Honorary Scientific Secretary of the Chester Society of Natural Science,. Literature and Art, in the last Annual Report of the Society, 1903.

physical training, but to watch the competition of professional athletes. Compare the numbers who are willing to sacrifice time and money in witnessing these matches with the numbers willing to attend a meeting for intellectual improvement. Yet I venture to say that a party of naturalists on a field excursion will secure more physical benefit—not to put the matter on a higher level—than those spectators can pretend to secure while spending hour after hour in the exciting atmosphere of a professional contest.

The love of sport at the present day is, however, only part of a larger subject—the excessive love of pleasure in general. Look, for instance at the crowds outside the doors of the theatres and musichalls with which London now abounds—waiting sometimes for hours before the doors open—and then think of the mean attendance at a scientific lecture. Such a contrast tends to support the opinion of those who are disposed to place the present generation, notwithstanding our schools and polytechnics, on a rather low intellectual level. But this only emphasises the importance of the work of our scientific societies in attempting to bring about some improvement by moulding the habits of the young.

The character of a man's amusement is absolutely a matter of habit. By constantly indulging in excitement he loses relish for all sober intellectual pleasure. He will be delighted at the representation of a house on fire, with a real fire-engine rushing across the stage, but can take no interest in a chemical lecture on combustion. Nor must it be forgotten that the degraded taste for sensational pleasure, like other cravings, will grow with feeding. Every time we indulge in unhealthy excitement we become less capable of rational enjoyment.

Mindful of these tendencies of our day, it is clear that a fine field of activity lies open to our scientific Societies. It is within their power to do much to correct unwholesome tastes; they mayaid greatly in forming, guiding and elevating the tastes of the young if only they will undertake to provide recreation of an intellectual order. Let it not be said that this is beneath the attention of a Natural History Society or a field-club. It is surely the business of such a body to do all it can to develop a taste for natural history within the area of its activities. Probably the most hopeful way of accomplishing this is by means of attractive lectures. Some people are apt to sneer at a Lecture Society, but the sneer is very ill-deserved. Such societies may be of inestimable social service in developing healthy tastes in the youth of our land.

My early remarks made it sufficiently clear, I hope, that I desire to encourage our societies to have the highest possible aims, to maintain an exalted standard of work, to seek primarily the advancement of natural science. But whilst the special meetings should keep such objects steadily in view, the general meetings must assume an altogether different character. It should be their purpose to create and develop a taste for scientific pursuits, to encourage intellectual intercourse, and especially to provide healthy recreation of a scientific character for the younger members.

There is no denying the fact that the scientific study of Nature demands some mental effort, and it commonly happens that people untrained to habits of severe study are indisposed to such exertion. It therefore behoves us to see that our intellectual pabulum is served up in a palatable form, yet without loss of nutritive value. Much has been done of late years in this direction by the free use of lantern illustration. The lantern and the camera have come to be valuable helpmates to our societies. Rut let us remember that the lantern, after all, should take its place only as the servant of the lecturer, assisting him in his explanations, yet not daring to take the lead in a lecture. There may of course be evenings devoted to lantern-illustrations, when the instrument reigns supreme, but I refer, at the moment, simply to ordinary lectures on science. Here it is not that the lecturer has to explain the slide; the slide has to illustrate the theme of the lecturer. Above all let us remember that the lantern does not relieve the lecturer from the responsibility of preparing his discourse with the utmost care.

It is the careless preparation of lectures that has so often brought science into ill-odour with the audience. Scientific men are usually so absorbed in their own pursuits that they fail to understand why other people are not interested in them. The matter which forms the subject of a scientific lecture is considered by the enthusiast to be so absorbingly interesting that he fails to see that it needs any embellishment. To be candid he is rather contemptuous of style. It would almost seem as though some scientific men deemed the cultivation of literary grace incompatible with serious work and robust thought. So long as the matter is sound, the manner in which it is presented counts for little. A science lecture is consequently too often a thin and crude composition, suffering much by comparison with the graceful style of good lectures on history, or literature, or art. So far, however, from such crudity being in any way justifiable, care of quite an exceptional character ought to be taken in the preparation of a scientific discourse. Just because a scientific subject is usually rather difficult for an audience to follow, there is the greater need for its presentation in as lucid, as attractive, as picturesque a form as possible.

It is all very well for scientific professors to hold that science itself is so valuable a thing that it needs no adventitious adornment, and that people ought to be attracted to its study by its inherent worth. No doubt this is what ought to be, but as practical men we have to deal with things and people as they are. It must not be forgotten that the members of our local societies, who attend our evening meetings and listen to our lectures, are usually not people who can devote their whole energy to scientific study, or can command leisure for investigation whenever they please. Our audiences are composed of people in various walks of life, including it is true some of the leisured class; but for the most part, I take it, they consist of people who have to devote the bulk of their time to some profession or trade.

Such people come to our meetings not with the freshness of college-students, whose studies are, for the time being, the engrossing business of life; but they come more or less exhausted by daily duty, perhaps from the office, from the school-room, from the shop— let me add even from the workshop, for I should like to see our societies popular enough to attract the artisan. As mdst of our members thus come to the meetings more or less work-worn and weary, exhausted in greater or less degree by mental and physical labour, jaded after the moil and toil of the day, they can bring to us in the evening only the dregs of their energy. If, therefore, we are to interest them in science, the science must be presented in an attractive form. This means work. It is so easy to make a scientific subject dull, so difficult to burnish it effectively.

It should be borne in mind that our members are not like collegestudents, bound to attend a definite course of instruction for some specific end. On the contrary, their leisure—scant as it may be— is at their own disposal, and if our meetings are too dull they discontinue their attendance; they either resign membership, or remain members simply for the sake of the field-meetings. It is therefore a matter of the first moment that those who have the management of our natural history societies should take care that while the special meetings maintain an exalted tone for carrying out those higher objects which all scientific men have at heart, the general meetings should be made as attractive as possible, especially by sound popular lectures, well illustrated and well delivered.

As to the field-meetings nothing need be said, for every society recognises in them an element of strong attraction. A judiciously prepared programme of work blends outdoor demonstration with in-door meetings according to season and local conditions, the great point being to interest as large a proportion of the members as possible—some in this way, others in that.

Turning to the programme which our Honorary Secretary has been good enough to draw up for our present meeting I think it will be admitted that he has set before us an intellectual bill of fare which is singularly attractive. On comparing it with last year's programme I find that we are to have fewer papers and more excursions—a change which, considering the season, and considering above all the attractions of the beautiful district which we are privileged to visit, seems to be a change that reflects much crediton the judgment of Dr. Abbott.

The Congress is to be congratulated on meeting, this year, in a part of the country exceptionally rich in scientific and archaeological interest. If I might be permitted to refer to that department of natural knowledge in which I happen to take special interest, I would remind you that Maidstone is a place of no mean importance in the annals of Geology. Exactly seventy years ago, a discovery was made in this neighbourhood which carried the name of Maidstone into every land where the science of geology was cultivated. It was in 1834 that the late Mr. W. H. Bensted discovered, in his quarries of Kentish Hag, those remarkable remains of the Iguanodon which are now preserved in the Natural History Department of the British Museum, and which have been studied and figured by such men as Mantell and Owen and Huxley. It is true the bones were in confusion and bad been scattered by the workmen in blasting the stone, yet at the time of the discovery, and, indeed, until the great revelation made in Belgium in 1H78, the Maidstone Iguanodon was unparalleled for its comparative completeness, and even now it remains a specimen of unique interest as having teen obtained from the Lower Greensand.

Maidstone was fortunate in having in Mr. Bensted a townsman ever on the alert to observe and chronicle any geological facts of interest which might be brought to light in his locality--and this at a time when geological science was much less cultivated than it is to-day. Would that every district were equally fortunate! Individual observers of his type are a source of constant strength to a provincial society, securing at once local success and general reputation.

Mr. Bensted's name is honourably associated with several fossils, which paleontologists gratefully dedicated to him in recognition of his discoveries. Moreover, students were indebted to him for a series of excellent papers on "The Geology of Maidstone," contributed more than 40 years ago to the pages of *The Ireoloijut.*" It is a matter of congratulation that during our present meeting we are to have the advantage of the guidance of Mr. Bensted, the son of the discoverer of the Maidstone Iguanodon, to a quarry of Kentish Rag in this neighbourhood, as well as to the famous gravel-pits of Aylesford which have been so rich in the relics of pleistocene mammals.

A geologist ought not to visit Maidstone without being reminded that this town was the birth-place of that accomplished master of the hammerman's craft-Alexander Henry Green. As I had the privilege of enjoying Professor Green's friendship for something like five-and-thirty years, I may be permitted to recall his memory with some touch of feeling.

Green's professional life dropped into three parts: first, a

"Notes on the Geology of Maidstone. " By YV. H. Bensted, Esq. *The GeologiKt,* vol. V (1862), pp. 294, 334, 378 and 447.

period of 13 years as an officer of the Geological Survey, marked by much work in mapping very diverse formations, but especially those of the Yorkshire coal-field—a district on which his masterly memoir is still the standard authority; secondly, a period of 14 *yearn* as professor of geology, and part of the time professor also of mathematics, in the Yorkshire College of Science, ut Leeds, where he was the colleague of such distinguished men as Sir Arthur Rvicker and Professor T. E. Thorpe; and thirdly, a period of 8 years in the L'niversity of Oxford, where lie occupied, with conspicuous Ability, the Chair of Geology in succession to such eminent men as Buckland, and Phillips, and Prestwich. Acute as a critic, rigorous in his reasoning, a vigorous writer, and a lecturer of exceptional power— Professor Green was a man of whom Maidstone may fairly be proud. Suddenly removed from us before he had reached the normal term of life, he still speaks to us by his writings— especially by his admirable treatise on " Physical Geology," written while he was at Leeds—a work which was justly de scribed last year by Professor Watts, in his presidential address to the Geological Section of the British Association as " the best and most eminently practical text-book on physical geology in this or any other language."

Assembled as we are in the valley of the Medway, it is well that we should keep in mind Green's views on questions of physical geology—clear, broad and sound as they always were. With regard to the origin of the scenic features of the district around us I may recall, with much fitness, a memorable paper read before the Geological Society nearly forty years ago by two friends and former colleagues of mine—now, alas! both passed away—Sir Clement Le Neve Foster and Mr. William Topley. The notable paper I refer to was one "On the Superficial Deposits of the Valley of the Medway, with Remarks on the Denudation of the Weald."

This subject of the denudation of the Weald has now found its way into all our text-books, and has therefore become familiar to the merest tyro in Geology; but when Foster and Topley wrote, the state of geological opinion in this respect was far from settled. The marvellous effects of subaerial denudation in sculpturing the surface of the land were realized only by a band of far-seeing geologists, mostly of the younger school; and in looking back at the controversy across this vista of time, it is a pleasure to note that the geologist who did perhaps more than any other man to secure due recognition for this type of erosion was a former President of this Union—our distinguished member Mr. W. Whitaker, F.R.S.

Few subjects are more fascinating than those which deal with the causes that have been operative in moulding the surface of our country into its beautiful and diversified features. Cobbett regarded *Quarterly Journal of the Geological Society.* Vol. xxi. (1865), p. 443. the country between Maidstone and Tunbridge as the finest for "fertility and diminutive beauty in the whole world." Thecharacteristics of this beauty spot on the map of Kent naturally find their explanation in the geological structure of the district. Such works as Sir Archibald Geikie's fascinating volume on "The Scenery of Scotland," or Lord Avebury's attractive work on " The Scenery of England," serve to show us in a striking manner how close is the connexion between the form of the ground and itsgeologic build. This is a subject well worthy of the attention of our local scientific societies. Each society should study the physiographical features of its own district, seeking to understand not merely what its physical characters now are, but by what seriesof natural processes they have come to be what they now are: in other words, they should trace the physical changes which have passed over the district and which constitute literally its natural history.

Every local society and field-club will, no doubt, possess a set of the (ieological Survey maps of its own district; but it is satisfactory to note that the recent introduction of colour-printing now places the newly-issued maps within reach of all the members individually. To buy a sheet of the one-inch map, however complicated the colouring, for eighteen pence, is a boon which has been given to us only in recent years, and one which every student of nature ought to appreciate.

It would be well, however, if every local society not only possessed the official maps, but would construct a model, on a rather large scale, showing the physical features and the geological structure of the area within its influence. Such a model might appropriately be contributed by the society to the local museum.

It is clearly desirable, wherever possible, that a bond of union should be established between the society and the museum of the district. The two institutions have much in common. Both aim at teaching us something about the multitude of natural objects besetting us on all sides, so that we may not walk through the world, surrounded with its garniture of surpassing beauty, and yet remain as ignorant of our surroundings as "a blind man in a picture gallery."

It was Professor Blackie who advised a young man on entering a strange town to seek at once the museum. This certainly is my own practice. One of the first places I make for, on visiting a town, is its museum; and I must confess that from the character of the museum 1 am apt to draw conclusions as to the character of the inhabitants of the locality. A good museum at once suggests people of thought, and taste, and general enlightenment.

If any stranger coming to Maidstone followed this course and applied my test, he would immediately form a very high opinion of the intellect and culture of the people of this town. Eor I cannot hesitate to rank the Maidstone museum among the best of our provincial museums. Members of this Congress who may not be already familiar with it will have ample opportunity in the course of our meeting to make themselves acquainted with its treasures, and with the admirable manner in which these are classified and displayed. Under the wise administration of the Corporation, and with so excellent and energetic a curator as Mr. Allchin, aided by his accomplished colleague Mr. Elgar, the Museum has come to be an institution of which any town might indeed be proud. Nor should I forget the advantage which the museum enjoyed in having had for many years the services of Mr. James, who, trained under one of the most scientific archaeologists that this country ever produced—the late General Pitt-Rivers—brought the antiquarian collections, which form so important a feature in the museum, into an eminently satisfactory condition.

Who will dare to assess the value of such an institution as this museum, to the town and surrounding district! It is not a mere place of amusement—a gallery in which a spare hour may be pleasantly passed; it is an educational engine of high potency, helping, like our scientific societies, to direct the tastes of the young into healthy channels; ministering in its scientific departments to the instruction of those who are seeking acquaintance with the works of the Creator; enabling us in its archaeological section to recall with vividness the forms of the past, and thus deepen outveneration for antiquity; and finally, in those rooms which are dedicated to art, introducing us to all the refining influences of artistic beauty. Would that such institutions were far more numerous! But it is not every town that has the good fortune to secure such enlightened patrons as Dr. Charles and Mr. Brenehley.

Formerly it was the common practice

of a local society to form and maintain a local museum. Such a course was not without grave difficulties. Both institutions usually suffered. In most cases the income of the society was quite inadequate to bear such a burden, and as a consequence, whilst the society seriously impoverished itself the museum struggled along in a chronic state of semi-starvation. Happily in these days the Natural History Society is usually relieved of such a charge. Under the enlightened policy which now prevails, the local museum is generally supported by the municipal authorities, and therefore enjoys the advantage of an income which, if not large, is at least secure.

It is still, however, desirable that the Natural History Society in any district should keep in close touch with the local museum. Mutual good-will should animate the two institutions. In most cases it will be found desirable that the society's collections should be deposited in the museum; whilst it is not without advantage that the society should hold its meetings in the museum buildings. The Free Library, the Science Museum, the Art Gallery and the Natural History Society, are all institutions with aims more or less akin, and might well be included where possible under a single roof.

Now that our local societies are generally relieved from the care of the museum, now that most districts possess catalogues of their flora and fauna, and maps and monographs fully illustrating their geology, it has been said that the day of the Natural History Society is past. A spirit of pessimism has been abroad, and it has been whispered that as the Society has become worn-out and effete the best thing it can do is to gather its robes around it, and prepare as decently as possible for dissolution. Such a gloomy view has probably found some countenance in the facts at which I have previously hinted—the unfortunate facts that many societies find their roll of members diminishing, and the attendances at their meetings steadily dwindling.

lint despite this discouraging state of things I am indisposed to take so dark a view of the future. Depend upon it the work of the Natural History Society is not yet finished, and so far as I can see never will be. Turn over the successive reports of the Corresponding Societies' Committee of the British Association, and note the work which is suggested year after year, and which still lies to a large extent unaccomplished. It is humiliating to think how little we know after all about any group of natural objects or any set of natural phenomena. Fresh subjects, too, are frequently cropping up for the attention of local societies. Only recently we have been reminded that though the plants of a district may have been completely listed there are still vegetation-maps to be made. Photographic surveys, too, are only just being started in most districts. How few counties, again, can produce maps showing the distribution of the prehistoric remains within their boundaries. Then there are meteorological phenomena calling for unceasing observation. Surely there is no lack of original work to be accomplished in the future by our provincial societies. No scientific man need ape Alexander and weep that there are no more fields to be conquered. On no line of work has the locomotive of science yet run into its terminus.

But the Natural History Society should not be confined to observational work. It is beginning to be felt that our societies, as I previously hinted, have still another mission to fulfil. Educational work is now-a-days recognised as coming well within their scope. The society ought to be, and if wisely directed will be, a potent agent in the education of its youthful members—leading out their faculties of observation and reasoning; and it is obvious that this part of the work must needs be perennial.

It is sometimes said that all the localities of interest in a particular neighbourhood have been visited by the field-club, and consequently its resources are exhausted. But surely old scenes should be re-visited for the sake of new members. A new generation is steadily springing up around us; and, notwithstanding the opposing tendencies which I referred to, each generation ought to contribute a fresh stream of members to the society. Any place visited, or any subject discussed, say ten years ago, will be practically forgotten by most of the older members, and be absolutely novel to those who have entered since. There never need be in a society any lack of material that will interest the majority.

The way in which opportunities are utilized must depend much upon the honorary secretaries of the several societies. Those members w:ho are so unselfish as to sacrifice their time in secretarial work deserve our profound gratitude. The government of the society is of course in the hands of its Committee, yet the practical administration is largely due to the honorary secretary. Hence a local society usually takes its complexion from its secretary, and necessarily reflects to a large extent his tastes. It is obviously desirable to induce the secretary to hold office for a considerable period, so that the work may be carried on without interruption or dislocation; and it is well to have tw7o secretaries whose tenure of office will close at different dates. In this way the continuity of the society's traditions is ensured, since the senior secretary carries on the work in accordance with established precedent, whilst the younger officer brings in new ideas and fresh energy to revivify the work. Their co-operation, if cordial, will bring about a happy blending of the old and the new elements, the conservative and the progressive tendencies, which must exist in every society. Stability is secured without stagnation; progress without precipitancy.

But, after all, too much must not be expected from the directing Spirits of a local society. It must never be forgotten that the Society's success remains practically in the hands of the general body of members. It is on thir personal exertions that the future of the society depends. Let them rise to the responsibilities of membership, not content with being merely sleeping partners in the concern; let them show a living interest in the society and its work, reserving the evening of meeting as a time set apart for a specific purpose, not to be interfered with by other engagements, and making it a point of honour to attend the meetings, whether the subjects to be discussed are personally interesting to them, or not.

Attendance at the meetings of a society no doubt frequently means a sacrifice on our part. If we were sitting at home all day, we might find it refreshing to go out in the evening to a scientificmeeting or a lecture. But, as a matter of fact, we have probably gone forth in the morning to some occupation and come home in the evening more or less fatigued. Small wonder then that we often find a book at the fireside on a winter's evening more attractive than a lecture or a meeting in a distant hall. Yet our attendance is of more value to the society than many of us are disposed to admit. It is hardly too much to say that the character of a lecture is often determined more by the audience than by the speaker. Be the lecture never so good intrinsically, it is likely to fall flat in an empty hall. Few lecturers are able to shake themselves free from the natural abhorrence of a vacuum.

The multiplication of small societies may tend to draw members away from the larger and older societies; but there ought to be sufficient enthusiasm for science in the twentieth century to support quite a multitude of these bodies. Notwithstanding what I was led to say—and I said it painfully—about the pleasureseeking tendencies of this age, I have yet sufficient faith in the power of science to believe that it must ultimately assert its proper influence, and that the day will come when our natural history societies will receive support more adequate to their importance.

Can we forecast the character of the Natural History Society of the future? It seems to me that it need not differ markedly from our existing societies; but while working generally on the same lines it will need perhaps more effective organisation, and certainly much more generous financial support.

We may safely conjecture that in any ideal society the Sectional Meetings will reach a higher and higher standard of merit, keeping pace with the progress of science, and growing with the general growth of natural knowledge.

Then the General Meetings will be characterized by a class of lecture superior to the ordinary type. Scientific men of high authority will not consider it beneath their dignity, or unworthy of their vocation, to deliver attractive lectures, so that scientific knowledge may become popularized without being in the least degree vulgarized. There will be no fear of meeting that horrible thing—unscientific science.

The area to be worked by the society will be mapped out and allocated to members or groups of members; and in this way ultimately the natural history of the whole district will be thoroughly known and faithfully delineated.

The excursions of the society will become more strictly definite in aim, and perhaps less picnic-like. By the increased use of photography, the records of the field meetings will be amply illustrated, so that what is collected during an excursion may be permanently recorded by figures.

Colour printing will be employed in the society"s publications, and in this way natural objects and natural scenes will no longer be represented with the infidelity of monochrome, but Nature will be faithfully mirrored in all her gaiety of colour. At present most of our publications suggest that flowers and insects, and birds and beasts, are nil living in a world of black and white. Yet every naturalist knows that colour rs often a valuable means of diagnosis, so important that we have sought to give it exactitude of description by such schemes as Hadde's *b'arbennkala.*

Such increased and well-organised activity as I have sketched will require greatly-increased expenditure on the part of the society. In this sordid world where money measures, if not all, at least most things, the activity and usefulness of a society must depend largely on the magnitude of its income. Most of our societies are haft-starved. To censure a society for not doing better work is like upbraiding a poor man for not faring sumptuously every day. The best way, because the most independent way, of securing additional income is, of course, by increasing the membership of the society.

It is not, perhaps, easy to do this in a dignified manner. Let us not importune a man to join, in the tenacious way in which he might be worried into buying a Cyclopwdia. Rather let us appeal to the sympathy of intellectual people by quietly pointing out the need of cultivating habits of seeing and thinking, by explaining the social and scientific value of the society, by insisting on the importance of that special kind of knowledge which the society cultivates.

"Surely," said an old writer on minerals, " we live not in the most unknowing times of the world."" If this could be said truthfully when it was written two centuries and a half ago, it can be said with far greater force to-day. The days in which our lot is cast are indeed not " the most unknowing times "; and it is just because "that Angel Knowledge" has visited us so beneficently that we feel assured of the future welfare of our societies. Whether they are really manufacturers of new knowledge or merely distributors of old knowledge, they have a grand function to perform in the elevation of the people. "Knowledge," to quote Bacon's words, " is not a couch for the curious spirit, nor a terrace for the wandering, nor a tower of state for the proud mind, nor a vantage ground for the haughty, nor a shop for profit and sale, but *a Storehouse for tlie Glory of God and the Endowment of Mankind.*
" " A Lapidary: or the History of Pretious Stones." By Thomas Nicols, Cambridge: 1652.

Allington Castle.

By DUDLEY C. FALCKE.

With the short time at our disposal I propose merely to give the briefest possible summary of the history of the castle, which will allow more time for wandering round the grounds. A castle is supposed to have occupied the site in Saxon times and to have been demolished by the Danes. The first name connected with the ownership is Ulnoth, 4th son of Earl Godwin, a younger brother of Harold. At the conquest it fell to the share of Odo:, Bishop of Bayeux, who appears to have been the " official receiver" in those days, for he seems to have "owned" most of the castles in Kent and elsewhere. After the conquest Odo raised a rebellion in Normandy which was suppressed by William (Earl Warrener), who became the next possessor. He rebuilt the place. It was then transferred to Lord FitzHugh, whose daughter married Sir Giles de Allington, from whose descendants it passed to Sir Stephen de Penchester. Henry de Cobham, who married Sir Stephen's daughter, was the next owner, from whom it passed to the Brent family. From the Brents it came to Sir Henry Wiatl in Henry VH's reign. His son, the first Sir Thomas Wiat, succeeded him, and was in turn succeeded by his son, the second Sir Thomas Wiat. At his execution it escheated to the Crown, but in the 11th year of Elizabeth the Queen granted it to Sir.lohn Astley, whose descendants sold it to Sir Robert Marsham (Lord Romney) in 1720, to whose family it still belongs. From this list of owners I select a few of special interest.

The building, as it exists to-day, was the work of Sir Stephen de Penchester, Governor of Dover Castle, who obtained leave by royal licence to fortify and embattle! bis mansion house of Allington. This was in the reign of Edward I, in 12H1. All the Early English work is of this period, the additions of the Tudor period having been made during the ownership of the Wiats. There is small record of the intervening owners till we come to Sir Henry Wiat. He was attached to the fortunes of the Earl of Richmond, and at his master's accession to the throne as Henry VII came into high favour. There is a curious story related of him. He was imprisoned by Richard III, and would have starved to death had not a favourite cat supplied him daily with a pigeon. This fact is duly recorded It may perhaps be interesting to note that in 1279 there was an Odo, Rector of Allington.

t Wiat is sometimes spelt Wyat or Wyatt on his tomb at Boxley Church, and a pigeon appears in two well authenticated portraits of him. Henry VIII visited him at Allington, as he did also his son when he succeeded. Sir Henry died in 1588.

It is interesting to note in passing that the Friars, which you visited yesterday at Aylesford, was given by the King to Sir Henry at the dissolution of the monasteries. He was succeeded by his son, the first Sir Thomas Wiat, poet and statesman, who was born at Allington in 1503. He is well-known as the unhappy lover of Anne Bullen (Boleyn). They became acquainted as children at Blickling Hall in Norfolk, the seat of Sir Thomas Bullen. When the latter removed to Hever Castle the intimacy between the families was renewed, and Wiat's early affection ripened into the love which Hoods his sonnets and poems. Sir Thomas was a man of great accomplishments—a gifted poet, a courtier, a linguist, an able diplomatist, and he was sent by Henry VIII as Ambassador to Paris and on an Embassy to Charles V, and acted as ewerer at Henry's marriage. He died in 1542 on his way to fulfil another duty for the King. He was succeeded by his son, the second Sir Thomas Wiat, a distinguished soldier and a favourite of Edward VI. He is perhaps even more widely known as the ill-fated leader of the Kentish Rebellion, which was started to oppose the marriage of Mary with Philip II of Spain. Wrhether Wiat meant to head the rebellion is uncertain, but what is certain is that he was on the eve of going abroad, but stayed at the urgent entreaty of his wife, who was near her confinement. When the rebellion broke out shortly after and he accepted the leadership, he seems to have had a presentiment of failure, as he is reported to have said on bidding farewell to his wife and child, " Thou mayest yet prove a very dear child to me." He marched from Penenden Heath to London where, after gaining considerable success, he was ultimately deserted by most of his followers, taken prisoner and beheaded on Tower Hill. The custodian of the Tower of London showed me the cannon balls used when Wiat besieged the Tower, but they are as authentic as the Roman rings I was once offered at Cannre, for Wiat never got as far as the Tower till he lost his head. In connection with Wiat's rebellion it is interesting to recall that Tennyson lays the first scene of the second act of Queen Mary in the Court where we now are. I have said that the estate was then forfeited to the Crown, and from that date, 1554, the place gradually fell into ruin. The next owner, Sir John Astley did not reside here, and when his descendant, Sir Jacob Astley sold the castle to Lord Romney, it was no doubt in ruins, as an engraving I have 15 years later, 1785, shows the Anne Boleyn is believed to have stayed at the castle when she was a young girl. (?) Did Henry VIII meet her for the first time on his visit.

building in exactly the same state as it appears to-day with the exception that the Columbarium then had a roof and the moat completely encircled the castle.

There is an entry in the Parish Register, under date 1681, "Elizabeth, daughter of John Bert of AUington, baptised." It may also be advisable to add that the present inhabited portion was fitted up in 1829, and that the other was dismantled in 1840.

The Friars, Aylesford.

By (rev.) C. H. FIELDING, M.A.

By the pretty meandering Med way near the picturesque village, of Aylesford, with its fine old church on a hill rising from the stream, and its quaint old bridge almost too narrow for the road traffic above, and the water traffic below, stands the old home of the Carmelite Friars. In those days, besides that still renowned delicacy, the Medway smelt, the river could boast of its salmon and trout, so that, though the rules of the Carmelites were hard, no doubt fasting on fish was a treat, especially if they were allowed eels. As for contemplation, the banks of the quiet stream, fringed in summer with yellow tansy and pink feathery loosestrife, must have inspired it, while, when evening fell, the calm that settled on the waters must appear to have promised a gentle rest to the Friars. We might tell of the fierce battles of Briton, and Saxon, and Dane, to gain the important ford of Eccles or Churchford, of which the name of Aylesford is a corruption, but instead of doing so let us ponder awhile upon the Monastery of the White or Carmelite Friars of Aylesford.

To understand Aylesford Priory thoroughly, we should first know something about the Carmelites, who held it till the Reformation. This Monkish Order derived its origin from a set of Hermits, whom the Crusaders found living as solitaires on Mount Carmel; they claimed Elijah as their founder, and professed that they were first converted to Christianity by St. Luke, and heard the story of our Lord's Passion from the mouth of the Virgin. In consequence, they claimed to be the oldest Christian Order of Monks, and carried on a word war with other Orders on this subject till the Pope forbad any more discussion. After the peace that Frederick II. made with the Saracens in 1229, Alan V. General of the order, declared the Virgin had bidden them quit the Holy Land, and they passed from Cyprus to Sicily and thence overspread West Europe. It is at this period that they certainly found Europe a better place for their Order than the wilds of Carmel. In England, two houses were founded by Crusaders, one by Lord Vesey of Alnwick, in Northumberland, and this one by Lord Grey of Codnor, then owner of the Manor of Aylesford. It should be mentioned that the Charter of the Northumberland Brethren and their rules, survive in Dugdale. In their charter it is curiously mentioned that Read at the visit paid to "The Priory " by the members of the SouthEastern Union of Scientific Societies. June 10th, 1904.

they were given a monopoly of the wild bees and honey on Lord Vesey's estate, that they might be provided with a perpetual supply of wax for their candles to light the church, a rebuke upon the waste and improvidence of these so called more enlightened days. As regards their rules, I will just mention one or two. The Friars' cells were to be separate, and they were not allowed tochange without permission, I believe some of the curious little stable-like buildings in the Courtyard to have been these cells; they had to remain and watch in them, except when lawfully occupied on the Canonical Hours. They had to fast from the Exaltation to Easter, and abstain at all times; they had to work with their hands and to keep silence from Yespets to Tierce. They had tobe obedient to their superiors, poop, and chaste. When they were in Europe their convents were to be in solitudes, as they were hermits; they might take meat, and they were to dine in the refectory. When they were dissolved at the Reformation, forty or fifty houses were found in England, and many of those were situated in various towns, and not in solitudes. Whether the Hermitage of Longsole, near Banning Station was an outpost of theirs, I am unable to determine. At Aylesford was held the first European Chapter of the Order, at which was elected Prior Simon Stoke, or de Stock, in 1245, an Aylesford brother, to be General of the Order. This meeting of the White Friars must have been very picturesque as they wound their way along the banks of the Medway to their meeting place, or came down from the hills above through the forest. I may mention that their garments were striped with grey or orange, and from this they were called Les Barres. Simon Stoke, or Stock, is said to have lived long in the hollow of a tree, whence he got his name. He died at Bordeaux in 1266, at a great age, and was canonized, and the Cathedral still has a shrine to his honour and believes in his bones. His day is May 16th and not to-day, so we cannot honour him now. The Church was dedicated to St. Mary, and in it, and around, the founder Lord Grey and many of his descendants were buried. The only notices we have of the Friars after this time are principally bequests. We give you therefore a chronological order of the history. 1818.—Three Acres were given by Lord Richard Grey of Codnor, to increase the Friars' property founded by his father. 1319.—Adam dictus ad aquam Maidstoniie, R. of Ditton, was probably a Member of the Order. 1394. —Richard II. granted a spring of water called Haeley Garden in the next parish of Burhain, for the purpose of making an aqueduct. This is possibly Cossington Stream. If this was the case the water for the Friars must have been carried by pipes under the present street to the Friars. 1896.—Richard Maidstone author of several works, died a Carmelite brother at Aylesford, and was buried in the Cloister. A modern mural painting represented him perhaps near the spot. In the wall near was found a skeleton, thought to have been that of a disobedient brother, but perhaps it was that of Richard; however, the skeleton has given rise to a ghost, which disembodied creature at midnight stalks the earth, and frightens brains overcome by indigestion and bad livers.

In 1500, Bro. William Arnold was Prior (Will of William Halle).

1588.—June 18th, Richard Ingworth, Suffragan Bishop of Dover, received the house of the Whitefriars into his hands for the King. The Church at that time was stripped of all its valuables, and razed to the ground, but the doorway of the Almshouses appears to have been part of the Friars Church, as also two windows in the boardroom there, which bear the Arms of Lord Grey of Codnor and his son Henry.

These Almshouses were founded by John and William Sedley, brothers in succession, Masters of the Friars, whose father held it by grant of Queen Elizabeth, for, having been granted, on the dissolution, by Henry VIII to Sir Thomas Wyatt, the elder, of Allington Castle, on the attainder and execution.of his son Sir Thomas Wyatt, the younger, it had again become Crown property. The Sedleys, like the Wyatts, produced another bard for Medway's stream, the gay and witty Sir Charles Sedley, born here in 1639. He was a copy of his King, except that he wrote poetry, which we do not believe Charles II ever rose to. Pepys says he was bound in £5000 on one occasion for his good behaviour in future, and he further tells us that he saw his play "The Mulberry Garden," in 1668, and thought the whole play had nothing extraordinary in it at all, either of language or design. His plays I have not read, but of his poems, 'tis only fair to give one of the Medway bard—

'THYRSIS, unjustly you complain,
And tax my tender hen rt,
With want of pity for your pain,
By sense of your desart.

By secret and mysterious springs,
Alas, our passions move,
We women are fantastic things
  That like before we love.

You may be handsome and have wit,
Be secret and well bred,
The person love must to us fit,
  He only can succeed.

Some die, yet never are beloved,
Others we trust too soon,
Helping ourselves to be deceived,
  And proud to be undone.

Sir Charles' brothers sold the estate in 1657 to Sir Peter Ricaut; his voluminous works us a great traveller are not only recorded on a tombstone in Aylesford, but have given valuable documents to the British Museum. His son sold to H. Caleb Banks, Esq., whose son's second daughter, Elizabeth, married Heneage Finch, created Baron of Guernsey in 1703, and Earl of Aylesford on the succession of George I, in 1714. A description of the Friars 1780-1790, may close this paper. "The Priory here commonly called the Friars, is pleasantly situated on the banks of the Medway, above the town. Phillpott calls it a skeleton, but certainly without reason, for even at this time the major part thereof remains very fair and the least demolished of any conventual edifice in this part of the country, owing to its having been after the suppression the residence of several eminent families. The great gate from the road is still extant; and the apartments over it, when I was last there, served for the residence of the steward and his family It opens to a large square court, in which are seen all the doorways to the cells. The side where the high buttresses are on the left hand within the gate, was the great Hall or refectory, now divided into rooms. The kitchen was likewise on the East side of the square, as appears by the large fireplace in one part of it. The Chapel was that part of the building which stands East and West, the North side fronts the garden as the South does the river; the East window of it was where now is the dining room or gallery door with iron balcony which faces the town: (Mr. Evans says, what is meant by the Chapel may have been the infirmary chapel, as the church stood where the lawn now is). The principal parts of the Convent, the Halls, Chapels and Cloisters were covered with plaster and converted into stately apartments by Sir John Banks, and the Cloisters were by him enclosed and paved with black and white marble. There is a fair high stone wall which fronts the road on the North and West sides, leading down to the gate and enclosing the present garden, as it did originally the Convent. The large ponds at the mill above belonged to the estate, but the monks did not require them to supply fish as the Medway gave them all kinds, including sturgeon and smelts, perhaps these worthies first found out the delicacy." There is a plan No. 38KH8 in the British Museum MSS. of Friars of Aylesford in 17H8. There are some slight errors in this description, one of which Mr. Evans has pointed out, the other is the square to the right of the gateway. Concluding with thanks to Mrs. Hunter foi permitting our rummage, 1 would only add that, I believe the single stables' stalls to have been the Friars' cells, and, if so, certainly, in their case, the ass knew his master's crib, for they chose a good, fat and productive home, where there was plenty to get and little to do while the river flowed on, until Henry VIII spoiled their calculations.

Ox THE MERIDIONAL POSITION OF MfiCiALITHS IN Kent COMPARED WITH THOSE OF WlLTS, AND ALSO WITH THOSE OK Earth-works
And Churches.
By F. J. BENNETT, F.G.S.

So very little is really known about megaliths that any further information, however small, needs no apology, I think, on my part in bringing it before this Congress.

My object also is to promote discussion. This may result, perhaps, in steps being taken to restore the fallen stones. To do this some excavation will be necessary, and this, if carefully done after the plan advocated by the late General Pitt Rivers, should result in some important information.

The work of this nature recently done at Stonehenge afforded definite information on matters only conjectural before. A most careful survey should be first made of the surrounding area, with contour lines at frequent intervals, with accompanying plans and models such as may so well be seen in that most admirable Pitt Rivers' Museum at Farnham Royal.

I would suggest that our great Kentish megaliths at Coldrum be first taken in hand. A trench cut through the ditch and mound might reveal the age of this.

All the megaliths that I have seen seem to be composed of unhewn, naturally detached, masses of local stone, left often by denudation after the softer portions had been removed by various denuding agencies.

This is certainly the case in Kent" and also in Wilts. Stonehenge is the only exception as far as I know where stones had been dressed on the spot or imported from a distance, so that megaliths are very much dependent upon the local geology.

The larger supply and larger size of the sarsen stones in Wilts, compared with those in Kent, may perhaps have something to do with the fact that the Eocene strata furnishing them thin away there to the west.

But the prehistoric folk of Kent did their best with the smaller stones, and the smaller supply, and we are justly proud of our Kentish megaliths. The area also for their erection was more limited.

It would also be an interesting matter of observation to note, especially in the case of the older churches, the use of naturally shaped surface stones used in their construction, thus connecting them with the megaliths. I have seen this in some of our Kentish churches, and Mr. Benjn. Harrison of Ightham informs me that he had remarked to the late Sir Joseph Prestwich on the fact of so By the words *Meridional Line* I mean one with a *general* N. anil S. trend— not an absolute N. and S. line—as *petition* must count both in the eases of Churches and Camps.

many pieces of Oldbury stone and Ironstone being found in the oldest part of Otford church (Anglo-Saxon), and states that the old churches (near Ightham) are usually built of surface picked stones from a drift and not quarried ones. He also informs me that Kemsing church is said to be on the site of a Pagan temple and proved so by finds in later years. But we must now come to the subject matter of the paper.

Some ten or twelve years ago, when engaged on the Geological Survey in the neighbourhood of Marlborough, Wilts, I was much interested in noting a certain relative order and distance in the position of Avebury, Silbury, and a stone circle south of it.

Jast about that time I had been told that someone had stated as a remarkable fact that Avebury, Silbury, Stonehenge, Ogbuvy camp, Old and New Sarum, were all along a north and south line.

The only published reference that I have found about this is in *The Drititlical Temple «/' the County of Wilts,* published in 1845 by the Rev. E. Duke, M.A.', F.S.A.', F.L.S. The book is a very discursive one, treating of ancient mythologies and seeking to connect these with Stonehenge, &c, it has naindex, and the subject matter is most difficult to follow.

On p. 6 he speaks of the meridional line 16 miles in length including, as he considers, seven temples, *riz.:*—to Venus, the Earth, the Sun, Mercury, Mars, Jupiter, and Saturn. He commences this line at the north with a stone circle at Winterbourne-Bassett, and takes in Avebury, Silbury, Walkers Hill, Easterly camp, and Stonehenge, and gives a diagram to illustrate this, and 1 understand considers he was the first to call attention to this meridional line.

I thought no more about this till I came to reside in this district some few years ago, and then, on visiting Coldrum and Addington, I was much struck by their meridional position, and also that they were one mile apart, and that they seemed at one time to have been connected by a possible *ria sacra,* marked out by stones of which some six or seven along a path south of Coldrum or Addington still remain.

I then noticed that Kit's Coty House and the Countless Stones were similarly situated. This led me to re-examine the Marlborough sheet of the Wilts map to see if the megaliths north of the Devil's Den, close to, and west of Marlborough, followed the same line. 1 found that the two Kistvaens, as they are termed on the map consulted, were both north of the Devil's Den, and the last one was two miles from that well-known megalith.

So that here we have this interesting fact, that the Kentish megaliths follow the same meridional line as those in Wilts.

The next thing I noticed in the Marlborough map was that north of Avebury Church were three other churches, all a mile apart, and that north again was the Winterbourne-Bassett stone circle and PinVnoll rrrnp. So that there you have four churches separated by the same distance that Avebury is from Silbury, and this from the stone circle to the south of it.

So that the meridional line, 81 miles long, starting from Binknoll camp, takes the four churches—Avebury, Silbury, the stone circles, Easterly camp, Stonehenge, Ogbury camp, Old Sarum, New 8arum, Clearbury camp—certainly a most interesting and remarkable fact.

I then looked to see if some of the other churches in that sheet also followed meridional lines and included camps and tumuli, and most of them certainly seemed to do so. I also noted that several of the churches were one mile apart.

Turning again to Kent, and especially to our area, I found that from the foot of the chalk escarpment and north of this and the Thames, and to the east and also the west of Maidstone, the churches seemed to follow meridional lines, including castles, camps, and tumuli, and also that many of the churches again were one mile apart, as in Wilts. Maps were prepared and shown to illustrate this paper; megaliths were indicated by blue, camps and tumuli by green, and churches by red, circles. These meridional lines depart sometimes a little from the north or south (as might be expected) where the position of the church or camp had to be considered.

In that part of the map comprising the Wealden Forest area, the churches seem rather to follow east and west lines, a fact noted in the discussion and dealt with in my reply.

As to why a north and south line should be chosen we can only conjecture. It seems easier to suggest a reason in the cases of the camps and churches than in that of the megaliths. Camps and churches were used in former days for signalling purposes by means of Beacon fires. The great highway for invaders was the sea and the Thames, and this latter is to the north of the chalk escarpment in our area, and it was in that direction that the landings would take place, and from which the signals would first come, and so *to* the north they would naturally look for such warnings of danger. It is also noteworthy that the parishes run up from the foot of the chalk escarpment in long, narrow parallelograms, with *A* north and south direction. Mr. Topley was the first to notice this. Something of the kind, too, may have governed the directions of the megaliths, but we must not forget that these have been so much mutilated, and have even been entirely removed in some cases, so that caution is needed in dealing with them.

Then we may say a few words about the mile interval that separates some of the churches. I have found this to be the case in several maps dealing with other areas than those touched on in this paper. Of course, there is always an element of uncertainty in comparing weights and measures of the present day, when the standards have become scientifically fixed, with those of other times and countries, with a varying and by no means fixed standard.

Still it seems worthy of note that we find megaliths separated by the same interval as churches. Sometimes these churches occur in groups of three or four, and sometimes also the distances are a little under and sometimes a little over our mile of to-day. Further investigation on these points seems needed.

It may also be found that the churches along north and south lines may be of older date than those that seem to follow east and west lines.

Another matter of much interest to us in Kent is that in thecase of some of the churches near our megaliths we find sarsen stones associated either in the foundations or in the walls.

In the case of Cobham church—just north of which are the remains of a stone circle—we rind lying on the ground just outside the north porch a large sarsen of a grotesque outline, another leans against the wall of the west end, and another is built into the south wall.

Just outside Meopham churchyard are two very large sarsens. These are found associated with the ruined churches of Maplescombe, Punish, and Paddlesworth, and notably in the foundations of Trottersclirl'e Church. Broken up they enter largely into the walls of many others. Outside the churchyard wall of BirlingChurch are some good-sized sarsens. This leads to the suggestion that some of our churches occupy the sites of the stone circles. We know the early Christian bishops forbad the worship of stones, trees, and wells, but no doubt for some time without much effect. Compromises may afterwards have been arrived at and the stone circles utilised in the building of the Christian churches, so that this seems to point to a continuity of worship right down from pre-historic to Christian times.

The Rev. II. N. Hutchinson, in his work *Prehistoric Man ami li'east*, pp. 258-259, says:—" Many churches have been built on the sites of stone circles. There is a common Gaelic phrase—'Am bheil thu dol don clachan (are you going to the stones)'? '— that is to say, ' Are you going to church?" And, again, the "Gaelic word *clarhan* signifies both a circle of stones and a place of worship. " All this seems to closely connect our churches with the megaliths, and must add much to the interest of both.

We will now take some meridional lines in Kent, including in some cases both megaliths, camps, tumuli and churches, and shall begin in the west.
1. Woolwich And Edenbrihgk, Meridional. *Woolwich,* this now occupying a most important spot on the river, may have done so in long past times. *Shooter's Hill.*—A most marked elevation and of much strategic importance from the earliest times, no doubt. *Chislehurst Church.*—Here, also, are numerous Dene-Holes and the now famous eaves. *Farnborough Church, Cudham Church, Westerham Church, Knockholt Church. Kdenbridge Church.*—Just south of the church are marked on the map the words *Devil's Den,* in antique type, what this is I do not know. We also have *Hever 'astle* to the south.

We might include in this line, as so close to it, the following churches a little west of it, viz.: *Eltham Church* and *Palace, Mottingham Church,* the name also suggests a *moot,* an old meeting-place; *Kenton Church,* and *Hutcircles,* with an old *camp* north of it; *Down Church, Cudham Church.* 2. Erith And Penshurst Meridional.—*Frith Church,* this was no doubt an important river post; *Crayford Church, Rcxley Church, The four Cray Churches, Chelsfield Church, Famingham, Kynesford* and *Lullingstone Churches, Shoreham Church, Otford Church* (? Anglo-Saxon) and *Palace, Checening Church, Thinton Green Church, liirerhead Church, Sundridge Church, Serenoaks Church, Knoirle House, lde Hill Church, Weald Church, ('lnddingstone Church, Penshurst Church* and *Castle.*

Some considerable deviations will be seen, the disturbing causes are here the valley courses.
8. Purfleet To Bidborowh Meridional —Here also might be included with this the churches in the Darenth Valley as so close to it on the west. *Purfleet Church,* and the *Deacon Hill,* a little to the east of it, *Dartford Church, Wilmington Church, Darenth Church, Sutton-atHoue, South Darenth Church, Horton Kirby Church, Maplescombe Church,* in ruins and with large sarsens in the interior; *Kemsing Church, Seal Church,*

Under Hirer Church, Hildenborough Church, Bidborough Church. 4. West Thurrock And Ti;nbridge Wells Meridional.—Here also we have to include other churches close to it. *West Thurrock Church.*—Here we also have a *Beacon. Greenhithe Church, Stone Church,* and *Stone Castle.* Suggestive names:— *(ralley Hill, Sicanscombe Church* (and *earthworks), Southjieet Church, 'Umgfield Church, Fawkham Church, Hartley Church, Ash Church, Ridley Church, Stanstead Church, Wrotham Church, Ightham CourtHouse* (with two *moated mounds), Oldbury Camp* and *Hock Shelters,* and on the well-marked ridge there is a possible *serpent,* whilst *serpent-worship* (according to Dr. Phene, the authority on this) was formerly celebrated there. *lledwell Tumulus.*—One mile south of the church, with springs at the base of the tumulus. *Rosewood Pit dwellings, Plaxtol Church, Fairlawn Park* (with possible *stone circles* there), *Shiphorne* If we include this we must also include *(Irays Church, Little Thurrock Church,* and *tforthfteet Church,* and the north of it. *Church, Tonbridije Castle* and *Church, Tunbruiife Wells rocks.* These are of most remarkable form, and so could hardly have escaped *worship* in prehistoric times. The Toad-Rock suggests a totem.

This surely is a most interesting line. At Galley Hill, too, a possible Paleolithic skeleton was found at some depth in a gravel pit containing implements of that age.

5. Gravesend And Brenchxey Meridional.—Here again we have to include churches a little outside the direct line. This line also includes rive megaliths; these, however, are along a north and south line.

The megaliths in question are, starting from the north, the stone circle, a little north of Cobham Church, the hardly-known remarkable group of fallen stones in Cockmlams Shaw-wood near Harvel. Here the stones lie touching one another in a line, as if they had once formed part of a structure; then the Coldrum group just two miles south of the former; then the two megaliths at Ad-

dington.
(t*raresend*.-Perhaps the two churches across the river to the north should also be included, *ciz., Chadnell Church*, associated with which is St. Chad's well, and *West Tilbury Church, (frarescnd* must he a place of great antiquity. *I field Church*.—This and the four others that follow are almost equidistant from each other. *Xunited Church, Cobham Church*, with the sarsens noticed before, *Meophatn Church*, also with sarsens, *l. nddesdoicn Church*. Here in the farmstead close by are some very large sarsens used for gateposts, and others, smaller ones, used in the walls of the buildings, here also are the remains of a very old manor house. *Punish Church*, in ruins, enclosing a large sarsen, the *ijroup of sarsens* in the wood above mentioned near Harvel. The *('nldrum uuujuliths, liirlimj 'hitreh*, and to the east of this, *Trotterscliffe Church*, the *Adilinijton Stone Circles, Addini/ton Church*. This is said to be built on an artificial mound. I can see no proof of this. The hill on which it stands has an artificial appearance, but the Folkestone Bed Sand composing it tends often to form conical detached hills, due no doubt to the ironstone it contains in lenticular beds, and one of these may cap and so protect the summit and give it an artificial appearance, causing it perhaps to have been selected for a sacred site probably from prehistoric times. *Kyarsh Church*, and, to the east, *Off ham Church* and *West Mallini) Church*. These three are veryclose together, Addington, Ryarsh, and If ailing being only very little more than one mile apart, while Mailing is exactly one mile distant from Ryarsh..St. *Leonard's Toner and Chapel*, and also the *Spring head* there, the deciding factor for locating both, no doubt. Roman urns, etc. , have also been found close to the spring-head. *Mereuorth Church*, HV.it *Peckham Church, Hast Peckham ('hurch*. The churches in this group are almost equidistant, *Mereirorth* being exactly one mile away from *East I'eckham Church;* south of this are two *Earthirorkxone* mile apart, the first at *Moat Wood*, the last at *Hale Street, Hale Chinch*, and *Paddock Wood Church*, while again south is *Rrenchley Church*, with an earthwork bank north of it. 6. East Tilbury And Yaldino Meridional. — Here again, owing to the grouping of churches on either side of the line, we have to include these:— *Kant Tilbury Church, Higham Church*, where also many antiquities have been found; *Higham I'pshire Church, Shorne Church*, with ancient *earth/corks* there; *Cuxton Church*, here also Paheolithic implements have been found and there are here prehistoric *cultiration-terraces* and *Roman remains. Lower Hailing Church, Wonldham Church*, just across the river, a *tumulus* at *Holborough*, and again just across the river the remains of a *Mithraic Temple, Snodlanii Church, Ilurham Church*, with a *Roman Villa and Road* to the south of it. *Xeir Hythe*, a ruined church used as a dwelling house; *Leybourne (hurch, Leybourne ('untie, Ditton 'liurch, East Mailing 'hurch:* these three are almost equidistant, one of them less, and one more than one mile apart. The Churches of *Wateringbury, 'lesion, Harming, West Earleigh*, and *Xettlestead*, of which *Wateringbury, Xettlestead, Teston*, and *Harming* are one mile apart. *Vailing Church*, with *Hunton* on the east of it. 7. Cliffe-coolinu And Maidstone, Meridional.—This is an important one as including three megaliths, again north and south, as well as churches and castles.
The *megaliths* are those of *Horstead, Kits Coty* and *Countless Stones. Horstead* is the traditional burial-place of Horsa; a megalith once stood here.
Kits Coty.—So well known, but very different now from what it once was, having been connected with a long tumulus apparently. Near it to the northeast are other recumbent sarsen stones.
Countless Stones.—These were thrown down some 150 years ago. A little to the west of these are other sarsens by the spring-head at Tottington Farm. South again of this are *Aliington Castle* and the *tumulus* once in the churchyard. *Clifi'e Church, Cooling Church and Castle, Frindsbury Church, Rochester Castle* and *Cathedral, Horstead, Kits Coty, Countless Stones, Aliington Castle, Church* and *Tumulus, Maidstone Church, Castle* and *Palace*, while due south of them you have *Tovil, Loose*, and *Linton Churches*. 8. High Halstow And Bouuhton-monchelsea Meridional. — *High Halstow Church, Hoo Church*, and *Roman remains, Gillingham Church* and *Roman remains, Luton Church, Huxley Church* and *Abbey, Penenden Heath*, with the ancient *Moot Hill* there, and *British camp* three miles south of the former, *Houghton-Monchelzea 'hurch*. 9. Ktoke And Chart Sutton Meridional. — This is interesting as including two camps and a tumulus. *Stoke Church*.—The word *stoke* is supposed to indicate an old fort or stockade. *liainham Church, Hredhnrst Church, Hinbury Camp, Thornham Camp, Castle* and *Church*, while close to this is *Vetting Church*, with *Heneholes* close to it, *Bearstead Tumulus, Bearstead Church, Olham Church, Langley Church, Chart Sutton* and *Sutton Valence Church*. 10. I'PCHURCH AND Ulcombe MERIDIONAL. *Upchlirch Church*.
This place is also celebrated as being the locality where the well-known Roman Ipchurch ware was made.
Ilartlip (lunch.—Here also Roman remains have been found. *Stoekburij Church* and *Castle*, a prehistoric one, *Bicknor Church, Hitching Church, Hollingbourne Hill*, here I have seen what may be earthworks and not all attributable, I think, to tip, etc., from the old chalk-workings there; here also appear to be *1eneholes*, and the old *cultiration terraces* there are well-marked. *Hollingbourne 'lunch. A tumulus* also is shown on an old map at *F.ghorne Street, Leeds Castle, Hast Sutton*, and *Clcoinbe Churches:* these are one mile apart.
Having now given the meridional lines, or what appear to be so, east and west of Maidstone, including nearly all the churches in the area, this brings the paper to a close. The persistence of these among the older erections, such as megaliths, tumuli, and camps, and the inclusion with them of so many churches and the apparent meridional position of many of these, and their equidistance in many cases, seems more than a mere

coincidence: and though it may not prove anything, it warrants, I think, further investigation along these lines.

It also seems to suggest that where you have one megalith, tumulus, or camp, you may look for others in the former traces of them, along a north or south line. The separation, too, of megaliths and churches apparently by almost equidistant lines of division sometimes equal to our mile, or only differing a little from it, and perhaps by multiples of a mile in places, with unhewn stones used in their construction, also seems a further connecting link between them, and suggests an interesting line of enquiry.

The Use Of 25in. Ordnance Maps For Estate And Agricultural
Purposes.
By F. J. BENNETT, F.G.S.

There is no doubt that the ordnance map, on the scale of 25in. to the mile, might be used as a basis for recording many important facts connected with the soil, geology, and the agriculture of each farm or landed estate in the country.

In a way this would amount to a new Doomsday Book Survey, and this leads us to remark that that work, accurate, stupendous, and remarkable in so many ways, may be said to have had no successor—a fact hardly to our credit. Of course the Doomsday Book was compiled under pressure, for a specific purpose, and mainly for the personal benefit of an autocrat, with no thought or intention that it would become a reference book of the utmost historical value to succeeding centuries.

The survey of the land that is now suggested is of a different nature. It would have a practical and utilitarian side of the utmost value, we conceive, for all landowners and farmers. Let us take the case of a person purchasing an estate. To a large extent he would, in a usual way, be very much in the dark as to the real nature of the property he had purchased. He would, of course, have all the information the seller could afford him and that would vary very much according to the way in which the estate had been managed. He might be able to obtain one inch, or even six inch, maps of the Geological Survey both Solid and Drift, with, in many cases, the accompanying descriptive Memoirs; and, according as he was able to understand them, they would give him much or little information. Yet to most this would be of a superficial or vague nature on many points, and perhaps, could not give the details most useful to him. But if he had followed the plan adopted, I believe, in the best Estate offices, the 25in. Ordnance maps would have been used, and the estate maps brought up-to-date by marking on them any alteration subsequently made. There would no doubt be a schedule of the amount of arable and pasture and woodland, with the kinds of trees, etc., and there might be a rough division of the soil into heavy and light.

Soil.—Now, let us suppose that the late owner had made these maps in the way this paper would suggest. Say, that on each field division should be noted the nature of the soil and subsoil, whether clay, sand, loam, gravel, or chalk, etc. , and the qualifying character of these. Of course, difficulties would arise as to how this should be done. Here, then, I would suggest that a visit should be paid to the Geological Survey Office, to ascertain what information was available. As a very useful preliminary to this visit, trial holes, or trenches preferably, should be dug, especially in the pasture lands, so that the subsoil could be exposed. In this way a kind of soil map could be made and recorded on the map or schedule accompanying it. Place and field names should also be recorded on the maps with first and last ways of spelling these, and the dates. *Wells.*— These should be all marked on the map, whether in use or not, and all measured and their total depth given, and that of the water and the variation of this, and where possible, a record of the soils met with when this well was sunk and the name of the sinker. *Sprinijs.*— All these should be marked and their variations and highest point in any special year, going as far back as possible. *Quarries.* —All these should be noted and characters recorded on the map. *1'itx.*—Where old pits exist, often, of Course, grown over, it will be found of the utmost importance, where all record has been lost, that they should be cleared and their true character ascertained. *1 train.*— Now, perhaps, the most important detail has been left to the last. 1 am informed that, in most cases, where land has been drained the courses of the drains have not been laid down on the estate maps, so that very often much of the money thus expended has, for practical purposes in after years, been lost, and where the drains have ceased to work much time and expense have been absorbed, sometimes to no purpose, in seeking the outlets, ivc. All this would have been avoided had their courses been laid down on the maps. To record the nature of the soil dug out, when drains are being made, is of the utmost importance to the agriculturist and also to the geologist, and this should he especially noted on the map. As the Government lends large sums of money for land drainage 1 would suggest that the Government stipulate in the future that the courses of all land drains should be laid down on the estate maps and the nature of the soil recorded, and that a copy and tracing of the drains be deposited with the Board of Agriculture. *Agricultural Maps.*—We would here suggest a further use of these 25in. maps for the recording on them by farmers of certain agricultural notes relating to crops, Ac. On each of the Held divisions year by year, and in one line if possible (so that the records of several successive years might be placed on the same division for reference, especially if contractions were used), should be noted the amount of seed sown, the kind and quantity of manure used, and the weather at the time; also the result of the crop, such as weight of grain, length of straw, &c, and the same with other crops. If the results of seven years were thus recorded they could be taken in at a glance and the reason often seen for success in one year and failure in another, and the varying results where different manures had been used could also be noted in a Field Book. The different kinds of trees and their growth in relation to the soil should be noted both by farmers and landowners. Many farmers, no doubt, would object to all this as an additional and useless labour on their part; but we would suggest that such information would be of the utmost value to the incoming farmer, and would of course be the private property of the late occupier. The incoming tenant should be glad to pay a very substantial sum for this accumulated information, as, without this, he might have to spend years and lose much valuable time and money in finding out these facts. Thus the late occupier would find that he had not only been getting together much valuable information for himself, but information of such a nature that the incoming tenant would be glad to buy it as so much goodwill. All this would tend to a systematic and thus to a scientific method of farming and its adoption seems of the utmost importance.

The Practicability Of An Artificial International Language.

By P. MATHEWS, M.A.

President of the New Prompton Naturalists' Society.

Py an "Artificial International Language" is meant a Language, which has been invented, which has not grown by a natural process. The goal aimed at by those who think this idea practicable, is, that in course of time, such a language may be adopted by, and taught in, the schools of every civilised country in the world. There is no desire to supplant the mother tongue, or to hinder those, who wish to study the literature of any country in the language of that country, from doing so to their heart's content.

As mathematicians frequently attack a problem by considering it solved and examining the results, let us for a moment suppose that in travelling about Europe (to say nothing of any other part of the world) we were always sure to rind that we could understand what was said to us, and to meet with people who invariably comprehended what we said to them. Would that not be a great benefit from the social point of view? Then again, suppose that all scientific work in every country were published in a language easily understood in every civilized land. Would that not be a great benefit to the scientific world? Persons engaged in scientific investigation are constantly finding, especially if their work is of a highly specialised nature, that important contributions to the subject which they have in hand are to be found in the Transactions of a Foreign Society. Even if the language be such a fairly familiar one as French or German, many will find some difficulty in translating, still more so if it be a language (such as Swedish or Japanese), which is rarely understood except by natives. Once more from the educational point of view—how much time would be saved if the schoolmaster were obliged to teach to all the pupils one easy exception-free language merely, leaving particular foreign tongues to be studied only by those who have special aptitude or need for them! Would not this be a great benefit to education?

The idea of inventing a common tongue for all the world is a very old one. I suspect that the story of the Tower of Pabel is a hint nt such a yearning. Like all human ideas, this idea has become considerably modified with the lapse of time. Three hundred and fifty years ago, the notion was to replace

Since (owing to unforeseen circumstances) the time for this paper at the Congress was somewhat shortened, the writer hopes he may be excused for having slightly expanded the discourse delivered on that occasion.

all the diverse languages of the world by a single tongue—an obviously unrealizable aim. Any one curious about these early attempts will find plenty of them in the British Museum Library. The earlier ones are in the nature rather of polyglot vocabularies, or of ideographic characters, the shape determining the meaning, while each person assigned the sounds of his own tongue to the signs, just as when we see the character 5 we say "five," but a Frenchmen says "cinq," a German "fiinf," an Italian "cinque," a Russian "pyaht," and so on. With Europeans this kind of thing is confined mainly to numbers, but in China, a man from the north seeing a manuscript will be able to understand it as perfectly as a man from the south, though, ask each of them to read it, and you will evoke sounds as different from each other as possible.

John Wilkins (Bishop of Chester) invented the most notorious of these early attempts. He avowedly intended his language for the use of the learned only, and his main object was the spread of the knowledge of the Christian scriptures. His language was at first simply ideographic, though he afterwards put sounds to the signs.

As an instance of failing to perceive a beam in one's own eye while wishing to remove a mote from one's neighbour's, I may just notice that another universal language inventor objected to Dr. Wilkins' sounds as being uncouth. I present, in juxtaposition, the Bishop's rendering of a phrase and that of his critic. It does not matter what the phrase means (each means the same), but I should not like.to decide upon the point of uncouthness.

"Hahee eobali owow reel dayd."—(Dr. Wilkins).

"Ho fonzipeerton dyoo pontoo inzoeeter."—(The Critic).

More than one attempt represents its root-words by numbers, each figure representing a sound (derived from the name of the figure in the language of the reader), *e.g.*, 84H1 to an Englishman would sound "threeforeighton," to a German "dryfeerachtein," etc.; the inflections are indicated by letters and diacritical marks each + representing a sound, *e.g.*, r 8= ricrocetre.

Sir Thomas Urquhart, about 1650, projected, a 'universal' language, but his preliminary prospectus is the only trace of it I have been able to find, and I felt disappointed at this, for he promised 12 parts of speech, 10 genders, 10 synonyms for each word, 4 voices, 7 moods, and 11 tenses. The ten genders especially would have interested me much.

But enough of these ancient matters—they are justly relegated to the limbo of forgotten fads. Let us come to modern attempts.

The idea has changed. Now there is no longer a hope of an "Universal Language," this unrealizable ideal is replaced by that of a *second* tongue commonly understood in all civilised countries. Several methods of arriving at this result have been suggested:— *(a)* Have an existing language adopted as the International means of communication, and taught in every country—Objections —National jealousy and the difficulties which every language, even the least complicated, presents to the foreigner, in the way of pronunciation, spelling, grammar and idiom.

(b) Take some existing language, strip it of its irregularities, simplify its grammar, and adopt it in its modified condition as the auxiliary dialect.

The languages which it has been proposed thus to improve (?) are Latin, English, French, German, Italian, and even Welsh. The main objection is, I think, the offence to the national pride of the nation whose language is so mauled—nay, one (" Anglo-franca*"*) combined to distort both English and French, the only result being, one would fear, seriously to endanger the *entente conlialv.* Here is a specimen (French pronounciation to be used).

"The navigateurs would trouv avantages not less grands in pouving to communiqu facile with un-auother, either upon mer, or at the grands stations of the ocean." What would we or the French think of this? No! I do not imagine that Anglo-franca will " deven the nouveau plan for the facilitation of international communication." () An entirely original language seems the only plan with the faintest possibility of success.

Such a language should be extremely easy to linguists, and at the same time not difficult to those who are not linguists, and it should not (from considerations set forth above) be based entirely upon any existing language. Many attempts of greater or less ingenuity have been made on these lines.

Schleyer's " Volapiik " held the field some 10 or 12 years ago, and attracted many adherents. It however carried the seeds of its own dissolution.

The very name of the language illustrates the reasons of its failure in two directions. The language was harsh and difficult of pronunciation, owing to the retention of the German, modified vowels, ii, o, and ii, whose precise pronunciation is by no means easy to everybody. The word "Volapiik" means "Speech of the world." Schleyer took his roots mainly from English. One would not think so from this word, for "vol" suggests to a linguist "thief "or "Hying," rather than " world," of which it is a modified form. (The letter "r" is not used in "Volapiik" because Chinese and Red Indians find a difficulty in pronouncing it). Piik is the English "speech " modified. In consequence of such changes Volapiik root forms take a lot of learning. The grammar, though minus exceptions, is complicated, and the number of inflections to be learnt enormous, at least some hundreds. Consequently Volapiikists soon quarrelled among themselves, and in trving to improve the language, killed it,

The attempt which at present holds the field is Zamenhof's "Esperanto;" this is far superior to Volapiik in every particular, to which allusion has previously been made. It is pleasant sounding, and the roots are obtained in most instances by choosing (without alteration) those which are common to several European languages, consequently one who knows Latin, French, and English readily recognises them, and finds no trouble in quickly acquiring an extensive vocabulary. The inflections are few, and case forms, except the objective, are replaced, as in English, by prepositional phrases. Verb conjugation is very easy, there being no changes for number and person, the pronoun being always used.

It is no exaggeration to sny that two hours is sufficient for a complete mastery of the whole grammar. But the chief improvements on all previous attempts are (1) the system of word-building by means of affixes and suffixes, and (2) the writing of hyphens between the various constituents of a word, dividing it into parts, each of which can be looked out in the dictionary. Thus it is possible, as I have myself proved more than once, to send a letter in Esperanto to a person who has never perhaps heard of the language, certainly never seen it, and he will be able, by the aid of an enclosed sheet vocabulary, to translate the communication without much trouble. These vocabularies, costing from Id. to 3d. apiece, are now printed in most civilised languages. Of course the writing of the language requires a little more time, but a well-educated person will write an intelligible letter in a day or two, and one would not expect that (even with a genius) in any existing language. For conversation again, one requires practice before one becomes fluent, but it is immeasurably more easy to become fluent in Esperanto than in a language which has evolved in the ordinary way. The accent is always on the penultimate syllable, and every letter has a fixed and invariable sound, so that no difficulty arises from there being any marked difference between the pronunciation of an English Esperantist and that of one from another country.

As an example shewing the use of the vocabularies which I have just mentioned, we will suppose a Russian wants to write to me. He writes in Esperanto as follows (enclosing an Esperanto-English vocabulary):—

Kar-a Sinjor-o, Chu vi komplez-e vol-os send-i al mi vi-a-n fotogmf-ajh-o-n'?

fidel-e la vi-a, X.

The vocabulary gives the following information: -*a* = adjective, *ajh*=a thing made of, oZ = to, *chu* is used in asking a question, -*e* = adverb, *jidel* — faithful, *fototjraf*= photograph, -*i* marks the infinitive mood, Aw=dear, *kompler* = oblige, mi = I, me, "marks objective (accusative) case,-o = a noun substantive,-« indicates future in verbs, *send* = send, *sinjor*=*Mr.,* sir, gentleman, n' = you, roi =vrish. It is unnecessary now to append the translation. When one writes to an expert one omits the hyphens. By the aid of syllables having a fixed meaning wherever they occur, word-building is very easy in Esperanto; to take a simple example the word " staff." This word may mean a stick *(baxtono)* or, as in the expressions, the staff of an army, of a school, of a business; each would be a different word in Esperanto. The staff of a school means the whole body or collection of the teachers, hence *in-ntru-ist-ar-o,* meaning the collection (*-ar-*) of persons engaged in (*-ist-*) teaching (*-imtrii-*); but the staff of an army would be *ofieer--ar-,* and of a business *sen--ist-ar-o,* the collection of the servants.

To show that Esperanto is not a harsh sounding language I quote the Esperanto version of the first verse of an old English sea song, which runs as follows:—

One night came down a hurricane,
The sea was mountains rolling,
When Barney Buntline turned his quid
And said to Billy Bowline,
A strong nor' wester's blowin', Bill,
Hark! don't ye hear it roar now?
God help 'em! how I pities all
   Unhappy folks ashore now!
   En nokt' okazis uragan'.
La maron montigante;
Tabakon machis Barni Bunt,
Al Bolin Bil dirante,
Nordokcidenta blovas. Bil,
Chu vi ghin audas brui nun?
Helpu ilin di'! bedauras mi
Bordanojn malfelichajn nun!

This quotation shows than Esperanto *can* be written rhythmically, though, of course, to produce poetry is no necessary function of an auxiliary language, which is intended primarily to facilitate mutual intercourse between those Who would otherwise understand each other only with great difficulty. It is the opinion, which is shared by the writer, of many thousands of people that the general adoption of Esperanto is practicable, and would be a real benefit to humanity. The worst criticism of it has been that of a Russian diplomatist, who called it so absurdly easy that it would enable all the rascals of the world to communicate with each other much more unrestrainedly.

"The letters in this vocabulary which are followed by "ft" have a special typeform, and are single sounds.

Ice Streams And Ice Caves."
By A. B. HARDING, F.Phys. Soc. Lond.
In France, Switzerland, and Styria, as well as occasionally elsewhere, are found caves, chiefly situated in limestone formations, wherein large quantities of ice and snow may be found at all times. Lantern views were shown of such-Ice Caves, (1) At St. Georges, in the Jura, 110ft. in length, the floor a sea of ice which is constantly replenished: the man who farms the ice chiefly draws his supplies from the north-west wall, which is 70ft. long by 22ft. high. (2) Glaciere of the Pre de St. Livres, near Aubonne, which has a sloping floor of snow and ice, leading to a lower slope of very smooth slippery ice, which, when cut with an axe, showed no ordinary vitreous fracture, but rather separated into prismatic nuts of limpid ice, free from air and surface-lines, dark grey in colour, and suggesting the idea of coarse internal granulation. The ice wall is 72ft. long and 22ft. high, while facing this are icicles forming a natural Gothic arcade. (3) Second and lower cave of same Glaciere: principal feature, a remarkable fan-shaped cascade of ice, 27ft. wide at its base; a smaller fan *pours* from a vertical fissure in the rock, looking as though water had been frozen while falling; the floor consists of a sort of shore of stones, leading to a lake of *black* ice. (4) Ice columns in Glaciere of La Genolliere, near Arzier. Three curious columns of ice, 11, 184, and 15ft. long, issue from fissures in the rock; between the rock and the ice is a space of one or two inches. The prismatic structure of the Glaciere ice is very marked here; some of the nodules could be pushed out, like knots in wood. (5) Entrance to the main cavern or Glaciere of Surtshellir, Iceland, described by Forbes. The cavern is 50ft. wide, 40ft. high, and 700ft. in length; the floor is composed of ice and sand.

In none of these cases is a sheet of water found superficially frozen, but the ice is always solid, when on the floor filling up the interstices between the stones, while the walls and columns of ice are often of great thickness.

Several theories have been advanced to explain the formation and persistence of the ice. Pictet and others ascribe the production of ice and snow to cold currents and evaporation. Prevost regards the caves as natural storehouses, in which the heat of summer is unable to dislodge the cold of winter and its consequent crop of ice. Deluc and Browne adopt an explanation which embodies both these views, *viz.,* the prduction of ice and snow by evaporation, and the retention of the cold heavy air of winter in the lower parts of the Summary of a lecture given at the Mayor's Keception, June 10th, 1004.

caverns, even during the hottest summers, particularly if the summer be *dry.* The lecturer illustrated the production of cold by evaporation by freezing a quantity of carbon dioxide into a carbonic acid snowball; partially dissolving this snow by strong sulphuric ether he showed that the temperature was reduced to — 80F. or 112 of frost. With this extreme degree of cold he froze a quantity of mercury into a solid resembling lead, and obtained a thick rime of hoar-frost on the containing vessel by the freezing thereon of the moisture of the atmosphere.

The latter portion of the discourse was concerned with Ice Streams (Glaciers), of which he briefly explained the nature and leading features, illustrating the various points by very beautiful views of moraines, ice-falls, pinnacles, seracs, etc., as seen on the Bossons, Mer de Glace, Aletsch, Rosenlaui, and other Swiss glaciers. He also showed several fine slides of snow-flakes (English aud Arctic), hailstones, and frost plumes, which were much admired.

Notes On The Lepidoptera Of Mid-kent.
By Capt. SAVILE G. BEID, H.E., F.E.S.

The following notes refer to a limited portion only of Mid-Kent, and I will ask you kindly to look upon them as merely a contribution to the study of the lepidoptera of this interesting district. I think 1 may fairly call it interesting, including as it does a portion of the chalk range of the North Downs, of the lower greensand "ragstone " range, and of the deep clay valley known as the " Weald" of Kent. With such varied topographical and geological conditions we may expect to find a somewhat extensive list of lepidoptera occurring, even within the small area worked by my friend Mr. Edward Goodwin, of Canon Court, Wateringbnry, and myself, which I may confidently say is one of the most productive in England.

The county of Kent, taken as a whole, is doubtless one of the richest in lepidoptera, a fact due to its geographical position as well as to its peculiar geological formation.

In this paper I propose to confine my remarks as much as possible to those species met with within our restricted boundaries which are most numerous and characteristic, or of most interest to collectors, and I am adopting the specific nomenclature of Mr. C. G. Barrett in his work on the *Lepidoptera of the British Islands,* now in course of publication. At the same time I hope I shall be excused for occasionally making use of the familiar English names of well-known species.

These notes principally apply to the districts round Wateringbnry and Yalding, but many observations refer to the chalk range between Lenham on one side and Eynsford on the other. Personally I have collected chieflyat lamp light, having agood windowin my housewith a favourable stretch of open grass-land in front of it, but my friend, Mr. Edward Goodwin, has done a great deal of work in every other way, more particularly in breeding from collected ova and larva1, and from ova laid by captured females. His collection is very fairly representative of the district concerned, and contains many fine local varieties.

I think the best plan will be for me to take the various species in the order of Barrett's list, and make a few remarks on those which seem to call for mention, the very common species being, as a rule, passed over in silence.

Before commencing this detailed account I might mention that, out of the total of 825 species of macro-lepidoptera included by Barrett in the British list, not counting the Psychids, which we have not embarked upon yet, no less than 478 are, or have been, to our personal knowledge, taken within the area treated of in these remarks, so that we may fairly claim that our hunting-ground is a happy one.

Commencing with the butterflies, I think it worthy of mention that a specimen of *Papilio machaon* was taken some years ago, at Snodland I think, and therefore almost within our province. For a time it was looked upon as a rare straggler of great historical value, but it transpired that many perfect insects bred from Wicken Fen larvsp had been liberated by an enthusiastic naturalist in this neighbourhood, and the incident soon ceased to excite interest. I do not think *I'. machaon* is likely to occur in a truly wild state in Kent. *Aporia erataeyi* does not come into our district, the only locality for it at present known being in another part of the county. The Common Whites are all numerous, including *Euchloe eardaniines. Leptidia xinapix* used to occur commonly in the Wateringbury district, but is not now taken there, and I fancy it is rare in all parts of Kent. The " Clouded Yellows," *Colias hyale* and *('. ediixa,* including var. *helire,* appear in fair numbers in their usual irregular way; 1900 was a phenomenal year for them, and it is worthy of note that *C. hyale* also occurred again in some numbers in 1901. *(ionepteryx rhamni* is common, and both ova and larva can readily be found on the buckthorns. Of the Hairstreaks only *Xephyrux querent* and *'allophryx rubi* have come under our notice, though *'/. ephyrux betidae* and *Tliecla iv-alhiim* occur in other parts. *Callophryx rubi* is apparently more plentiful on the chalk than elsewhere.

As a whole the Blues are well represented; *Polyommatux ayextis, P. ale.rix* and *I'yanirix aryiolux,* occur numerously throughout, while *I'lebeius (lei/mi, Polyommatux adonis, I'. corydon,8,nA Cupido alxiix,* are found more or less abundantly but locally on the chalk range. We have not taken *Xemeobiitx lueina,* and *Apatura iris* must be looked upon as a great rarity. *I.imenitix xibqlla* has been taken in the Weald.

Among the Yanessids the Maidstone Museum has two good examples of *Fnranexxa antiopa,* taken in 1HH9, one in Earl Street, in the town, and another at Hunton, near Yalding, while Mr. ((loodwin has a very tine specimen captured at Hast Farleigh, about six years ago, and has seen two others on the wing at various times near Wateringbury.

F.uyonia pulyrhloros is frequently common, larva' having occasionally been found on sallow, elm, wych-elm, and cultivated cherry. Iioth this species and *Aylaix urtieae,* oddly enough, have come into my house at times in the evening, attracted by light. *Vanexxa in* and *I'yrameix atalanta* are always common. */'. carttiii* occurs annually; last year (1908) it was in hundreds. We have never, unfortunately, had the luck to meet with *Polyyonia c-album,* though Newman alludes to its having been reported commonly from the hop-districts round Maidstone many years before his time.

We are not well off in Fritillaries, though *I'ryax paphia* in the big woods, *Aryynm adippe* on open rough ground and heath, and *A. at/laia* on the chalk downs are fairly common. I have seen *A. tifilaia* in numbers flying over the grassy slopes at the foot of the Uoxley Hills, and on one occasion, when looking for the silverspotted Skipper without success, was much astonished and interested when I observed the hitherto invisible Skippers rising suddenly out of the long grass to attack the passing Fritillaries. Thanks to the latter I was able to catch one or two very good specimens of the pugnacious little *Pamphila comma. Brenthix euphroxytie* is, of course, numerous, but we do not take *H. xelene* except on the Weald. We regret that we cannot include *Ixxoria latmut* in our list, though it occasionally turns up on the coast, while *Melitaea athalia, M. riii.ria* and *M. artemix* do not occur.

The Marbled White *Melanaryia yalatea* is local, but very common on the chalk, the same remark would apply to the Grayling, *tiipparchiaxemele.* Curiously enough the familiar *Para rye eyeria* has never been seen by us in our province. Two specimens in the Maidstone Museum were taken at Boxley about 1875, where, in a very restricted locality, the species formerly occurred, but it is no longer to be found there. It is still common, I believe, in the Chattenden Woods. *Para rye meyaera* is a familiar species. *Epineplwle ianira, Knodia hyperanthux,* and *'oenoitym/dta pawphilttx,* also, but the Large Heath, *Kpinephele tithonux,* is not so plentiful, being somewhat local in its distribution. *Syriekthtix aleeolux, Thymelicux thanmax, Sixoniadex tayes,* and the large *Pamphila xyleanux* are common everywhere, *l'. comma,* is locally numerous on the chalk. The Essex Skipper *T. linenla* has not, unfortunately, reached us yet. The Maidstone Museum has a well-marked specimen of *S'. alveolus* ab. *tarax,* and Mr. Goodwin took another fine one in the middle of last month near Wateringbury.

Hbterocera.—Among the Sphingids we find, chiefly in the larval state, *Smerinthiix ocellatux, Amorpha popnli, Mimax tiliae,* and *Sphinx liyastri,* not uncommon—also *Sexia xtellatartim,* which is a frequent visitor to garden flowers. *Manduea utropnx* is scarcer, though occasionally larva-are found plentifully in potato fields. *Atjriux conrolruli* is at times fairly numerous at the flowers of the white tobacco plant. In the autumn of 1898 they were extraordinarily common in the neighbourhood of Wateringbury and Yalding, and I think it worth mentioning two rather curious facts in connection with this unusual immigration of the species. Nearly every specimen captured in a gentleman's garden near Wateringbury, where they were especially numerous, was found to be very much torn and damaged in the wings, and Mr. Goodwin found, after carefully watching the place, that the big moths in their rapid flight towards the tobacco flowers dashed headlong through the wire netting enclosure round the tennis court and thus knocked themselves about badly. Mr. Blest, the owner of this garden, and Mr. Goodwin, the same autumn, paid a visit to Mr. Henry White of Wateringbury, who possessed a cat which was in the habit of catching and bringing into the house specimens of *A. rnnrol villi.* During their visit this cat was let out into the garden at the proper time and soon returned with one, going out again and bringing in a second!

A. ronvolvuli has a habit of falling to the ground, if struck, and feigning death, so doubtless the wily pussy took advantage of this—otherwise I fail to see how she could have managed to secure such a strongHying moth. A very fine specimen of *Hippotimi celerio* in the Maidstone Museum was captured in a backyard in St. Faith Street, Maidstone, at as late a date us October 9th, in 1903. *Kumarpha elpenor* is common at honeysuckle, the larva along the river Medway on *Kpilobitim.* *'l'heretra pnveellux* larva? are found on the chalk-hills on *Galium,* but the perfect insect does not seem to be common; one very tine one obligingly sat on my drawing-room wall near a lamp two years ago; L bad never taken it at light before, though *S/diiii.r liijuxtri, Amorpha populi,* and *Kitumrplia elpenor* have paid me visits. I reported to the entomological journals the occurrence of *Daphnix nerii* at Yalding in September 1H9H. Unfortunately my small friends who captured it had no proper net or apparatus and it was badly mauled. They brought it to me the next morning for identification, and 1 made what I could of it for their collection. It is worthy of note that the moth, apparently attracted originally, like *A. ronrolndi,* by white tobacco flowers, was further influenced by a light in the room adjoining the garden, and came into the room. *Hemarix furiformix* is taken at *l.i/rhnix* flowers, the larva' being numerous on honeysuckle, easily found through their habit of making small round perforations in the leaves. We have no authentic record of *//. hnwbyliforinix.*

The Clearwings are not strongly represented. *.Kijeria tipulil'orinix* is sufficiently common to do some damage to currant-trees, especially black currants. *.K. cnliii/'oniiix* larva' and pupa? are very common in stumps of birch-trees cut down the previous year, but the Titmouse family probably destroy 90 per cent, of them. *.'.. rj/nipif'ormix* occurs on oak stumps, but is not so common..*/.'. iiii/oprieforinixlx* frequent in apple orchards. *Troehilinm beinbeeiformix* larva1 common in sallow stems, so much so that nearly every hop pole cut from sallow trees shows traces of their work.

Among the Foresters and Burnets the Museum has examples of *Adxcita xtaticex* from Detling, but we have only taken the ubiquitous *Anthrocera tilipendidae,* though several other species are found in Kent.,

Of the Zeuzerids, *Xeicera aexcuii.* and of the Cossids, *Coxxitx liijniperda,* are rather too common, for the comfort of the fruitgrowers, in the larval state. The perfect insects are seldom taken. J have bred (*'oxxux liijniperda* from larva? taken out of fine healthy large oak-trees at Yalding. Most of the Hepialids are common, and of the Chloephorids, the common "Silver-lines," *Hylojihila praxiiittna* and the scarce *II. qiiercana* are both taken, the larvte of the latter being difficult to dislodge from oak. Of the Nolids, *Nola riirullatflla* comes freely to my lamp-light. Only eight of the "Footmen" have come within our ken. The "Tigers" are well represented, both *Aretia raja* and *Phraymatobia fuliijinom* come freely to light, while *Xrmeophila ritssula* occurs at Boxley.

Of the Liparids, I have taken several males of *Porthexia chryxmrhoea* at light at Yalding. This was in 1898, but as I had just previously lost a batch of small larvte (hatched from eggs sent me from another part of Kent), which escaped through the meshes of the muslin sleeve in which I wjas feeding them, 1 came to the conclusion that I was merely recovering some of my own lost property. All the same it is quite possible that the " Brown-tail" may be found in this neighbourhood, as it is common in other parts of Kent. Of the "Tussocks" *Dasyrhira pudibunda,* the larva of which is well-known locally as the "Hop-dog," is naturally very common. *1 tenia eoryli* and *Lymantria monacha* are by no means rare. Of the Lachneids only the commonest are met with. Males of *Kntrirha quern folia* occasionally come to my lamp.

My friend, Mr. R. H. Kremlin, informs me that he found a batch of 15 to 20 eggs of the "Kentish Glory," *IHmorpha rersicolora,* near Wateringbury, many years ago, all of which were successfully bred, the perfect insects being given away to friends. There is, I believe, no subsequent record of the occurrence of this fine species in this neighbourhood, and Mr. Goodwin has recently taken freshly-emerged females up to the woods, where the eggs were found, without any "assembly " resulting, so that it is probably extinct at the present date. Of the Attacids, *Saturnia car/rini* occurs on Banning Heath. All the ' Hook-Tips," except, of course, the scarce *Drepana xinda,* are found in fair numbers.

The family Notodontidie is present in sufficient variety to make their study most interesting. Excluding, of course, *Dicranura hicns/iix,* and the rare visitors *Xotodonta tritophux, X. to ma,* A", *bicolora,* and *(ilnphixia crenata,* Mr. Goodwin has bred the whole of this family from eggs which, with the exception of those of *Vrymonia dodonea,* were all obtained in the district. *I'tilodontis plumiyera* and *Lop/toptt-ry. i-ntadlina* deserve special mention on account of their rarity. 1 have taken males of many of the species, including *I. dodonea,* at light, so that I think we may claim for our district that it is rich in Notodonts. *Axteroxcopiix caxxinea* is fairly common, and Mr. Goodwin has found the larva feeding on cultivated cherry, and has bred a fair number. Males come to light. The only "ChocolateTip " we get is *Pyyaera recluxa,* which is not common. Larvre of *Dilnba coernleocephala* numerous on hedgerows and many fruit-trees. Passing to the Noctuina group, which, having due regard to the length of my notes, I fear 1 must only touch lightly upon, we find among the C'ymatophorids, *Thyatira batix* and *T. deraxa,* both common, *Cymatophora tlnplaris,* ('. *tiiluta,* ('. *or,* and *As/dialia riaricornix* come to sugar, larva-of '. *or* on aspen, common, while Mr. Goodwill takes the rarer '. *ductuoxa* in the Wateringbury district.

We take nearly all the common Acronyctas, but the Agrotids are somewhat poorly represented. All the Yellow Underwings occur except *Triphaena xubxeyua,* and most of the ordinary true Noctuie, *Xovtua ilahlii* coming freely to sugar. *Apleeta tiiutu* at times is numerous on tree-trunks and at sugar, *Ilatieiui yenixtae* I have often taken at light. Lame of *('leocerix riminalix* common on sallow. 1 fear, with the exception of a single *Leiieama albipunrta* taken at light on September 10th, 189G, 1 have nothing of interest to record till we come to *'hortotlex arciwsa,* which is very plentiful at Yalding, the larvie feeding on the extensive patches of *Aira* grass near Twyford Bridge. *Tethea xubtiixa* and *'/". retuxa* occur, and come to light occasionally; (*'alymnia di/fhiix* and *'. a (finis* are also taken, they both visit my lamp rarely.

Females of *Hoporina rroeeayo* are often taken at sallow-bloom, and we have bred fine series from them, the best plan being to sleeve the females on oak boughs, adding *ileail* oak-leaves for them to lay their eggs upon.

Xylena xemihrunnea we come across occasionally at light, at sugar, and ivy-bloom, and I have taken it once on sallow-bloom after hibernation. (*'alueainpa e.eoletn* comes to sugar not uncommonly, but (*'. vetnxta* is rare. Of the "Sharks"' the larvw of *Cneullia rerbaxci* are common on various mulleins, and those of ('. *asterix* on golden-rod, on which plant Mr. (ioodwin has occasionally found and reared the rare *'. ynapltalii.* ('. *umbratica* is common. Of what we may term the "Burnished brasses " the commoner species are found, and the now ubiquitous *I'lnxia moneta* occurs in numbers, its larva' being taken both on *Delphinium* (Larkspur) and Aconite, in gardens. Last spring in my own garden they were all feeding on *I delphinium,* none being found on *Avonitum:* this spring, oddly enough, they avoided the *helplu'niiim* and were all on *Aconitinn.' Aeimtialnctuosa* occurs abundantly in various lo calities on the chalk; the beautiful larva-of *Anarta inyrtilli* can be found on heather where it grows. *t'atocala nnpta* is our only "Red I'nderwing." and is common. *F.neliilia ylyphirei* and /*'. mi,* common generally, on the chalk-hills especially. *Ilreplwx partheninx* abounds in some years, but we have no record of *II. notha.*

Omitting the rather uninteresting Deltoides, or rather merely remarking that *Arentia rie.rnla* visits my lamp occasionally, 1 will now pass on to the group (ieometrina, to my mind, and that of Mr. Goodwin also, the most fascinating of all the Macro-Lepidoptera. I regret that, as in the case of the group Noctuina, my remarks on our local species must be very brief. We have not come across *Alenrix jiictaiia,* but *Mararia notata* is common in woods, and the local *Scoria dealbata* may be met with in our area, though we have not yet taken it ourselves. *Aspilatex ijilraria* is common on the Boxley Hills above the Pilgrim's Road. *Knrymene dolobraria* may be met with in the woods, the pupw being found in moss on oaktrees. We have bred all the "Thorns," and all of them from this district, except *Ennomox autumnaria, /'.'. eroxaria,* and (of course) *Kpione rex/wrtaria. K. eroxaria* males come often to my lamp-light, however, and many /*'. fuxcantaria* also. A male *Bixton hirtariiix* came to light as late as May 22nd, in 1898, it was in excellent condition. *Amphidasyx prodroinarinx* is to be found, *A. betidarittx* males come to light, *Bixton hixpidaria* is rare, (*jnophox obxcurata* has been taken at Boxley. *'leora lichenaria* often visits my lamp. All the Tephrosias occur, and the commoner Boarmiids. Mr. Goodwin has a fine lot of these, including some splendid dark forms of several of them, and his series of this group is probably unequalled. I have taken several males of *Boarmia roboraria* at light. The Hyberniids are all taken, *Ilybernia defoliaria* being sometimes a great pest in fruit orchards. Of the " Fmieralds " (*reometra rernaria* is decidedly common, it comes often to light, as does *Phorodexma bajidaria* rarely. Many species of the Acidaliids, a most interesting family, occur to us. the rarest being, perhaps, *Aridalia xtraminata, A. promiitata,* and *A.xiibxericeata,* all three of which I occasionally take at light. Mr. Goodwin has bred a fine pink form of *A. avertata,* five examples out of a batch of 120 eggs laid by the same female. *A. ornata* is common on the chalk range. *A. imitaria, Timandra amataria,* and *A. emarginata* come to the lamp, especially the two first. *Melanthia riibiijinata* and *Antiiiea rubidata* are common, and the handsome *'omnia quadrifasciaria* is often met with, though very local. *Asthena sylrata* is found in the woods; I have taken a single *F.upixteria obliterata (heparata)* at light, it appears to be a scarce insect with us, for Mr. Goodwin does not meet with it. *Emmelexia nnifasciata* occurs, I have taken it at light, as I did years ago in Hampshire, to the great surprise of a critic of my notes in the " Victoria" History of that county, who apparently considered it unlikely to occur in the south of England. It is now known, however, to be far from uncommon in Kent, and to occur wherever its foodplant, *Bartsia odontites,* grows in any quantity. I have nothing special to say about the C'idarias, or any other subsequent families, full of interest as they are, and contributing, as they do, very many species to our list, until we come to that undesirable pest *('lieimatobia brumata.* It happens, fortunately not very often, that the larva: of this "winter" moth, aided by a less number of those of *Hybernia defoliaria, Oporabia dilntata, Phigalia pilosaria,* etc., do immense damage to cherry and other fruit-trees in this district, whole orchards being occasionally seen stripped of every leaf and flower-bud. Farmers have had, at times, to resort to beating their nut-trees, bushels of *('lieimatobia brumata* and other lame being collected and destroyed. The use of the grease-bands round the fruit-tree trunks, and the gradual development of the spraying processes, have, however, made a great change in the old order of things, and we hear much less now-a-days of wholesale damage by lame. *Cheimatobia boreata,* a much more interesting and aristocratic species than his brother, occurs commonly on birch in the Wateringbury district. And now I must pass over several genera in silence, and bring my uotes to an end with a few brief remarks on that numerous and formidable genus *Eitpithecia*. On the whole, the "Bugs" are fairly represented in our area, and would well repay further investigation, for Mr. Goodwin and I cannot claim to have done them justice in the past, though he is now working at them systematically. They are indeed a terrible puzzle to the collector, and unless he breeds them all he is likely to be in difficulties in properly identifying many of his captures.

Owing to the quantity of Golden-rod growing in this district we take several very interesting Pugs, *Ku/iithecia exjialliilaUi* being about the best, I have taken it at lamp-light. /,'. *snrrenturiatu* is found, /,'. *renosata, hi. lariciata, /-'. absi/ntliiata, hi. sobrinata,* and many others. Both /.'. *linariata* and /''. *fmlrhellata* visit the lamp.

In conclusion, 1 hope I shall be pardoned for bringing in my "lamplight" so often; as l said at the commencement of my notes T have done little collecting by any other means, and I have taken a great number of insects during the last few years. It may lie worth mention that I once caught and released a skylark, which came into my lighted room about 1-30 a.m. on a summer's night, also that I have taken numbers of *Amitm/ms nirciis* and *I'arajumyx stiatiotata,* also a single */ atarlysta Irmnata*. Rather strange visitors these last, though doubtless the existence of the little stream called the Beult. and the river Med way. not far oft", account for their appearance.

With a last word of gratitude to my friend. Mr. Goodwin, for his invaluable assistance, I will bring my notes to an end, and express the hope that, in spite of their brevity and imperfections, they may prove of interest to lepidopterists in this part of the country.

On Blacklown, Near Haslemehk.

By E. W. SWANTON.

The objects depicted in the adjoining plate were found, together with many Hint arrow-heads, scrapers, knives, etc., on Blackdown, near Hasleiuere. All belong to the late neolithic period. No pottery has, as yet, been found.,

Blackdown is a bold promontory of the Lower Greenland (Hythe Beds), about 918ft. above sea-level. It is the highest point in the county of Sussex, the spot in which the 'finds' have been made being a few hundred yards removed from the Surrey border. Jt consists of wild moorland, covered with heather and gorse, inter spersed with Scots pine and birch trees; a great deal of it has been quarried for road metal.

For several years past we have picked up odd pieces of Hint in the neighbourhood of two little ponds on the moor, and old villagers have told us that their grandfathers were accustomed to visit that part of Blackdown in search of Hints for "strike-a-lights." Some of our Hints showed very evident traces of human workmanship, and Mr. Allen Chandler, J.P. (an enthusiastic local collector of flint implements), decided, after obtaining the necessary permission, to have the ground around one of the ponds excavated.

He has secured a very large series of arrow-heads—some exhibiting a high degree of workmanship--scrapers, hammer-stones, pot-boilers, etc.

All were found about a foot below the surface. Amongst the objects shown in our plate (from a photograph by Mr. Roger Hutchinson, M.R.C.S.), the three grindstones are of great interest. Sir John Evans in his classic work on "Ancient Stone Implements" remarks— "The grindstones on wrhich stone celts and axes were polished and sharpened were not like those of the present day, revolving discs, against the periphery of which the object to be ground was held, hut stationary slabs on which the implements to be polished or sharpened were rubbed. Considering the number of polished implements that have been discovered in this country, it appears not a little remarkable that such slabs have not been more frequently noticed, though not improbably they have, from their simple character, for the most part escaped observation, and even if found, there is usually little, unless the circumstances of the discovery are peculiar, to connect them with any particular stage of civilization or period of antiquity."

That on the left in the illustration is of local ferruginous sandstone; the others are quartzite, and with the Hints must have been brought to the summit of Blackdown from a considerable distance. There is no flint occurring in natural beds nearer than Petersfield, eight

Memoranda Respecting Some Late Keltic Pottery Found At Hasi.emkre, Surrey, In November 1908.

By E. W. SWANTON.

The vessels represented in the adjoining)late (pi. v.), were discovered by a gardener whilst planting trees in a field about a quarter of a mile from the Town Hall, on the road to Grayswood. They were not more than 18 inches below the surface, in greensand soil. Fragments of several other vessels, including a large cinerary urn, were found at the same spot, together with some bits of calcined bone. The gardener asserts that the vessels were arranged around the broken urn. 1 did not have an opportunity of verifying this. The pottery was brought to the museum the day after the discovery; the shape of the bottle or jug led me to think the ' find " a Roman one, but upon taking some of the vessels to the British Museum I was told that they belonged to the Late Keltic Period (Early Iron Age), probably about B.c. 200, an opinion confirmed by Sir Henry Howorth who happened to be passing through the gallery at the time of their examination.

The ware was of two qualities, one a coarse clay of blackish colour with large grains of silica in it; the other a sandy clay of a reddish tint. The jug (black ware) shown in the middle of the plate stands almost six inches high, and has a maximum diameter of five inches. It is very imperfect, and was with much difficulty partially restored to allow of its being photographed. Below the bottle is a shallow vessel of the red ware. It much resembles a modern flower-pot saucer of four-inch diameter.

The shallow vessel in the top left hand corner of the illustration is of dark ware; its diameter is seven inches, and it is a little less than two inches in depth. Immediately below it is a cup-shaped vessel of the reddish ware, four inches in diameter and two inches in depth. The pot in the top corner on the right is of an elegant shape, made of dark ware, measures five inches across, and is three inches high (fragments of another vessel of similar size and design were also found).

The vessel below it, and a similar one in the lower left hand corner of the plate, are of great interest. They are made of the red ware, and are four inches across. Three years ago Sir John Hrunner presented to the British Museum some fine examples of Keltic pottery from the Ticino valley, Switzerland; amongst them may be seen some vessels which differ only from the above in having a much deeper rim.

On two fragments there were traces of linear ornamentation, thus:—

Fragments of Late Keltic Pottery with ornamentation resembling that on vessels from the Lake Dwellings, near Ulastonhnry, Somerset.

It is very similar to that on some pottery obtained from the British Lake Dwellings, near Glastonbury, Somerset.

The ground around the site of the discovery was carefully excavated under nry supervision. We found six cinerary urns, placed about three yards apart, just below the surface. The field had been many times ploughed and also used for allotment gardens. The urns being so near the surface were very much damaged, the shallow vessels described above were better preserved because they were a foot or more below the surface.

All the urns contained calcined bone; in two of them I found flint flakes. Several worked flints were found in the soil around, but no metal could be traced. The urns were placed in an upright position. The more nearly perfect specimen is 8 inches high, 3 inches across its base, and 9 inches in maximum diameter.

At one spot about three or four yards from the outside burial, we found many stones with pieces of charcoal and calcined bone amongst them; here in all probability was the funeral pyre.

Perhaps the most interesting discovery was a rude pavement of small flat pieces of ferruginous sandstone. It resembled a huge saucer, was two yards across, and three feet below the surface. It contained a large quantity of burnt wood and sand with fragments of pottery; no bone was found, though very careful search was made. The nearest burial was nine yards away. It may represent the kiln in which the vessels were baked.

The urns and flints were found in that part of the field owned by Mr. Eollason, of Hindhead, and were kindly presented by him to Dr. Jonathan Hutchinson's Educational Museum. The pottery shown in the plate was found in another part of the field, and has been kept by the owners.

The Teaching Of Nature-study.

By WILFRED MAKK WEBB, F.L.S.

At the outset I am assailed by two memories. The first is that the great majority of those whom I have heard speak upon NatureStudy, have said or inferred that there is no need to explain what the words mean as " every one knows all about it." The other is the advice offered to the popular lecturer that he should assume no knowledge whatever upon the part of his audience. While it would be far from complimentary upon the present occasion to adopt the attitude last suggested, I am sure that it will be useful if I indicate what those who have been publicly promoting the movement in its favour, mean by Nature-Study, and what, as a result of the efforts which have been made, the more enthusiastic teachers have recognised it to be.

What Nature-study Is.—Briefly considered, Nature-Study is a natural method of education through personal consideration of the ways and works of Nature. In it curiosity is made to play its proper part; by the spontaneous observation and deduction that ensue, the mental powers are developed: and by the interest aroused, memory is cultivated.

From what I have said and from the title of my paper it must follow that it is mainly concerned with education, but it is possible for us, as we shall see, to look at this question somewhat from a biological point of view.

What Nature-study Is Not.-Let me, before going further, say that while 1 am an ardent advocate of science teaching in schools, the value of which, luckily, has been recognised for some time, I must emphatically point out that Nature-Study is not science teaching, however accurate the observations may be, and however valuable they may lie found as a preparation for scientific work. At one time there was a great outcry upon the part of certain teachers with regard to what they called "the new subject" of Nature-Study. I have already shown that it is not a subject, but a method of teaching, and now 1 am going to say that it is not new. What is novel about it is the name which conies from America, and the fact of its general recognition and adoption.

The History Of Nature-study.-1 shall probably make myself clearer if 1 briefly trace the history of the movement and show what smoothed the way for its progress. The most important step was made when the child mind was studied; when it became evident that young human beings acquire their information in a very different way from that adopted by adults: when it was recognized that through their inquisitiveness, which they possess in common with all young animals, though in a higher degree, they are enabled to collect together a host of isolated pieces of knowledge through their own experience for their mind to classify and co-relate, as time goes on, into a harmonious whole. The work of Pestalozzi, and more particularly of Froebel, did much to prepare the way, and in some of the work of the kindergarten we have the beginnings of what we call Nature-Study. We can trace also in the work of the older naturalists the observational spirit, the desire to know intimately the creatures amongst which they dwelt, and this at a time before the classification and arrangement of dead specimens on the one hand, or detailed examination of their anatomy in the laboratory on the other, had in turn monopolised attention. A few years ago, however, it was recognised that ecology, or the study of creatures in their environment, had equal claims for the biologist with morphology and classification. In support of these contentions I will give three short extracts. The first is from the original tale of "Eyes and no Eyes,"' by Aiken and Barbauld, published towards the end of the 18th century.

"One man walks through the world with his eyes open, another with his eves shut; and upon this difference depends all the superiority of knowledge which one man acquires over another. I have known sailors who have been in all quarters of the world and could tell you nothing but the signs of the tippling houses and the price and quality of the liquor. On the other hand, Franklin could not cross the Channel without making observations useful to mankind. While many a vacant, thoughtless youth is whirled through Europe without gaining a single idea worth crossing the street for, *the observing eye and inquiring mind will find matter for improvement and delight in every ramhle."*

The second was contributed fifty years ago to the *Westminster Review* by Dr. W. B. Hodgson.!

"It is a fallacy to regard memory as a vessel which receives and retains impartially what may happen to be poured into it; it is only what has awakened a child's interest thHt it remembers tenaciously and recollects quickly; and only those impressions awaken a child's interest which are adupted to the stage and conditions of its mind, which excite while they gratify its appetite for knowledge. Now can it he doubted that it is external objects which most attract and fix the attention of children, and which are consequently, most naturally, easily and permanently remembered! *This rust field, which has been partitioned, as it has been, ainony very many xiemes, for which collectively we want an adequate title, and of which we would i ow mention only one, though a very comprehensive division—natural history—affords most ample materials through the longest school course, for developing as well as storing the youthful understanding, and for arousing the young wonder and sense of beauty."*

The third is from an article which 1 wrote for the same review in 1H99, before the word Nature-Study was known in our midst.

"One would have to go a long way before one could find anything better calculated to train boys and gills to use their powers of observation—which is a piece of education not always provided for even now—than a properly conducted In *Evenings at Home,* or *The Juvenile Hudi/et Opened,* bv Aiken and Barbauld (1792 1796). t October, 1868. study of plant and animal structures easily obtainable. Indeed, such an interest can be aroused that the work may be unconsciously continued during walks and rambles, from which the idea of task is far removed. A general knowledge quickly becomes a special one. The gleeful urchin blows away the miniature parachutes from the ripened head of a dandelion to learn the time o' day in accordance with the nursery myth. A question as to the use of the flying down may set him thinking, and a search for other contrivances by means of which plants give their children a start in the world, will show him the hooks of the burdock, the slings of the broom, and will let him see that the juicy fruits he loves so well were intended, in the first place, for the birds he hitherto has looked upon only as thieves."

I said just now that Nature-Study is not science teaching, and I may at this point explain what I mean. It is a great deal more informal, it considers things just as they occur, or as the seasons bring them round. It is scientific in its method, but it does not consider matters in any logical sequence or inter-connection, though one thing may be observed in its successive stages. The idea is based, as we have already indicated, on the way in which the tiny child gains its knowledge in its earliest years. It is not strange perhaps that critics who have not followed the Nature-Study movement, and who are ignorant of the sound common-sense and scientific principles on which it is built, fail to find any method in it, and spend their energies in an attempt to ridicule it.

What actually brought Nature-Study to the fore, and obtained official recognition for what was advocated in the books our fathers read, was the need for an education suitable for dwellers in rural districts, and calculated to give the people a love for the country.

At the time that this happened there were other needs to be satisfied in regard to general education. The time had gone by when the mental digestion by the pupil of second-hand information was assisted by the outward application of the strap or cane to his body. A period had been reached when the scholar was crammed, even as chickens are crammed, with carefully prepared food, cut and dried, of course, first, but softened and made ready for assimilation by the overworked teacher. Small wonder was it, therefore, that children grew up into machines incapable of acquiring any knowledge for themselves, and uninterested in the things around them, while the disastrous results of this system made themselves felt, particularly in the army, and among the officers who had never been taught in a way that cultivated their powers of observation. For these, among other reasons, Nature-Study was seized upon to improve the education of all children, and to lead them to learn for the pleasure of the thing, and through the exercise and satisfying of their own instinctive activities. It is only just to mention the great part which the Agricultural Education Committee took in obtaining official recognition in this country for observational

"Biology as a branch of Education," *The Wentmintter Hevieic,* December 1899, pp. (1(15 and tj(6.

teaching on Nature-Study lines in elementary schools. We must allude also to the influence of The Parents' National Educational Union, and to the various bodies which have encouraged and directed the movement by the holding of exhibitions and conferences. We must not, however, forget that little could have been done if the efforts of individual teachers and schools had not furnished evidence of the value and feasibility of Nature-Study. It is obvious that outdoor observation under various conditions must necessarily take place either in school hours, or in the pupil's own time. We can point to such an elementary school as that of Stanbnry, in "Bronteland," where Mr. Jonas Bradley, years ago, introduced outdoor school, or to one of the old Scotch parochial schoolmasters, such as James Shaw, who used to accompany his pupils on their way to their moorland homes. It was he who was delighted when one of his pupils described a donkey as il a wee horse with a tail like a coo's." Then again there is the rather more defined natural history work that has been going on for a hundred years in the schools of the Society of Friends:—at Bootham, for instance, where the first number of the school Natural History Magazine appeared in 1838.

Wherk And How Nature-study Should Be Taught.—In discussing the question of where and how Nature-Study should be taught, it will be of interest to approach it from a biological point of view. We are gradually coming to to live under more and more artificial conditions, in which it is not easy to satisfy several mental instincts and to acquire thereby keen powers of observation, which are of no less importance to us than they were to our primitive ancestors. The principle is well established that the individual creature in its bodily development passes through stages which often furnish a brief epitome of tbe changes that have taken place during the history of its race. The frog when a tadpole, favours its fish-like progenitors. The egg of the hydra is an amtvhoid cell, and young lions are covered with spots.

It seems (mite possible to apply this theory to thc mental development of man. and, when determining the best methods of training in early years, to see what we can learn from looking at the past history of man and from utilising his hereditary tendencies. There was a time when man was almost "in a state of nature," and practically depended for his existence upon the knowledge which be obtained through his own observations, that is to say. be had to rely on his personal acquaintance with the habits of animals, with the properties of plants, and upon his power of making weather forecasts. It follows, therefore, that the ideal place for NatureStudy teaching is in the open air, on the wild moorland, on the mountains, or in the woods, which of all spots, in our country are the least changed, and where we meet least the things that are artificial. It is, of course, impossible in many cases, at present at least, to attain to the ideal, but of this more later. One of the first advances which man made on the road to civilization was the domestication of animals. How he came to do this we do not know; possibly the dog to which he seems first to have turned attention, was distinctly chosen to help him in the chase, and we might digress here a moment to say that Nature Study satisfies the hunting instinct which we still possess, even though we may use the field glass and camera instead of the spear or more modern fowlingpiece.

It is possible, however, that we may owe many of our domestic animals to the interest or, who shall say otherwise, to the pity which the prehistoric hunter felt for the young ones whose mother he had slain. No doubt to his helpmate and children we owe something, for the care of the young creatures would probably full to their share. In our Nature-Study, therefore, it would be no bad beginning to interest our children in pets, before they are old enough and strong enough to roam in the country. It is surely one of the very best ways to teach children that animals have feelings and that these should be respected. Is it not often because an old lady has a pet dog or cat which she loves like a child and which she could not bear to see hurt, that she takes an unreasoning hatred to vivisection, even under the present restricted conditions? After rearing herds, man went on to grow crops and to acquire property, and similarly the modern child should have his little garden and his own plants and shrubs. Ky having things of their own, people learn through their own feelings to respect the property of others. While at first the observations made, must necessarily be quite informal, later on, records of gradually increasing accuracy may be kept. Our prehistoric ancestors were no mean draughtsmen long before we learnt the art of writing, and our children can represent what they see with brush or pencil at an age when they do not know a letter. Many were the stories told in cave or lake dwelling, and some teachers get their young pupils to describe what they have seen and write it down for them, but as time goes on the young naturestudents will keep their own notebooks and calendars, will make their own rough maps of their neighbourhood on which to record their discoveries in general or particular directions.

The school garden should primarily be to provide material and a place for Nature-Study near at hand. It is not a bad thing to pay special attention to a certain field or copse, or even to an individual tree. It should hardly be necessary to emphasise the great value of continued observations, except in the case of very young children, and to point out that a drawing should not merely be a picture from an artistic point of view, but that it should show some stage or habit in a creature's life, and in like manner a photograph should do the same.

Last of all, when some branch of Natural History has been chosen to be considered scientifically as a result of Nature-Study, the student may turn to books, for in the progress of man nothing has had greater weight than the accumulated experience which his predecessors have written down.

It now behoves us to dwell for a moment upon the cases where the ideal of pursuing Nature-Study out-of-doors cannot be attained. There are here and there schools where three parts of the work is carried on in the open air, but everything cannot be done at once, and where it is not possible during school hours for scholars to see living things in their natural surroundings, some of the creatures may be brought to the class. It is not difficult to watch the growth of plants, to discover their peculiarities, and the uses of their parts, within the four walls of the school-room. Many living animals may also form the basis for observations, and their life-histories can be followed without difficulty. On the one hand such work may encourage its voluntary continuation out of school hours, and it forms a useful supplement on the other, for it is impossible to complete some observations out-of-doors. The creatures must be near at hand if they are to be watched in any detail, and it might be impossible to discover them again if they were left to roam at large, or their enemies might make short work of them. At this stage we might point out that there are some interested in Nature-Study who would permit no animals to be kept in captivity on the score that it teaches children to be cruel. At a Conference" held last Tuesday, Sir George Kekewich, in the course of a very suggestive address, took this view. The majority of speakers who followed him, and who are engaged in Nature-Study teaching, were, however, entirely at variance with him. They said that no cruelty at all need ensue, and that to leave out the study of such animals as could be watched in captivity, would be to stop the greater part of the Nature-Study work in many town schools. My own opinion expressed on the matter is that children have so many opportunities of seeing cruelty to animals, that to omit to keep them in captivity, and to show that they are worthy of the kindliest consideration, is to wilfully throw away a great power for good. It may be apt at this point to discuss an objection which has been made to another means of Nature-Study, and by some unthinking persons, in consequence, to all Nature-Study. The matter at issue is the making of collections. Now the chief harm that can be done by making of collections is, on the scientific side. by the rooting up or extermination of rare species, and on the (esthetic side by the lessening of the numbers of pretty Mowers and insects or singing birds. Now the object of Nature-Study is the making of observations in many directions, and although some few specimens may be kept as records, and got together to illustrate June 7th. 1!H)4. At the Horticultural Exhibition in the Royal Botanic Gardens.
special habits or life-histories, or some economic points, the making" of systematic collections by pupils is not one of its aims and does not fall under its scope. When Nature-Study, however, yields to the scientific consideration of some special branch of Natural History, then collections may Income necessary, but with this I am not here concerned, though it is hoped that Nature-Study will then have given the student such an appreciation for nature, and such a desire that others may enjoy what he has come to know, that the spirit of the mere collector may have evaporated. On all sides I meet with cases where the cause of animal and plant protection is being happily advanced by the nature-teacher and the nature-student.

I have a word to say with regard to pictures and dead specimens. In Nature-Study, real living things should be put before the student in preference to anything else, but there are cases where the work may be helped on by illustrations and museum collections. For instance, we may like to contrast the appearance of a tree in summer and in winter. After we have followed the life-history of the silkworm, it may be valuable to be able to compare the various stages by means of preserved specimens. The museum, whether concerned with the immediate neighbourhood of the school and belonging to it, or whether it be the more extensive collections of a county local institution, may help in showing what pupil and teacher may look for, or in telling them what they have found. A good sign is shown by the fact that museums are beginning to include live creatures as well as dead specimens in the sections intended for nature-students.

At present many schools have to rely entirely on the voluntary work of teachers and pupils out of school hours, in the directions which we have indicated.

Preparation for examinations, overcrowded time tables, the failure to recognise that Nature-Study should form part of general education--and should precede formal scientific teaching, in the case of those who are to be given such a training have contributed to prevent progress. The large number of children in the classes of elementary schools in many cases, and the expense incurred and time taken in going from town schools into the country, are difficulties still to be solved in the pursuit of ideal Nature-Study. The schools which have most opportunity at present are private and village schools, where the whole teaching can be devised from the Nature-Study standpoint, where the essay is an account of a ramble, where the geography lessons deal first with the neighbourhood of the school, where the germinating seedling is a model as it grows, for one drawing lesson after another. It is surprising, however, that town schools, and those often in very difficult circumstances, produce some of the best results. Here, we find masters who encourage their boys to save their pence and who take them from South East London for a fortnight among the mountains in the Easter holidays; there, we see a school in the Borough, where pets are a feature of the life of the infants, whose happiest hours are those spent in school. While it is obvious that preparatory schools can do a great deal, it is of interest to see that some of our public schools are introducing Nature-Study, and that in others, Natural History is not overlooked. As the conditions at a public school are somewhat special, and out-door school, except possibly in connection with a school garden, is at present not possible, it may be of interest to outline what has been clone at Eton. My friend, Mr. Matthew Davenport Hill, who took the initiative in the matter, and I, have prepared a course of lessons to be carried on in school and which we call " observational lessons," which, as they tell the pupil very little or nothing and leave him to carry on his own investigations under the guidance of the teacher, might, perhaps, claim to be called Nature-Study, and in order to encourage the out-door consideration of country lore in the small time that can be snatched from games or found in the holidays, we have interpolated a number of chapters, suggesting rather than informing as consonant with the true spirit of Nature-Study, and pointing also to any number of fascinating branches of Natural History which, as time goes on, may be taken up as a hobby." One kindly critic has been so good as to say that we "have accomplished the very difficult task of producing a valuable text-book for a study, one of the leading characteristics of which is that it is independent of books."

The Results Of Natuke-study.—In turning to the results of Nature-Study, it should be pointed out that it is quite as needful for us at the present day, as it was for primieval man, to see what we look at, to appreciate what we hear, and to be able to judge how to act for ourselves in every and any emergency. Nature-Study properly conducted will enable us to do this. It has one or two special points about it which we may particularise. It interests practically all pupils, if they are introduced to it before their curiosity is dulled, and their desire to find out things for themselves has been crushed out of them by the old didactic methods. It is remarkable, too, how certain children who are not successful in ordinary school subjects may outshine their fellows in work of this kind. NatureStudy should incidentally give a love for beautiful things. Mere sentimentality should, however, be rigorously excluded. If NatureStudy provides a definite hobby or even if it does not, much enjoyment will be added to life. In the hands of good teachers it should develop culture, refinement, and humbleness of judgment. It is the ideal stepping-stone to science, and when pursued out-of-doors *'Eton Nature-Study and Observational Lexsons,* by Matthew Devenport Hill and Wilfred Mark Wcbb, with a foreword by The Rev. Edmond Warre.— Part I., 1903; Part II., 1904.

«().-.,. ,,i ,/. i'»" cannot fail to improve tbe health. It is just as easy to deal with plants and annuals of economic importance, and with the weather, as with other matters, and many contend that it will stay the rural exodus. Opinions differ as to whether Nature-Study will have any such effect. It may he claimed, however, that it will give children a love for the country which ordinary education does not, before it is1 too late, but it cannot be expected to do everything by itself. An opportunity should be given to boys at least, to learn something of the economic side of Nature-Study before their school career is over. Earl Carrington, on Tuesday", was able to give figures showing how, by the provision of allotments and small holdings in a district of Lincolnshire, the rural exodus has been greatly lessened. The number who left the country for the towns between 1K82 and 1892 was 2500, while in the succeeding ten years less than fifty took the same course. If the provision of school-gardens and Nature-Study, such as Mr. Jesse Collings seeks to make compulsory by his Hill for "promoting Agricultural Education and Nature-Study in Public Elementary Schools," prepares the hoys, and the holdings are forthcoming, much may yet he done to check a great social evil. Before passing from the results of Nature-Study there is a point which I should like to touch upon.

"'We all look to Nature-Study to teach us to see things, but there is,'I think, a limitation.' D6 Ave in aftei life want to he bothered with every detail that is around us? Do we not need in the end so to direct our mental powers that, although the subconscious part of our intelligence may take count of much, it shall only force upon our notice what is of Value to us?''

'Towards "this I1 fhjrik the purlsuir,"as time goes on, of some particular branch'of NathrpiStudy should greatly help.

1!ltow 'the'. Movemekt' Mst Be Pvshei Forward. — I have reeetUlyf had occasion to briefly Summarize the methods that may be adopted to push forward the Nature-Study movement. The majority of theni deal tyitbJ the existing teacher, though, of course, it'is bftantatnotiiit importance that the'futnre exponents of NatureStu4y in1 but'schools should'1 be properly trained, and in a way that makes them enthusiasts. The methods alluded to are as follows:— 1 (i)' *Hy hMiiiij' t;yltibitjons nn ititirt and more restricted lines in tilit'chrtfie s/icfidl, iHcftnint/ atyi'iittMiti'nilx 'of nature-study are made clear'-*This'Mil no 'doubt be done, at not too frequent intervals, by the!Natin-e-Stii(l'y'Society, arid' it' wbuld.be advisable afterwards to showed' r'eWre'sentttfVc selection of exhibits at a number of local centres.

'(il) *J'j tlrrainfinii con t'e fences at ichich teachers, trim hare recot/nised* 1 Ul".i: -",.. i.'/', i ',!

At the Conference referred to. "'M/filrt *Shirty Y Itls Pmrret rtni httvrprt-tiitton,* "Record of Technical (tmt .Sront'(ii"J/Kiri»crfo»i."a-A. piiUTunp,'11904:, fftigfe'22&.

irlmt Nature-Study is, should describe their work. It has been customary to organise conferences in connection with exhibitions, and this is likely to be done in the future. The School NatureStudy Union has already carried out some excellent work in this direction. County Education Committees may well follow the example set in the past by some of the more thoughtful of the Technical Instruction Committees. (iii) *By yiriny direct help to individual schools anil teachers.* The programme of the Union, to which allusion has just been made, shows, among its objects, that of giving useful advice and of offering suggestions and specimens to schools and to teachers who have a difficulty in introducing Nature-Study. The periodic exhibitions will give the Union an opportunity of seeing the results of its work. (iv) *By providing literature of special use in Nature-Study, both to teachers and to pupils.* For some years the Agricultural Education Committee has successfully circulated nature-knowledge leaflets, and it seems that the time has now come to develop this branch of the work.

What The Field Clubs May Do.—I remember very well, and with considerable gratitude, the help that the members of the Homesdale Natural History Club gave to me when I was a schoolboy. I was by no means an isolated case, and 1 have no doubt that, by encouraging young people who get no opportunities for NatureStudy in the ordinary way, much good may be done in the future. Perhaps the clubs might set themselves to give help and advice to teachers by inviting them to their rambles, and by directing their attention to the various branches from which to choose a special line of work, for, be it said, many pupils owe much to some particular Natural History hobby taken up by their masters, and, again, it is quite possible, as has been abundantly shown,:: for a teacher to begin wrork with his or her pupils knowing less at the beginning about the matters in question than those whom he or she is called upon to teach. Certain individual naturalists and others, have sought to help the teachers by circulating specimens, but unless these are sent round with the special object of suggesting actual work on the part of the pupil, it seems likely that they might encourage the old information lesson which ought now to be extinct.

Summing up the matter, instead of being a passive recipient of dry facts, which the teacher has learned in similar fashion, the child, through Nature-Study, becomes an active seeker after knowledge with the teacher as his guide. His note-books are no longer rilled with second-hand information, but his diaty, calendar,

Miss Edith Aklerton's paper in the *Record of Technical and Secondary Education.*—April to June, 11)04, pp. 200-211.

and map, are crowded with records of work which he has himself done, of discoveries which he has himself made. In places " Exempt from public haunt" he, like the exiled Duke in *As You Like It*—"Finds tongues in trees, books in the running brooks,

Sermons in stones and good in everything,"

and to complete the quotation—

"I would not change it,"

Would you?

The Abbey And St. Leonard's Tower At West Malling.

By the Rev. G. M. LIVETT, B.A., F.S.A.

The Abbey Of St. Mary The Virgin At West Malling.— The Abbey (pi. ii) was founded for nuns of the Benedictine Order in or about the year 1090 by Gundulf, second Norman bishop of Rochester, who appointed the first Abbess thereto shortly before his death in 1106. The priestly functions of the establishment were performed by a succession of priests, who were known as "the prebendary of the high mass." There still exist in the High Street, behind a shop now in occupation of Mr. Harrington, grocer, the remains of a middle-Norman building which may have been the prebendal house. From the time of the surrender in the 10th century, when the Abbess Margaret Vernon and eleven nuns were dispossessed and pensioned, the Abbey-buildings remained for three and a half centuries in the hands of lay-owners. In the closing years of the nineteenth century they were, by private munificence, again devoted to their original use as the home of an Anglican Sisterhood. The sisters endeavour, as far as possible, to keep the Benedictine Rule. They occupy their time in doing Church embroidery work, in educating a few poor children who are received into the establishment, in attending to the wants of a few boarders, and in endeavouring to maintain a ceaseless *rota* of intercessory prayer in the beautiful little chapel in the south transept of the old church.

All that now remains of the old church is the fine west front, the south wall of an aisleless nave, with signs of the processional doors that communicated with the cloisters, and the south transept with its great arch of communication now blocked with masonry.

The west front shows work of three distinct periods of architecture: early-Norman, middle or later-Norman, and 15th-century. The. contrast between the early-Norman work along the lower part of the front and the later-Norman above it is very instructive. The early Norman work is very rude in character, with plain wall-arches turned in calcareous tufa or *travertine,* a local material which characterises early-Norman work in the neighbourhood generally. The later-Norman work is executed in Caenstone and shows all the fine shallow axe-finished carving that is characteristic of the period. The earlier work runs up to the first string-course, and here and there a little above it. The irregularity in the junction of the two works suggests the theory that the building of the front was left incomplete for some years.

The original design of the front is somewhat obscured by alterations. The early doorway has disappeared, and in its place one sees a late and poor segmental-headed opening, and above it a two-light window apparently of late 14th-century date. The flanking pinnacles now appear to rise from the first string, adorned with wallarcading in their lowest stage. It is probable that originally thisarcadingran right across the front along this stage,perhaps slightly recessed along the central portion now destroyed, and there pierced in two or three places for windows. Therefore, in the original design the base of the pinnacles must have appeared as forming part of the next stage, viz., the third stage of the front. In this stage the Norman arcading remains, boldly recessed within the bases of the pinnacles, which assume an octagonal form by means of a bold chamfer on the angles. A central window in this stage has been blocked and its semicircular head removed. Perhaps tbe head of this window ran up into the gable that originally rose up between the Hanking pinnacles, But the gable has given place to the base of an octagonal tower of 14th-or loth-century date. The upper part of the tower has disappeared and left a ruin that is very massive, plain and ugly; but if, as seems likely, it was surmounted by a lofty spire, the resulting design must have been grand in the extreme. The Norman pinnacles are lofty well-proportioned examples, octagonal in form, adorned with arcading with banded shafts. The stone caps of the pinnacles have lost their top.

There is a marked parallelism between the history of this front and that of Rochester cathedral. Both were founded by Gundulf. The remains of the early-Norman front of Rochester exist only underground. The later-Norman front was completed probably in the fourth decade of the 12th century, and, judging from the similarity in workmanship and design, one would imagine the Rochester masons must have come straightway to Mailing to rebuild or to complete the front of the Abbey-church.

If the great perpendicular window that now fills the centre of the Rochester front were removed and the original design restored according to the signs revealed during the recent repairs—not that such vandalism is for one moment to be thought of—its similarity to the design of the Mailing front would be apparent. The chief difference lies in the fact that the Abbey with an aisleless nave required only a single pair of flanking pinnacles, whereas the Cathedral, having aisles, was given two pairs of pinnacles, one pair to form abutment for the nave-arcades, and a second pair at the end of the aisle-walls.

Behind the front there are buildings attached that must have been erected at the time of the surrender or somewhat later, when the nave and eastern parts of the church were destroyed or allowed to fall into ruin. In their walls there are windows that must have come from the destroyed parts of the church.

The cloister-garth of the Abbey remains. Its western range of buildings, no doubt the cellarer's buildings, have disappeared. The north boundary-wall is the original south wall of the nave, and ..

Livett—Tin' Abbey anil St. Leonard s Tower at II est Mailing. 71 ;-r-i! -i,;- ..I: V.' !.'!!' aflords to the student an excellent example of the early-Norman materials and method of building. The east side comprises the original south transept, now the sisters' chapel, and a building (with an excellent timber roof) that stands on.the; site, of the chapter house and is now used as a dining-ball. Remains of the dorter to the south of this building would probably be revealed by an expert survey.

On the south side of the garth there are the remains of an Early English cloister-wall of unparalleled interest and much beauty. The arcade of trefoiled arches (now glazed) has good foliated capitals and a series of flower-ornament of unique design running round the head of the openings on the inside. At intervals there are external buttresses of fifteenth-century date, cleverly bonded into the older work, the old caps adjoining the buttresses having been replaced by new caps which were worked on to the stones of the buttresses. The range of buildings on this side of the garth, on the site of the frater and kitchen, is for the most part work of postReformation date, In the garth there have been placed two or three sepulchral slabs of early thirteenth-century date.showing crosses of unexceptional design. In the south Wall of the south transept there is an arched recess of some interest. We believe that a full account of the abbey and its buildings from the expert pen of Mr. W. H. St. John Hope is to appear in the forthcoming *Victorian County History,* for the publication of which archaeologists are anxiously waiting.

The So-called St. Leonard's Tower. —J. H. Parker was doubtless right in assigning the tower (pi. iii) to bishop Gundulf, who received a grant of land at Mailing. Perhaps it was scarcely correct to say that it was the first Norman keep erected in England: for towers of nearly identical design and possibly of as early (if not of earlier) a date were built by the same bishop at Rochester and at Dartford. Like the Rochester tower (and probably like that of Dartford also) St. Leonard's Tower had been built in close connection with a church. It was a keep in the sense that many a Norman churchtower must have been a keep. Such towers doubtless served a double or triple purpose: they were bell-towers, church-treasuries, and places of refuge, possibly of military defence, in times of war. Like the Rochester tower, St. Leonard's tower seems to have been detached from the church or chapel to which it belonged. The remains of the chapel, which was a cell to the Abbey, may still be seen in the wall that runs east from the north east angle of the tower. The western part of that wall for some 80 feet seems to be modern. At a distance of BO feet in the wall may be detected the signs of the return of the west wall of the chapel. Further east there is a change of direction in the line of wall that indicates the line of the cross-wall of the chancel-arch of the chapel. Further east still may be seen the remains of one of the chancel-windows. The rest of the chapel has been destroyed, but its dimensions are preserved in Thorpe's *Antiquities*.

At the side of the road, again to the east, there is a spring which forms one of the sources of the Lille Bourne. It is covered with an arch and Hows along a paved course for some feet fiom the arch. These works are of early-Norman date. The arch, originally all of tufa, has recently been rebuilt in a way that betrays ignorance of its archieological value.

The Tower is very massive, and in every way a splendid example of the buildings of the Gundulf school. A careful study of its masonry would enable a student to master the peculiarities of local work of the period. There is scarcely a church for many miles around that does not contain some evidence of contemporaneous building. But perhaps the chief interest of the Tower lies in its detached relation to the chapel. In this respect it seems to have suggested the plan of the parish churches of East and West Mailing. For in both these cases there is (or *was*) a space between the tower and the west end of the nave proper, though in both cases that space seems to have been occupied by an intervening building, which perhaps served in the original design as a baptistery. But this, as well as a detailed description of the Tower, is a matter which merits a fuller treatment than the writer of this paper was able to give it on the occasion of the Society's visit. He hopes to find an opportunity of dealing with it elsewhere.

List Of Localities, Not Recorded In Recent Floras Of Kent And Surrey, For Some Comparatively Rare Plants'".

By W. H. GRIFFIN.

Except where Surrey is stated, all the localities mentioned in the following list are in Kent. The nomenclature is that of *The London Cataloi/iw of British Plants*, and, in the case of aliens not included in the *Catalogue*, Mr. S. T. Dunn's *List of the Alien Flora of Britain*.

Ranunculaceje.—*Ranunculus drouetii*, Dartford marshes; *11. hetcrophyllus*, Southend, Catford; *11. peltatus*, Chaldon, Surrey.

Berberide.e.—*Berberis vulyaris*, wood near Layham's Farm, West Wickham; hedge at foot of Meentield Wood, Shoreham.

Crucifer.e.—*Nasturtium sylrestre*, Orpington Common;-V. *palustre*, Ravensbourne, Beckenham; *Arabis pcrfoliata*, Hayes Common; *Cardamine amara*, Crofton, near Orpington, Eynesford; *('ochlearia anylica*, bank of Dartford Creek; *Hesperis matronalis*, Keston Common, appears now to be well established; *Erysimum cheiranthoides*, Catford, Lower Sydenham, West Wickham, chalkpit off Brighton Road, Purley Oaks, Surrey; *Brassica elonyata*, near the prison, east of Dover Castle; *Hunias orientalis*, Keston Common, Sevenoaks Road, Halstead; *lHplotaxis muralis*, footpath above Eynesford on the East. Ditto var. *babinytonii*, Stoat's Nest, Coulsdon, Surrey.

Violari.e.— *Viola permixta*, Down, very abundant; *V. ericetorum*, Hayes Common.

Polygale.b.— *Polyijala serpyllacea*, meadow opposite "The Fox," Keston; *P. calearea*, Keston, Cudham.

Caryophyllee.—*Saponaria raccaria*, West Wickham, near the Railway Station; *Cerastium semidecandrum*, new road off Bromley Hill; *Armaria serpyllifolia* var. *leptocladog*, Petts' Wood, Pauls Cray.

Geraniace.e.—*Geranium sanyuineum*, chalk hollow in a meadow near Down, probably bird sown, but several plants well established.

Leguminos.e.—*Genista tinctoria*, Cudham, abundant in two localities; *Medicayo denticulata*, gateway of strawberry field, between Chelsfield and Parkgate, Lullingstone; *Trifolium subterraneum*, Joyce Green, near Dartford; *T. medium*, Hayes Common, The Scrubs, Bromley Common; *T. resupinatum*, many plants in the place stated above for *Medicayo denticulata*; *T. Jiliforme*, Hayes Common, Keston Common; *T. striatum* var. *erectum*, near Reigate Found or verified by W. H. Griffin, Hon. Sec. Catford und District Natural History Society, in the years 1901-4.

Hill, Surrey; *Lotus tenuis*, Stone Marshes; *Galeya officinalis* var. *albiflora*, gravel pit, Hayes Common, an escape; *Lotus corniculatus* var. *rillnsiis*, Dungeness; *Hippucrepis comosa*, near "The Salt-box," Westerham Road, Cudham, Poll Hill; *Lathyrus nissolia*, ditto, and the Warbank, Keston.

Rosace.— *Primus cerasus*, Hayes Common; *Habits iilaeus*, Keston Common; *Fratjaria elatior*, Farley, Surrey; *Potentilla norreyiea*, roadside Iwtween Farnborough and Down; *I'. aryentea*, Hayes Common; *Poterium poly/jamum*, Keston, Down; *Rosa ruhiijinosa*, Warbank, Keston; *Pyrits tormitialis*, hedges and copses, Whitefoot Lane, Southend, Catford; *J', communis*, Fairchild, Chelsham, Surrey.

Saxifraoe.e.— *Sa.rifrat/a tridactylites*, Down, Shoreham, Addington, Surrey;.S. *yranulata*, Holwood Park, Keston; ut side of public footpath, Lullingstone. *I hrysospleninm nppiisitifulium*, Crofton, near Orpington; *Hibes i/rnssularia*, hedge, Down Road, Keston; *Sedum telephiitm*, Crofton, wood between Down and Cudham.

llAiMKMiEX. — *Myriiipltyllniii rerticillatum*, Dartford marshes; *Vallitriche staynalis*, Joydens Wood, Bexlev; *('. hamulata*, Crofton; *('. obtnsanyula*, Dartford Marshes.

Onaurarie/e.— *F.pilnbium anyiistifoliitin*, wood near Down; *I.'. bradi year pit m*, old limekiln near Otford Station; *I'.'. atlnatum*, Greenhithe; *(Knotliera biennis*, roadside, Shortlands, an escape.

I'mhkllifer.e.—*Sinyrniiiiii olusatrum*, near Dartford; *Apirnn ijrareoleus*, Swanscombe; *('arum petroseli mini*, Farnborough Common; *C. carui*, bank of Medway, Cuxton; *Silaits tlarescens*. The Scrubs, Bromley Common, Dartford Marshes; *Caucalis nodosa*, Swanscombe.

Cai'rikoi.iack.e. *Ailn.ra museliatelliua*, Down; *Lmiicera xylosteum*, hedgebank, roadside near Keston Church; one old bush; as it hasnot been hitherto recorded elsewhere in Kent, it is a pity that in the last two seasons all the flowers have been gathered, thus preventing the extension of the species by seeds.

Rubiace.k.—*Galium tricorne*, arable

field, West Wickham; (/'. *antjlicum,* garden wall, Lullingstone Park; *Aspernla odnrata,* roadside below Holwood Park, Keston, wood between Down and Cudham; *A. rynancliica,* meadowbanks between Keston and Down. Yai.kkiane.e.— *Valeriana tlioica,* Lullingstone; I. *Mikanii,* Down, Ciidham, near Addington and Fairchild, Surrey. *Valerianella olitmia,* Hayes; I. *dentata,* Keston, Down.

Dipsace.t:.—*Pipsaeus pilnsns,* Vale of Cudham.

Composite.—*Calendula officinalis,* roadsides, Keston and Down, escapes; *Aster saliyuns,* roadside, Crofton, an escape; *A. laeris,* gravel-pit, Ravensbourne, an escape; *Filayn minima,* between Addington and Chelsham, Surrey; *Achillea ptarmica* var. *cartilaijinea,* Hayes Lane, Beckenham, an escape; *Setiecio risensus,* gravel-pit near Hayes Station, Southborough, near Bickley; *Antliemis nobilis,* Kenley Common, Surrey; *Chrysanthemum parthenium* forma *jiorepleno,* roadside, Farningham, an escape; *Matricaria chamomilla,* waste ground near Hither Green Railway Station; *Tanacetum rulyare,* between Hayes and Farnborough, Leves Green; *Petasites rnlyaris,* Southend, Catford, Horton Kirby; *Picris hicracioides* var. *arralis,* Cudham; *Crepis foetida,* Swanscombe; *Hieraciiim murorum* var. *pellucidiim,* Godstone Road, near Oxted, Surrey; *Taraxacum officinale* var. *erythrospermum,* Shoreham; *Lactuca rirosa,* Dartford Marshes; *L. scariola,* Greenhithe; *L.saliijna,* Dartford Marshes; *L, muralis,* Keston Common, Farnborough, Shoreham.

Cami'anulace.e.—*Jasimie montana,* gravel-pit, near Hayes Station; *Phytenma orbicnlare,* Farthing Down, Coulsdon, Surrey, abundant; *Campanula i/lniiierata,* Cudham, rifle range, Woldingham, and chalk escarpment above Westcott, Surrey; (*'. traclielium,* Keston Common, Chelsfield; ('. *rapiinculoides,* roadside, Locks bottom, Farnborough, an escape, plentiful; *Speculaia hyhrida,* arable field, Down Road, Keston, abundant.

Ericaceae.—*(alluna erica* var. *incana,* Hayes Common.

Monotkope.e.—*Hypopitys monotropa,* Fairchild, Chelsham, on Kent side of county boundary.

Primulace.e.—*Samidus ralerandi,* Stone Marshes.

Poi,emoniace/e.— *Pidennmium caeruleum* form *album,* bank of Darenth, between Eynesford and Shoreham, an escape.

Gentiane.:.—*Gentiana amarella,* Down on upper ridges of Vale of Cudham, finer than seen elsewhere in Kent or Surrey.

Boragine.e.—*Cynot/lossitm officinale,* Down, Lullingstone; *Borayo officinalis,* Bromley Hill; *Myosotis sylratica,* Titsey, and beech copse near Tatsfield, Surrey.

Convolvulace.e.—*Cusciita epithymum,* Hayes, Keston, and Farnborough Commons, Shirley Hills, Surrey; generally on (*'alluna Erica,* shifts its position year by year.

Solanace.s.—*Atmpa belladonna,* meadow-bank between Keston and Down, lane opposite The Salt Box, Cudham, near Farley, Surrey; *Hyoscyamus niyer,* Lullingstone Park.

Scrophularine.— *Ycrbascum thapsus* x *V. lychnitis,* Keston; 1'. *nii/rum,* roadside between Addington and Hayes, abundant; *Linaria cymbataria,* on and at foot of garden wall, West Wickham Court; *Antirrhitnum ornntium,* arable fields between West Wickham and Keston; *Veronica anayallis-aquatica* var. *anai/alliformis,* Southend, Catford, Northfleet meadows; *Euphrasia officinalis* (subspecies or varieties as defined by Mr. F. Townsend in *The Journal of Botany*)—*E. rostkoeiana,* meadow opposite The Fox, Keston, also a hybrid between this and *E. breripila;* /,'. *stricta,* near Farthing Down, Coulsdon, Surrey; *K.nemorosa,* very common on top ridges of many valleys in the chalk; *E. kerneri,* Leves Green, Down, hill west of Shoreham; *Bartsia odontites* var. *scrotina,* old chalk-pit Northfleet; *Orohanehe minor,* same place, in abundance and very robust; old limekiln near Otford Station. *Lathraea squamaria,* roadsides, Keston, and Down, plentiful and increasing; I have traced it from Keston, through Down, and across the Vale of Cudham to Knockholt; it is known, locally, as " Butcher-boys."

Labiat.e.—*Thymus chamaeilrys,* Hayes Common; *Calamintha officinalis,* bank of road from Shoreham Station to the village; *Salvia rerticillata,* West Wickham, roadside bank of arable field between Wellington and Stoat's Nest, Surrey; *T. sylrestris,* bank near Knockholt Station; *Nepeta cataria,* Westerham Road, near LevesGreen, Chelsfield, Sevenoaks Road between Greenstreet Green and Knockholt Station; *Scutellaria yalericulata,* South Darenth, Otford; *Marrubium valyare,* Mitcham Common, Surrey; *Teucritim botrys,* between Addington and Farley, Surrey, abundant and now well established; *Ajuya chamaepitys,* in three spots on land relapsed from cultivation near Fairchild, Chelsham, on both sides of the county border; on such land this rare species appears suddenly, but seldom recurs in the same spot for more than one or two seasons.

Chenopodiaceje.—(*'henopodium polyspermia*)!, West Wickham; also var. *cymosnm,* Ravensbourue, Crofton; *V. ficifoliitm,* roadside, Bromley Common; ('. *bonuslwnrieus,* roadside between Farnborough and Down, Lullingstone, Otford; *Atriple.r littoralis* form. *serrata,* close to the river, Greenhithe.

THY.MEr-.SACE.E. — *Daphne lauivola,* coppices, Whitefoot Lane, Catford, in hedge of arable field, Keston, Meenfield Wood, Shoreham.

Euphorbiace.t;.—*Euphorbia esula* var. *pseudo-cyparissias,* roadside, Woldingham, Surrey.

I'kticace.t:.—*I'arietaria officinalis* var. *falla.r;* Farningham.

Orchide.e.—*Xeottia niilus-aris,* Down, Fairchild, near Chelsham, Surrey; *Spiranthes autumnalix,* meadow near The Fox, Keston, two roads, with houses, are being made through this meadow, which will result in destroying the only station for the species known to us in West Kent; *'ephalanthera pallens,* Keston, Farnborough, Shoreham; *Epipactis latifolia,* Chaldon, Surrey; *E. media,* Fairchild, and near Reigate Hill, Surrey; *E. ciolacea,* near Down; *Orchis pyramiilalis,* Down, Cudham; . *morio,*

Down, Cudham, Crofton; (. *viascula,* same localities; . *incamata,* Lullingstone; . *latifolia* Cudham, Crofton; *L macnlata,* same localities; *Aceras anthropojiloo-a,* Polhill; *Ophrys apifera,* Down, Cudham, but becoming more rare every year; . *mitscifera,* the like; *Herminium monorchis,* has been almost, if not quite, exterminated on Darwin's famous "*Orchis* bank" at Down; *Habenaria eonopsea,* Down, between Romney Street and Otford; *H. chloroleuca,* Cudham, Fairchild, Surrey.

Lii.iaceje.—*lluscus aculeatus,* Hayes Common, roadsides, Keston, and Leves Green, the plant does not produce fruit in these localities; *Museari raeemosiim,* lane near Chelsfield, an escape; *Ornithoyalum umhellatum,* one plant on chalk-bank near The Salt Box, Cudham; *Paris quailrifolia,* Crofton, abundant.

Juncace.—*Luzula* Foisted, West Wickham Common, Crofton.

Typhack-b.—*Typlia latifolia* var. *media,* Southend, Catford.

Lemnaceje.—*Lemna polyrhiza,* Farnborough Common.

Naiadace.t.—*Potamoyeton polyyonifolius* var. *erieetorum,* Keston Common.

Cyperace.b.—*Seirpus sylratiens,* Lullingstone.

Gkamine;.—*Glyceria plicata x&r. petlicellata,* Southend, Catford; *Ayropyron jmnyens* var. *littorale,* Greenhithe; *Hordmm sylraticum,* coppice on chalk escarpment above Westcott, Surrey.

Filices.—*Asplemum Kuta-miiraria,* Hint wall at Down, bricktomb Cudham Church-yard; *Oplnoylossum milyatum,* close to public footpath, HolwoodPark, Keston, meadows about Down and Crofton.

Eqi;isetace.k.—*Equisetnm maximum,* in a farm lane, half-mile south of Chislehurst Common.

A Few Notes On The Corporation Museum, Maidstone:
Formerly Chillington Manor House.

By J, H. ALLCHIN, Chikf Cituatdr Ani Librarian.

The large amount of interest in the Museum (pi. vi) and its contents, expressed by many of the visitors at the recent Congress in Maidstone of the South-Eastern Union of Scientific Societies, has prompted the writer to put together the following notes concerning the past history of the building, as a tribute to all those members of the Union who favoured Maidstone with their presence; and as a memento of a visit which was a source of pleasure to those whose privilege it was to act the part of hosts and hostesses during the three days, June 9th, 10th, and 11th, 1904.

The first recorded mention of Chillington (or Chillingdon, as it is spelt in some of the deeds) is, that in the fourteenth century the manor of that name was in the possession of the great Cobham family, who were barons in the time of Edward I, and in A.D. 1343, Sir John, Lord Cobham, Justice Itinerant, received from Edward III a grant of free warren of all his lands in Kent, including the Manor of Chillington. Sir John was connected, by marriage, with William, Archbishop Courtenay, a member of the old Devonshire family of that name, and who was so closely associated with the building of the Collegiate Church of All Saints', Maidstone, towards the end of the fourteenth century, when he obtained a licence from King Richard II to convert the then existing Parish Church of St. Mary into a Collegiate Church.

The Manor of Chillington was subsequently held in possession by the College of All Saints', and afterwards by the Maplesden family, of Digons, in Maidstone, now called the Priory, and at the present time the residence of the Vicars of Maidstone.

The Maplesdens held Chillington until the reign of Queen Mary, when George Maplesden forfeited it to the Crown, as one of the penalties for his participation in the ill-fated Wyatt Rebellion (a.d. 1551). In A.D. 1501 (4th of Elizabeth), the estate passed into the possession of Nicholas liarham, Sergeant-nt-law to Queen Elizabeth, and Recorder and Member of Parliament for Maidstone. To him is attributed the building, or re-building, of the central portion of Chillington House, now known as the Museum; the oldest portions of which, i.e., the Cloister and the Long Gallery (pi. vii), which project from the back of the building, are apparently the remnants of a still older structure, and are supposed to be of about the time of Henry VII or Henry VIII. Barham died in 1577, and was succeeded in the estate by his son Arthur, who, early in the seventeenth century (circa A.D. 1009), sold the property to Haule, or Hall, a

Museum, Victoria Library, And Lientlif Art Gallery At The Present Time. member of an ancient family settled at Wye, in this county; from him it descended to his grandson, George Hall, who, dying in A.D. 1650 without issue, left the estate to his sister and heiress, Elizabeth, wife of Sir Thomas Taylor, Bart., whose son Thomas, the second baronet, married the heiress of Sir Thomas Colepepper, the last of the Colepeppers of Preston Hall, Aylesford. Sir Thomas and Lady Taylor sold the estate to Sir John Beale, Bart., who died in 1684, and from him the property passed through the hands of various owners, until in 1801 it was purchased by Mr. William Charles, a manufacturer of felt, who adapted some portions of the old building for felting and blanketing.

He died in 1832, and left the property to his sons, Thomas and William; the latter died in 1840, and Thomas, who was a general medical practitioner, but retired from practice, and a bachelor, continued to reside in the house until his death, in 1855. During his lifetime he amused himself with the study of archaeology, and in tlie course of numerous excursions in the county he made several interesting pencil drawings of many old buildings and other objects of antiquarian interest: he also formed a collection of various objects of antiquity, mostly Romano-British pottery, found in the town and surrounding neighbourhood, and, at bis death, he bequeathed the whole of his collection, including several oil-colour paintings, to the town, and thereby gave the nucleus of what is now one of the largest of the provincial museums of England.

In the same year as Mr. Charles'

tleath the Corporation adopted the "Public Libraries Act, 1855," then known as the "Ewart Act," and in 1857 they secured, by purchase, Chillington Manor House (pi. viii) with the adjoining garden, now almost entirely covered by the Municipal Schools of Science and Art. In the following year (a.d. 1858) the old building was opened as a Public Museum, with Mr. Edward Pretty, F.S.A., as the first Curator. Mr. Pretty, who was a native of Hollingbourne, a small village between Maidstone and Ashford, was an old and intimate friend of Mr. Charles, and had held the position of drawing master at Rugby School for many years; he also was of an antiquarian turn, an accomplished draughtsman, and an excellent miniature painter. In the Reference Library—in the Museum—there is a collection of very skilful pencil drawings of a large number of the old buildings of Maidstone and district, and of several of the Kentish Churches; nearly all the former have altogether disappeared, and many of the latter have been "restored" beyond recognition, so that the drawings themselves possess a great value from an archaeologist's point of view, quite apart from the artistic merit which they possess; for they show that Mr. Pretty was endowed with a keen artistic talent, and that he possessed a rare feeling for detail and its truthful delineation. He died in 1865, aged 73, and left his library of books, chiefly on art and archeology, and his large collection of prints, and pencil and water colour drawings by himself and other artists, to the Museum.

Mr. Pretty was succeeded by Mr. W. J. Lightfoot, from the British Museum, who held the appointment until his death in 1874. During his Curatorship many important alterations to the old building, and extensions in the shape of new wings were effected. In 1868-9 the original east wing of the building, which was in the occupation of a separate owner, and was used as a coal and straw store, was purchased, and a new building consisting of two large rooms was erected, mainly through the liberality of Mr. Alexander Randall, a banker and native of Maidstone. In 1870-8 the west wing, which also was in separate occupation, was purchased through the generosity of Mr. Alexander Randall, Messrs. Samuel and Richard Mercer, and Mr. Julius L. Brenchley; and by means of a public-subscription, and a loan on liberal terms from the late Mr. Samuel Mercer, an entirely new wing was erected. In February, 1878, Mr. Julius L. Brenchley, one of the most liberal benefactors to the Museum, and the town, died at Folkestone, and bequeathed, with very few exceptions, the whole of his valuable collections of ethnographical objects, oriental pottery and porcelain, pictures, and many other works of art, also an extensive library, to the Museum, with an endowment fund for the maintenance of the collection. Shortly before his death Mr. Brenchley had purchased the old St. Faith's Green, together with some adjoining land and the houses that stood on it, and converted the whole into a public garden immediately adjacent to the Museum on the north and east sides; he bore the entire expense of laying out and planting the garden, and enclosing it with a wall and railings, and then presented the whole to the town as a free gift, now known as the Brenchley Gardens.

In 1878-4 the Chapel was built on to the east side of the Long Gallery and Cloister, and in the latter year the south wing of the Court Lodge, East Farleigh, an interesting half-timbered building of the time of Henry VIII, was carefully taken down, carted into Maidstone, and re-erected as an annexe to the Cloister; the cost of removal and rebuilding was defrayed by a member of the Tyssen family, to whom the building originally belonged.

In the following year (1875) the south front of Chillington House was restored, and the present iron railings, gates, and paving in the fore-court were provided by the late Mr. W. Laurence, J.P., who discharged the entire expense of the same as a parting gift on his retirement from public life. About the same period the present main staircase, designed according to the style of the Elizabethan period, was supplied as a gift from the brothers Messrs. T. and J. Hollingworth, and Mr. Richard Mercer; and the two former benefactors also bore the expense of rebuilding the large chimney stacks of the " Elizabethan " and " Anne " periods, in 1885.

In 1890 the annexe known as the "Bentlif" wing was erected by the late Mr. Samuel Bentlif, J.P., in memory of his brother, Mr. G. A. Bentlif, an old Maidstonian.

The building consists of four spacious rooms, one of which on the ground floor, is occupied by the Public Reference Library, and the adjoining room is devoted to exhibitions of various works of art. The two upper rooms are reserved as a Picture Gallery, and contain the Bentlif collection of oil and water colour printings, bequeathed by Mr. Samuel Bentlif at his death, in 1897.

He also left a liberal endowment for the maintenance of the Bentlif wing, the government of which is in the hands of a body of Trustees separate from the local authority.

In 1897-99 the latest addition to the Museum was effected by the erection of the Victoria Library and County Room, for "which the necessary funds were provided by public subscription, in commemoration of the Diamond Jubilee of her late Majesty, Queen Victoria. This building udjoins the beforementioned west wing of the Museum, and consists of two large rooms; the one on the ground floor is set apart for the Public Lending Library, and the upper one, which is approached from the Bird Room, is reserved exclusively for collections of the Kentish Fauna and Flora.

In concluding the foregoing very abridged account of the history of this interesting building, the writer ventures to express the hope that, amongst those members of the Union who attended the Congress, there may be some who were inspired with sufficient interest in the Museum and its contents to be moved to render assistance towards filling up some of the gaps in the collections in the County Room. Specimens are still desired for the collections of mammals, fish, insects, Crustacea, shells, and plants; and the donation of any of those desiderata will always be welcomed and gratefully acknowledged by the Curator.

In the course of his address, the President of the Union dwelt upon the importance of Museums as educational institutions, and it is the one aim of the Staff of the Maidstone Museum to place it in the front rank of the educational factors of the county of Kent, so that none who come within its walls shall go away without gaining some item of knowledge.

Summary Of The Principal Collections Contained In The Museum. 1.—Brenchley Room (West Wing):—English, Chinese, and Japanese Pottery, Japanese Bronzes and Enamels, Chinese and Japanese carvings of Ivory, Crystal, and Jade, and many other examples of the Far East. Oil Colour Paintings by Canaletto, Pannini, N. Poussin, Northeote, Opie, G. Morland, T. S. Cooper, Albert Goodwin, and other Artists of the English and Foreign Schools.

2.—Ethnographical Boom:—The Brenchley collection of objects illustrating the Ethnography of New Caledonia, New Hebrides, Solomon Islands, Friendly Islands, the Fiji and Sandwich Islands, Australia, and New Zealand. 3-4.—Entrance Hall and Great Hall (Portion of Nicholas Barham's structure):—Armour, and old Furniture of the seventeenth and eighteenth centuries. 5.—Cloister (Henry VIII period):—Geological collection.—Holocene and Pleistocene, Kent.—Pliocene, Suffolk and Essex.—Miocene, France.—Oligocene, Isle of Wight.—Eocene, S.E. England and N.W. France.—Cretaceous and Neocomian, Kent and Sussex.—Jurassic, Triassic, Carboniferous, and Silurian. 6.—Long Gallery (Henry VIII period):—A large General collection of Minerals. 7.—Drawing Room (Portion of Nicholas Barham's structure):—Antiquities of the Bronze, Romano-British, and Anglo-Saxon Ages, the majority of thein having been found in Kent. Egyptian Pottery, including several specimens from the Beni Hassan excavations; and other antiquities from Egypt, Greece, and Italy.

Collection of Portraits of the Hausted or Hasted family, predecessors of Edward Hasted, the Historian of Kent (1782-1812).

8.—Bird Room (Upper Floor of West Wing):—Brenchley collection of birds from Australia, New Zealand, South Pacific, and North and South America, Kentish collection of birds, nests, and eggs. 9.—County Room (the latest addition to the Institution):—This room is reserved exclusively for collections representing the Fauna and Flora of Kent. At the present time it is only partially furnished, but will ultimately contain the large collection of Kentish birds now in the adjoining room. It is also proposed to arrange in this room a type collection of fossils from the various Kentformations.

The collections now here are:—Some cases of Birds arranged in groups with natural surroundings, Birds' Nests and Eggs; Mammals; Fish; Land, Marine, and Freshwater Shells; Crustace; alnsects, including an extensive collection of Bees found in the county, the majority of them from the immediate neighbourhood of Maidstone; a Kentish Herbarium—some of the specimens are exhibited in the wall case in the Gallery; and the Harrison collection of Eolithic, Palaeolithic, and Neolithic Stone Implements from the Chalk Plateau of Kent, the Oldbury Cave Shelters, and the Medway Gravels.

10.—Shell Room 'Upper Floor of East Wing):—An extensive general collection of Shells and Corals, and a general collection of British and Foreign Lepidoptera. 11.—News Room (Lower Floor of East Wing):—This room is set apart as a News Room and for General Readers. It also contains a large portion of the Reference Library. 12.—Victoria Lending Library:—This, the lower room of the latest addition to the Museum, was erected by public subscription in commemoration of the Diamond Jubilee of Her late Majesty, Queen Victoria, and was opened to the public in June, IB'.)!), by the Lord Mayor of London, Sir John Voce Moore, during the first Mayoralty of the present Mayor, Mr. Alderman W. Moiling.

Bentlif Art Gai.leky. 13.—Upper Floor:—The Bentlif collection of Oil and Water Colour Paintings, including examples by David Cox, Turner, Copley Fielding, Clarkson Stanficld, J. Varley, Aaron Penley, Samuel Prout, John Brett, Henry Bright, Albert Goodwin, Harry Goodwin, T. S. Cooper, Arthur Hughes, Walter Shaw, and other Artists. 14.—Ground Floor:—Oil Colour Paintings by deceased Artists:—Salvatore Rosa, Snyders, Steenwijck, Loutherbourg, Wijnants, Van der Neer, and others.

A valuable collection of old English Needlework of the seventeenth, eighteenth, and nineteenth centuries.

Illuminated MSS., Books of Hours, and early printed English Books, including a copy (imperfect) of the Golden Legend, dated 1527, printed by Wynkyn de Worde, William Caxton's Assistant and Successor.

15.—Vestibule:—Plaster Casts from the Antique, and the Five Orders of Architecture. 16.—Reference Library:—An extensive collection of Books on Archaeology, Numismatics, Topography, Genealogy, History, Biography, Art, and Natural History; and a special collection of works relating to the county of Kent generally, including the Topographical Drawings by Edward Pretty, F.S.A., and Thomas Charles. 17.—On the walls of the staircase leading from the Vestibule to the Upper Floor is a collection of Engravings and original Drawings by William Woollett, one of England's greatest Engravers, a native of Maidstone (1735-1785). One of the drawings is a Chalk Portrait of himself when a youth.

Note:—The illustrations to this article are introduced by permission of the Museum Committee.

LIFE MEMBERS. *(Instituted at the Canterbury Congresn. 1902.)*

Adkin, Robert, F.E.S. 4, Lingards Road, Lewisham, S.E. (D).

Adkin, Mrs. B., ,,,,

Bennett, F. J., F.G.S. "The Acacias," West Mailing, Kent.

Bullen, Rev. R. Ashington, F.L.S., F.G.S. Pyrford Vicarage, Woking, Surrev.

(Hon. Sec.) (D).
Bullen, Mrs. Ashington. Pyrford Vicarage, Woking, Surrey.
Coomaraswamy, Ananda K., F.L.S., F.G.S. Walden, Worplesdon, Guildford.
Foran, C. Elm Grove, Southsea (D).
Gray, H. Norman, P.A.S.I. Newlyn House, 131, Earlham Grove, Forest Gate, E. (D).
Howorth, Sir H. H., K.C.I.E., F.R.S., F.G.S. (V.-P.). 30, Collingham Place, Earl's Court, S.W. (Ex-Pres.).
Meeson, F. 98, Sutherland Avenue, Maida Vale (D).
Merrifield, F., F.E.S. (V.-P.). 24, Vernon Terrace, Brighton.
Neate, P. J., J.P. Watt's Avenue, Rochester.
Rudler, F. W., I.S.O., F.G.S., &c. 18, St. George's Road, Kilburn, N.W. (President)..
Stebbing, Rev. T. R. R., F.R.S., F.L.S. (V.-P.). Ephraim Lodge, The Common, Tunbridge Wells. (Ex-Pres.).
Stebbing, Mrs. Ephraim Lodge, The Common, Tunbridge Wells.
Stebbing, Miss Grace. Catton, Southborough.
Stirling, Sir J., Bart., F.R.S., Finchcocks, Goudhurst, Kent.
Turner, Miss E. L. Langton Green, Tunbridge Wells.
Vardon, Rev. S. A., M.A. Langton Green, Tunbridge Wells.
Walker, A. O., F.L.S. Ulcombe Place, Maidstone.
Walmisley, A. T., M.I.C.E. Atherstone, Castle Avenue, Dover.
Whitaker, W., F.R.S., F.G.S. (V.-P.). 3, Campden Road, Croydon. (Ex-Pres.).

MEMBERS, ASSOCIATES, AND DELEGATES FOR 1904.

Abbott, Miss M. 59, Oxford Street, Whitstable (M).
Abbott, George, F.G.S. 33, Upper Grosvenor Road, Tunbridge Wells (M). (Treasurer).
Abbott. Mrs. G. 33, Upper Grosvenor Road, Tunbridge Wells (M).
Abbott, E. W. B. Redan, Reginald Road, Maidstone (A).
Allchin. J. H. The Museum, Maidstone (M). (Local Sec.)

Allport, Lieut-Col. The Barracks, Maidstone (A).
Arkcoll, J. Foley House, Maidstone (A).
Baldock, J., F.C.S. St. Leonard's Road, Croydon (D).
Bannennan, W. Bruce, F.L.S., F.G.S., F.S.A. The Lindens, Sydenham Road, Croydon (M).
Barker, W. Cobbett. Bryant House, Bryant Road, Strood (D).
Baker, F. H. 141, Haverstock Hill, N.W. (M).
Barton, Arthur. Sunnycroft, Holland Street, Maidstone (M).
Barton, Mrs. A. Sunnycroft, Holland Street, Maidstone (M).
Beach, Mrs. 11, Parkli'ill Road, Haverstock Hill (M).
Bedford, E..1. "Anderida," Gorringe Road, Eastbourne (D).
Bennett, A. 143, High Street, Croydon (Referee).
Bensted, W. H. "Longfield,'" London Road, Maidstone (M).
Bensted, Miss. "Longfield," London Road, Maidstone (A).
Bird, C, B.A., F.G.S. Mathematical School, Rochester (M).
Bishop, A. 3, Earls Road, Tunbridge Wells (M).
Bloomfield, Rev. E. N., M.A., F.E.S. Guestling Rectory, Sussex (V.-P.).
Boulenger, G.A., F.R.S. 8, Courtfield Road, S.W. (V.-P.j. (Ex-Pres.).
Boulger, Prof. G. S., F.L.S., F.G.S. 11, Onslow Road, Richmond, Surrey (V.-P.). (Ex-Pres.).
Box, J. W. 25, Henry Street, Chatham (M).
Brackett, A. W., F.S.I. 51, Queen's Road, Tunbridge Wells (M).
Britton, C. E. 25, Victoria Road, South Lambeth (A).
Burr, Malcolm, B.A., F.Z.S., F.E.S. 12, Fitzjames Avenue, West Kensington, S.W. (M).
Bullen, Miss E. A. Pyrford Vicarage, Woking, Surrey (A).
Carpenter, Miss E. 41, Quarry Road, Hastings (D).
Carre, Rev. A. A., M.A. Headcorn Vicarage, Kent (A).
Carre, Mrs. A. A. Headcorn Vicarage, Kent (A).
Chambers, Miss. 8, Albion Place,

Maidstone (M).
Chapman, T. A., M.D., F.E.S. Betula, Reigate (M).
Charles, F. Bedford Lodge, South Parade, Oxford (M).
Cole, Miss E. 53, London Road, Canterbury (M).
Collis, Rev. Canon H., M.A. St. Philip's Vicarage, Maidstone (A).
Cook, G. F., Junr. 62, Earl Street, Maidstone (M).
Cooper, J. H. Dashwood, Gravesend (A).
Cooper, Mrs. J. H. Dashwood, Gravesend (A).
Cornell, G. Gabriel's Hill, Maidstone (M).
Crafer, Mrs. M. H. 102, Beaconsfield Villas, Preston Park, Brighton (M).
Crisp, F. G. Buckland House, London Road, Maidstone (M).
Day, W. H. St. Michael's Road, Maidstone (M). (Local Sec.)
Day, Mrs. W. H. St. Michael's Road, Maidstone (M).
Dobie, Miss. Bedford Place, Maidstone (M).
Donisthorpe, H. St. J. K., F.Z.S., F.E.S. 58, Kensington Mansions, South Kensington, S.W. (Referee).
Downing, F., F.S.I. 165, Brigstock Road, Thornton Heath (D).
Edwards, S., F.L.S., F.Z.S. 15, St. German's Place, Blackheath, S.E. (M).
Eele, J. W. Shottermill, Haslemere (D).
Elgar, H. J., The Museum, Maidstone (D).
Elliott, E. 6, Albion Place, Maidstone (M).
Elvery, Mrs. The Cedars, Maison Dieu Road, Dover (M).
Enock, F., F.L.S., F.E.S. 13, Tufnell Park Road, N. (M).
Fielding, Rev. C. H., M.A. West Mailing, Maidstone (Mj.
Fitzgerald, Rev. H. P., F.L.S. Wellington College, Berks. (D).
Fleay, F. G., M.A. 27, Dafforne Road, Upper Tooting, S.W. (D).
Fremlin, H. S. Government Lymph Laboratories, Chelsea, S.W. (M.
Fremlin, R. J. (»). Heathfield, Maidstone (M).
Frisby, G. E. Fengate's Road, Redhill

(D).

Frost, R. C. 11, St. John's Road, Plumstead (M).

Froft, Mrs. R. C. 11, St. John's Road, Plumstead (M).

Gilbert, E. J., M.D. Madeira Park, Tunbridge Wells (D).

Goodwin, T. 2, Canning Street, Maidstone (M).

Goodwin, Mrs. T. 2, Canning Street, Maidstone (M).

Goodwin, E. Canon Court, Wateringbury (A).

Grant, R. P. Maybank, Bearstead (M).

Grant, Mrs. R. P. Mavbank, Bearstead (M).

Green, L., F.C.S. 3, St. Michael's Road, Maidstone (M).

Green, H. The Godlands. Tovil, Maidstone (A).

Green, Mrs. H. The Godlands, Tovil, Maidstone (A).

Griffin, W. H. 6, Rutland Park, Perry Hill, S.E. (D).

Groom, J. B. St. Paul's School, Maidstone (M).

Groves, Jas. 55, Jeffreys Road, Clapham Rise, S.W. (Referee).

Gruner, Miss Joan F. Oakhill, Hindhead (M).

Hailey, Miss. Girl's Grammar School, Maidstone (M).

Halliday, J. Linden House, St. Mary's Road, Hastings (D).

Hammond, Miss. 48, Upper Grosvenor Road, Tunbridge Wells (Ml.

Hannen, Hon. H. The Hall, West Farleigh, Maidstone (A).

Harding, A. B., F. Phys. Soc. Bond, lielmont, Catford, S.E. (M).

Harris, P., M.D. 63, Lower Addiseombe Road, Croydon (M).

Harris, J. C. 23, Knightrider Street, Maidstone (M).

Hembry, F. W., F.R.M.S. Langford, Sidcup (D).

Hepworth, J. Linden House, Rochester (D).

Hill, T. 75, Trafalgar Street, Gillingham (D).

Hill, F. V., Junr. 75, Trafalgar Street. Gillingham (M).

Hills, E. Lenworth, Ashford Road, Maidstone (M).

Hills, Miss. Bedford Place, Maidstone (M).

Hoar, W. Norfolk House, Bower Street, Maidstone (M).

Hoar, 11. Hawthorndene, Buckland Road, Maidstone (M).

Holmes, E. M., F.L.S. Ruthven, Sevenoaks (V.-P.).

Howes, Prof. G. B., LL.D., F.R.S. (V.-P.). Inglcdene, Chiswick, W. (Ex-Pres.).

Hubble, Mrs. Grove Lodge, Hunton, Maidstone (M).

Hutchinson, Jonathan, F.R.S. (V.-P.) Inval, Haslemere (Ex-Pres. 1.

Hutchinson, R. R. 28, Princes Street, Tunbridge Wells (Sec. N.H.S.).

Inniss, F. J. 1, Adelaide Road, New Brompton (M).

Inniss, Mrs. F. J. 1, Adelaide Road, New Brompton (M

Jeffcoat, Rev. W. R. 1, Lower Stone Street, Maidstone (M).

Jenner, J. H. A., F.E.S. 209, School Hill, Lewes (M).

Kearton, 1!., F.Z.S. Ardingly, Caterham Valley (M).

Keeton, H., Mus. Doc. Thorpe Road, Peterborough (A).

Kensctt, Miss. 106, New Street, Horsham (M). (See. Museum Soc.).

Lambert, F. 2, Borneo Street, Putney, S.W. (A).

Legg, Rev. W. A. H., M.A. Knottingley, Maidstone (M).

Legg, Mrs. W. A. H. Knottingley, Maidstone (M).

Lincoln, J. E. 80, High Street, Northfleet (D).

Liven, Kev. G. M., B.A. Wateringbury Vicarage, Maidstone (M).

Loblev, Prof. Logan, F.G.S. 28, Palace ltoad, Buckingham Gate, S.W. (M).

Lowne, B. T. Bromley Road, Catford, S.E. (D).

Manwaring, Miss. Stonewall, Hunton, Maidstone (M).

Marten, C. J. 3, Mason's Avenue, Croydon (M).

Martin, E. A., F.G.S. 23, Campbell Bond, Croydon (D).

Masters, Mrs. 2, Bower Place, Maidstone (M).

Mathews, Paul, M.A. 32, South Avenue, Bochester (D).

Mathews, Mrs. P. 32, South Avenue, Rochester (D).

McDakin, Ciipt. Gordon. 12, Pencester Koad, Dover (D).

McDakin, Mrs. G. 12, Pencester Boad, Dover (D).

Measham, Miss C. E. C. 39, Star Hill, Rochester (D).

Mitton, W. Hurstpierpoint, Sussex (Beferee).

Monro, W. 138, Britton Street, Gillingham (M).

Morgan, J. 12, Moreton Road, South Croydon (M).

Morris, J. 17, Throgmorton Avenue, E. C. (M).

Neate, P. J., J.P. Belsize, Watts Avenue, Rochester (M).

Newmarch, Major-Gen., R. E. (i, Norfolk Terrace, Brighton (M).

Nicholson, E. S., F.E.S. 22, Crouch Hill Road, Crouch End, N. (M).

Nicholson, W. E., F.E.S. Lewes (M).

Nottidge, A. J. Yardley Lodge, Tonbridge (M).

Oliver, C. P., M.D. West Dean, Tonbridge Road, Maidstone (M).

Oyler, T. H. Queen's Avenue, Maidstone (M).

Paine, Miss B. Chart Place, Maidstone (M).

Paine, Miss M. Chart Place, Maidstone (M).

Pannell, C. East Street, Haslemere (M).

Parkinson, S. T. South-Eastern Agricultural College, Wye (M).

Pavne, Mrs. Linden Gardens, Tonbridge Wells (M).

Payne, E. S. 45, Boslvn Hill, Hiunpstead, N.W. (D).

Peirson, H. 57, Castle Hill, Hertford (A).

Pollock, Sir F., Bart., LL.D. Hindhead Copse, Shottermill (V.-P.).

Potter, G. W., M.D. 19. Molyneux Park, Tunbridge Wells (M).

Bayner, A. J. 32, Dover Road, Northfleet (M).

Beid, Capt. Savile G., B.E. The Elms, Yalding (M).

Richards, Bev. S. Kilworth, Buckland Boad, Maidstone (M.

Robarts, N. F., F.G.S. 23, Oliver

Grove, South Norwood (D).

Rogers, J. T. Athelney House, Gillingham (Dj.

Rogers, Rev. W. Moyle. Grosvenor Road, Bournemouth (Referee).

Rogers, G. H. J., F.R.M.S. King Street, Maidstone (M).

Rogers, Mrs. G. H..1. King Street, Maidstone (M).

Roods, Alfred. 67, Thornhill Road, Croydon (I)).

Rowlstone, Miss. 78, Darnley Road, Gravesend (A).

Ruck, W. 11, High Street, Maidstone (M).

Ruck, F. W. Westcombe, London Road, Maidstone (M).

Sargant, Miss E. Quarry Hil!, Reigate (M).

Saunders, Sibert. Bank House, Whitstable (M).

Semark, Miss. 4n, Ashford Boad, Maidstone.

Sharp, Miss. 225, Boxley Road, Maidstone (M).

Sharp, J. 225, Boxley Road, Maidstone (M).

Silcock, P. H., B.A. Rocky House, London Road, Maidstone (M).

Silcock, Mrs. P. H. Bocky House, London Boad, Maidstone (M).

Smith, Miss. 29, Albion Place, Maidstone (M).

Soames, Rev. H. A., M.A., F.L.S. The Hawthorns, Otford (M).

Sparks, H. "Anderida," Gorringe Road, Eastbourne (D).

Stansell, L., F.I.S. 7, Albion Place, Maidstone (M).

Stallworthy, Rev. G. IS. The Manse, Hindhead (D). (Local Sec. 1901 Congress).

Stallworthy, Mrs. G. R. The Manse, Hindhead (D).

Starling, Dr. E. A. Chillingworth House, Tunbridge Wells (M).

Steadman, O. F. Dover Lodge, Northrleet (M).

Steadman, W. H. Dover Lodge, Northrleet (M).

Stebbing, W. P. D., F.G.S. Plavfair Mansions, Queen's Club Gardens, S.W. (M.I.

Styles, M. Grassington, Terrace Road, Maidstone (A).

Styles, Mrs. M. Grassington, Terrace Road, Maidstone (A).

Swanton, E. V. Rrockton, Haslemere (M).

Sweetland, Miss A. K. 3, Park Road, Southborough (D).

Tapsfield, Miss, c/o Dennis, Paine ct Co. High Street, Maidstone (M).

Trumavne, L. J. 29, Cockspur Street, London, S.W. (M).

Treutler, Dr. (V.-P.) Goldstone Villas, Hove (D).

Trollope, V. T., L.D.S. Camden Park, Tunbridge Wells (D).

Trollope, Mrs. W. T. Camden Park, Tunbridge Wells (D).

Turner, J. W.. B.A., R.Sc. Lindtield Lodge, Folkestone (M).

Tutt, J. W., F.E.S. 119, Westcombe Hill, Rlackheath, S.E. (Editor).

Vincent, W. T. 189, Barrage Road, Woolwich (D).

Walton, G. C, F.L.S. 80, Guildhall Street, Folkestone (D).

Walker, Mrs. A. O. T'lcombe Place, Maidstone (M).

Ward, H. Snowden. Hadlow, Kent (D).

Ward, Mrs. H. Snowden. Hadlow, Kent (M.

Ward, W. R. Sutton Valence, Maidstone (M).

Waters, F. 3. 98, High Street, Northteet (M).

Webb, Sydney. 22, Waterlow Crescent, Dover (D).

Webb, Wilfred Mark, F.L.S. Odstock, Hanwell, W. (D).

West, W. 15, Horton Place, Bradford (Referee).

Williams, J. A. Wheelside. Hindhead (M).

Wilson, Rev. F. R. Prospect Place. Old Brompton (D).

Wilson, II.. M.A., F.S.A. Farborough, Kent (D).

Window, Miss. Howbcrry, Haslemere (Local Sec. 1901 Congress).

Wright, T. H. St. Michael's Road, Maidstone (M).

Young, W. P. 251. Lavender Hill, S. W. (D).

Young, G. W. 82, Bridge Road West, Battersea, S.W. (D).

Youngman, G., J.P. "Lenlield," Ashford Road, Maidstone (M).

Youngman, Miss. "Lenfield," Ashford Road, Maidstone (A).

TITt.KH AND ARRANGEMENT OF PLATES.

To be placed opposite Pack. ate i. Members and Delegates present at Altington Castle (Title page) ii. Mailing Abbey G9 iii. st. Leonard's Tower, West Mulling.. 71 iv. Neolithic Stone Implements 55 v. Keltic Pottery found at Husleuiere...... 78 vi. The Corporation Museum, Maidstone.... 78 vii. The Long Gallery and Cloister.. 78

Tiii. Chillinglon Manor House 79 ix. The Wtthdrawing-Room H2 THE SOUTH EASTERN NATURALIST BEING THE TRANSACTIONS OF THE SOUTH-EASTERN UNION OF

SCIENTIFIC SOCIETIES

ALSO THE PROCEEDINGS AT THE TENTH ANNUAL CONGRESS HELD AT REIGATE.

JUNE 7th, 8th, 9th and 10th, 1905. EDITED BY

J. W. TUTT, F.E.S.

"President for 1905-6:

Professor W. M. FLINDERS PETRIE, D. C.L., LL.D.,

F.R.S. &c.

LONDON:

ELLIOT STOCK, PATERNOSTER ROW, E. C.

1905,

The objects of the Union are to systematise Scientific Work among the different Societies composing it, to give greater impetus to Scientific research, and, in general, to promote the study and advancement of Science by Cooperation. In view of these objects, School Natural History Societies receive special consideration, aad are admitted on payment of a nominal fee.

Authors are entirely responsible for the facts and opinions contained in their papers.

Readers of Papers are requested to send to the Editor, J. VV. Tutt, Rayleigh Villa, Westcombe Hill, S.E., a list of any *errata* they may detect in the present volume.

The Congress for 1906 will be held at Eastbourne, on

June 6th, 7th, 8th, and 9th, under the presidency of

Professor FRANCIS DARWIN, LL.D., F.R.S.

The Editor would be glad to exchange Transactions with other Unions and Natural History Societies. All communications relating thereto should be addressed to J. W. Tutt, Westcombe Hill, S.E.

Objects of the South-Eastern Union of Scientific Societies
  Date and Place of Congress for 1906
  Places where Meetings have been held, and names of Past
    Presidents
  Officers and Council for 1905-1906..

  Reigate Local Committee
  Rules
  Bye-laws
  Botanical Research Committee
  Original work done by Members of Affiliated Societies, in
    Archaeology
    Botany
    Geography...
    Geology
    Hygiene
    Miscellaneous
    Photography
    Zoology
  List of Affiliated Societies
  List of Lecturers
  Ninth Annual Report...
  Balance Sheet, 1904
  Letter sent out by British Association
  S.E.U.S.S. Lantern Slides
  Referees
  Museum Notes...
  Report of Delegate to British Association
    Proceedings of Congress, 1905
    Presidential Address...
    The Extinct Postpliocene non-marine Mollusca of the South
      of England
    The Law of Treasure Trove as it affects Archaeologica
      Research
    Antiquities of Reigate...
    Gatton in the Past
    Lepidopterous Ova
    Mendel's Theory of Heredity...
    The flora of the Reigate district
    Report on Wild Plant Protection
    Notes on an Address...

Obituary
Life Members...
Delegates
Members and Associates for 1905
3lelgate local Committee:
  His Worship the Mayor of Reigate (Alderman W. H. Bagaley).
  Alderman F. E. Barnes, J.P., C.C.
  The Venerable Archdeacon Daniell.
Rev. E. J. Baker, M.A.
  Miss K. Baxter.
Dr. vv. A. Berridge.
'Rev. R. Ashington Bullen, B.A., F.L.S., F.G.S.
Mr. G. W. Butler, B.A., F.Z.S., F.G.S.
Dr. T. A. Chapman, F.Z.S., F.E.S.
Mr. J. J. Colman, M.A., D.L., J.P.
  Mr. W. Conolly, J.P.
Miss Crosfield.
Mr. A. J. Crosfield.
'Mr. J. B. Crosfield, President of Holmesdalc Natural History Club.
Dr. F. Curtis.
  Rev. F. C. Davies, M.A.
  Rev. Walter Earle.
  Rev. Selwyn Evans. Mr. G. E. Frisby.
  Rev. J. Gardner.
  Rev. J. M. Gordon, M.A.
  Mr. H. Gough.
  Mr. P. N. Hasluck.
Those marked are Officers and Council of the Holmesdalc Natural HistoryClub.
Hon. Local Secretaries—
  Mrs. G. R. Taylor, Clear's Corner, Reigate.
G. E. Frisby, 9, Fengates Road, Redhill.
Hon. Local Treasurer—
Rev. E. J. Baker, M.A., Nutley Lane, Reigate.
As revised at the Fourth Annual Congress held at Rochester, May 27th, 1899, with subsequent amendments.
1. Objects.—The objects of the Union shall be to systematise work among the various Societies composing it, to give a greater impetus to research, and to promote the interests of the Societies by co-operation. 2. Management.—The affairs of the Union shall be managed by a Council and a General Committee. 3. The Council shall consist of a President, Vice-Presidents, General Secretary, Treasurer, Editor of the Transactions, and seven other persons, three to form a quorum; all to be elected annually,' and none except the VicePresidents, Secretaries, Treasurer, and Editor, to be eligible in the same position for more than two years in immediate succession. The filling of casual vacancies to be at the discretion of the Council itself. *i*. The General Committee shall consist of the Council, Past-Presidents, and the Delegates. 5. Affiliation.--All Scientific Societies in Hampshire, Kent, London, Middlesex, Surrey, and Sussex, shall be eligible to join the Union, provided that the Society claiming to join comprises at least 10 members. 6. Congress.—A Congress for the furtherance of the general work of the Union, and for the reading and discussion of papers, shall be held annually in June, at such place and at such date as may be decided on by the General Committee at the preceding Congress, or, failing such decision, by the Council. 7. Delegates.—A minimum Annual Subscription of 5s., *payable in advance at least a fortnight before the Congress,* shall entitle a Society to affiliation and a voting ticket for one Delegate at the Annual Congress. Societies with more than 50 members, exclusive of honorary members, shall if they so desire, be entitled to voting tickets for additional Delegates in the proportion of one for every additional 50 members, and one for the number (not less than 10) in excess of every multiple of 50, on payment of 5s. for each ticket. 7a.—School Natural History Societies may affiliate for a subscription of 2s. 6d. (Rule added June, 1900). 8. Members.—Members of Affiliated Societies shall be admitted to the Congress on payment of 2s.' 6d. 9. Associates.—Persons unattached to any Affiliated Society may, at the discretion of the Council, be admitted to the Congress on payment of 3s. 6d. 9a. Life-Members.—Members, Associates, and other persons, at the discretion of the Council, may compound for the Annual Subscription by a single payment of £2 2s. for LifeMembership. (Rule added June, 1902.) 10. General Meetings.—The meetings at the Congress shall

be for the reception of reports of work and the reading and discussion of papers. 11. The General Committee shall, at some time during the Annual Congress; receive a Statement of Accounts, appoint an Auditor, elect the Council for the ensuing year (by ballot if demanded), appoint such Sub-Committees as may be required, decide on the next place of meeting, and, when necessary, revise the Rules. 12. Executive.—All other affairs of the Union throughout the year shall be managed by the Council. 13. Transactions.— Such Transactions of the Union as may be published shall be issued free to all affiliated Societies, Members, and Associates. 14. Local Receptions.— Each Society or Town inviting a visit of the Union shall appoint a Local Committee and Local Secretary to assist the General Secretary in drawing up the Programme of the Congress, which shall be arranged at least a month before the said Congress. 15. Expenses.—The expenses of printing and general management shall be paid out of the funds of the Union; those of providing rooms for the meetings of the Congress by the Society or Town issuing the invitation. 16. Changes in the Rules may be proposed and discussed at any meeting of the General Committee, but cannot be passed until the following year, unless they have been submitted to the General Secretary at least three months in advance, so that he may report the proposals to the Affiliated Societies before the Congress.

I.—The Union shall have the right, at its discretion, of printing *in extenso* in its Transactions all papers read at the Annual Meeting. The copyright of a paper read before any meeting of the Association, and the illustrations of the same which have been provided at his expense, shall remain the property of the author; but he shall not be at liberty to print it or allow it to be printed elsewhere, either *in extenso* or in abstract amounting to as much as one-half of the length of the paper, before the 1st of November next after the paper is read.

2.—The author of any paper printed in the Transactions shall be entitled to receive 25 separate copies of it gratis, and to have any further number printed at his own expense by private arrangement with the printers to the Association. 3.—If proofs of papers to be published in the Transactions be sent to authors for correction, and are retained by them beyond four days for each sheet of proof, to be reckoned from the day marked thereon by the printers, but not including the time needful for transmission by post, such proof shall be assumed to require no further correction. 4.— Should the extra charges for small type, and types other than those known as Roman or Italic, and for the author's corrections of the press, in any paper published in the Transactions, amount to a greater sum than in the proportion of ten shillings per sheet, such excess shall be borne by the author himself. 5.—A time limit of 25 minutes is prescribed for each paper, with 5 minutes for each speaker, and the discussion of the subject is to be closed at the end of one hour. March, 1900. BOTANICAL RESEARCH COMMITTEE.

This will be added to from time to time, so that it may embrace 2 or 3 representatives from each county in the district.

Prof. G. S. Boulgek, 11, Onslow Road, Richmond, S.W.

W. H. Beeby, Hildasay, Thames Ditton.

Jas. Groves, 58, Jeffreys Road, Clapham, S.W.

E. Chas. Horkell, 58, Copleston Road, Denmark Hill, S.E.

Thos. House, Cholmondeley Cottage, Riverside, Richmond.

Rev. E. N. Bloomfibld, Guestling Rectory, Sussex.

Wm. Mitten, Hurstpierpoint, Sussex.

W. E. Nicholson, Lewes, Sussex.

E. M. Holmes, *Chairman,* Ruthven, Sevenoaks.

ORIGINAL WORK DONE OR BEING DONE BY MEMBERS OF AFFILIATED SOCIETIES. (Furnished by the Secretaries.)
ARCH/EOLOGY.

Croydon N.H. and Sc. Account of Bermondsey Priory N. F. Robarts, F.G.S. Soc.

BOTANY.

Dover Sciences Soc. Verification of botanical records Rev. J. Taylor, and

Capt. McDakin, R.E. Photography of plants, and particularly as to the underground life of British *Orchiddcei Tc* S. Webb *Hepaticae* and other low orders of plants... T. Moring

Eastbourne N.H. Soc. The Lady's slipper orchis and other plants N. S. Whitney, M.B., of the Caldecott herbarium M.R.C.S.

South London Entom. Orchids H. J. Turner, F.E.S.

and N.H. Soc. Morphology of Rritish plants E. Step, F.E.S.

Tiffin's Boys' School, Ferns P. Varley

N.H. Soc.

Tunbridge Wells N.H. Varieties of Cereals F. S. Munford, F.E.S. Soc.

GEOGRAPHY.

Brighton and Hove N.H. A tour in Spain (illustrated with slides in

Soc. natural colour photography) Ernest Payne, M.A.

Eastbourne N.H. Soc. Some hill towns of N. Italy Rev. G. H. West, D.D.

Notes on the Cinque ports and their harbours Rev. W. Hudson, M.A., F.S.A.

Tiffin's Boys' School, Norway Rev. G. L. Swinnerton,

N.H. Soc. M.A.

GEOLOGY.

Croydon N.H. and Sc. Descripton of some fossils from a Croydon

Soc, garden. *Porasphacru* (plates) G. J. Hinde, Ph.D.,

F.R.S.

1. Notes on New Cross Gate section, Woolwich and Reading beds, L.B. & S. C.

Railway N. F. Robarts, F.G.S.

2. Notes on New Cross Gate section, Woolwich and Reading beds, L.B. & S.C. Railway,,,,,,

Further notes on some Surrey hills W. Whitaker, B.A.,

F.R.S.. F.G.S., &c.

Dover Sciences Soc. 1. Sea temperature and coast erosion Capt. McDakin, R.E.

2. Geological investigations,, ,,

Pakeontology Maj.-Gen. Cock burn Haslemerc M. and N.H. Geology of

the Haslemere District (with
  Soc. section) Rev. J. B. Fowler
  Tiffin's Boys' School, 1. The making
of rocks and fossils, C. J. Grist, M.A.
  N.H. Soc.
2.,,,,,, scenery ,,,, HYGIENE.
Brighton and Hove N.H. Social evolution and public health A. Newsholmc,
  Soc. M.R.C.P., M.D.H.
  Weekly articles in " The Kentish Mercury"
newspaper on the Flora, Fauna, Geology,
and Archaeology of N.W. Kent, and the neighbouring parts of Surrey. New stations for rare plants, and superficial deposits (old river-gravels and brickearths) not denoted on the Geological Drift maps, have been recently recorded. Marine and Estuarine fossils from the Woolwich Beds at Bromley Hill, Belmont Hill, Lee. and Well Hall, near
Eltham, have been recently sent to the
  Public Museum, Maidstone W. H. Griffin
  Oxshott, its beasts and flowers W. J. Lucas, B.A.
  South London Entom and N.H. Soc.
Tunbridge Wells N.H Soc.
Catford N.H. Soc.
PHOTOGRAPHY.
  Eggs of Lcpidoptera A. E. Tonge, Noad Clark.
Nature photography in relation to Bird Life Miss E. L.
F.L.S.
Five full plate photographs of Lower Tertiary sections at Belmont Hill, Lee,
  S.E., on descriptive mounts W. H. Griffin (Photo. Survey of Kent.)
School, Lantern slides to illustrate shells W. Bridger
and F. Turner.
ZOOLOGY.
  South London Entom. and N.H. Soc.
  City of London Entom.
and N.H. Soc.
  Dover Sci. Soc.
Eastbourne N.H. Soc.
  Holmcsdale N.H. Club.
  Tunbridge Wells N.H
Soc.
  Marine Fishes;E. Step
  European Lcpidoptera (particularly

life histories—eggs, &e T. A. Chapman, M.D.,
  F.E.S. and J. W.
Tutt, F.E.S.
  Revision of Alucitides(Pterophorides)
  Revision of life histories of British butter-,,,,,,,, flies—eggs, larva;, &c
  Orthoptera of the world M. Burr. B. A..V.E.S.
  Life histories of Coleophondes H. J. Turner. F.E.S.
  Geometrida: of the World L. B. Prout. F.E.S.
  Revision of Alucitides... A. Bacot, F. E.S., A.
Sich. F.E.S., J. W.
Tutt, F.E.S.
  West Indian Lepidoptern W. J. Kaye
  The life-histories of the Emerald moths Rev. C. R. N. Burrows
  Palasarctic Lepidoptera Dr. T. A. Chapman
  Variation;ind Death in relation to Natural
  Selection A. W. Bacot. F.E.S.
  Life histories of the Pterophorida: T. A. Chapman, M.D.,
  F.E.S.
  The genus *Emtitctesia* L. B. Prout, F. E.S.
Polyoinntittus Corydnn (variations and abnormalities) C. B. Pickett. F.E.S.
Verification of animal and insect records... S. Webb
  The aquarium at Naples Miss M. E. Vinter. M.A.
  B.Sc.
Phinisftini zebru var. *holleri* n.var. and *Chloritis inuhinfiensls*, n.sp., and varieties of *Xestti citrina*. L. from the Isle of Gisser(coloured plate). (*Proc. Malac. Soc.*, vol. vi.. pt. 4, 1905) Rev. Ashington Bullen, F.L.S.. F.G.S.
South African Crustacea, Part iii., II plates Rev. T. R. R. Stebbing.
M.A., F.R.S., Sec.
L.S.
"Crustaceans" in Victoria History of the counties of England *(continued)*
  Zoological Nomenclature. International ,, ,,,.
rules and others *Journ. Linn. Soc,* vol. xxix)
  Mamaia and Mamaiitlie *(Proc. Biol.*

Soc. Washington, vol. xviii) Rev. T. R. R. Stebbing,
  M.A., F.R.S., Sec.
  The Amphipoda and Cladocera (G. H. L.S.
Fowler's *Biscuynin Pliinkton,* 2 plates),,,,,t,,
  British Hymenoptera F. S. Mumford F.E.S.
  Tiffin's Boys' School, Insects in my garden P. E. Towell
  N.H. Soc. Four insect life-histuries F. C. J. Hawkins
  Preserving caterpillars C. T. Casey
  Reptiles R. Cleary
  Note.—The object of this List is not only to form a record of work done, but also to assist workers to communicate with each other—sec also last year's list. Mere lists of Lectures given before Societies are not wanted for insertion, but only particulars of *original* work, whether published or not.
  Where no fee is mentioned it may he assumed that none is expected, hut the travelling expenses must he paid and accommodation for the night provided if required. For Lantern Lectures the Society will please provide lantern and operator. In future issues of this list names of ladies or gentlemen *recommended* by Affiliated Societies will be inserted. It is expected that many of them will be able to give at least the name of one capable Lecturer willing to repeat his or her lecture before other societies in this Union. Lecturers.
Mr. J. H. Allchin, The Museum, Maidstone. Private address, Chillington House.
Fee, 2 guineas and travelling expenses.
  Professor G. S. Boulger, F.L.S., F.G.S., Kd. of *Nature Notes,* 11 Onslow Road, Richmond, S.W.
  Fee on application.
  Mr. A. W. Brackett. F.S.I., 51, Queen's Road, Tunbridgc Wells.
  Mr. Ed. Connold, F.E.S., 7, Magdalen Terrace, St. Leonard's-on-Sea.
  Fee, 2 guineas, and travelling expenses.
  Dr. Vaughan Cornish, F.G.S., F.C.S., F.R.G.S., 72, Prince's

Sq., London, W.
 Fee on application.
 Mr. Martin Duncan, South Park, Reigate, Surrey.
 Mr. F. Enock, F.L.S., F.E.S.
 Salisbury Road, Bexley. Fee on application.
 Subjects. 1. A Glance at the Early History of Kent and the life of William Caxton.
2. Some Kentish Celebrities—Wm. Harvey, Wm. Shipley, Wm. Alexander, Wm. Woollett. Both with lantern illustrations. 1. Insectivorous Plants. 2. Seed-dispersal. 3. Geological Photography—all with lantern slides. 1. A simple calculating machine. 2. Phenomena outside our apprehension. 3. Ether. 1. The Galls of the British Oaks. 2. Wasps and how they live, and other subjects. 1. Waves in Water, Sand, and Snow. 2. Shifting Sands and Drifting Snow. 3. The Snows of Canada. 42, 1.
 Mr. R. R. Hutchinson, Hon. Sec. 1. Tun. Wells N.H.S., 28, Prince's 2. Street, Tunbridge Wells. 3.
 Mr. A. B. Harding, F.Phys.Soc. 1. Lond., Belmont, Catford, S.E. 2. Fee on application. 3.
 A Naturalist's Ramble on the seashore
Flesh-feeding Plants.
Devil Fish and Kraken some Long-armed Monsters of the Deep.
 Wonders and Romance of Insect Life.
British Trap-door Spiders.
Insect marvels in a back garden.
 All with specially prepared lantern slides.
 Rust Fungi.
Mycetozoa.
Dispersion of Seeds.
 The Romance of a Snow Flake.
From Noise to Music.
A Flash of Lightning.
1. Haunts and Habits of British Birds. 2. Peeps into Nature's Secrets. 3. Wild Life at Home: how to study and photograph it. Each lecture is illustrated by photographs taken direct from nature by the lecturer. 1. Marvels of the Subterranean World (Jenolan Caves, N.S.W.) 2. A Tour through the Crystal Caves of
 New South Wales.

3. Through the Mammoth Caves of Kentucky in search of Eyeless Fish and other Blind Fauna. The dazzling nature of the Stalactites is conveyed by means of a patent crystalline screen. 1. Buried Cities, or Northern England 2,000 years ago. 1. The History of Valleys, submerged and exposed. 2. The Physical future of the Earth. 3. The Coal Problem, geologically and economically considered. 1. Aims of Natural History Societies. 2. The Classification of Animals. 3. Artificial Languages. 1. Plumage.
 Mr. W. H. Griffin, 6, Park, Perry Hill, S.E.
 Rutland 1. Plant Folklore.

NINTH ANNUAL REPORT.

It is pleasant to be able to report at this Tenth Congress, that the S.E.U.S.S. has again had a successful year, financially and otherwise.

The number of affiliated Societies remains the same as in 1904. Wellington College N.H.S. has left the Union owing to distance from our places of meeting, but Epsom College N.H.S. having rejoined, the number remains unaltered at 43.

The Ninth Congress at Maidstone was very well attended; the local museum was much appreciated, and the kindness of the Mayor and Mayoress (Alderman and Mrs. Morling), the museum officials, local Societies, and others, the successful excursions in delightful weather, the interesting papers and discussions, will not soon be forgotten.

A special feature of the Congress was the crowded meeting of teachers, pupil teachers, and elder scholars, to listen to Mr. W. Mark Webb's paper on the "Teaching of Nature Study." This was a successful innovation, and whatever the views expressed in the subsequent discussion, for and against, such a meeting is an evidence of another useful purpose that may be served by our Union.

This year an extra day has been added to the Congress in view of the remote meeting place of the British Association in South Africa, but the proximity of the Whitsuntide holidays prevented larger extension. Had this been feasible the Congress might have assumed a more important character, and Botanical, Zoological, and other Sections, taken the place of our General Meetings. But perhaps this scheme would have been too ambitious; it certainly would have been a great tax upon our generous Reigate friends.

No autumn meeting has been held during the last year, but the scheme has not been by any means abandoned. It is hoped to organise this part of our work under a separate Secretary, and Mr. H. Norman Gray has consented to see to this special work.

There are now 23 Life Members, and the fund thus subscribed, together with a part of the surplus funds, amounting in all to £62 2s. 0d., is placed at deposit.

The Union has to chronicle this year, in common with many other Societies, the loss which it has sustained in the decease of its Vice-President, Professor G. B. Howes, LL.D., F.R.S. A vote of condolence with Mrs. Howes was drawn up and proposed by the President (Mr. F. W. Rudler) at the Council Meeting on April 14th, and unanimously adopted. A portrait and short biography will appear in the coming volume of the *Sontli-Eastern Naturalist,*

The *South Eastern Naturalist* or 1903 and 1904) has been forwarded to the Library of the British Museum (Natural History).

Branch proceedings, programmes, occasional papers, &c, have been received from the Croydon N.H.S., Rochester Naturalists' Club, Fulham Field Club, North London N.H.S., Balham Antiquarian and N.H.S., Haslemere N.H.S., East Kent N.H.S., Horsham Museum Society, South London Entom. Soc, City of London Coll. Sci. Soc, and Epsom College; also the report of the British Association (Cambridge meeting, 1904), and the account of the Delegates' meetings at Gonville and Caius, last August. These are put out for consultation in the Public Hall, and to use a nautical expression, one Society may "pick up a wrinkle" from another, and in the language of the village politician in *Sylvester Sound,* thus lead to the "fructifying concatenation of concurrent ideas." BALANCE SHEET, 1904.

RECEIPTS.
Brought Forward
 Maidstone Congress—Proceeds of Tickets and Contributions by Local N.H.Soc
  Laurence Green, Esu_. (don.)
  Delegates' Subscriptions, 1904...
  Delegates' arrears. 1903
  Members' Subscriptions
  Associates' Subscriptions
  Sales of-S.E. Naturalist
  Reprints
  Interest on Deposit at Birkbeck Bank
EXPENDITURE.
£ s. d.
 Printing Congress Programmes, &c 4 17 8
 Printing S.E. Naturalist, 350 copies and Reprints 41
 Expenses re museum at Congress
 Postages
 Put on deposit at the Birkbeck Bank
  Balance in Hand
£ s. d.
2 8 4 6 0 4 i n 3 Id 15 18 0 16 1
June 6th, 1905.
Audited and certified correct with vouchers produced.
W. T. TROLLOPE, *Auditor.* LETTER SENT OUT BY THE BRITISH ASSOCIATION AND APPROVED BY THE CONGRESS.
 British Association For The Advancement Of Science.
  Burlington House,
London, W.
 April, 1905. To The General Manager.
 Sir,
 I am directed by the Corresponding Societies' Committee of the British Association to submit for your consideration the following statement:—
 This Committee represents seventy Societies for the Promotion of Science, a list of which is enclosed herewith. It often happens that the Members when engaged in carrying out the objects of their respective Societies are not associated in parties, and consequently at present have no advantage in respect of railway fares at reduced rates.
 It is understood that members of many Angling and Golfing Clubs enjoy the privilege of railway travelling at a reduced scale of fares on production of their tickets of membership.
 It seems not unreasonable to hope that a similar concession may be granted to the members of the Societies represented by this Committee, and I am, therefore, requested to solicit your favourable consideration of this appeal, so far as applies to the Societies travelling on your line.
 The privilege, if granted, might be safeguarded from abuse by such conditions as your Company might think proper to enforce.
1 am, Sir,
 Your obedient Servant,
 This appeal has been submitted to the S.R. Union of Scientific Societies, and is approved on its behalf by W. M. Flinders Petrie, President.
 S.E.U.S.S. LANTERN SLIDES.
 The following sets of Lantern Slides are available for use by affiliated Societies on application either to the General or Photographic Secretary: 1.—*Some British Orchids* (50 slides) contributed by Mr. S. Horsley,
M.I.C.E., with explanatory lecture. 2.—*The Gitult and lower Greensand* (about 80 slides), with lecture. 3. — *The Wealdett Formation* (about 50 slides), with explanatory notes. 4.—*Ice Flowers and Crystals* (small set), with explanatory notes by G. Abbott.
 No charge is made except for carriage both ways. The orchid slides are a new and interesting set, dealing with the general and detailed structure of many British species and their adaptation to insect-fertilisation.
 The Society for the Protection of Birds, 326, Holborn, W.C., also lend *to their subscribers* very beautiful Lantern Slides relating to Birds, which are well worth the attention of Secretaries and Lecturers.
 The notice of Secretaries is particularly called to the suggestions made in the Photographic Secretary's Report that the Union should solicit loans or gifts of *small* sets of slides (a dozen or so) illustrating *any* particular scientific phenomenon or limited branch of scientific work. Such sets, with full explanatory notes, to occupy about half-an-hour for exhibition, would doubtless be much appreciated for the purpose of soirees or other occasions in which tfme is necessarily limited, while two sets might furnish material for an ordinary evening meeting. Many members of our Societies possess sets of this character which they have prepared for their own use, and which they would be willing to lend for the use of the affiliated Societies. Secretaries are hereby asked to furnish the Photographic Secretary as soon as possible with the names and addresses of any of their members who, in their opinion, might be induced to co-operate.
 Contributions are still solicited towards the following larger sets that are in course of formation: 1.—*Pre-Hiitoric man in S.E. England. 2.—English Wild Flowers,* with special reference to forms of capsules and their dehiscence. 3.—*Photomicrographs.* 4.—*Coast Erosion in S.E. England.*
 Contributions of lantern slides may be sent to the Photographic Secretary, Mr. E. A. Martin, F.G.S., 23, Campbell Road, Croydon (West), who will be glad to give any information as to this branch of the work of the Union.
REFEREES. BOTANICAL.
(Additions to this list would be welcomed).
The following gentlemen have kindly consented to name a limited number of specimens for our Members and Associates.
(A stamped directed envelope should always be sent, or no replies need be expected).
Cryptogams (not microscopic).—Thomas House, Cholmondeley Cottage, Riverside, Richmond.
 Freshwater Algae.—W. West, 15, Horton Lane, Bradford.
 Specimens of species of *Zygnema, Spirogyra,* and *Motigcotia* should be fruiting. They are best sent in small tubes in water. Habitat must be always stated. Permanent reedy ponds and ditches yield the best results, especially those where *Utriciilaria* occurs. Plants

like *Utriculariii,* leaves and peduncles of *Nymphaea, Nuphar,* &c., might be sent in tin boxes; and Mr. West will examine these for minute forms. Gelatinous or slimy coverings of damp, shady, or trickling rocks should also be sent in small tins.

Marine Algae (excluding diatoms and desmids).—E. M. Holmes, F.L.S., Ruthven, Sevenoaks.

Fresh Alga: should be rolled separately in old muslin or calico, so that one plant does not touch another; then packed so as to be free from pressure in tins or boxes.

Mosses.— W. Mitton, Hurstpierpoint, Sussex.

Phanerogams.—A. Bennett, F.L.S., 143, High Street, Croydon, and Rev. E. N. Bloo.mfield, Guestling, Sussex.

In some orders like *Cruci/erac, Cyperaccae, Umbelliferae,* the fruit is almost a necessity.

W. H. Griffin, 6, Rutland Park, Perry Hill, S.E.

Cyperaceae.—A. Bennett, F.L.S., 143, High Street, Croydon

Hieracii;—Rev. W. R. Lynton, Shirley Vicarage, Derby.

Rubi.— Rev. vv. Moyle-rogers, Chetnole, Grosvenor Ror.d, Bournemouth, West.

The work of the *Rubi* referee would be very greatly lightened, and his determinations and suggestions proportionately more satisfactory, if correspondents would send only good *representative* specimens, and place them always on *paper stout enough and large ennugh to bear them safely.*

There is, of course, least room for uncertainty when panicles show both flower and fruit, and the stem pieces mature leaves from about the centre of their length; other less satisfactory pieces *in addition* may sometimes help a referee. But if in any case a correspondent can feci justified in sending only such pieces, he should at least press them carefully and supply them in extra quantity.

In explanation of the term "stem" and "panicle," it is to he remembered that all the fruticose *Rnbi* throw out long leafy shoots (the "stem" or " barren stem ") directly from their roots, which, normally, produces no flowers in the first year. From these spring, in the following year, the flowering shoots, and it is to the flowering part of these, including all the branches and branchlets, that the term " panicle " is applied.

No "specimen" can be determined with certainty, unless it consists of both panicle and stem pieces; and it is often of great assistance to the referee when the label accompanying the specimen contains—in addition to the usual memoranda of locality,county or vice-county, date. &c.—further notes made from the living plant, of the colour of the flowering organs and the comparative length of styles and stamens, with any other conspicuous character lost in the process of drying.

ZOOLOGICAL.

Aculeate Hymenoptera.—G. E. Frisby, 9, Fengates Road, Redhill.

Diptera.—Rev. E. N. Bi.oomfield, Guestling, Sussex.

Tenthredinidae.—Ditto.

He will also be glad to hear of any " finds" in either the Fauna or Flora of his part of Sussex.

Coleoptera.—H. St. J. K. Donisthorpe, F.Z.S., F.E.S., 58, Kensington Mansions, South Kensington, S.W.

Lepidoptera.—J. W. Tutt, F.E.S., Rayleigh Villa, Westcombe Hill, S.E.

Orthoptera.—M. Burr, F.Z.S., F.E.S., Royal Societies' Club, S.W.

Galls.— E. Connold, F.E.S., Hon. Sec. of Hastings Nat. Hist. Soc. 7, Magdalen Terrace, St. Leonard's.

Hydroida (Calyptoblastea).—Rev. H. A. Soames, F.L.S., Hawthorns, Otford, Sevenoaks.

MUSEUM NOTES.

Edited By The HONORARY CURATOR.

It is gratifying to report that the interest in our temporary museum is steadily increasing. Last year at Maidstone, apart from the Nature study exhibition, ve had eleven exhibitors, this year there were twenty-two. The value of the museum is very obvious, it brings us more into touch with one another, and causes us to realise perhaps more fully than anything else could—how very closelv one branch of natural history is linked with others. In the following pages the majority of the notes have been kindly contributed by the exhibitors. It would materially assist the curator if, in future, all exhibitors would bring (or send) notes upon their exhibits for insertion in the *Transactions,* without waiting to be called upon for such. It will be observed that the arrangement of the notes is alphabetical.

Dr. George Abbott, F.G.S., Tunbridge Wells.—(1) Twigs with leaves from a copper beech-tree on Mt. Ephraim, Tunbridge Wells, gathered in May, which showed that the leaves which had been protected by others from the effects of frost were *green,* and not purple; even when part of a leaf was thus protected, the same difference was shown quite clearly, the leaf being partly green and partly purple. (2) Galls on male catkins of oak, gathered in Reigate, caused by *SpatJiegaster baccarum.* This gall is unusuallyabundant this year, hundreds of them lay on the ground under some of the trees. The alternating generation is the spangle gall *(Nenroterns lenticularis)* which occurs, often in large numbers, on the undersurface of oak-leaves in August and September. (3) Photographs of the Stonehenge monoliths showing the effects of the growth of lichens on their weathering, taken at Easter, 1905. Extensive patches appearing white in the photographs were parts from which the lichens had peeled off, and had carried away with them some of the sand grains, leaving a glistening surface of white sarsen. In this manner, during the three or four thousand years since their erection, considerable changes must have been effected. (4) Chrysalids of a small moth *(Solenobia liclienella)* found amongst the lichen of the Stonehenge monoliths. (5) Album of photographs illustrating many different forms of weathering in sandstone, limestone, brickwork, mortar, etc.

Mr. Francis H. Baker: Photographs Of Rural Villages And Hamlets Of Middlesex. "Middlesex in general," wrote Leigh Hunt, "is a scene of trees and meadows—of trees and nestling cottages." This was a hundred years ago, before the "Great Wen" had spread itself over half the county, and threatened to drive all rusticity out of the other half. Yet there are still many little out-of-the-way hamlets in Middlesex—assemblages of rustic cottages nestling among trees and bordering brooks and meadows, and even some of the villages, which still in some degree retain their rural character, though this must inevitably be soon destroyed in the progress of railway and tramway extension and the operations of the speculative builder.

Photographs recording the present appearance of these places would, it is thought, be interesting in view of future changes. Those exhibited illustrate about sixty of the villages and hamlets in the county which have not yet been absorbed into the metropolis, or converted into outlying residential suburbs, and include

Early Gravestone In West Mallinc Chirchvard (See " History of West Mailing Church," Chap. XII.) *Photatt by Edvtin Filkins, Oravetend,* Gravestone Of The Profile Type In West Mallinc Churchyard. *The South-Eastern Naturalist, 1905.*

Ickenham, with its pretty little green and village pump; Hayes, with its old swing gate; Perivale and its Church with the wooden tower; Bentley Heath, Northolt; the Hall, Harmondsworth; Longford, Poyle, Yeading, etc. In selecting the point of view, preference has been given to the oldest portion of each village or "hamlet, as being not only generally the most picturesque, but the part most liable to change.

Mr. F. T. Bennett, F.G.S.: Photographs Oh Gravestones.—During the past two or three years I have been working up the hitherto unworked ground of the gravestones at and around Mailing, of the 15th, 16th and 17th centuries, and was much interested to find that Mailing Churchyard contained some of the best examples of almost every variety. These stones are divisible into three broad groups, *viz.* :— (1) the upright, (2) table tombs, (3) flat ones.

We are mostly, if not entirely, concerned with the upright variety, as it is among those that the quaint types to be noticed appear. Some of the types, such as the Doll's Head and Profile, die out within a radius of a few miles from Mailing, and are peculiar to that area. Of these upright stones, which are plain and have no faces cut on them (though there are some exceptions to this), the oldest that I have found is dated 1623, and the latest 1690. I have termed these of head and shoulder pattern, as that fairly well describes the shape of the stone. They are massive, about 2 feet high by 15 inches broad and 7 inches thick, and sometimes have a simple moulding; they have a broad border, and the face of the stone is deeply recessed. The back of the stone is towards the west, so that the lettered face is protected from the prevalent wind and weather. The lettering is deeply cut and well executed, and appears in many cases as if newly cut, especially in the parts that have sunk in the ground. A few good examples of this type are to be found at West Mailing, 1657; Ryarsh, 1648; and East Banning, 1623. The exceptions to the plain early stones are these very few which show a trace of rude faces, attempts apparently at likenesses, that have been carved on them. Examples of these occur at East Mailing (two) and Offham (one), 1675 to 1680.

Doll's Head Type Of Skull.—The early face type seems to have been abandoned almost as soon as it was attempted, and was succeeded by conventional patterns, the earliest type of which is a curious delineation of the skull that I have designated the doll's head type, as it resembles that (if a Dutch doll's head, The illustrations in connection with Mr. Bennett's paper arc from Photographs by Edwin Filkins. Gravesencl, represented by a circle in which are two smaller ones, and beneath these a triangle and below the triangle a kind of portcullis, crossbones also appear under the skull. These small circles are the orifices of the eyes and the triangle that of the nose, while the portcullis stands for the teeth. On looking full face at a skull it will be seen that these small circles and the triangle, and the portcullis, do not at all inappropriately portray the skull of the conventional doll's head type. The dates of the doll's head are all within the latter half of the 17th century. The doll's head passes through an interesting series of changes and finally emerges into the skull proper much later on. The changes are the suppression of the rows of teeth, four being the greatest number, till only one row is left, then, the meaning of this having apparently been lost, a circle for the mouth is added, and finally the remaining row of teeth is omitted and the mouth only is left.

Face Type Of Skull: 1708 to 1730, about.—The next type of skull to the doll's head kind is what I have termed the face type; to the head are added some facial features, such as wrinkles under the eyes, eye-brows, and apparently a beard, and hairs on the face. The crossed bones in this type are crossed behind the head, only the ends of the bones appearing. Death's dart, too, is sometimes added, apparently transfixing the skull. Very many examples of these are to be found in the area around Mailing. Associated with this type of skull is a peculiar form of moulding in the stone of a late perpendicular type.

Profile Type: 1717 to 1760, about.—This type is most quaint, and apparently quite a local one. It dies out to the east, not ranging further east than Seal, or north than Luddesdon or Northfleet, or west than Charing, or south than Cranbrook, and passes through many phases. It would seem to have been copied from the head on some coin. The earliest example and the best is at West Mailing, 1717, and it falls to the right, as the King's head did in that reign. One very striking point about this is the way the hair is arranged in a series of stiff corkscrew kind of curls, and these may be seen to pass gradually into the full wig, the curls becoming more and more confluent. The latest phase of this is in high relief, all the faces before noticed being merely in the flat, and the outlines cut in shallow incised lines. Some of these profiles are most rudely executed. The best example of the group may be seen at West Mailing. Elsewhere archaic examples appear. Some thirty years later, the worst executed examples are to be seen at Seal, to the east of which the profile is not to be found.

Cherub Type: 1740 to 1810, about.- This type, in its latest forms, is almost universal, but the earliest form has a very restricted local area, with Mailing again as the centre. The earliest form is most quaint, and is associated always with the late perpendicular Gothic form of stone, and is formed apparently from the face-type of skull, to which wings at the side of the face are added squeezed into the narrow space that this form of stone afforded. In all these late Gothic form of stones it is noticeable, as the moulding and the lettering are so well executed, and the faces are so rudely done, that here we have the work of the mason pure and simple. The later and more ornate type of cherub is only found in stones of the Renaissance pattern. Here the stones are much broader and the designs are most varied and artistic, so that here, at least, the artist and the sculptor appear for the first time on the scene with most beautiful results. With the classic and renaissance pattern of stone we get many various designs. These have a very wide distribution, and they take us over all parts of the country, and thus away from the special and restricted types just described, that seem peculiar to Kent, and chiefly confined to small areas, of which Mailing," for some reason, seems to have been the centre.

Rev. E. N. Bloompield, M.A., F.E.S. —A few of the larger Diptera, some of which are of special interest on account of their life-histories. To mention a few of these—the Chamaeleon fly *(Stratiomys ripariu)*, in the larval state, lives in water, and is there remarkable for being very enlongate, while the fly itself is short and broad, hence the pupa only occupies a portion of the hardened larval skin. The Great Gad-fly *(Tabanus bovinus)*, and two species of *Chrysops*, "Golden-eyed flies," inveterate bloodsuckers. Wolf-flies, the most conspicuous was the Great Hornet-fly *(Asilus crabronifonitis)*, one of the largest of British diptera. The beautiful family of the *Syrphidae*, represented by eight species. The three species of *Volucella* are parasitic on the humble-bees, which some of them very closely mimic. *Scricomyia borealis*, when apparently at rest utters a sound something like singing at a distance (see *Ent. Mo. Mag.* , vol. xviii., pp. 159 and 189). Besides these were the exotic and handsome *Calliprohola speciosa*, and several other very fine species. The curious family of the *Conopidae* which, in the larval state, are parasitic on bees, represented by two species, *Conops vesiciilaris*, and the slender-bodied *Physocephala rufipes*. In addition to these, we may mention the *Qistridae*, containing *Gasteropliilus equi*, the larva of which lives in the stomach of the horse; and *Cephenomyia auribarbis*, which is viviparous, and deposits its larvae in the nostrils of the red deer; the young larva; make their way to the throat where they remain until full fed.

Vide " History of West Mulling Church. " (Oliver, publisher, West Mailing.) It contains Mr. Bennett's description of the churchyard, where the Kentish types described appear.

Professor G. S. Boulger, F.L.S., F.G. S.: Photo-micrographs Of Woods.—A series of sixty photo-micrographs of transverse sections of British and exotic woods, prepared and photographed by Mr. J. A. Weale, of Liverpool. These are but a selection from a far larger number. Each is from a section f of an inch square, magnified 30, or in a few cases 10, diameters. Some of them are to be utilised in a forthcoming new edition of *Familiar Trees.*

Rev. R. Ashington Bullen, B.A., F. L.S.: Holocene Shells From The Horseshof. Pit, Coi.lf.y Hii.l, Reioate.—This find has been fully described in the third volume of *Proc. Maluc. Soc*, p. 326. Thirty-five species were found, the most interesting being *Etia montaiia* and *Clausilia biplicata*, now restricted to only a limited area. All four British *Clausiliae* occur in the deposit. Another link in the chain of evidence that *Helix pomatia* is pre-Roman in this country is afforded by its occurrence beneath neolithic remains of a very early type. Some of these points are alluded to in Mr. Santer Kennard's admirable paper.

Mr. G. W. Butler, B.A., F.Z.S.: Eggs, Larvae, And Young Stages Of Tei. eostkan Fishes.—The eggs exhibited included the following types:— (A) Eggs that float (apparently confined to sea water) including (1) Eggs floating separately, illustrated by Eggs of Cod (from Captain Dannewig), Bass, Sole, Merrysole, Plaice, Flounder (all about the size of a pin's head) and other smaller eggs. (2) Eggs that float connected together, illustrated by the eggs of Fierasfer, fixed to the inner walls of cells of a remarkable transparent honeycomb structure (other eggs float more simply connected). All the above floating eggs are spherical. The eggs of the Anchovy, short cylinders with hemispherical ends, are quite exceptional among floating fish eggs.

(B) Eggs which do not float (found in both sea-and freshwater)—including (3) Eggs which lie loose at the bottom, illustrated by the eggs of the Trout (one-fifth inch in diameter). (4) Eggs which adhere *(a)* to each other in masses, or (/;) are attached separately side by side to foreign objects. The former *(a)* were represented by eggs of *Cottus scorpins* and *C. bubalis* deposited in clumps, as among branching roots of *Lamiiiaria* seaweed; eggs of *Liparis inontagui*; eggs of *Cristiceps argentatus* adhering byfibres in a tiny grape-like cluster; and eggs of Perch (spawned in the Zoological Society's fishhouse in 1897) adhering together in a lacelike network one egg thick. (The eggs of the Herring are peculiar among marine food fishes in that they sink and adhere at the bottom in masses.) The second group of adhesive eggs *(b)* was illustrated by eggs of the shore fishes, *Gobius pagaiicllus* (pear-shaped), *G. jozo* (cigar-shaped), *G. capita* (spindle-shaped), and other Gobies; eggs of *Blennius;* and *LcpaJogaster* (oval eggs with one side flattened against inside of empty *Pecten* shell). Eggs adhering in groups such as classed in section 4, are in many cases known to be tended by the male, or as in the case of the Gunnel, described by Holt, by both parents.

(C) Eggs which would not float by

themselves, but which are carried about by the parents; illustrated by eggs and parent fish of *Lophobranchii;*— *Ncrophis,* whose eggs are attached to skin of abdomen of male; *Syngnatlnis* and *Siplionostoma,* pipefishes whose eggs are carried by male in elongated subcaudal pouch formed by apposition of paired longitudinal folds; and the sea-horse, *Hippocampus,* with pouch in same position, but with small opening. (The males of the Siluroid genus, *Arius,* carry the large eggs in their pharynx, which is also recorded of *Chromis.* In only two genera of Teleosts, *Asperdo* and *Solenostoma,* according to Giinther, is the female known to carry the eggs outside her body; but we may also place in group C the eggs of viviprous Teleosts, *e.g.,* the British *Zoarces vivipara).*

The above mentioned series of eggs illustrated the association, that is generally observed, of parental care of the eggs with the allotment of relatively much yolk to each egg; comparatively little food yolk being stored in each egg, when, as in the case of freely floating eggs no parental care or protection can be given. (From the work of T. H. Milroy, 1898, it appears that the greater transparency and lower specific gravity of floating fish-eggs is due to a dilution of the yolk by watery fluid shortly before spawning. Therefore such eggs contain even less yolk than their size, small as it is, might suggest.)

The larval and young stages exhibited, included two series of Trout eggs and larvae in alcohol, mostly fixed on glass, showing to the naked eye the development during five weeks before hatching and during two months after hatching (these were preserved from eggs kindly given for study by the late Mrs. Andrews, from the Haslemere Trout farm in 1895-6); also series of larval stages of Perch, Bass, Plaice, Sole, Merrysole, Fierasfer, *Cottits bubalis, Gobius capito, Ceiitronotiis gunnellus* and *Hippocampus,* and, in addition, a series of larval Cod from Norway (from Prof. G. B. Howes), of Herring (from Mr. E. W. L. Holt) and larvae of Pilchard, Mackerel and Angler (from the Plymouth Laboratory, collected by Mr. J. T. Cunningham, M.A., F.Z.S.); also young or post-larvel stages of *Belone, Atherina,* and Flounder; also a series larval and young stages of Turbot (from the Plymouth Laboratory, collected by Mr. J. T. Cunningham).

The above series of larval and young stages illustrated, with the aid of camera sketches, magnified 40 diameters, (1) the undeveloped and larval conditions of the fish as hatched, especially in the case of the poorly yolked freely-floating eggs; (2) the development of special larval structures such as the dorsal filament in Fierasfer, which quite disappears in the subsequent development; (3) the exaggerated development in the larva of structures destined to persist, *e. g.,* the specialised fin-ray processes of the Angler and the elongated lower jaw of the Garfish; (4) the development of the tail and other medium fins within the continuous larval fin; and (5) in the Flounder and Turbot the course of head distortion in Flat-fishes whereby both eyes appear on the upper side.

In addition to those already mentioned, the exhibitor acknowledges his great indebtedness, first and foremost to his teacher in zoology, the late Prof. G. B. Howes, also the Zoological Stations at Plymouth and Naples, and to the British Association for the use of their table at the latter in 1890.

Mh. Hubert Elgar, Maidstone Museum.—A case containing a few of the larger *Syrphidae* or Hover-flies; also photographs to show the wing neuration of the fly, *Voliicella bombylaus.*

Mr. G. E. Frishy, Redhill.—British And Foreign Aculeate Hymenopteka. The British species included the following rarities: —*Formica exsecta, Myrmica sulcinodis, Stenamma westu'oodi, Methoca ichneumouides, Pseiidagenia pmictum, Pompilus unicolor, Sal ins obtusiventris, Astata stigma, Stigmus solskyi, Mimera shuckardi, Gorytes bicinctns.Cerceris quadricittcta,Crabro gonager, C. lituratus, Vespa austriaca, Odynerus melanocephalus, O. antilope, Colletes margiii(tta,C. cunicularia, Prosophis corntita, P. dilatata, Sphecodes reticnlatns, S. spimdosus, S. rubicundus,* S. *hyalinatus, Halictus bicinctus, H. malacliiinis, Andrena apicata, A. ambigna, A. lapponica, A. fasciata, A. cetii, A. proximo, Cilissa melanura, Nomada sathbiiriana,-N. lateralis, Megachile versicolor, Osmia pilicornis, O. leucomelana* and *Bombus smitliianus.*

Among the most conspicuous of the foreign species were *Kctatomma inbercuJata,* and *Oecodoma dentata* (Peru). The latter is one of the leaf-cutting ants. *Lobopelta septentrionalis* (Colorado), the handsome *Mutilia accidentalis* (New Jersey), *Scolia riibiginosa* (Pegu Hills), and *Dielis 4-uotata* (Florida), were particularly noticeable. The large *Macromeris violacea,* from Sikkim, provisions its nests with large hairy spiders, and *Parageina argent if ions* (Lower Burma), does the same, but with smaller specimens. Other striking insects were *Pepsis marginata* (Cuba), a very fine series of East Indian and North American species of *Spliex, Vespa cincta* (Rangoon), '. *oricntalis* (Egypt), V. *Carolina* (United States), *Masaris zonal is* (United States), *Enmenes petiolata* (Rangoon), *Rliyichium brnnneum* (Rangoon), *Halictus parallelus* (New Jersey), *Nomia nortoni* (Nebraska),*Megacilissa eximia*Venezuela), *Nomada grandis* (Colorado), *Megachile disjiuicta* (Lower Burma), *Anthopliora tarsalus* (south France), A. *go//i77*(Darjeeling), *Trigotia fulviveutris* (Para), *Cgntris Haviirons* (Nevada), many species of the Carpenter bees, /. *xylocopa,* from India, Burma, Egypt, Sierra Leone, Cape Colony, Transvaal, United States, and Brazil, *Bombus orientalis* (India), and *B. ternarius* (Colorado).

Mr. W. H. Griffin, Catford And District Nat. Hist. Soc.— (1) Mounted specimens of *Epipactis latifolia* and *E. media,* taken and pressed at the stage when the flower-buds were just appearing, shewing the marked difference in habit at that stage, which fully justifies their separation into two species. In *E. latafolia* the inflorescence stands clearly above the leaves, and is slightly bowed over at the top; whereas in *E. media* (and also *E. violacea*) the stem, about one-third of its length from the apex, is bent at a right angle, and sometimes downwards at an angle of 45, and the flower-buds are completely hidden in the foliaceous bracts. (2) Maps and sections showing the course of the Ravensbourne from Catford Bridge to its original sources, in two rivers now extinct, one of which ran from Hayes and through Addington to Chelsham, and the other from Hayes to the low chalk lands below Keston church; the special feature being a terrace gravel to the west of Bromley, not denoted on the geological drift map, and which is 25 feet above, and half-a-mile distant, from the present river.

Dr. Gerald Hodgson. Reicate.— (1) Living specimens of *Orchis laxiflora* (sent for the Congress by E. Marquand, Esq., of Guernsey) side by side with *O. incamata*, for comparison of the leaves. Three plants of *O. apifera* in each of these the new (1905-6) tuber had been removed, they were potted with chalk nodules and down mould under a thick mossy sod of turf. (2) Photographs of orchids, representing for the most part local species, shewing the first appearances above ground, and the underground appearance of some spring-appearing plants. (3) A series of the common species of our native "blue" butterflies, arranged specially for study of parallel variation, or otherwise.

Mr. Jonathan Hutchinson, LL.D., F.R.S., D.sc—(1) Terminal SHOOTS OF THE AUSTRIAN PINE BORED BY THE PINE-BARK BEETLE (hy Lurgus Piniperda.—A freshly attacked twig may be at once known by the white lump of hard turpentine which forms around each orifice made by the insect, there are usually two or three holes in each twig. Later, the attacked twig may be recognised by the withering of the needles above the point of attack. *H. piniperda* never breeds in the shoots, and the absence of excrement at once distinguishes the burrow from that of the larva; of the moth *Retinia buoliana*. The Scots pine is more usually attacked by this beetle. Schlich says the Weymouth pine sometimes suffers, the larch very rarely. In our experience the Austrian pine suffers quite as much as the Scots pine.

(2) A series of overgrowths and tumours from various trees accompanied by the following notes: On The Establishment Of Ekratic Growth-centres In Vegetable Tissues.—These specimens are exhibited as illustrations of local over-growth to which vegetable structures are prone under the influence of local irritation. The effect of local irritation of various kinds is in the first instance to attract and accumulate sap, ubi irritatio ibi fluxus. From this, overgrowth, often with extreme rapidity ensues. The type of the overgrowth and the quality of the new structure produced are modified according to those of the part attacked. Galls illustrate these facts in a very remarkable manner. In many cases, the overgrowth quickly comes to an end when the cause of irritation ceases, as when a larva becomes encapsuled. In others, however, a new growth centre having been initiated, the hypertrophy is continued long after the irritation has ceased. The specimens shown bear upon this last mentioned fact. No. 1 is a large growth developed upon the branch of a vine. It is as big as a large orange, although the branch which it wounds is not so thick as a little finger. Nos. 2, 3, 4, etc., are from English oak-trees, in which they are very common. The growth of these knobs is continued for several years, although the source of irritation ends its existence within the first year. A section of these growths shows great increase in thickness both of bark and of the wood. The bark which shows no signs of disease is simply greatly thickened, being four times that of the opposite side of the branch. The wood is similarly thickened, whilst it shows no signs of disease, and no trace of a parasite. The conditions are those simply of overgrowth. A notable consequence of the continued overgrowth of the wood is that the bark is unable to enlarge itself sufficiently and cracks under the distending force; to the deep rugged fissures which result, the name " canker" is popularly given. In the early stages, however, the bark is usually quite smooth and unbroken. The wood in these nodosities, although at first rather soft, may subsequently become exceedingly hard. This hardening is well seen in the growths which occur on crab-trees. These growths have no true analogy with cancerous and other new growths as observed in animals. They approach more nearly, though not closely, to the hypertrophy known as "elephantiasis." (3) Thistles (carduusarvensis) Attacked By A Fungus (puccinia Suaveolens) Contrasted With Healthy Ones To Show That Disease May Induce Premature Orowth.—The healthy specimens were robust and short, and the diseased much taller and weaker. The difference is only noticeable in the early stages of growth. The fungus does not spread by aerial infection; an infected plant may be closely surrounded by others which are without any trace of the parasite. The following note upon the biology of this interesting fungus is taken from Dr. C. B. Plowright's *Monograph of British Uredinaeae and Ustilagineae* (p. 183):— "The first generation consists of roundish uredospores, spermogonia, and sometimes a few stray teleutospores. The mycelium invades the whole of the infected plant, and hybernates in the upper part of the rootstock, so that every shoot sent up from this root contains the mycelium, and bears spermogonia and the primary uredo (*Uredo suaveolens*, Persoon). The affected plants appear sooner than the healthy ones, have a sickly pale-green colour, and do not bear flowers. The second generation consists of uredo spores, and teleuto spores, and has only a localised mycelium (--*P. obtegens*, Fckl)." (4) A LARGE PIECE OF CORAL WITH SPECIMEN'S OF THE CORALDWELLING Shell (magilus Antiquus.)—This mollusc takes up its abode within the crevices of living coral in the Red Sea, and Indian and Pacific Oceans. It is soon compelled to enlarge its home (a spiral shell about the size of that of a banded hedge snail), to keep pace with the coral which is always growing around it and threatening to seal it over. This it does by adding to the apperture of the shell, which is sometimes prolonged into a tube (seldom straight), nearly twelve

inches in length, the animal leaving the spiral part and living just within the mouth of the tube. (5) Glass-rope Sponge (hyalonema), showing the long tail consisting of opaque glass-like hairs, by which it is fixed in the mud of the deep-sea bottom, and a parasitic zoophyte attached to the upper part of the rope. This sponge is fished up with hooks from the deep seas off Japan. (6) Venus' Flower Basket (euplectella Aspergillum).—The skeleton of a fixed glass sponge; showing the delicate trellis work of glass, the lid resembling lace in appearance, and the glass strands by which it is anchored to a stone. The living sponge is soft, gelatinous and brown. It occurs off Japan and China.

The Photographic Record And Survey Of Kent had on view a selection of about 50 prints from their permanent collection, which was founded at Maidstone at the last Congress. Amongst the number were some architectural studies at Yalding, by Mrs.

Savile Reid.

Remains of Boslcy Abbey Rev. Gardner Waterman, Loose.

Prints of the Dutch Invasion... Benj. Wade, Sheerness.

Weathering of Rocks G. Abbott, AI. R.C.S.

Studies of Canterbury Cathedral... F. J. Argles (Hon. Treas., Maidstone).

Orchids *in situ* H. J. Elgar, Maidstone Museum.

Barreston Church T. Nottage, Ashford.

Evolution of carving on tombstones.. . F. J. Bennett, West Mailing.

Groombridge place......... H. Gower (Surrey Survey).

Views on the Darenth J. H. Baldock (Surrey Survey).

Old Rochester............ H. F. Wingent (Hon. Sec.) and members of the Rochester Naturalist's Club Photo. Section.

Mr. Eustace Large, Catford Nat. Hist. Soc.—A series of stereo-photographs to emphasize the value of stereo-photographs for scientific purposes. *The South-Eastern Naturalist, IUU5.*

Captain Mcdakin, Dover Sciences Society.—Examples of *Helix elegans* from the colony, found by Mrs. McDakin, near Dover in 1891. This colony is still flourishing; specimens were presented by Mr. Stanley Cox to the British Museum in the same year. See note by T. Carrington, in *Field,* August 29th, 1891).

Mk. E. Penfold, Reigate.—Original coloured sketches, also prints and photographs depicting ancient Reigate.

Mk. C. E. Salmon, F.L.S.—A series of rare local plants from his private herbarium, about a dozen sheets, including *Althaea hirsuta, Cerastium puinilum, Tcuciium botrys* and *Mentha gentilis.*

Photographic Survey And Record Of Surrey.—The collection of Photographs exhibited by the Photographic Survey and Record of Surrey, comprised examples of the different sections of work which is being done, viz.:—Architecture, Art and Literature, Anthropology, Geology, Natural History, Topography, and Passing Events. The collection which is stored at Croydon, and now numbers considerably over 1700 prints, bids fair to become one of extraordinary interest and value.

Mr. E. W. Swanton, Mem. Assoc. Economic Biologists, Brockton, Haslemere,—(1) Photographs of Potatoes from a crop almost totally destroyed by slugs of four species:—*Agriolimax agrestis, Milax sowerbyi, Arion hortensis* and *Arion subfitscns.* The tubers were honey-combed, and in section resembled a bath sponge.

(2) The early state (egg-like) of the stinkhorn fungus *(Ithyphallus impudicus).* Specimens were sent to the exhibitor by the Rev. E. N. Bloomfield, with a notification that they were brought to him as truffles. They were probably mistaken for the White Truffle *(Choiromyces meandriforniis),* a fungus which belongs, however, to quite a different order; being an Ascomycete. The *Ithyphallus* is one of the *Gustcromycetes*—Dr. Cooke and other writers have remarked that it is edible in the young state. (3) Photographs and coloured drawings shewing fasciated stems in the great valerian (*Valeriana officinal is).* These fasciations usually result from injury to the growing buds; they are not infrequent in daisies and other plants met with on lawns, but are seldom seen in such an exaggerated form as generally obtains in the valerian. The condition is produced by the uniting side by side of twenty or more stems. The union begins very early in the formation of the stems, whilst their tissues are soft and succulent, and before the budding stem makes its appearance above ground. Each of the united stems retains its life endowments and produces flowers.

Two Views Ok Ironstone Implement Prom Near Haslembrr. (Slightly less than natural size).

(4) Coloured drawing depicting leaves of *Viburnum lantcma* with pustular galls caused by an unknown Cecid, also photographs of the same by Mr. Edward Connold, F.E.S. This gall is apparently a new record for Britain. It was found on the chalk downs near Maidstone by the writer, on Sunday, June 12th, 1904. The pustules were about-Jineh in diameter, reddish at first, purplish in specimens that had been placed in a vasculum for two or three days. A few of the leaves were simultaneously galled by *Eriophyes (Phytoptus) viburni.* (5) A scalarid example of *Hygromia rufescens* found by Mrs. E. W. Swanton, at Inval, Haslemere, and a Melanistic form (var. picea) of *Helicigona arbustorum,* found near Marnhull, in Dorset. (6) Galls on the common spruce caused by *Chermes abietis,* showing two well-marked forms, *(a)* That usually described and figured, in which the pseudo-bracts are brightly coloured at their margins; *(b)* a form lacking the brilliant coloration, more globular, and usually terminal. It always arrives at maturity earlier, the flies emerging at least a fortnight sooner than those from the other form. (7) Ironstone implement found in the railway cutting below Haslemere station about the year 1850. It measures four inches in length, and is two inches in width. (The illustrations are a little less than natural size). The hole tapers towards the centre. Sir John Evans remarks that this implement is of a considerable interest, and is not unlike fig. 124 in his

Ancient Stone Implements, which represents an unfinished green-stone implement found near Stourton, Wiltshire. It was probably used as a kind of hoe. Iron-stone implements are rarely met with; there are only four references to such in the work mentioned. On page 84 it is stated that " Celts, merely chipped into form and unground, occur also in other kinds of stone. One of iron-stone from Sussex, 8 inches long and 3£ inches wide at the broad end, is in the Blackmore Museum." On p. 184 there is a reference to an axe-head formed of some kind of iron-stone 5 inches long, " found with the remains of a skeleton, an amber cup, a white-stone, and a small bronze dagger with two rivet holes, in an oaken coffin, in a barrow at Hove, near Brighton" (see *Coll. Sussex Arch.* vSoc, vol. ix., p. 120). "The hole is described as neatly drilled. A weapon of the same kind (3 inches), blunter at the ends and described as a hammer, was found with a deer's-horn hammer, and a bronze knife-; in a barrow, at Lambourn, Berks. " The fourth and last reference (p. 222) is to the stalagmitic breccia in Rcbin Hood Cave. "In this were found implements ofquartzite and iron-stone, eighty-six in number, more than those of flint in the breccia." Fig. 413n depicts one of the iron-stone implements.

Mr. A. W. Vennf.r, Reio.ate.— Portion of Roman bonding tile from the Reigate tilery; Roman roofing tile, showing imprint of the foot of a dog.

Mr. R. H. Welch Man, B.A., Assisted Dy Mr. Mitchener.—A series of living plants to illustrate the paper on "The Flora of Reigateand District,"includingthe Ground Pine*(Ajugachai)iaepitys),* Man Orchis *(Aceras atithropophora).* Fly Orchis *(Ophrys muscifera).* White Hellcborine *(Cephalanthera pollens),* Deadly Nightshade *(Atropa belladonna),* Corn Cromwell *(Lithospermum arvense),* Hound's Tongue *(Cynoglossnin officinale),* var. Field Clover *Trifolium pratense* var.*parviflornm),*Crimson Vetchling *(Lathyrus nissolia.)*

Mr. W. P. Young, F.R.M.S., Battersea Field Club.—Thirtylantern slides from negatives taken for the Photographic Survey and Record of Surrey. Photomicrographs showing *(a)* stinging hair of stinging-nettle; *(b)* growing point of the stem (longitudinal section) of *Hippuris vulgaris;* (c) diatoms; *(d) Pinnularia nobilis; (e)* stem of umbrella plant (transverse section); *(/)* transverse section of mid-rib of leaf of allspice; (g) leaf of *Deutzia,* etc.

REPORT OF THE REV. R. ASHINGTON BULLEN, DELEGATE OF THE S.E. UNION TO THE BRITISH ASSOCIATION AT CAMBRIDGE, IN AUGUST, 1904.

Your delegate has to report as follows:—

A. Recommendations Of B.A. Committee.— 1. The Corresponding Societies' Committee, in their report, recommend Mr. William Cole's suggestion that local societies should make it a part of their systematic work to enter upon the 6-inch ordnance maps of their respective districts, any natural features and archaeological remains which are not indicated thereon.

2. That Corresponding Societies should assist, by scientific, advice and otherwise, those teachers in elementary and secondary schools who are taking up Nature Study. Certain societies have already undertaken such work, e.g., the Halifax Scientific Society has given lectures to children and scientific aid to teachers, and the Croydon Natural History and Scientific Society is furnishing loan collections of natural history objects to schools' and has assisted educational work by means of addresses. 3. Only twelve of the Affiliated Societies have reported that they have been able to do any original work during the past year, most of the work undertaken has been botanical, relating chiefly to local botanical surveys. 4. The Committee desire to urge upon the representatives of the various local societies the desirability of taking up some of the subjects set forth in former Annual Reports of the Committee.

B. First Conference Of Delegates, August 18th, 1904 (58 of the Corresponding Societies appointed delegates). —1. The first Conference was presided over by Principal Griffiths, F. R.S.,who summed up his speech in the following proposals.

(1) That any society which undertakes local scientific investigation, and publishes the results, may become a Society affiliated to the British Association. (2) That the Delegates of such Societies shall be members of the General Committee.

(3) That any Society formed for the purpose of encouraging the study of natural knowledge, which has existed for three years, and numbers not less than fifty members, may become a society associated with' the British Association. (4) That all Associated Societies shall have the right to appoint a Delegate to attend the Annual Conference, and that such Delegates shall have all the rights of those appointed by the Affiliated Societies, except that of membership of the General Committee. (5) That all Affiliated or Associated Societies shall contribute annually the sum of at least 5s. for each 50 members, and that the funds thus obtained be utilized for the purposes of a "Journal of Corresponding Societies." (6) (In case proposition (5) is approved) That the Council of the B.A. be requested to make an annual grant towards the expenses of such a journal, on the understanding that such grants shall cease if the journal become self-supporting.

Discussion.—Mr. W. Whitaker instanced a Society in an outof-the-way country place that published nothing but a bare Annual Report, and yet had a splendid museum. It did not publish because it spent all its money on its museum. He thought that such work ought to be encouraged, and that the B.A. Committee of Corresponding Societies should reconsider the terms on which societies are affiliated.

Sir Norman Lockyer explained that a " British Science Guild" was being started, quite independently of the British Association, and that it could work shoulder to shoulder with the various societies in the extention of their interests. He felt that it was necessary, in the interests of science, to influence the man who has to vote, not only in the County Councils, but in the House of Commons.

The Rev. G. B. Stallworthy, expressed his faith in the value of local societies and their connection with the British Association. He was glad that headquarters urged them to do something. He hoped that a Central Committee would consider whether it were possible to appoint a dozen responsible gentlemen to visit local societies, to report upon their nature and the kind of work done, and the report of such responsible inspectors, quite apart from the question of museums or publications, might determine the admission of such societies to connection with the British Association. Dr. Abbott urged the importance of Unions in different parts of the Kingdom on the lines of the S.K. Union.

The Rev. T. R. R. Stebbing thought the attendance of the President of the British' Association at the Conference of Delegates was a precedent of the highest value. He thought societies should be associated with the British Association, independently of their publishing or not.

Mr. A. 0. Walker thought the payment of 5s. per 50 members would prove burdensome in practice.

Mr. Reunert and Mr. Reid (delegates from South Africa) instanced the generosity of Cape Colony and the Transvaal Colony in providing ample funds, and heartily welcoming the British Association to South Africa in 1905, as an encouragement to Science.

On the enquiry of Captain Dubois Phillips, R.N., it was stated that the question of the reduction of railway rates for members of scientific societies was under discussion, but no definite result had yet been reached.

2. The Rev. W. Johnson, B.A., B.Sc, introduced the subject of " The utilization of Local Museums, with special reference to schools." In this valuable paper he urged:— (1) That a great amount of material lies buried in local museums. (2) That it needs proper description and exhibition to make it available for the use of young students. (3) That it is very desirable that local natural history rather than general science should be illustrated and studied in that connection.

He proposed that each town should have a strictly limited collection of the objects commonly found in that area; that the district of each should be indicated; that thei months in which plants may be found should be added; that there should be a large geological map of the area, with suitable vertical sections showing the connection between the underground conditions and the variety of life on the surface. He advocated taking the classes to the specimens in the museums. This would accustom the young student to resort to a definite place for the solution of his difficulties. The museum is too often a swamp in which the streams of knowledge lose themselves. In his opinion one of the firsa improvements by our museum trustees will be the provision of rooms for demonstrations with lanterns, and the needful lecture room appliances.

Discussion.—Mr. G. P. Hughes dwelt on the importance of getting the minds of the younger generation trained to the industrial interests they must follow in later life.

Rev. G. Capell instanced the practical training given to French and German children.

Mr. F. W. Rudler had suggested in 1891 (Museum Association at Cambridge) that demonstrations should be given in a separate lecture room, followed by an adjournment to the museum. He was glad to hear that it was successfully carried out at Leeds. He thought that the museum might also be taken to the children by circulating small loan-cabinets of simple specimens.

Mr. Hopkinson found that more than 100 school-children per week visited the Herts County Museum; they quickly detected any additions to the collections, which certainly indicated their interest. C. Second Conference, August 23rd, 1904.— (Principal Griffith, F.R. S. (Chairman) followed by Dr. Tempest Anderson.)— 1. The Chairman asked the Delegates to support the following resolution.

"That a Committee be appointed, consisting of members of the Council of the British Association, together with representatives of the Corresponding Societies, to consider the present relation between the British Association and local Scientific Societies.

"That the Committee be empowered to make suggestions to the Council with a view to the greater utilisation of the connection between the Association and the affiliated Societies, and the extension of affiliation to other Societies, which are at present excluded under Regulation 1."

The resolution was adopted unanimously.

Mr. John Hopkinson dealt with the following subject: "On the Conformity of the Publications of Scientific Societies with certain Bibliographical Requirements." He urged the necessity of— (1) Uniformity of size in scientific papers.

(2) Proper and adequate indexing. (3) Proper and accurate dating of parts of each publication. (4) Correct pagination of reprints.

Air. W. Whitaker supported Mr. Hopkinson's statements. A good editor required technical knowledge; agood writer and reader might make the very worst possible editor. Delegates should try and get their journals printed in regular orthodox fashion, contents in front and index at the end.

Mr. T. vv. Shore thought that the British Museum authorities should be advised that they were losing some out-of-the-way information of a very valuable kind by refusing the publications, however small, of any society.

The Rev. T. R. R. Stebbing thought the British Museum had no room for its books and pamphlets. The case was different at the South Kensington branch, however, for every possible care was taken there to collect pamphlets, reports and transactions. He thought the large London societies might advise as to the best form to adopt, and if an uniform size were adopted it would be a great advantage to our book shelves.

Dr. Tempest Anderson thought that the size of the *Century Magaziiie* 'woud be a good standard to adopt, as it would be an advantage, in the publication of photographic plates, to have the increased size.

D.—Reports Of Delegates From Various Sections.—Section A.—Dr. H. R. Mill reported that the Royal Meteorological Society was endeavouring to increase the interest in meteorological observation and to direct it systematically.

Section C.—Mr. W. Whitaker stated that it was the joint wis,h of the Geographical and Geological Sections to get a Committee appointed to determine and record the exact significance of local terms applied to topographical and geographical objects.

Section D. —Rev. T. R. R. Stebbing presented the sub-committee's report. The subjects proposed for the co-operation of local societies had not produced any substantial results.

The subjects are (1) Cave faunas, (2) Zoological changes on a given plot of land during the year, (3) Compilation of local faunas, (4) Systematic observations of micro-organisms of a pond or ditch, (5) Over-land lines of migrations of birds, (6) Slugs from all parts of the British Isles. (For further information see p. 400, Report of British Association for 1904).

Section E.—Dr. Herbertson desired that information as to papers dealing with questions of distribution should be sent to the Royal Geographical Society that they might be noticed in the Bibliography of Geography.

Section K.—Miss Ethel Sargant stated that she had only received one communication with regard to orchids, but none about seedlings, information concerning which she still solicited.—R. ASHINGTON BULLEN. PROCEEDINGS OF THE CONGRESS OF THE SOUTH-EASTERN UNION OF SCIENTIFIC SOCIETIES, 1905. Wednesday Afternoon'. *June 7th.*—The proceedings of the Tenth Congress commenced on the afternoon of June 7th, when Miss Ethel Sargant entertained the Delegates to lunch. The Mayor of Reigate was present. Mr. F. W. Rudler, I.S.O., expressed the thanks of all present for Miss Sargant's kindness and congratulated her on her election as a Fellow of the Linnean Society.

After lunch 40 members joined in a botanical and geological excursion to the North Downs, first visiting Messrs. Hall's Lower Greensand Pit in the North Albert Road, where the junction of the Gault and Lower Greensand can be seen. Then, in rapidlyincreasing damp, the party went up the Pilgrim's Road, across the Gault Plain and Upper Greensand Terrace, up to the summit of Colley Hill, being welcomed to tea at Margery Hall by Mrs. G. Taylor and her family. Miss M. Crosfield conducted the geological section, and Professor Boulger, F.L.S., &c, and Mr. C. E. Salmon, F.L.S., the botanical section. In such pouring rain, however, work was almost impossible, and the 29 survivors who reached the summit were glad of Mrs. Taylor's hospitality. Mr. and Mrs. Taylor were thanked by Mr. Rudler (the President) in a speech that was warmly applauded.

The Local Arrangements.—Naturally the local arrangements have involved a considerable outlay of labour on the part of the executive committee. The names of members will be found on page (v).

The Delegates.—The names of the delegates present and the societies represented are appended:-.Messrs. R. Adkin, F.E.S. (South London Entomological), J. H. Baldock, F.C.S. (Photographic Survey of Surrey and Croydon Natural History), E. J. Bedford (Eastbourne), F. Blackburne (Hastings and St. Leonard's), F. Campbell Bayard (Croydon), Dr. Chapman (City of London Entomological and Holmesdale), Messrs. A. T. Comber (Horsham), H. J. Elgar (Maidstone), S. Edwards, F.L.S. (West Kent), J. W. Eele (Haslemere), F. G. Fleay, M.A. (Balham), R. C. Frost (Woolwich), H. Norman Gray (City of London College), W. H. Griffin and E. Large (Catford), T. Haldane Harrison (Photographic Survey of Surrey), F. W. Hembry, F.L.S. (Sidcup), E. Gane Inge and E. W. Swanton (Haslemere), F. J. Inniss (New Brompton), F. Merrifield, F.E.S., and Dr. E. J. Spitta (Brighton and Hove), Capt. McDakin, R.E. (East Kent), Mrs. McDakin and S. Webb (Dover), Messrs. F. Meeson (Woking), C. P. Nicholson, F.L.S. (North London), J. L. Otter and W. Mark Webb, F.L. S. (Selborne), E. S. Payne (Hampstead), N. F. Robarts, F.G.S. (Croydon), W. T. Symons (Northfleet), H. Sparks (Eastbourne), Miss A. K. Sweetland (Southborough), Rev. T. N. H. SmithPearce, M.A. (Epsom College), Mrs. G. R. Taylor (Holmesdale), Messrs. W. T. Vincent (North Woolwich), W. Mark Webb, F.L.S. (Fulham), H. F. Wingent (Rochester Naturalists), H. Wilson, M. A., F.S.A (Bromley). W. Plomer Young, F.R.M.S. (Battersea), Mrs. Plomer Young (Polytechnic); Mrs. Adeney, Rev. and Mrs. T. R. R. Stebbing and Mrs. Dodd (Tunbridge Wells). Those also attending the Congress include the following ladies and gentlemen resident in Reigate and Redhill:— The Misses Allfrey, Mrs. S. Allen, Mr. and Mrs. J. W. Ashby, Miss Bennett, Rev. E. J. Baker, Miss Baxter, Miss A. Bowyer, Mr. and Mrs. G. W. Butler, Mrs. Burt, Dr. T. A. Chapman, Misses Chapman, Misses Campion, Misses Cooper, Miss Clayton, Mr. A. J. Crosfield, J.P. Mr. J. B. Crosfield, Miss Crosfield, Mrs. Cudworth, Mr. and Mrs. C. Davison, Miss Dickens, Mrs. Eumorphopoulos, Mr. and Mrs. G. E. Frisby, Mr. and Mrs. K. R. Fletcher, Miss L. M. Gibb, Mrs. Grant, Mrs. Heisch, Mr. and Mrs. F. Hughes, Mr. F. H. Hughes, Miss Home, Mr. A. Hyde, Dr. Hewetson, Mrs. T. Halsted. Mr. W. Johnson, Misses Johnson, Mrs. Johnson, Mr. E. Dunkinfield Jones, Mrs. Latham, Mrs. F. E. Lemon, Mr. G. L. Merriman, Mrs. Mould, Misses Mould, Mr. and Mrs. P. Mordan, Miss G. Nicholson, Mrs. J. Powell, Misses Payne, Miss Phillips, Miss Pilleau, Miss Roberts, Mr. and Mrs. R. S. Ragg, Dr. Stone, Mrs. Simpson, Mr. S. Salmon and Miss Salmon, Mrs. Stone, Mrs. Sargant, Miss Sargant, Mr. and Mrs. Sambrook, Mr. C. E. Salmon, Mr. E. B. Sargant, Miss Taylor, Miss C. Taylor, Miss A. Taylor, Mr. and Mrs. W. F. Taylor, Miss Trollope, Mr. and Mrs. A. E. Tonge, Mr. A. W. Venner, Mrs. Ward, Miss Whitehead, Dr. and Mrs. Walters, and Mr. A. W. Yeo. The Company also included: — Mr. C. E. Britton (South Lambeth), Miss Bostock (Nuffield), Mr. W. H. Day (Maidstone), Mr. and Mrs. H. Warren (Dover), Mr. P. Harris (Croydon), Mr. R. Kerr, Mrs. Cole (Canterbury), Mrs. C. Beach (Hampstead), Miss Kerr, Mr. H. J. Turner (New Cross), MajorGeneral Newmarch (Brighton), Dr. Griffiths and Miss Griffiths (Horley), Rev. J. Thornton (Betchworth), Miss A. C. Klaassen and Miss E. F. Klaassen (Croydon), etc.

The following delegates were unable to be present:—Mr. Paul Matthews, M. A. (Rochester), Mr. and Mrs. Halliday (Hastings). Messrs. W. Whitaker, F.R. S., V.P., and Mr. J. H. Allchin, of Maidstone, also wrote regretting their inability to attend the Congress.

Wednesday Evening. (F. W. Rudler, I.S.O., F.G.S., in the

Chair).

There was a large attendance at the opening meeting at the Public Hall, on Wednesday evening, at the inauguration of which Mr. F. W. Rudler, I.S.O., F.G. S., (the retiring President), occupied the chair. He announced that his duty was fortunately a duty of the simplest possible character, and he could assure them it was as pleasurable as it was simple, for he had simply to resign this chair formally in favour of his distinguished successor. Introduction was unnecessary, for Professor Flinders Petrie was a man who had a reputation which was not European—it was world-wide. Wherever archeology was cultivated, wherever men where sufficiently intelligent to appreciate the value of the application of scientific principles to the interpretation of the past, there the name and work of Professor Petrie were known, and being known were held highly in honour. It was not simply that the Professor by his remarkable and long continued researches had thrown a flood of illumination upon ancient Egypt, but they, assembled there, representing scientific societies in the South of England, were not unmindful of the work which he did manyyears ago in connection with Stonehenge; nor could they forget the interest Professor Petrie had always taken in the question, which all local societies had greatly at heart—the great question of museums.

Professor Petrie then took the chair and with a few felicitous words to the retiring president, the S.E.U.S.S. members and Reigate friends present, proceeded to deliver his presidential address (see page 1).

Sir H. H. Howorth, K.C.I.E., in a eulogistic speech, moved a hearty vote of thanks to Professor Petrie for his able paper. He said he was perfectly astounded at the versatility of his old friend, and more than astounded with the surprise he had given them that night. For years past he had been investigating the antiquities of Egypt by a scientific method previously unknown, and everybody was expecting him to say something about these wonderful discoveries, and something about its old histories, but he was determined to spring another surprise upon them. He (the speaker) could only hope that after this interlude in his career he would spend many hours in these dark and smelling Egyptian tombs, out of which he had dragged so many secrets, and if he would give them the benefit of the great work with which his name would always be associated, they would be very grateful. With a kindly reference to Mrs. Petrie and her assistance to her husband in his work, Sir H. H. Howorth sat down amid loud applause.

The Rev. T. R. R. Stebbing said there are certain recognised formulae for seconding a vote of thanks. It can be done by modestly rising and bowing to the chair, or more boldly by saying "I beg to second this resolution," or, with still greater ceremony, by remarking that "it is impossible to add anything to the eloquent terms in which the proposer has moved it." The last is especially applicable when one has to follow such a speaker as Sir Henry Howorth. Nevertheless, there are this evening one or two facts, sufficiently obvious though they may be to the eye, which I shall venture to put into words. This is, I believe, the largest audience, that has ever been present at the Congress of this Association to hear the opening address by the President of the year. Certainly, never before has there been so large a gathering of past presidents as that which has so agreeably assembled this evening to listen to their distinguished successor. There is much encouragement to be derived from the address itself. It has, I think, been calculated that when Bessemer developed his process for the preparation of steel he practically made the world a present of some millions a year. The advice which our president has given us to-night opens up a pleasant vista of a saving of forty millions a year to Great Britain alone. When that occurs, a grateful nation surely cannot do less than settle a handsome pension on every member of this Association for their services in calling Professor Flinders Petrie to the chair, from which so brilliant a prospect has been unfolded. I shall ask Sir Henry Howorth to include in his resolution a request that Professor Petrie will allow his very able and suggestive address to be printed in our *Transactions*, in full assurance that the motion so amplified will be carried with and by your unanimous applause.

The motion was then carried by acclamation and acknowledged from the chair.

Thursday Morning. (Professor W. M. Flinders Petrie in the

Chair.)   June   8th.—Delegates'Meeting.—There was a good attendance at the delegates' meeting, which was held at the Public Hall, Reigate, in the morning, when the Annual Report of the Council to the delegates, for 1904-5, was read. It stated that the last year had been financially a success. Forty-three societies belonged to the Union. The Maidstone meeting last year was much appreciated, and would not easily be forgotten. A special feature of the Congress was the crowded meeting of pupil teachers and elder scholars to listen to Mr. Webb's paper on "The Teaching of Nature Study," the meeting being very successful. Mr. Norman Gray was appointed secretary to manage the autumn meetings. There were reported to be 23 life members in the Union. The death of the late Prof. G. B. Howes, LL.D., F.R.S., was deeply regretted. The report appears in full on page xvi.J

The Rev. T. R. R. Stebbing, being called on to move its adoption, observed that it was a pleasure to move that the Report by the Hon. Sec, the Rev. R. Ashington Bullen, be adopted and printed. Financially, and in all other respects except one, it is a most satisfactory and agreeable report. The single exception to our pleasure in it is the unavoidable note of sorrow occasioned by the death of our lamented vice-president and past president, Professor G. B. Howes, F.R.S. The motion was seconded by Mr. Meeson and carried.

Dr. Abbott was elected a delegate from the S.E.U.S.S. to the British Association, for the Conference to be held in London in the Autumn, on the proposition of Dr. Boulenger, F.R.S., seconded by the Rev. T. R. R. Stebbing, F.R.S.

The British Association's suggestion that the members of natural history societies should receive the same travelling privileges as members of golfing and fishing clubs was unanimously adopted. See page xviii. for the form accepted by the Union and signed on behalf of the Delegates by the President.

Mr. E. Penfold then read an exhaustive paper on "Old Reigate," which will he found on page 33. The members then adjourned to a darkened room, where Dr. Hodgson read an interesting paper on "The Orchids of the District." Having described the district as an area covered by a walk of twelve miles out and twelve back, lie gave a list of the resident natives, remarking that very frequently the orchids in this district corresponded with the London catalogue's British station number, the exceptions being the Bee orchis, which is much more abundant here than according to the London catalogue; the Man orchis, which is more abundant here, the Musk orchis, which is commoner, and the Marsh orchises, which are almost unknown. Dr. Hodgson, having given a list of the orchids, some new localities since the *Flora of Surrey* in 1863. and some disappearances, mentioned that Reigate Hill was no longer a locality for the musk or the dwarfwinged varieties, nor to any practical purpose were many others likely long to be localities for any but a few chance specimens. All the hills east of Redhill were tending to a pandemonium of unorchis-like plants. Two or three good localities for broadleaved and other Helleborines, owing to building, had ceased to be so. Disappearances from extant localities had occurred at Reigate, Colley, Buckland, and other hills to and beyond Boxhill; and in marshy lands such as Merstham pools, Reigate Heath, and thence on to the south side of Betchworth, and in other marshy ground formerly round Redhill station, the marsh orchids were increasingly difficult to get. Looking at the district generally, there was a terrible tendency to degenerate, and not the least criminal were Sanderstead, Coulsdon, Epsom, and Walton-on-theHill. Dealing with the variations, he said observation showed that the plant was not necessarily produced in the same form year after year. Occasionally, as in the four-leaf butterfly orchis, it would have four leaves instead of the usual two in two consecutive years. With respect to Bee orchis he remarked that the variety with imperfectly reflexed appendage was probably to be met with wherever the plant was at all abundant. With regard to the Helleborine known as *Epipcictis media*, it was most interesting to note, that in a locality on Boxhill, it appeared some years as *media* and others as *violacea*, perhaps well marked with cream and paler violet variegation of leaves. Wide variations appeared in the same plant, and from the typical form of a plant nothing could be absolutely predicted as to its form another year. In fact, they would find it was self-fertilising one year, and another year largely not. The green-winged meadow orchis had narrow leaves in 1895, the same plant having broad leaves in 1894. Having dealt at some length with these variations, Dr. Hodgson said the absence of the small butterfly orchis from Walton Heath and similar places seemed remarkable, and the presence of Bee orchis in fir copses in one or two places struck one as strange. The marsh orchids were perhaps exceptionally abundant now-a-days, hundreds being obtainable in a three or four miles' walk. The green-winged meadow orchis, more than in other parts one had come across, showed an extraordinary contrast between the meadows fed over by cattle, or cut yearly for hay, even in the case of opposite sides of a hedge and opposite sides of a road. The paper also dealt with many points of interest suggested by a close study of orchid.s.

The paper was illustrated by lantern slides, the lantern being splendidly manipulated by Mr. Brooks.

The papers being lengthy there was no time for discussion.

Professor Petrie briefly thanked Mr. Penfold and Dr. Hodgson for their able papers, and the Congress then adjourned.

Thursday Afternoon. Upwards of 100 members took part in the afternoon ramble, which proved very enjoyable. Under the leadership of Mr. E. Penfold various spots of archaeological and historical interest were visited and inspected. At the Parish church the old library, mentioned by Mr. Penfold in his paper, was on view, while the old dungeons in the Market Place, the ancient prison, the chapel of St. Lawrence, and the caves, which were illuminated, were also visited. Tea was served in the Castle grounds, by kind permission of the Mayor, and the afternoon proved a most delightful one.

Thursday Evening (Professor W. M. Flinders Petrie in the Chair). In the evening another meeting was held at the Public Hall, Professor Petrie occupying the chair.

Mr. A. Santer Kennard (member of the Geologists' Association and the Malacological Society), read a paper jointly prepared by himself and Mr. B. B.Woodward on "The Land and Fresh Water Shells of the South of England." The paper dealt with the various extinct species of well-marked varieties of land and fresh water shells which had been found in the pleistocene and holocene beds of the South of England. It was pointed out that while twenty extinct forms occurred in the southern area, only two occurred in the Severn area, and it was possible that these western examples had reached England from the West of France, since both species now occur there in a living state. The authors pointed out the difficulties there were in correctly determining the various species, owing to the chaotic state of the nomenclature, the needless naming of so-called species, especially on the Continent, and the fact that often the form which bore a certain name in this country was not the same as the species which Continental authors called by the same name. In their opinion the correct determination of the pleistocene fossils was a matter of great scientific importance, since it was clear they would throw considerable light upon the climate of the period. It was noted that the situation at the present time with regard to the so-called glacial period was one of conflict between two opposing schools. The geologists, in order to account for certain deposits, had postulated an ice age, during which the larger part of these islands must have been covered by ice, and the fauna and flora practically exterminated. On the other hand, the palaeontologists and students of distribution maintained that there could not have been this wholesale extinction of life, and consequently no great ice age. All the post-pleistocene land and fresh-water shells were now known to occur on the Continent, either living, fossil or both recent and fossil. One species, *Pisidium astartoides,* which was for a long time supposed to be confined to England, had now been found in two pleistocene beds in Denmark. As an instance of the imperfection of the geological record, it was noted that one species *Neritina gratcloitpiana,* which had hitherto been unknown in any deposit later than the Upper Miocene, had been found in countless myriads in a small patch of ground belonging to the 100 feet terrace of the Thames, at Swanscombe, all the specimens retaining their coloration. Whilst admitting that species had become extinct in this country, it was urged that this had been brought about by the operation of natural laws, and not by hypothetical cold periods. The co-operation of all observers was necessary in order that these pleistocene and holocene problems might be solved. The President expressed the thanks of the Congress to Mr. Kennard for his paper, and a brief discussion followed. In this discussion, Mr. Griffin, Mr. W. Mark' Webb, Professor Boulger, and the Hon. General Secretary took part. Mr. Kennard in replying said it was to be desired that Professor Flinders Petrie would record all land and freshwater shells from his Excavations in Egypt, at whatever level found, or of whatever date.

Friday Morning (Professor W. M. Flinders Petrie in the Chair).

June 9th.—The meetings were continued at the Public Hall, Reigate, on Friday morning, the session being opened with a very clever demonstration by Miss' Saunders (botanical lecturer at Newnham College) on "Mendel's Law. " At the conclusion of the demonstration, which was frequently loudly applauded, a brief discussion ensued, and Miss Saunders was heartily thanked. The President said their thanks were due to Miss Saunders for her extremely clear and lucid statement upon this very important question. It would be very desirable, if anyone wished to ask any question or to make any remarks, that they should do so immediately, as unfortunately the time remaining for discussion was very limited.

Mr. Griffin said comparisons were odious, but having attended this congress for several years, one felt bound to say that the best papers they had were always from the ladies. They were invariably the most original and instructive, and Miss Saunders' paper was the best and most instructive they had had at the Congress, after the Presidential address. Speaking upon the title of the paper, " Mendel's Law," he declared they were a tremendously long way from any settled law. Anyone experimenting with this system would know very well that they had two forces to contend with, one was the tendency to reproduce the form of the parent, and the other, and it seemed to him the stronger, was an inherent tendency to discover fresh types and make new laws. Apparently, some attempt had been made to reduce these two opposing forces to law. Unfortunately, experiments were absolutely necessary, and if any innocent friends, who had heard this lady say it was such a very easy matter to do, intended to try their hand, they would find it was not such an easy matter to secure any results that would he of any value, because they would find that the insects would upset all experiments. They would have to be very careful in making their records, and instead of it being easy, they would find these experiments one of the most difficult things to carry out. He would not advise them to embark on this, unless they were prepared to exercise infinite patience, and carry on their labours for a long number of years.

The Rev. E. J. Baker, inquired the name of the manual on the subject that had been quoted in the lecture.

Miss Saunders replied that the manual she referred to was "Mendelism," by Punnett, published by Messrs. Macmillan & Co.

Mr. G. C. Walton, F.L.S. (Folkestone), said that doubtless t+ie occurrence of natural hybrids lay outsidethe scope of Miss Saunders' lecture. He would, however, as a practical botanist, like to hear a few words on the question of hybridization as they saw it in several of the genera of British wild plants, *e.g.* , in the species of *Primula, Verbasciim, Caniitus,* and *Salix*. He asked could any light be thrown on the tendency to hybridization, in a few cases, whilst it was not observed in the great majority. He would be glad if the lecturer would elucidate the point.

Miss Saunders said she knew nothing of the point that had been raised. That hybrids did occur they all knew, but as to the conditions that resulted in the hybrid occurring in some instances and not in others she had no knowledge, and she was afraid she could not offer any opinion or throw any light upon the subject.

Professor Boulger said he would like to add his testimony anil congratulate Miss Saunders upon her admirable exposition of this new law. He did not see that it was of any use talking about inherent tendencies. He never yet had met anyone who could tell him what a tendency was. It seemed to him to be pure evasion and ignorance, and Mendel had set them a good example, by coming away from speculations as to inherent tendencies, and endeavouring to get at facts. He had one rather important question to ask as a matter of information, and that was as to whether Mendel and his followers had found any or no difficulties as to the matter of reciprocal crossing, and whether the results were affected in any way by arranging which was to be the pollinating parent and which was the pollinated parent. Miss Saunders did not mention the fact, and he wanted to know whether the results were the same or not.

Miss Saunders said she should have pointed out that the same result was obtained in reciprocal crossing in each case, and this was based upon the results of the experiments of the past five years.

Mr. VV. M. Webb remarked that Professor Boulger had said he did not know what a tendency was, but he thought the tendency of most lecturers was to fog their audiences as much as possible, but in the present instance the lecturer had gone the length of the opposite extreme. He would like to ask whether Miss Saunders could give them any idea as to the way in which Mendel's law assisted in the development of new species?

Miss Saunders replied that Mendel's law, as formulated by Mendel himself, was a practical law of inheritance only, and did not deal with the evolution of new forms. It was a law for the inheritance of such characters as were fixed, or characteristic of the race or species.

Mr. E. Large asked what was the best practical method of preventing interference by insects?

Miss Saunders said that was a point upon which workers in this field were disagreed, and *the* thought that was accounted for by the fact that the best possible means depended upon the particular plant they desired to protect. In some cases they could use a particular form of parchment bag, but on the other hand there were certain plants which would not stand being covered any length of time by a material so thick as a parchment bag. In other cases a very fine muslin or fine wire gauze was recommended, but she always preferred to use muslin. In reply to further questions by Mr. Large, Miss Saunders said it would be probably difficult to carry out experiments with bottles because she would imagine the plant might possibly rot with accumulation of moisture. Dealing with plants in glass houses, was to deal with them under the easiest of circumstances, because they could be left uncovered. The President proposed a hearty vote of thanks to Miss Saunders for her admirable paper, and this was cordially agreed to.

A brief account of Miss Saunders' address will be found on pp. 57 *et seq*. This was kindly supplied by herself.

Dr. William Martin, M.A., LL.D., then read a paper on "The law of treasure trove as it affects archaeological research." The contention in the paper was that archaeological research was facilitated by the existence of a law of treasure trove, and that the denunciations to which the law had been subjected by antiquarians and others were scarcely deserved. Such strictures should be reserved for faulty administrations. See pp. 26 *et seq.* for full report.

Sir H. H. Howorth, having thanked Dr. Martin for the lucid manner in which he had prepared his paper, touched upon the importance of improving and reforming the present law with regard to treasure trove. He contrasted the custom that prevailed in Norway, Sweden, and Denmark, with that of England where every object of antiquarian interest was impounded by the Government, and where it w-as impounded unfairly, because an adequate price was never paid. He could not help reflecting on the public spirit which existed in those countries, and he thought we might introduce into this country the system in practice thereof posting notices in the national schools, which stated that every child or peasant finding anything of value should bring it to the schoolmaster. The ridiculous action of throwing guineas into the roadway, found at the residence of Mr. Justice Phillimore, could not have occurred in any other community than England. They wanted to encourage people to bring these things to the Treasury. He thought it would be well if the distinction in the law as to whether money lost inadvertently or not should be given up, should be done away with. These distinctions were purely academic and perfectly ridiculous.

The President said he held entirely opposite and discordant views on the subject. In the first place he would like to see everything 500 years old and over—buildings, cathedrals and private houses —the property of the State, which could not be made away with, destroyed, or dealt with in any way whatever except by the consent of the State. That was what he might call the extreme view. Of course, he would be told that was impracticable, but something must be done to put the law on a more satisfactory footing. He dwelt upon the difficulty of persuading people that they would get full valuation when the goods were handed over to the care of an agent of the State probably the authorities at the British Museum: while, at the same time, he held it was more disastrous for things to go into the melting pot—as happened to such a large extent when they were liable to be claimed by the State —than into the collections of private individuals. His remedy was that all relics, coins, etc., found should be the property of the finder, subject to two conditions being imposed—first, that he should inform an officer of the State within twenty-four hours, and, secondly, that the articles thus found should be submitted to public auction, and the amount realised given to the finder. Commenting upon the awful destruction that had taken place with respect to priceless relics, the President stated that he knew of one case in which the architect of a certain cathedral receiving £300 a year, opened up certain recesses in that cathedral, and died worth £80,000. Having dealt with other cases, he said as an archaeologist, he desired to see the preservation of all ancient things, and he would like to see an army of inspectors whose duty it should be to draw attention to these ancient buildings, whatever they might be, whenever they might be in danger of being either removed or interfered with. The police were certainly not ideal persons for this particular work, but he thought if they could enlist the sympathy and aid of land agents, who were more in touch with these matters, some good might be accomplished. Adverting to his proposals, he argued that if the finders of treasure trove could be assured that they would receive full value, upon allowing the State to sell by public auction, it would ensure that these relics would not go into the melting-pot. It would be seen that he felt rather irreconcilable and held very opposite views on the subject, but they must avoid dealing with this subject purely by logic. Everything was illogical, and the law more than anything else.

Dr. Abbott suggested the Union might send a resolution to Parliament, asking for an amendment to the scheme.

In reply, Dr. William Martin said that he did not think that the President's views differed materially from his own. In the paper, he had shown that to avoid the danger of the melting-pot and to secure for the nation relics of antiquity, a stiffening of the law of treasure trove was required, together with its uniform application, an improvement in its administration, and, what was most important, a dissemination of a knowledge of the liberal offer of the Treasury to remunerate finders of treasure trove. There were undoubtedly faults in administration, but so long as scientific societies were content to let matters rest, it was unlikely that the government officials would depart from their customary course of action. Officials did not lead public opinion, but were, nevertheless, always ready to follow it, provided opinion was certain and indicated a practicable scheme. It behoved archaeological and other societies to take the amendment of the law seriously in hand, and, in conjunction, evolve a working arrangement which, commending itself to the Crown, would also secure, to a far greater extent than had hitherto obtained, the preservation of old-time relies.

Friday Afternoon.

According to the programme, the delegates should have driven to Gatton to inspect Gatton Hall and its lovely surroundings, but this pleasant excursion was frustrated by the wet weather; consequently a meeting was held in the Public Hall, Reigate, when the Rev. E. J. Baker read a paper written by Mr. S. W. Kershaw, M.A., F.S.A., dealing with " The history of Gatton." Sir Henry H. Howorth occupied the chair in the temporary absence of the President. See pp. 50 *et seq.*

In commenting upon the paper, Professor G. S. Boulger, a native of Bletchingley, pointed to the great statesmen who had been returned from these " rotten boroughs." He was curious to learn how it was that Gatton became a borough at all, and why it was made one as early as 1451. What influence was it that determined the writ?

The chairman added a few general observations, noting how England stood out prominently ahead of other European countries so far as the history of the smallest village was concerned.

Following this a short musical entertainment, kindly arranged by Mrs. G. R. Taylor, was thoroughly enjoyed and appreciated, the following being the contributions and contributors:—" Mendelssohn's songs without words," Mr. Meeson (flute) and Mrs. Adeney (piano); song, "Irish folk song," Mrs. Treutkr; violin solo, "Reverie" (Vieuxtemps), Miss Furze; song, "All Souls' Day" (Lassen), Miss Coles; song, " Indian desert song," Mrs. Treutler, Mrs. Adeney and Miss Majorie Furze acted as accompanists to the songs.

Professor Petrie then occupied the keen attention of the audience whilst he related the results of his labours in Egypt, pointing out that it was a very convenient ground from which to study successive periods of the civilisation of man. He gave a general idea of the geological appearance of the neighbourhood of Sinai; the meaning of the enormous masses of burnt offerings, and, lastly, he dealt with the results of the civilisation of the whole people in the Mediterranean, which he thought affect-

ed them more than any other branch of the history of Egypt. He expressed the hope that students would be able to continue their researches before the remains of ancient history were wiped out— and that was being done rapidly.

Professor Flinders Petrie's address is to be issued as a supplement to the *Transactions,* the address was illustrated with large photographs of Sinaitic scenery, and recent discoveries there. Sir H. H. Howorth expressed the thanks of those present, that Professor Petrie had delighted and instructed them with such a succinct and luminous account of his 25 years' explorations, and a vote of thanks was warmly accorded for this addition to his Presidential labours.

Friday Evening. One of the most enjoyable functions in connection with the visit of the Union to Reigate took place on Friday evening, when the Mayor (Alderman V. H. Bagaley) held a largely attended reception at the Municipal Buildings, Reigate. A large number of residents were invited to meet the members of the Union, and the gathering proved as successful as it was pleasing to those who accepted His Worship's invitation. The whole of the upper suite of rooms were thrown open to the guests, who numbered about two hundred. His Worship, who wore his chain of office, received the company in the mayor's parlour, where the borough mace was displayed in a prominent position. During the reception a delightful programme of music was discoursed by the Anglo-Viennese band under the conductorship of Mr. Howard Aynstey.

A most interesting programme had been drawn up and the evening proved to be one of instruction as well as of enjoyment. Among other objects of interest were a number of microscopes manipulated by Mr. F. Hughes, Mr. J. B. Crosfield, and Mr. E. Larmer. Considerable interest centred in a photo-micrographic exhibition by Miss Ethel Sargant, F.L.S., in the Aldermen's retiring room. Quite one of the most interesting experiences of the evening was the remarkable and singularly successful demonstration of Tesla's high voltage experiment. So much interest did this arouse that the police court in which the demonstration was given, and its approaches, were thronged while Mr. Richard Kerr, F.G.S., was giving his demonstration. Drawings with the harmonograph and geometric pen, also by Mr. Kerr and his daughter, proved a great attraction. He also exhibited Chladni's figures, produced by the grouping of sand on a steel bar in response to the vibration caused by musical notes caused in the bar by friction. The figures varied with the note sounded.

A few minutes before the demonstration took place the company assembled in the council chamber, where the Mayor, in a brief speech, extended a hearty welcome to his guests. His Worship said he did not wish to detain them many minutes, but he wished to say he had much pleasure in welcoming the South-Eastern Union of Scientific Societies there that evening. He was sure the borough of Reigate felt highly honoured by the Union having held its tenth annual congress in the borough. He was sure that if they had been favoured with better weather they would have been delighted with the natural beauties of the place. He considered it was his duty as mayor of the borough to wefcome them there that evening. He did not consider himself a scientific man; he Was a business man, and, therefore, would not talk to them upon matters connected with their Union in the presence of such able men as their president and vice-presidents, but he hoped they had found their visit to the borough in many respects a pleasurable one (applause). He appeared before them that evening at a certain disadvantage as being a bachelor, he had no lady to assist him at his social gatherings, but he hoped they would make themselves at home.

The President briefly acknowledged the welcome extended by the Mayor, expressing the great pleasure it gave the delegates and members of the Union to be so cordially welcomed by His Worship. He acknowledged the extreme kindness they had met with from all the inhabitants, and he could only express his thanks to those who had enabled them to spend such a delightful afternoon, despite the inclemency of the weather, which had caused them to have to abandon their trip to Gatton.

Sir H. Howorth, who was called upon to say a few words, said that he had nothing to add to the very graceful words that had been spoken by their President. There was nothing so pleasant as to be welcomed upon the occasion of their annual congress by the chief magistrates of the country towns in which their association met. It was a peculiarly English institution, and the chain worn by his Worship represented a great chain of patriotic men who had devoted themselves to making their neighbours happier, more healthful, and more prosperous, without any pay and without any favour, and they owed them great gratitude, and it was very pleasant to come there and be welcomed in such simple and such genuine terms as the Mayor had expressed.

Following Mr. Kerr's demonstration, Mr. A. E. Tonge read a brief paper in the council chamber, "On the Eggs of Lepidoptera," the lantern illustrations, which were beautiful, eliciting frequent outbursts of applause. See pp. 54 *ct seq.*

At the close of a pleasant evening a number of reproductions of old Reigate prints were thrown upon the screen, the lantern being ably manipulated by Mr. V. Brooks. A brief and interesting description of the various slides was given by Mr. E. Penfold, A.R.I.B.A., who expressed indebtedness to Mr. Wenman and Mr. Brooks, whose kindness had enabled several of these prints to be preserved and shown that evening.

During the evening light refreshments were dispensed, and the comfort of the guests was studied in every way. The company departed at about half-past ten, expressing the heartiest thanks to the Mayor for the pleasant evening he had afforded them.

Saturday Mornino. (Professor W. M. Flinders Petrie in the

Chair.) *June 10th.*— On Saturday morning Professor Petrie presided over a delegates' meeting, when the election of officers, etc., took place. The Rev. R. Ashington Bullen, as delegate for theS. E.U.S.S., read a report of the delegates' meeting of the British Association held last year at Cambridge. (See pp. xxxv. *ct seq.)*

Dr. Abbott submitted the balance sheet, which was passed on the motion of Rev. T. R. R. Stebbing, seconded by Mr. Haldane Harrison. On the motion of the President, seconded by Mr. Tutt. Messrs. G. W. Bulman. W. E. Nicholson, W. H. Griffin, W. T. Vincent, and Rev. V. Hudson were elected to fill the vacancies in council, subject to their consent. 'Thanks were heartily Note.— The retiring members of Council were Messrs. F. G. Flcay. Lawrence Green, H. Norman Gray,V. H. Griffin, and C. J. Martin, but as by an unfortunate slip Mr. Griffin's name was omitted last year, he has not attended all the council meetings, and consequently not served two full years, and, therefore, in justice to him, he was again nominated for another year.

accorded the retiring president and councillors, on the motion of Mr. R. Adkin and Mr. N. F. Robarts, and in accepting the expression of thanks, Mr. F. W. Rudler, I.S.O., proposed a vote of thanks to the Mayor of the borough for his reception on Friday evening and for permission to visit the Caves (which were illuminated by his order), and also the Castle Grounds, and take tea therein. He also expressed the indebtedness of the Union to the friends who favoured them with demonstrations on Friday evening, including Mr. Brooks, Mr. Penfold, Mr. R. Kerr, and Mr. A. E. Tonge, to Messrs. F. Hughes, J. B. Crosfield and E. Larmer, for their microscopic exhibitions; also to the Literary and Scientific Institution for treating them as temporary members; and to the local committee who had made such perfect arrangements for reception and entertainment, especially mentioning Rev. E. J. Baker, Miss Ethel Sargant, Mrs. G R. Taylor, and Mr. G. E. Frisby; also the president of the Natural History Club, Mr. J. B. Crosfield, for upon these, the bulk of the work had fallen. He mentioned that everything was so pleasant, there was nothing to grumble at, not even the weather, for their loss at Gatton had been their gain at Reigate. This vote was seconded by Mr. G. A. Boulenger, F.R.S., and carried with acclamation. Mr. E. J. Bedford and Mr. Sparks invited the Association to Eastbourne for 1906, and this was accepted by Prof. Petrie with warm thanks. Being called on by Professor Petrie to move a resolution, the Rev. T. R. R. Stebbing said:—"Mr. President, in your presidential address you showed that the current of your reflections had flowed widely and deeply over many subjects of human—and even superhuman—knowledge. Yet there is one point on which you have possibly never reflected, and that is the terrible anxiety which presidents cause to those who may be called their foster-nurses. There is ope point, indeed, to which you have yourself just made allusion, the question of successorship In the presidential chair. For scarcely is an in-coming president well seated, before some unfortunate victim, like myself, is selected for the invidious task of proposing to his face that we should choose some one else to reign in his stead. But the anxiety is intensified during the period of incubation. I remember a case in which the president-elect, though a man advanced in years and unaccustomed to travelling, after his nomination flew off to the uttermost parts of India to consort with lepers, exposing himself to the chances of being interned for life, and consequently of never coming back to address our Congress. Similarly, in another instance, the man of our choice was spirited away to Mount Sinai, in search of invisible temples, so that our Association, like the Israelites of old, might have been excused for saying, 'As for this president, we wot not what has become of him.' Up to the very time of meeting we cease not to feel a well grounded concern. On one occasion the president, instead of being at hand to deliver his address on the first evening, could not deliver it till the last morning of the congress. Another time, when the audience had especially been invited to hear the highest authority on the prehistoric civilisation of Egypt, the authority himself decided to deliver a stimulating address on the universal equation and the boundaries of space. But, as you say, Sir, there is a balance in all things. That president, who in his year of office, came unavoidably late, in the following year, when he had ceased to be president, made up for his previous tardiness by giving us a most attractive lecture on a most unattractive subject, that of leprosy. On this a question of order was afterwards mooted, since our presidents are allowed to discourse on any subjects they please, but other members are expected to confine themselves to subjects connected with the South-Eastern counties of England. It was, however, pointed out that Mr. Jonathan Hutchinson's paper dealt with the old leper houses of Canterbury and other south-eastern towns, and was therefore quite in order. May we not hope, then, that the familiar presence of Gypsies or Egyptians in this part of the world will encourage our distinguished and honoured Egyptologist to give us a paper on his special subject at our next year's Congress? For that Congress I have been empowered to propose that Professor Francis Darwin shall be our president.

Most of us must be acquainted with the fascinating life of his father, Charles Darwin, which Francis Darwin partly wrote and partly edited. He was associated with his father in many elaborate botanical experiments. Together they got, I believe, to the bottom of that conspiracy in which so many vegetables have engaged, for sending the tap-root downwards instead of directing it sideways or upwards. They investigated together many parts of that scheme of conduct by which plants show themself to be, on the whole, very sensible creatures. Professor Darwin's scientific activity is still in no way abated. He is also foreign secretary to the Royal Society. Notwithstanding the pressure of many engagements he has consented to place his services at our disposal, on the faith of an assurance that he would be relieved as far as possible from the routine business of the Association by one or other of our vicepresidents. It is then with great confidence that I invite you to ratify the Council's nomination of Francis Darwin, Esq., M.A.. F.R.S., F.L.S., F.Z.S., as president-elect of the S.E.U.S.S. for next year's Congress at Eastbourne.

Miss Ethel Sargant seconded, that Professor Francis Darwin, F.R.S., should be the president in 1906, and this was unanimously agreed to.

A paper on " The Flora of Reigate and District," prepared by Messrs. R. H. Welchman, B.A., and C. E. Salmon, F. L.S., was read by-Mr. Welchman. The account of the flora of Reigate was given for comparison with the flora of other districts. The enumeration of species well known to botanists, coupled with the names of localities unknown to strangers to Reigate, necessitated a description of the neighbourhood topographically. (For full report see pp. 64 *et seq.*)

The paper concluded with a few remarks on the old records, and the disappearance of species, a short list of aliens also being given. One of the causes assigned for to the disappearance of some species was the disturbance constantly due to building operations, and in other cases the causes were not so obvious.

Captain McDakin stated that the Lizard orchis was growing in the neighbourhood of Wye Downs. It had been photographed by Mr. Hammond, of Canterbury, and recorded in the East Kent transactions.

Rev. T. R. R. Stebbing believed that in the more generally known writings of J. S. Mill, no reference occurred to local botany, so that if, in addition to his works on logic, liberty, and political economy, he had ever published original researches on botany, it would be interesting to know the title of the book.

Dr. G. Hodgson mentioned *Symphytum tauricum* as growing wild within five miles of Reigate Station, but as it was a Russian plant it was not easy to see how it got there.

Mr. G. C. Walton thanked Messrs. Welchman and Salmon for their very interesting paper and list of plants found in the Reigate district. He thought that a few typically chalk species seemed to be wanting, and that it would have been a good thing if the authors had given approximately the number of species in their list, including, of course, the grasses and sedges. He said that it was interesting to contrast the flora of this district with that of Folkestone and Dover. Did the authors include *fiianthus armeria* and *Silene nutans,* or *Lepidium draba,* a plant introduced some time ago, and now rapidly spreading over Kent, and probably the?djoining counties? Which species of *Orobanche* are found here, and is there any record of the occurrence of that rarity, *Orchis hircina,* the Lizard orchis? Mr. Walton was not sure if the authors mentioned the Viburnums, the White Helleborine, *Hypericum, humifusum, H. elodes, Genista anglica,* and the Sundew.

Mr. Welchman said that Mr. Walton had enquired if *Silene nutans. Genista anglica, Orchis purpurea, Cephalanthera grandiflora. Viburnum lantana, V. opulus, Hypericum elodes, H. humifusum,* were found in the Reigate District. The reply was in the affirmative in all cases except *Silene nutans,* and the localities were given. Dr. G. Hodgson enquired if any particular reasons could be given for the presence of an east of Europe alien, *Symphytum tauricum* in the neighbourhood. The writers of the paper could only suggest that it had probably been introduced with foreign seeds, as seems to be the case with most aliens. Oulton Broad was mentioned as an example where large quantities of Mediterranean grain are used at the maltings, and upwards of eighty aliens have been found growing within a mile of the place.

In conveying the thanks of the meeting to the writers of the interesting and instructive paper, Professor Petrie expressed the opinion (which met with manifestations of cordial approval) that more organised resistance was required against the vicious habit of destroying flowers. It was not only vicious, but extremely unscientific, and it was one of the most obvious duties of all societies similar to theirs to prevent the destruction of flowers. Personally, he never took a single flower, for he would rather see them growing. He thought the growth of photography might tend to the preservation of the plants.

The Congress then dissolved, the excursions arranged for Saturday afternoon being abandoned in consequence of the wet.

TRANSACTIONS OF THE SOUTH EASTERN UNION OF SCIENTIFIC SOCIETIES. 1905. PRESIDENTIAL ADDRESS. By PROFESSOR W. M. FLINDERS PETR1E, D.C.L., LL.D. F.R.S.

When considering upon what subject I might venture to address you, with some possibility of profit, I could not hope to deal with a special science in the presence of specialists, nor would I impose my own corner of the field of science as the point of view for many different minds. It seems more profitable to turn to some broad underlying considerations with which we all have to deal, to look at some general attitude of mind, and mode of treatment in science, which we may all perhaps lay to heart with advantage.

There is one such aspect of science which seems hardly to have been viewed as a whole, but only yet in details of separate instances. This is the study of the balance in each subject; the broad conception that in every subject the incomings must balance the outgoings, and that the most feitile mode of study lies in tracing out this quality.

This conception of balance is almost entirely a growth of modern science. The ideas prominent in the ancient philosophies were those of the regular rotation and recurrence of nature; and this point of view is by no means out of date, but is continually applied with good results to subjects where cause and effect are the less easily traceable. But the idea of balance was almost impossible to grasp in the absence of accurate measurements of various physical kinds. Yet it was felt after, when the principle was stated, that no object comes from nothingness or disappears into nothingness. This statement required a large amount of scientific faith, faith in uniformity, at a time when men were yet incapable of tracing the source of carbon of a growing plant, or the disappearance of fuel in burning. This is a lesson for us, that we must not be deterred from general laws because we cannot yet explain all the exceptions and difficulties.

The first accurate enunciation of balance seems to have been by Newton, when stating that action and reaction are equal and opposite. Here was a clear view that the action in one direction, however produced, must be found equal to the action in the opposite direction.

A century more passed before another such step was reached, in announcing the balance of matter in whatever states it mayexist. The demonstration that where the weight of solids was changed during a process the difference was made up by gases, which entered into combination or were liberated, was the assertion of balance in the chemical world. Caloric was slowly dispossessed, and the fundamental idea was grasped that the total mass of any compound precisely balances the separate masses of each of its component elements.

Mechanical force and matter were now on an exact footing, but the disappearance of force in friction defied balancing, until Romford compelled the recognition of what every workman vaguely knew, that friction turned force into heat; and a generation later, Joule gave precision to the balancing, by showing that equal forces produced equal heat. After this it was an axiom that the force and heat put into the beginning of a mechanical process must exactly balance that given off at the end, whether it be a steam engine or a coffee mill that is considered. The broad conceptions of the Conservation of Energy, and the Correlation of the Physical Forces now followed. But you will observe, these views entirely rested on the precise balancing of these forces in the first place, and the consequent belief that whatever changes may be produced, matter or forces are bound to balance at whatever stage we may examine them in any given system of action.

Fortunately, general principles had advanced thus far before the great development of the application of electricity. Hence that study has been based entirely on the balance that may be found in every transformation; and the whole system of terminology is uniform, and linked with the definitions of force and heat.

So far, this may seem to be very familiar to us; but we need to remember that all these ideas, which are now part of the mental stock of educated persons, are entirely due to the general principle of balance; and that principle must be valued as one of the main methods of research and advance in other subjects, for which it has not yet been adopted as fundamental.

One of the most obvious deficiencies in our knowledge of physical equivalents at present is in the relation of light to heat and force. The mechanical equivalent of light has not been at all accurately ascertained; and the difficulties of absorbing or reflecting it, apart from an amount of heat which represents far more energy, stands so much in the way of measurement, that theory will probably outstrip observation. The amount of force perceptible as light is millions of times less than that perceptible as heat; for the heat of a candle may be barely felt at a couple of feet, the light of it is seen at four miles distance. And this difference in scale is not only in our senses but also in the physical effects that can be measured. When the equivalent of light is known, and the balance can be made out taking it into account, there will be a great opening for research.

In geology the principle of balancing has scarcely been used yet, but it will be the key to some of the greatest questions. The equivalence of metamorphic rocks to sedimentary deposits is a question of the balance of their components. What types of sedimentary rocks could be resolved into the metamorphic, or so-called igneous, rocks? The composition before and after metamorphism must balance. And this must be a first consideration in resolving the question of whether there be any rock known to us which is the product of primary fusion, or whether all rocks are metamorphic.

Linked with this is another question. Is the present heat of the earth due to its first gravitation from a diffuse condition, or due to its crushing by present gravitation? This depends on balancing the loss of heat by conduction against the production of heat by the crushing due to contraction. Given the known conduction and loss of heat by the earth, some think that this may be balanced by the heat produced by a certain rate of shrinkage. But if this theory breaks down in the balancing, then we must accept the usual idea of the heat being due to first gravitation, and increasing continuously to the centre. It has lately been pointed out that on this view the earth must nearly all consist of gas compressed to a solid, and, as we are wholly ignorant of the properties of matter under such heat and pressure, it is almost useless to speculate on the contraction of the earth. It may be that the structure of solidified matter may be more bulky than that of intensely compressed gas; much as a house occupies more space than do its ruins when crushed together. If so, the earth might expand by cooling, or remain of constant volume. The balancing of this great account seems hopelessly beyond us, but attempts at this balancing of actions are the only possible road to understanding them.

Another balance that needs striking is that of the sum of all the sedimentary rocks, as against the composition of a primitive mixture from which they have been sorted out by washing and solution. What is the total of silica, alumina, lime and other material in the whole pile of strata? And how will that balance against any supposed condition of the material before its degradation? The usual lists of strata and their thickness are of no use in this question, as those state what is traceable as the maximum thickness known of each rock, the nearest approach that we can make to its original thickness. By taking this we should count the same material more than once; the silica in all its successive states as siliceous rock, Devonian, Permian, flint, sand, etc.; the lime both as silurian and chalk. The strata therefore must not be credited with more than the average of what can be found, leaving the eroded parts to be counted in their later forms. What average composition of primitive stock rock this total represents has yet to be worked out.

Another question of balance in geology, for which the data have still to be sought, is the difficult consideration of climate. The presence of remains showing a far warmer or colder climate than at present, have been variously explained. The flora of the extreme north, and the ice age in the temperate latitude, have suggested to observers various possible causes. The altered diffusion of heat by currents of sea and air, and by a different outline of land may make great variations, as we see by the temperatures of Irkutsk and the south of Ireland in the same latitude. Yet the greater warmth of one place must be balanced by greater cold elsewhere, as the total is supposed to be constant. The eccentricity of the earth's orbit which has been invoked, and the greater heat falling on one hemisphere than on the other, are again questions of balance. Which country was colder when another was hotter? Can we trace alternations in the temperature in a single locality during one geologic age; or can we identify as being of the same age two countries in the same latitude with very different climates? The balance must be found, and we ought to aim at writing out a balance sheet of so many square degrees at each different temperature in each of the various past ages, as we can state it for the present age.

The depth of the sea and the amount of submerged land are again matters in which a continual balance has to be struck. There has been much consideration of the direct causes of changes of level. The expansion of strata by heat, the changes due to a change of state in liquefying or crystallising, the effect of more or less weight on the crust of the earth, the shifting of the centre of gravity of the earth by a polar ice-cap, all these have been estimated and shown to be causes capable of making considerable change of level. But the secondary changes, which must result from the more obvious movements, are brought into view as soon as we consider the balancing of the matter. The rise or fall of strata beneath the sea must raise or lower the whole sea-level. And as the earth may be supposed to lose its heat more rapidly beneath a sea bottom which is almost at freezing point, than it does beneath an atmosphere which is much warmer, it may be presumed that ocean beds tend to sink by contraction more than land surfaces; hence the sea-level around the shore line must recede more often than it advances. In another case, of an area being submerged, such as the formation of the present Mediterranean Sea, a considerable fall of sea-level at all other shores must result, unless it was balanced by a rise of the ocean floor elsewhere. On the other hand, denudation of land must always raise the sea-level by filling up the space in which the water rests. The effect of a polar ice-cap would be greater by its abstraction of water from the sea than by its shifting the centre of gravity of the earth. It would make but little difference of sea-level in its own hemisphere, but it would have a doubled effect of recession of the sea in the other hemisphere. Thus it will be seen how essential the questions of balance are in all changes of level, and how incomplete our views must be till both sides of the account are made out.

When we turn to organic life, we see in every direction how imperative it is to understand the balance of materials and of forces. In the crudest view of nature there is the balance between vegetable and animal life—how many deer or sheep can be kept per acre; and this, if it were worked out, would be of some statistic value for past ages. The coarse feeder can live on a much greater range of vegetation and of ground than those of more limited appetite, and this must be taken into account as well as the mere size and voracity of the animal. Even under the economic management of man, the cow requires most, or all, of the proverbial three acres to keep it alive. Some actual statistics of the numbers of different species of animals that live in a limited extent of country in a wild condition, would give a basis for the relation of animal life to land area in geological times.

The balance between carnivora and herbivora requires to be defined. The destructive animals are necessarily but few in proportion to their prey; we may reasonably suppose that a single lion would be maintained by not less than the progeny of a hundred sheep or deer. The actual statistics of the relative numbers of carnivora and herbivora under unrestrained conditions would be of great interest for comparison with the proportion of remains of these two classes found in any fossiliferous zone, and we might get an entirely fresh light on some question by making even an approximation to this balance.

When we look closer to the single organism, the relation of gain and loss in the various parts of the body is of primary importance. How much of each kind of food is required, how it is utilised, and how its waste matter is removed, have been the subjects of a great mass of experiments and statements. The one certain fact is that the difference of state between the material as gained and as lost must be the equivalent of the energies of the body. How much carbo-hydrate goes to supply the lost heat, how much nitrogenous matter goes to keep the involuntary and voluntary muscular action in play, is a complex enquiry as yet far from settled. The working out of this balance is one of the long tasks before physiologists.

The maintenance of the condition of the body has also been shown to depend very largely on the balance produced by various obscure products of apparently inert glands and organs. These are the unheeded regulators and stimulators, the treasury clerks and the philanthropists of the body, which direct the various energies so that excessive or deficient action is prevented; and, broadly, most physiological discoveries of recent years show further and further how every activity is controlled by two opposing impulses of excitation and inhibition. Every non-infectious disease may be, perhaps, shown to be caused by the deficiency of one of the two opposing forces. The latest researches on the brain show how the fearful effect of tetanus is the result of a reversal of the control of inhibition in the brain, so that the will to relax the muscles is translated into greater contraction. The last suggestion on cancer is that it is connected with, if not originated by, the absence of hydrochloric acid in the stomach; the general alkalinity in the blood promoting the formation of the sexual type of cell-reproduction which is now known to characterise cancerous growth. When the balance of action is out of gear, and a false preponderance has been established, then the constant presence of even an apparently harmless secretion may set up great chronic dangers. Thus it is beginning to be now realised that we may even say that the balance of the controlling powers, in one form or another, may be in future generalised as the only self-originating cause of disease, as well as being often the induced effect of diseases of external origin, such as bacterial and other parasitic attacks. To study, therefore, the balance of the motive functions, to ascertain what is the seat and action of each excitation and its counter inhibition, is the true line of advance in understanding physiological action and in learning how to control it beneficially.

In the study of vegetation, the nature of the balancing of gain and loss, is being elaborately studied now. The absorption of water and transpiration of plants has been determined with the greatest care in the experiments of Dr. Horace Brown; and the theoretical relation of thermal emissivity, water transpired, and difference of temperature between the leaf and its surroundings has been worked out, showing how any two of these determine the third, and verifying this by experiments. The amount of radiation of a leaf proves to be as great as that of blackened copper. The effect of sunlight which causes the absorption of carbonic acid, and thus increases the weight of the leaf by formation of carbohydrate compounds, has also been minutely measured, and, at last, a complete account is rendered of the dissipation of the whole energy received from the sun on a growing plant. This is further shaped into a balance sheet of the total gains and losses of energy, and tables which give a complete account of the expenditure of the energy received by different plants. The amount actually used for construction is not more than S0th of the energy which the plant receives in full sunshine, which agrees with the vague impression that the heat produced by burning a plant cannot be more than a very small fraction of that which has fallen on it during its growth. But the full sunshine can be cut down to,.2th before it appreciably diminishes the assimilation of carbon by the leaf; so that under the most economic conditions the leaf can use i50ths or,..,th of energy falling upon it.

The extreme amount of solar radiation observed at Kew has been one calorie per sq. centimetre per minute; and if this amount should fall fully and squarely on a leaf, it can yet be dissipated in a moderate breeze without a rise of more than 10 C. in the leaf. This great power of radiation in leaves (as great as blackened copper) is of essential value in preventing their being heated by the sunshine above their vital limits. And further, in most cases, about half of the possible total of heat can be disposed of by evaporation. The vegetable basis of all animal life has thus been worked over on an exact system, which shows what energy is received, and the balance of how it is used or dissipated. Perhaps the most familiar idea of balance is that of balancing accounts. What is in the practical application of it the most dreary and unprogressive of employments, becomes, in the theoretical study of it in history, one of the most enlightening and stimulating researches. The work of Giffen in computing a national estimate of production and expenditure is of invaluable service to indicate where we are and what the trend of changes may be. But we need far more of this kind of research. A statistical department costing a ten-thousandth of our national income might well save us ten times or a hundred times its cost of shewing what effects various changes produce in the politic welfare.

The existing summaries of how we spend our wealth are of the greatest value to show where waste and losses take place. But a far more searching enquiry is needed to give the real balance in different aspects. There is in the first place what may be called the day-book balance. In the receipts (1) So much earned by direct manual labour. (2) So much earned by labour using large capital assets such as factories, ships, etc. (3) So much earned by direction, which is the increment gained by control and supervision, the difference in value between a system of individual work and a system of combined action. Thus if a man can earn 30s. a week, as an independent workman, while 20 such men can each earn 40s. a week with the assistance of skilled direction and combination of their labour, the earnings of the direction are 200s. a week, which is a national earning, quite apart from the physical labour of the workers, and (4) So much earned by the gains from past labour invested. On the opposite side we need to set down expenditure on so many men paid at £1 a week, so many paid at £2 a week, and so forth in different grades, and so many paid by the profits of their past labour accumulated. Thus we should see what were the direct sources of receipts and the direct expenditure.

But a very different balance sheet is needed if we are to gain a truer insight into the utility of our labours and expenditure. The ledger balance must be a much more searching matter in order to discriminate the true sources of national gain and the various uses to which it is put. To this end land and machinery stand on almost the same footing; the land is worth very little in a wild state of nature, just as the raw material of the machine is worth little. Labour has been invested in reclaiming and fitting the land to be the basis for the continuous toil of man, just as labour has been spent in framing the machine as a basis of toil. And the profit from the use of land and of the machinery must be divided into gain by interest on labour of the past and gain by labour of the present. Luckily this complex division is almost ready made for us by the separation of rent and labour charges. The sources of national income really are then:—(1) Manual and personal labour and thought with or without appliances supplied to us by the past, (2) Interest on past labours, as reclamation of land, building of railways, factories, machinery and houses. Each of these may be divided into what is done for the inhabitants of the country, and what is done for the use of other countries; thus interest on foreign loans comes under the class of interest on past labours. And the expenditure is (1) on the direct maintenance of the manual workers apart from that of their families; (2) on the upkeep of the past labours for the future, and on the laying by of gain for future labours, including all kinds of investments and construction; and (3) what might be regarded as a part of 2, the provision for renewal of the human material which wears out; that is to say the entire cost of children to replace the present generation, which amounts to about half of the national expenditure.

Now this form of balance of the position of a nation may seem very obvious, but it is certainly entirely ignored by nearly all economists. Not one of the standard writers seems to allude to the division between work for the worker's self, and work for the replacement of the worker. The only facts bearing on this that are of much use are in Rowntree's *Poverty*. We shall not be far out if we say that the ultimate purpose of national work is fourtenths for the upkeep of the worker, one-tenth for renewal and improvement of permanent material, and five-tenths for renewal of the human material.

If now we find difficulty in probing into our true condition in the present day, how far more difficult it will prove to reach any such estimate for the past ages. If we look at different conditions of civilisation in our own time in different cities, we see very different balances of labour; the lower the civilisation the less is the renewal charge, until at the lowest the cost of the next generation may not be more than a sixth of the labour. Hence we must regard the kind of civilisation as well as the national habits and system in framing any sort of estimate of the balance of a past age. Yet I venture to think that anyone who would frame the roughest estimates of the balance in any period of ancient history would soon find a great reward in understanding that history as no one yet has understood it. In the middle ages we have some guidance as to actual money expenditure; and a careful study of some wellrecorded village would be a first step, making out how much value was produced, and how that was used in support of the workers, in support of the lord with his feudal service which was the equivalent of our heavy fighting taxes, and in permanent works; also the balance sheet of maintenance and human renewal charges. For earlier times an attempt at a social balance sheet for the different ages of the Roman empire would compel us to recognise the great cost of the army, and the much greater loss by the idle mob of Rome who were only fed by draining the industrious provinces. The balance is wanted as between Rome and the provinces, and also the balance in Rome itself, and in a drained province such as Egypt, or an almost undrained province such as Britain was, which only had to rear fighters. Then for Greece we might try a balance of Athens before its expansion, when the main asset was the great mass of slave labour, with the free citizen living on the profits of his mastership; and Athens after its growth, with tribute from other states, and a great armament to keep up. We have some indications for a still earlier balance, as in the tomb of Rekhmara the vizier of Thothmes III., the tribute of each province of Egypt is recorded; and it shows that the amount of taxes in kind and in gold could not possibly have supported a great central government, but were only for the maintenance of the court, while administration was entirety kept up locally. Probably the great estate owners managed the whole routine of affairs, and were onlyresponsible to the king and his ministers, without any interference by officials.

We have just considered here the fact that about half of all the energies of the nation is used up in the replacement of the present generation by fit successors. This view of the balance of the economic life of the nation is of the first importance in many questions; yet, strangely, it has been hardly looked at, and certainly it has never been given its place as one of the leading features in the political economy of a civilised state. The first and most obvious bearing of this expenditure is that the more it can be reduced in quantity and increased in quality, the better for us all. Now the death rate has been steadily reduced by civilisation. In Roman times it seems to have been about 50 per thousand in Egypt, judging by the ages on the labels of mummies. In Cairo and Alexandria now it is 35, or just double that of London. Probably we may say that sanitary progress has halved the death rate and doubled the average length of life. Now that does not imply that we spend on human replacement only half of what was spent before, but that we are able to spend twice as much on the better quality of the life, health, and training, of each generation; and, at the present time and condition of public health, we may remember that a decrease of one per thousand in the death rate, implies a gain of one-eighteenth on the lives, and so much off the need of replacement. This means a saving of one-thirtysixth of the national expenditure, or about 40 million pounds per annum. The unhealthy towns in the kingdom are wasting life a quarter more than London, or wasting an eigth of their whole income in needless loss. To put an extra tax on all incomes in Liverpool of 2s. 6d. per pound, in order to bring its mortality down to that of London, would be true economy. The tropical countries cannot hope to quite attain the economy of more healthy latitudes. But the example of Rio Janeiro with a death rate of only 20'8, and New Orleans and Havana each better than Liverpool or Dublin, ought to put to shame the waste going on in Egypt and India with rates of 35 to 42 per thousand. There are also many cities better off than London yet is in longevity. Amsterdam, Antwerp, Brussels and Hamburg, all show that a saving of life in London is yet to be aimed at. The study of the balance of our expenditure on replacement shows, therefore, how primary a consideration improvement in general health must be reckoned.

Another conclusion forced upon us by seeing the heavy cost of replacement is that it is worth a little more to secure quality as well as quantity; that health and education are the last things to cheapen over, nationally as well as individually. When the nation is spending 700 millions or 800 millions yearly on the next generation, a matter of a few per cent, more on this account, in order to render the product as fit as possible, is not worth consideration. Nothing that can be shown to really conduce to the future ability of the nation should be sacrificed, while such an enormous cost is being incurred whether or no. The apportionment of burdens is not our business; but we must open our eyes to the fact that when 10s. in the pound of national income goes to the human renewal, the heaviest education rates, which never exceed 2d. or 3d. in the pound of income, form an absurdly small amount. The only true question is— What is really wanted, and what will make the most effective men and women? That education, whatever it is, should be secured at any cost.

What training is wanted in life is again also a matter of balance, in which we must see what the requirements are and how to meet them. In no subject is the balance of requirements and the provision for them more imperative. We cannot on this occasion touch the well-worn subjects which have been long debated; but in one respect our considerations here must lead us to look at the balance afresh. We have seen that human renewal costs half of the national energies. Yet could it be credited that a subject on which half our labour is spent is absolutely untaught except to a few professional people, and that the whole preparation of what costs so much is mainly left to chance and ignorance? Our colonies show us the road, as in Canada there is now a regular course of education in the rearing and training of children. It is not too much to say that a half of the education of women— the parallel to the technical education of men in business— should be given to enabling them to make the best use of the moiety of the national expenditure on renewal.

Each college for women needs an orphanage attached to it where the majority of those attending the college should be required to seriously study the physical development of children, the elements of physiology and medical care, the moral and aesthetic training of the mind, the psychological variations and growth of the intellect. By no means should this supersede general culture and studies, any more than the technical training of a man should supersede the general culture of an university course. We all wish that extraordinary ability should have the freest scope among both men and women, and that in the ordinary individual no technical training for the business of life should hinder a liberal education. But a woman who was not trained to give her children the utmost advantages for life that were within her reach, should be looked on like an incapaple doctor, an inefficient lawyer, or a bad carpenter, and it is not only for its immediate utility that this training is required, but for its indirect effect. The great influence of the present system has been lately pointed out; how entire separation from the knowledge and interests of family life, and isolation under the'guidance of exceptionally celibate women, has created ah ideal of interests which results in a great preponderance of celibates among the students in their after life. This is repeating with deadly effect the story of the terrible injury that monasticism did to the middle ages, when every gentle and thoughtful soul was weeded out of the life of the people, and left an ever increasing mass of violent and impulsive natures to carry on the nation. It is an old observation that able men have clever mothers, but, if all the cleverest women are weeded out, where is the supply of able men to come from? The teaching of heredity should have at least enough influence to enlarge a system which, as it at present stands, is in direct contradiction to the welfare of the nation. I would even say that this people will be better off a century hence, if every woman's college was closed in which the practical teaching of human development is not carried out.

Lastly, let us turn to the greatest action of all nature, an action in which we cannot yet discern the balance, though we may feel a conviction that somewhere it has to be sought, and that, when found it will teach us deep things. The scattered energy that every sun radiates into space, seems apparently never to be regained. But before we can think of this we must have some idea as to whether visible space—the extent of the ether—may have any limits within our conception. There is one line of argument which seems as yet not to have been fully considered. We know the average distance from us of that zone of brightest stars which we class as first magnitude. We also know the average brightness of the night sky apart from special stars. From these data we can say that the light of the stars beyond visibility is equal to 300 times the light of the zone of first magnitude stars. Or 300 such successive zones of stars would give a night sky as bright as that we see. And that zone of stars is at such a distance that light takes 50 years to reach us from the middle of it, or it is at 50 lightyears' distance. The total starlight that we see would therefore be given by stars extending out to 15000 light-years' distance. If these stars are coeval with our sun, they are almost certainly 100000 times as old as this; that is to say, if stars existed beyond the limits we have just named we should certainly see them. In other terms, if there was a continuous dispersion of stars outward as far as we could see them from their'birth, the whole night sky would be much brighter than the surface of the full moon. All of these data are as yet but vague, and we have no certainty that the diffusion of stars is uniform in the ether, or that a large part of that light may not be hidden by dark bodies. But when we find on balancing the light seen against the light possible to be seen, and reach a disproportion of 1 to 100000, there is a grave case for supposing that this cannot be explained by the uncertainties of the data, and that we can grasp some conception that about 15000 light-years is a possible limit of the universe of matter.

Now according to a somewhat generally tolerated theory, matter consists of variously formed knots of vortices in the ether, or some form of recurrent cyclical motion. If so it would be unlikely that this should only occur in one small spot of the ether, namely our visible universe, which is in proportion to what we might see as a pin's head is to an immense ball. If then we can grasp the limits of matter, the *onus probandi* lies on those who would suggest that the ether extends much farther than those limits.

It is therefore a possible question what becomes of the radiant energy of the sun when it arrives at the confines of the ether in which it can travel. I will only add what may seem a wild dream of impossibilities. May the waves of energy in ether be thrown back when they can proceed no further, so that in some fashion a portion of their energy is transformed into atomic vortices? May thus a constant shower of atoms be flowing in from the confines by gravitation to the rest of matter, and by their concentration yield up again a fresh stream of radiant energy? Such a system might be in perfect balance, and yield by its radiation and condensation of energy an eternal continuity of action. Dreams, fantastic dreams, this may all seem. Let us turn back to the essential view which I hope I have brought before you to-night, that in seeking the balance of all actions and conditions, we are taking the truest road to understanding what is before us; and that the most powerful means of tracing fresh causes, and fresh action yet unobserved, lies in studying the balance sheet of what we already know.

The Extinct Postpliocbne Non-marine Mollusca Of The South Of. England.

By A. S. KENNARD And B. B. WOODWARD, F.L.S.

It is only within the last few years that the fossil fauna and flora of our more recent deposits have received adequate attention. Although an enormous amount of literature has appeared dealing with the Glacial period, yet the fossils contained in our Postpliocene beds have been either ignored or misidentified. It has been assumed that they are of the same species as the forms now living in this country, and, therefore, of no interest. This neglect is indeed remarkable, since it must be apparent that the correct identification of the species would throw considerable light on the various geological problems of the period, and it is not too much to say that, had this work been satisfactorily accomplished years ago, much of the literature dealing with this period would not have appeared, and more than one beautiful theory would not have been propounded. The last few years have seen, however, a large amount of work done, many valuable papers have been published, and the extreme school of glacialists has not been so prominent. At the present time the situation is indeed noteworthy. On the one hand we have the geologists, who, in order to account for certain deposits and stria; on rocks, postulate a period of intense cold when the greater part of these islands was covered by glaciers and the fauna and flora must have been practically exterminated. On the other hand, there are the students of distribution and the palaeontologists who are of opinion that there has been no such wholesale destruction of the fauna and flora as an ice age demands, that many of our species were living in these islands in Pliocene and early Pleistocene times, and have continued to occupy these islands to the present day. It is quite true that some species have become extinct, whilst others have been introduced, but all such changes have been very gradual and are probably to be explained by other agencies than by the operation of hypothetical cold and warm periods. Whether these two schools will ever be in agreement is on the knees of the gods, and it is not our intention to attempt a reconciliation. Some years ago we endeavoured to trace the origin of the non-marine mollusca of these islands and we pointed out that they had reached England from various sources. We were of opinion that many forms were boreal, and had travelled hither from the north along the now sunken land to Scotland; some had journeyed through Siberia and the continent; a large number had come from southern and middle Europe; others had reached us from an extension of southwest Europe, whilst a few might be endemic Since then we have exploited many Pleistocene and Holocene deposits, and numerous contributions have been made on the same subject by J. P. Johnson, Gilbert White and the Rev. R. Ashington Bullen, but we have seen as yet no reason to alter our conclusions, but, on the contrary, all the evidence tends to substantiate our views. The principal factor in the distribution of our freshwater mollusca has been the Thames-Rhine river system. It is well-known that the Thames must, at one time, have been a tributary of the Rhine, and to this system must also have belonged many of the rivers of northern France, and the south of England as far as Selsea, and the rivers of eastern England, at least as far north as the Wash. All these streams had, in Pleistocene times, practically an identical molluscan fauna. We would, however, point out that, in some cases, when the deposits were laid down, as at West Wittering, the beds near Amiens and the Chislet deposit, this connection had already been destroyed by the inroads of the sea. On the other hand, the Severn river system, and the other rivers of the West of England exhibit a marked contrast in their fauna. Only two extinct species have been recorded, *Unio littoralis* and *Paludestrina marginata*, both from Cropthorne, near Worcester. These two species also occur fossil in the east of England, but since they both are to be found living in the west of France, in all probability these western examples were immigrants from that part and not from central Europe, where they also exist at the present day. Moreover, it should be pointed out that the Cropthorne examples of *Unio littoralis* represent a well-marked variety, differing so much from the recent examples,»and also from the other English examples, that they were described by H. E Strickland as a new species, *Unio antiquior.j* We have thus additional evidence of the former connection between southwestern England, western France, and Spain; the old land bridge by which the well-known Lusitanian flora and fauna reached these islands.

It is perhaps advisable to give a short account of the various deposits which have yielded the materials on which this paper is based. Excluding the Cromerian (Forest bed of Norfolk and Suffolk) the oldest deposit is Dartford Brent, which is probably rather older than the well-known section in the 100ft. Terrace of the Thames at Swanscomb. Much later are the Beds of the third Terrace at Grays, llford, and Crayford; Chislet is probably of the same age as Swanscomb. West Wittering, Orton Waterville, Proc. Geol. Assoc. Loud.. 1901, vol. xvii., p. 254.

t Memoirs of H. E. Strickland, 1858. Scientific writings, p. 97 and fig. page clvii. near Peterborough, Clacton, Grantchester, Barnwell, and Stutton, are not far removed in time from each other. The relative age of the Ightham fissure deposit is uncertain, though it is certainly Pleistocene, and probably belonging to a late stage of the period. Spring Gardens, Westminster, belongs to the fourth Terrace of the Thames, and is late Pleistocene. Under this locality are included the shells found by W.J. Lewis Abbott in the excavations for the foundations of the new Admiralty Buildings. Shoeburyness, probably includes deposits of two distinct ages, one the equivalent of the Crayford brickearths, and one much later, perhaps of the same age as Spring Gardens. The relative age of Copford is still unsettled, though it is probably late Pleistocene. The St. James' Square deposit is of the same age as that at Spring Gardens. The beds at the following localities, are undoubtedly Pleistocene, but their relative ages are uncertain: Biddenham and Harrowden, near Bedford; Gedgrave, near Orford, Suffolk. The Holocene deposits are extremely difficult to deal with as to their relative ages. Walthamstow is probably early, whilst Staines, Uxbridge, Wargrave, Westminster, and

Tooley Street, are much later. Under Westminster we include the specimens from the foundations of the Houses of Parliament, the site of the new Scotland Yard, and Spring Gardens. It should be noted that both Pleistocene, and Holocene beds occurred at Spring Gardens, and to prevent confusion we have listed the Holocene shells as from Westminster. The greatest difficulty in dealing with this subject arises from the chaotic state of the nomenclature of the European mollusca, as anyone who has worked at the subject knows. Whilst admitting that local races and well-marked varieties (not of colour and size) are worthy of differentiation, the tendency of European conchologists to multiply species is only a hindrance to science. Moreover, the confusion in the application of specific names is even worse. Thus the *Pisidimn fontinale,* Drap., of British authors, which, as a rule, comprises two, or even three, well-marked forms, is totally different from the *P. fontinale* of Continental authors, whilst it is only within the last few years that *Hygromia (Helix) sericca,* Drap., has been removed from British lists, that species being quite distinct from the shell which used to bear that name in these islands, and which is now known as *H. granulata,* Alder. *Vitrea rogersi,* B.B.W., formerly *V. glabra,* is another example of misidentification, whilst if anyone would like a pleasant piece of work, we would suggest the attempt to name correctly the shells formerly known here as *Hydrobia similis,* Drap. , and latterly as *Paludestrina confusa,* Frau., and the species w-hich has recently been found at Oulton Broad by J. Le Brockton Tomlin, and identified as *Pseudamnicola* W. J. Lewis Abbott, Proc. Geol. Assoc, vol. xii., 1892, pp. 346-356. *anatina,* Drap. That this state of affairs will be remedied in the future is certain, but it will only be done by careful work and co-operation with fellow workers on the continent. We would take this opportunity of again thanking our numerous friends who have helped us in our labours, and we venture to appeal for more assistance. These deposits are too often so fragmentary, and sections in them as a rule are only temporary, that, without the aid of friends, it is impossible to do much. We have been fortunate in the past in this respect, and we would ask anyone who is acquainted with any deposit, no matter how recent, yielding molluscan remains, kindly to communicate with us. Even the much despised hill washes are of extreme interest, for the Rev. R. Ashington Bullen obtained a most interesting series of mollusca from one at Colley Hill in this neighbourhood, and he was able to demonstrate that *Ena montana,* Drap., and *Clausilia biplicata,* Mont., once lived here, though they are now quite extinct for many miles round, and he conclusively proved that *Helix pomatia* lived in these islands prior to the Roman occupation.

List Of Species.

Pyramidula ruderata, Studer.—Pleistocene: Barnwell, Clacton, Copford, Ightham, Swanscomb, West Wittering. Except at Clacton, this is a rare fossil, and we doubt if more than thirty examples are known from all the other deposits. It is a widely distributed species at present, being recorded from Norway, Sweden, Finland, Siberia, north China, Amurland, Russia, and the whole of central Europe.

Eulota fruticum, Mull.—Pleistocene: Barnwell, Grantchester, Ilford, Stutton. This is by no means a common fossil, and, as will be noted, it is unknown south of the Thames. A single example only has been found at Stutton, and it was from this specimen that the species was first described as British (). This example is now preserved in the Norwich Museum. It is known in a fossil state from the Red Crag of Hollesly, Suffolk, and, on the continent, it has been recorded from the lower Pleistocene of Mosbach, and the middle Pleistocene of Cannstadt and Nussdorf near Vienna.

Hygromia umbrosa, Partsch.— Pleistocene: Ightham. This species is only known from the Pleistocene of the Ightham fissures. It has been freely stated that the remains obtained from the Ightham fissures are of various ages, but except at one spot, where there had been a modern intrusion, and it was easy to discriminate between the modern and the older relics, the bones and the shells from the Ightham fissures must be considered as undoubtedly Pleistocene. It is probable that the deposit covers a very long period of S. V. Wood, *Crag Moll.,* vol. ii., p. 308, pi. xxxi., fig. 19.

time, but there can be no question as to the great antiquity of the fossils. *H. umbrosa* is extremely rare at Ightham, less than half-adozen examples being known. On the continent its present range is southern Germany, Bohemia, Switzerland, Silesia, and the Carpathians. In a fossil state it is known from the middle Pleistocene of Leuben near Lommatzach, and Robschulz near Dresden, and from the upper Pleistocene of Weimar.

Hygromia montiyaga, Westerlund.— Holocene: Harlyn Bay, Cornwall. The discovery of this species from the grave level of the early cemetery at Harlyn Bay, by our secretary, the Rev. R. Ashington Bullen, is of the utmost importance. It is to a palaeontologist no great matter whether the cemetery is Neolithic, Iron Age, Roman, or even post-Roman. What is of the greatest importance, is that here we have a shell, which at the present day is only found living in Spain and Portugal, and is, up to the present, unrecorded in a fossil state on the continent, occurring on the north coast of Cornwall. We have thus an addition to the wellknown Lusitanian fauna, the importance of which was first pointed out by the late Edward Forbes. It is just possible that the term Lusitanian may be a misnomer, and that this group did not have its origin in the Iberian peninsula, but rather in the now sunken land which formerly extended far into the Atlantic, if not to America. We have ample confirmation of this in the old river which deposited so much of the Eocene and Oligocene of the south of England. We may perhaps he permitted to quote the words of J. S. Gardiner, "it drained a vast continent with an indefinite westerly extension, even connected in some mysterious way with America. The breadth of this river, in its purely freshwater reaches, is actually seen in Hampshire and Dorsetshire to have been at least 17 or 18 miles," and he adds that " Dollfus compares its bulk in France to that of the Amazons1." It is only when we admit the existence of this now destroyed land that we can account for the fact that our Oligocene mollusca have well-marked affinities with Central American species, whilst it furnishes an explanation of the fact that *PlcbccuUi bowdichiana*, a recently extinct form in Madeira, is known from the upper Oligocene of Wurtemburg, as well as the affinity there is between several of the non-marine forms from the Red and Coralline Crags of England and living Madeiran species.

Clausilia pumila. Pfeiffer.—Pleistocene: Barnwell, Grantchester. Though not uncommon at Barnwell and Grantchester, this species is quite unknown from any other deposit in these islands. According to Dr. Boettger, the Cambridge examples are near to var. *sejuncta*, A. Schm., which is considered by some to be Q.J.G.S., 1883, p. 199. adistinct species. These formsoccur living in southern Scandinavia, Jutland, the Russian Baltic provinces, France, Belgium, Germany, the Tyrol, and Austrian littoral, and the Caucasus.

Planorbis arcticus, Beck.—Pleistocene: Crayford. For the identification of this form we are indebted to Dr. A. C. Johansen. Though in all probability it is only a northern race of the polymorphic species *P. parvus*, Say, yet since it can be separated from that species as well as from *P. glaber*, it is worthy of being placed on record.

Planorbisstroemii, Westerlund.—Holocene: Clifton Hampden, Fulham, Kew, Staines, Tooley Street (Roman and pre-Roman beds), Uxbridge, Walthamstow, Wargrave, and Westminster. This is one of the most interesting forms that have lately been found in this country. It is quite unknown from the Pleistocene, but occurs in some of the Holocene beds in abundance. This is particularly true of those at Walthamstow and Clifton Hampden. In all likelihood it was a late arrival in this country, probably being introduced here when, at the end of the Pleistocene period, the land stood at least 90 feet higher than it does to-day. The exact date of its extinction is uncertain. In all the Roman and post-Roman beds it is either quite absent or else extremely rare. Moreover, there is always the possibility of these specimens being derived fossils from an older bed. In all probability the form was either extinct, or else nearing extinction, in Roman times. On the continent it now lives in Siberia, Finland, and northern Scandinavia, whilst in Denmark it only occurs in deposits of the Oak period (= Neolithic).

Planorbis Yorticulus. Troschel.—Pleistocene: Grays, Swanscomb, West Wittering. This species was first detected in these islands at Swanscomb by Dr. A. C. Johansen. Since then we have noted it as occurring also at Grays and West Wittering. It is an extremely rare form, not more than twenty examples being known. On the continent it has only been found in a fossil state, in the Pleistocene of Weimar, Burg, and Grafen in Thuringia, and from the Holocene (neolithic) of Refsnaesand Kareboek, in Denmark, whilst, as a recent species, it lives in Holland, middle Germany, Switzerland, southwest Germany, and northern Italy.

Neritina grateloupiana, Ferussac,—Pleistocene: Swanscomb. The discovery of this fossil form in the 100 feet terrace of the Thames, at Swanscomb, is indeed noteworthy. It was by far the commonest mollusc there, occurring in myriads, all the examples still retaining their coloration. A few examples are pure white, without any trace of markings, whilst in others, the ground colour is almost obscured by them, and every graduation between these is met with. We are again indebted to Dr. O. Boettger for kindly identifying this species for us. On the continent it is known from the Upper Miocene of Switzerland, Baden, Wurtemburg, Austria, Bavaria, Hungary and Piedmont. At the present time its nearest living ally is *N. danubialis*, Mlf. The truth of the observation of the imperfection of the geological record is well borne out by this species. Hitherto unknown in any deposit newer than the Upper Miocene, it occurs in abundance in a small deposit at Swanscomb, the total area of which cannot be more than three or four acres, and, but for a chance excavation, its existence in this country would be still unknown.

Valvata piscinalis var. antiqua, Sowerby.—Pleistocene: Grays. Holocene: Kelvedon, Essex. This form was first described and figured in 1838, by G. B. Sowerby, as *Valvata antiqua*'". Though the types are not known, yet, since the figures represent the extremely high spired form, common at Grays, there ought to be no confusion, but the tendency is to call all the high spired examples by this name. *V. piscinalis* is an extremely polymorphic species, and practically every deposit will yield elongated examples, but a careful comparison with Grays' examples, will at once show the differences. In the *Journal of Conchotomy*, 1892, p. 64, it is stated " that *Valvata piscinalis* var. *antiqua*, Sow., is now considered to be practically the form named by Dr. Jeffreys, var. *subcylindrica*, and as Sowerby as priority, it has been adopted." We can only conclude that the author of this statement had never seen the true *antiqua* from Grays. In describing his variety, Dr. Jeffreys states that it has a flattened apex which the true *antiqua* never has. The var. *subcylindrica* is not uncommon both in a living state and also in the Holocene alluvium of the Thames. *Valvata* var. *antiqua* is stated by Dollfus to be now living in the canals of northern France, but, in this case, it is uncertain whether it is the true form he is referring to. We have only seen it from the Pleistocene of Grays', and from the deposit at Kelvedon, Essex, of which we have practically no details and whose age is uncertain. We cannot but think that many of, if not all, the continental records of *V.* var. *antiqua*, are open to question. There is apparently some connection in this species between the height of the spire and the depth of the water in which the animals live, the high spired examples being usually found in deep water.

Valvata piscinalis var. naticina. Menke.—Pleistocene: Crayford,

Swanscomb. This is a well-marked form, and was kindly named for us by Dr. Boettger. It was rare at Crayford, but was fairly common at Swanscomb. It differs from the type in the proportion of its whorls, the last one being far more inflated, and its name is well merited. Amongst continental writers it is usually considered a good species, but it is perhaps better in our present state of knowledge to group it as a variety of *V. piscinalis.* On the continent, in Germany, it is known in a fossil state from the middle Pleistocene of Mauer, near Heidelberg, and from the lower Pleistocene of Mosbach, near Wiesbaden. The living animal occurs in northern and southeastern Germany, Galitzia and lower Hungary, and possibly in Bohemia and Moravia.

Vivipara diluviana, Kunth.—Pleistocene: Clacton, Swanscomb. This species was not abundant at either locality. In this country it was first noted by S. V. Wood, who, in 1878, described it under the name of *Paludina clactonensis*", but he noted that a shell from the Pleistocene of Templehof, near Berlin, *P. diluviana* Kunth, might prove to be the same speciesf. This identity we have been able to determine through the kindness of Dr. E. W. Wiist, of Halle, who sent to us examples of *V. diluviana.* On the continent it is apparently confined to the Pleistocene of Germany.

Paludestrina marginata, Michaud.—Pleistocene: Barnwell, Biddenham, Brentford, Copford, Clacton, Grays, Ilford, Stutton, Spring Gardens, Stoke Newington, Swanscomb, West Wittering. Of all these localities, there are only two where this species occurred frequently, namely, Swanscomb and Clacton. In all the other deposits it was rare. It is quite unknown from Crayford, though such is the sporadic occurrence of all mollusca in these beds that it may yet be found there. It is a common pleistocene fossil on the continent and still lives in France.

Paludestrina confusa, Frauenfeld. — Pleistocene: Stone, Hampshire, West Wittering, Sussex. Holocene: Plumstead Marshes. It is only within the last few years that this species has become extinct, its last known habitat being a damp ditch at Erith. It is indeed possible that it may still exist in the Thames' marshes below Gravesend, but that district is practically a *terra incognita* to the conchologist. Mr. Clement Reid was the first to detect it in a fossil state in the Pleistocene of West Wittering and Stone, and in both localities it was rare. The record for the Holocene is on the authority of J. Gwyn Jeffreys who stated that it was found by Mr. Prestwich and Mr. Pickering in peat in the main drainage cutting between Woolwich Arsenal and Crossness J. Mr. J. T. Marshall informs us that, in 1870, it occurred between Erith and Abbey Wood and also at Tilbury, but it must be looked upon as now extinct in these localities. The shells which were obtained by J. Brockton Tomlin from Oulton Broad, Norfolk, and which were identified by Dr. O. Boettger as *Pseudamnicola anatina,* Drap., S. V. Wood, *Cidg. Mollusca.* 2nd supplement, 1878, p. 69, and tab. i., figs. 4 a-b.

t *Zeitschrift Deutsch. Gcol. Gcsellschaft Berlin,* 1865, tab. vii., fig. 8, J *British Conchology,* vol. i., pp. 64-5, certainly differ from the Thames' shells, hut we have not sufficient material to state definitely whether this is due to local variation or whether they are distinct species.

Unio littoralis. Lamarck.—Pleistocene: Barnwell. Brentford, Clacton, Crayford, Grantchester, Grays, Harrowden, Orton Waterville near Peterboro, Stutton, Swanscomb, Spring Gardens. This species was not uncommon at Brentford, Barnwell, Clacton, and Crayford, abundant at Swanscomb, rare at Spring Gardens and Orton Waterville. The largest examples we have seen are those from Crayford and Clacton. The Swanscomb specimens are small. It still lives in France, Spain, Morocco, and Algiers.

Corbioula fluminalis. Miiller.—Pleistocene: Barnwell, Chislet, Clacton, Crayford, Dartfordbrent, Grays, Gedgrave near Orford, Grantchester, llford, Shoeburyness, Stoke Newington, Stutton, Summerstown near Oxford, West Wittering. This form has been recorded from the Red Crag of Waldringfield, but we are unable to state whether correctly or not. It certainly occurs in the Norwich Crag of Thorpe, near Aldeburgh, and is found throughout the Pleistocene, as far as the fourth terrace of the Thames at Shoeburyness, where the examples, were extremely small, and still retained their periostracum. It is not known outside the Thames-Rhine system in these islands. On the continent, it is a widely distributed fossil, occurring in the Pleistocene of Denmark, Germany, Belgium, northern France, Austria, and as far east as Omsk, in Siberia, and in the Pleistocene of Algiers.

Sphserium corneum var. moenana, Robert.—Pleistocene: Crayford, llford. This form has hitherto only been noted at Crayford and llford. It is not uncommon, but the typical S. *corneum* is perhaps more prevalent. On the continent it is a rare form, living in the river Main. One or two examples of *Sphaerium,* that we have seen from the Pleistocene of Barnwell, are near to this variety which was figured in 1895.'::

Pisidium amnicum var. danubialis, Boettger.-Pleistocene: Crayford, Grays. At the latter locality it was extremely rare, whilst at Grays it was very common. In shape it greatly resembles *P. supinum,* but is much larger. There is great variation in the examples of *P. amnicum* from Grays. The typical form there is a very large shell much swollen and strongly striated. It is, however, only an extreme form, for many of the shells from the Holocene of Walthamstow also possess these characters, though not in so marked a manner. The var. *danubialis,* which was identified for us by Dr. Boettger has been recorded from the Pleistocene of Denmark.+ A. S. Kennard, "Pleistocene Mollusca of Cravford," *Science Gossip,* N.S., pp. 39-40.

f V. Madsen and V. Nordmann. " Oct Interglaciale N'ematurella," *Medd Dansk, Geol, For.,* no. 8, pp. 21-30.

Pisidium astartoides, Sandberger.—Pleistocene: Barnwell, Clacton, Crayford, Ilford, Grays, Swanscomb, Stoke Newington. This species first makes its appearance in the Cromerian, where it is not uncommon. At Grays and

Swanscomb it was abundant, less common at Crayford, rare at Stoke Newington, Ilford and Barnwell. On the continent it is known from the Pleistocene of Gudbjerg Laget, Forslevgaard, and Kobenhavn Frehavn, all in Denmark.

Pisidium supinum. A. Schmidt.— Pleistocene: Crayford, Grays, St. James'Square, Spring Gardens. Holocene: Clifton Hampden, Staines, Vvalthamstow. This species, which has only lately been detected in these islands, was probably a widely distributed form both in the Pleistocene and more recent times. It was extremely abundant at Grays and Walthamstow. Recorded by Sandberger as living in Scotland; we have been unable to obtain any confirmation of this, though it is not improbable that it may yet be found living in these islands.

Notes.

We have thus twenty well-marked forms which have become extinct. Of these *Eulota fruticum* first appears in the Red Crag, it is unknown from the Cromerian (Forest bed of Norfolk and Suffolk)," reappears in the later Pleistocene and disappears before the close of that period. *Corbicula fluminalis* is first known from the Norwich Crag, and apparently lived in these islands from that period to the late Pleistocene.

Three forms *Pyramidula ruderata, Paludestrina marginata,* and *Pisidium astartoides* are first known from the Cromerian, but did not survive the end of the Pleistocene.

Eleven forms *Hygromia umbrosa, Clattsilia pumila, Planorbis arcticus, P. vorticidus, Neritina grateloupiana, Valvata piscinalis* vars. *antiqua* and *naticina, Vivipara diluviana, Unio littoralis, Sphaerium corneum* var. *moenana, Pisidium amnicum* var. *danubiulis* are restricted to the later Pleistocene.

Two forms, *Paludestrina confusa* and *Pisidium supinum,* first occur in the Pleistocene and survived into the Holocene, the first named having only become extinct in the last few years, whilst *Hygromia montivaga* and *Planorbis stroemii,* are only known from the Holocene. We thus see how the south of England was continually receiving new species from the continent; sometimes they were able to establish themselves here, and have continued down to the present day, whilst, as we have seen, a large number died out, though for a time they were abundant. One cannot speculate on the causes of these extinctions since there are no facts to guide us.

In some quarters this is considered to be late Pliocene, but in our opinion should be classed as early Pleistocene.

The past history of our non-marine mollusca certainly lends no support to the school of extreme glacialists. Thus *Pisidium astartoides* first occurs in the Cromerian, which is usually considered as preglacial, it also occurs in the Pleistocene of the Thames valley. Now, if this period of intense cold, which we are told occurred after the deposition of the Cromerian, really happened, this form must have become extinct in these islands, so that the problem arises as to where this species survived. It cannot be on the continent, because it is quite unknown there, except from the late Pleistocene of Denmark, and, moreover, glacial conditions are postulated for nearly the whole of Germany. Is it not more reasonable to suppose that this, as well as many other, species was a continuous resident? Until the last few years, it has always been held that the 100 feet terrace of the Thames was deposited under more rigorous conditions than now exist, and that this deposit marked the later stages of the great Ice age. Now we know, as the result of a chance excavation at Swanscomb, that an abundant fauna lived in that period, including the wild pig, a species which is quite unknown in cold climates. Moreover, this bed yielded a number of remains which had been derived from a still older fluviatile deposit, and the wild pig is amongst these derived fossils.

True conclusions as to the climatic conditions prevailing at the time of the deposition of a bed can only be inferred from the general facies of the contained fossils, and not from the occurrence or absence of a few species. Thus, at Crayford, the presence of the musk ox *(Ovibos moscliatiis), Spermopliilus erythrogenoides* and *Planorbis arcticus,* does not imply that much colder conditions prevailed than the present, for with them occur *Corbieula fluminalis* and *Unto littoralis,* indicating a temporate climate. The general facies of the mollusca points to a climate similar to that which we now enjoy. It has been stated, and there are good grounds for so doing, that more genial conditions prevailed in early Holocene times, but we find that deposits of this period yield *Planorbis stroemii,* a species which, at the present time, is decidedly northern. A distinguished geologist once called *Planorbis albus* a land shell, and drew conclusions based on the presumed absence of freshwater forms. But, perhaps more interesting is the fact that the shells from the Pliocene beds of St. Erth, Cornwall, were once adduced as conclusive evidences of the glacial period in Cornwall, the clay being called Boulder Clay. One is reminded of the words which Canon Jessopp attributes to a scientific friend—" Give me theories, I can understand them, but your facts, I do not believe them."

If recent work has upset theories relating to the Pleistocene, it has also materially altered certain views with regard to the supposed introduction of species into these islands. We now knowthat *Dreissena polymorpha* was not introduced about 1830, in spite of the beautiful chain of evidence that has been adduced. *Helicella cartusiana,* so far from being a modern arrival, is an old established form, whose area of distribution has been greatly reduced in recent times. *Helix pomatia* is not a Roman introduction, but lived here in pre-Roman times. *Sphaerium pallidum* is not a recent American importation, but existed here in the Pleistocene, and since *Helix aspersa* occurs in the Pleistocene of Ireland, as-well as in many pre-Roman beds in England, it cannot be considered a recent arrival. We do not pretend that we have exhausted our subject. Large areas, even in the south of England, have as yet received no attention whatever. The extremely interesting deposit at Stutton has not, so far as we know, been visited by any geologist for nearly thirty years, and there must be many most interesting beds still awaiting the pick of the geologist, deposits which will yield even more interesting discoveries than those of Swanscomb, Grays and Ightham.

The Law Of Treasure Trove As It Affects Archaeological Research.

By WILLIAM MARTIN, M.A., LL.D.

The gist of the remarks, which I have the honour to lay before the South-Eastern Union of Scientific Societies, is that archaeological research is facilitated by the existence of a law of treasure trove and that the denunciations, to which that law has been subjected by antiquaries and others, are scarcely deserved. Such strictures, indeed, should, in the author's opinion, be reserved for faults in administration.

There is not a more fallacious doctrine than that summed up in the phrase "Findings, keepings," yet there is scarcely a maxim which appeals more strongly to the uneducated. The reduction into possession of ownerless or derelict property, and securing that possession by brute strength, or by artifice, is to the primitive man almost the first law of nature. Consequently, an eradication of the idea, that the physical detention of ownerless articles does not necessarily carry with it the right of ownership, is even now proportionately the more difficult.

Scarcely is the feeling "Findings, keepings," more difficult to extirpate in any other connection than in the case of ownerless hidden treasure. Nor, in respect of treasure, is the instinctive feeling that " Findings" is, or ought to be, "Keepings," confined to the uneducated or the illiterate, if we may judge by the clamour of many scholars in their condemnation of a law that allocates to the Crown the ownership of treasure trove.

To the author, however, the condemnation appears to be founded on an inadequate perception of what would be the state of affairs if there were no law of treasure trove, while the reasons which are commonly put forward for its abolition, or for a drastic alteration of its provisions, are such as would ensure, in a heightened degree, a continuance of the very evils denounced.

For a full comprehension of the reasons leading to this conclusion, a knowledge of the law of treasure trove in its relation to the law of first finding, of which, indeed, it forms an important part, is almost a necessity. When this relation is considered, it will be found that the alleged evils are due rather to the spasmodic application of the law than to the law itself, and that the remedy is not so much in a lessening of its rigour, but rather in an increase of its scope, an uniformity in application, and an improvement in its administration.

With the view of showing that, by a maintenance of the law of treasure trove, the science of archaeology profits, national museums are the richer, and individuals the gainer, this paper has been written. Incidentally, the opportunity is taken of stating exactly what in England and Ireland constitutes treasure trove. A definition collated from many sources is of importance, since many writers have taken as their basis one or other of the numerous and often erroneous or misleading definitions which occur in the compilations of their predecessors, definitions which, in many instances, appear to have been propounded with but little regard to the real state of the existing law.

In the first place, then, it will be convenient to discuss very briefly what is the law of first finding, for then the relation to it of the law of treasure trove will readily be perceived. It will be seen that the law of treasure trove is in the nature of an exception to the law of first finding, but that, in common with the latter, it will be found to deny, speaking generally, the right of possession to the finder. It will also appear that the Crown promises to act far more benevolently towards the finder than could reasonably be expected from him to whom, as a rule, the results of first finding accrue.

According to the law of first finding, it by no means follows that the finder of an article becomes, in the absence of all knowledge of the owner, entitled to its possession. In some cases, he may become the possessor or owner, but, in many cases, ownership falls to others. Whether the one or the other is entitled to possession depends on a number of conditions which may best be indicated by reference to a few typical cases which have been dealt with by the Courts. The chief, perhaps, of these conditions is whether the place where the article has been found does, or does not, belong to the finder. This was put so succinctly, in 1896, by the late Lord Russell, when two gold rings had been found by a labourer on cleaning out a pool, that the passage may be quoted. The Lord Chief Justice said:—"The general principle is that where anyone is in possession of a house or land which he occupies, and over which he manifests an intention of exercising a control and preventing unauthorised interference, and something is found in that house or on that land by a stranger or a servant, the presumption is that the possession of the article found is in the owner of the *locus in quo."* (South Staffordshire Waterworks *v.* Sharman 1896 65 L.J., Q.B.D. p. 462.)

Manifestly, this ruling considerably limits the number of cases where a mere finder is entitled to his find.

Let us now observe the position of the lessee of land as against his lessor. On the question of the ownership of a find, it was settled in 1886, that a prehistoric boat, which had been exhumed during the course of certain excavations, did not belong to the lessee of the land where it was found, but was the property of the lessor; it making no difference whether the lessee knew, or did not know, of the existence of the boat at the time of granting the lease (Elwes v. Brigg Gas Co. 1886J 33 Ch. Div. 562.)

Evidently then a lessee is not entitled to buried relics of antiquity.

We may now direct our attention for a moment or two to objects found in public places. When a bundle of banknotes was found on the floor of a shop, it was held, in 1851, that the shop keeper had no right to the notes when claimed by the finder. (Bridges *v.* Hawkesworth 1851, 21 L.J., Q.B.D. 75.) Probably this case governs the finding of articles in public thoroughfares, and, perhaps, it is the case most favourable to a finder as regards possession of articles found by him.

In respect of quasi-public places, such as theatres, railway carriages, tram-cars, promenade-piers, and so on, the ownership of articles there found depends upon so many conditions, that a discussion of the subject would occupy more time than there is available. It must here suffice to say that not in every case, or rather in few cases, is a finder entitled to what he picks up in those places.

There are also reported cases which show that convictions for theft have followed upon appropriation by finders where the true owners of articles taken have been unknown. Notable instances are the abstraction of pieces of iron from the bed of a canal into which the iron had, from time to time, accidently dropped from vessels navigating the canal (Reg. v. Rowe 1859J, Bell's Crim. Cas. 93); the taking of money from a bureau which had been entrusted to a carpenter for repair, neither the carpenter nor the owner of the bureau knowing at the time of the presence of the money (Cartright v. Green 1803J, 8 Yes. 405); and the retention, by the buyer of a bureau at an auction, of money which he found secreted in the bureau (Merry v. Green 18411, 7 M. & W. 623).

These brief references will indicate that, in most instances, it is truer to say that a finder may not retain what he finds, even when the owner is unknown, since the articles found, in virtue of their situation and of the other circumstances in which they are discovered, usually belong to others.

To recapitulate, it is probable that an article found in a highway, or a banknote in a shop, belongs, in the absence of the owner, to the finder; but when an article is found on private ground, whether secreted below the soil or drowned in water or exposed to view, the finder, as such, has no right to it. It might belong to the lord of the manor, who with a show of legality could prefer a claim; certainly the owner of the soil would have a strong right to the find. The instances where the law is doubtful need not detain us.

Where the find consists of the precious metals, gold and silver, more powerful claimants may appear, viz., the Crown or the Crown's assignee. When gold or silver is surrendered at these instances, the action is not usually due to the law of first finding, but to the law of treasure trove.

As regards the constitution of treasure trove, and in what way treasure trove differs from other finds, it may be said that:— *Treasure trove consists of gold or silver advertently deposited anywhere witliout abandonment, the owner being unknown.*

This short definition, it will be perceived, differs considerably from those customarily given. It will also be seen that no express mention is made of the attribute "hiding" which is so usually associated with the idea of treasure trove. Now, " hiding" is an equivocal term and may mean either a depositing out of sight without the knowledge of others, or a depositing out of sight with their knowledge. In most cases the presumption is that the treasure was secreted, but in many instances the conclusion may fairly be drawn that it was not. The definition given avoids the use of the equivocal term, but yet will be found to include its meaning when the determination that a particular find is treasure trove necessitates a reference to hiding.

In the present paper the author does not conceive it necessary to show the steps by which he has arrived at the definition. A study of the law and of its administration during many centuries— so far as that knowledge has been available—leads him to suppose that the definition is an accurate generalisation. A few comments may be made upon the definition. Many definitions have included coins in treasure trove, without specifying the metal of which the coins were composed. For centuries, however, the Crown has laid no claim to coins found in England which are not of gold or silver. As regard's the "abandonment" of the definition, if the gold or silver had been cast away by its owner, that is " abandoned" with or without a thought of its ultimate destination, then, when the treasure is discovered, the law of first finding obtains, not that of treasure trove. Consequently, in every disputed case, evidence must be adduced towards proving the correctness of, or negativing, the presumption that the treasure was cast away and was not deposited advertently in the place where it was found, or deposited there with the intention of reclaiming the treasure at a seasonable period. An " advertent deposit" is, of course, the antithesis of "abandonment." Again, if treasure, e.g., a gold ring, is lost accidentally, clearly there has not been an advertent deposit. Consequently the gold ring is not treasure trove, and the law of first finding settles the ownership, provided of course the owner of the ring is unknown or is practically unknowable.

The question may now be considered as to what would happen were the law of treasure trove abrogated, and has been advocated in many quarters, and whether archaeological research would be benefited by its abolition. In the first place, let us see what theCrown does when it stands upon its right to treasure trove and secures for itself the results of a discovery. According to the official statements which, appearing in the Government Publication of 1899 relating to " Certain Celtic ornaments found in Ireland," the author believes to be still applicable, on the receipt of treasure trove, His Majesty's Treasury transmits the find to the British Museum, where its value is assessed. The find is then offered to the Museum at that price. If the Museum does not elect to retain it, or has chosen such articles of the hoard as it desires, the find, or the surplus, is offered to other national institutions, with the result that coins or relics of bygone ages, or whatever they may happen to be, are placed on view for the edification of all and sundry.

Now as regards the remuneration to the finder. In spite of the number of years since the Crown has, with exemplary benevolence, rewarded the actual finder of treasure trove, the idea is still occasionally to be encountered that the finder, on the delivery up of treasure

trove, obtains nothing by the transaction. To show how erroneous is this idea, reference need only be made to the amended notice which was circulated by the Treasury, in 1886, concerning the remuneration that the Treasury was prepared to pay finders upon treasure trove being handed over to the proper authorities. By this circular, which has not been recalled, the Treasury promises to return to finders, such treasure trove as is " not actually required for national institutions." As to the treasure trove which is retained and sold to institutions, the finder is to receive its " antiquarian value" less a deduction "at the rate either—
1. Of 20 per cent, from the antiquarian value of the coins or objects retained; or
2. Of 10 percent, from the value of all the Gbjects discovered, as may hereafter be determined."

This notice must, however, be read with the context which says that this "arrangement is tentative in character" and with the letter which the author received a few days ago from the Treasury, as follows:—" With reference to your letter of May 31st, I beg to say that the Treasury have not, so far as I am aware, expressed any intention of departing from the general principles on which they have hitherto acted in dealing with treasure trove. No precise rules are, however, capable of universal application in respect of treasure trove, and it is obvious that the Crown must reserve its discretion as to the manner of dealing with particular cases."

Let us now consider what might happen if the law of treasure trove were abolished. Would our institutions be the richer and facilities for antiquarian research increased? When rare articles appear on the market, purchasers are sought among the wealthy, with the result that the articles are usually relegated to private museums. How much treasure has found its way into private collections may be judged from the very small amount which, having been claimed by the Crown's assignees of the franchise of treasure trove, is to be seen in our public Museums, the assignees, of course, not being bound to follow the example of the Treasury. Considering, too, the difficulty in obtaining access to private collections, when such collections are known to exist, it is obvious that students of archaeology and, indeed, the general public, would be the loser by an abolition of the law of treasure trove.

As regards the individual finder, would he be in abetter position if the law were abolished? Such rights as he would then have would arise from the law of first finding. From what has already been stated, the instances where he would be entitled to his find are rare. It would not perhaps be too strong to say that in no instance, in practice, would the finder of a hoard of precious metal become the owner of the hoard, considering the situations, the conditions, and the circumstances of the hoard when discovered. On the other hand, the Treasury promises to act, as we have seen, with great liberality towards the finder of treasure trove.

It is not to be denied, however, that the law still propounds problems which have yet to be solved. As in other branches of our legal system, it is from cases which have been settled by the Judiciary, or from the intervention of Parliament, on comparatively rare occasions, that questions are answered, difficulties removed, and the law made more capable of comprehension. Parliament, however, has scarcely touched treasure trove, while the cases before the Court have been so extremely few, and a knowledge of the instances in which the Crown, acting administratively, has claimed finds, has been so difficult of attainment, that the public has been in the dark with respect to the real reasons for the Crown's assertion of its right. Undoubtedly, ignorance of the applications of the principles of the law of treasure trove has produced dissatisfaction in its administration. When the higher Courts of Law are invoked, their judgments are reported by trained lawyers. When, however, from the facts of a case the Crown concludes that a find is treasure trove, the minutes of evidence of the departmental investigation, and the steps which have led the Crown to its conclusion are so rarely published that the grounds upon which the official action is based are practically unknown. A removal of the prejudice with which the law is viewed might partly be effected by publication of the reasons for the Crown's conclusions in cases of discovered treasure. From such publication, those interested would become acquainted with the circumstances which led the Crown to claim particular finds of precious metal.

As regards amendments whereby the law of treasure trove might be made more subservient to a study of antiquity, it is desirable for the promise of the Treasury to accord remuneration to the finder to be put upon a statutory basis. In addition, the author considers that the law of treasure trove might well be extended by statute, so that it could be invoked, if possible, whenever a relic of antiquity, whether or no of precious metal, was found. In this respect it would then accord with what is understood to be the law of Scotland. An extension of this nature would go far towards preserving for the country many relics which are now scattered and lost. Steps also should be taken to bring a knowledge of the law, and of the Treasury remuneration, to every one, as for example, by a judicious "posting " of notices in villages and rural districts, as for instance, in post offices. Finders would then learn that it was more to their pecuniary advantage to obey the law and deliver up objects to the proper authorities than to dispose of the treasure to a passer-by. The employment of the police in the collection of alleged treasure trove is to be deprecated, for a knowledge that the police, who are usually associated with crime in its various forms, will intervene, is often sufficient to make the unthinking secrete a discovery. Civil functionaries should undertake the custody of alleged treasure trove. Until a crime is suspected or committed, it is a mistake, from the point of view of the preservation of priceless relics for the nation, for a finder of treasure to be visited by the police and to that extent to be treated as a possible criminal. Other small, but desirable, amendments in the administration of the law, will readily suggest themselves, all having for their object the preservation of old time relics and their public exhibition in accessible situations.

In conclusion, the author trusts that he has indicated, though briefly, how the evils attributed to the law of treasure trove would not be removed by a repeal of that law, but that a repeal would, in effect, intensify those evils; and that a more rigorous application of the law, with a possible extension of its provisions, together with improvement in its administration, would not only benefit the individual, but also materially assist in the promotion of archaeological research.

Antiquities Of Reigate.
By EDWARD PEN FOLD, A.R.I.B.A.

So much has been written about the history of Reigate that anything I can say must be to a very large extent a *resume* of what is well known to any one at all interested in Surrey, but I hope I shall be able to recall to those persons some of the associations of the past, while affording to others an opportunity of becoming acquainted with our old town.

Like many other towns in the south of England, Reigate, in early and mediaeval times, was called upon to play a not inconsiderable part in the events of the country, and, in this way, it bears much resemblance to the Cinque Ports, the small size of which seems inadequate to the great events in English history in which they played so important a part. Unfortunately, in the views of many persons, Reigate is so near the metropolis that it is fast losing its independent position, and is being grasped by the tentacles of that ever-growing city, and becoming one of its many suburbs and a mere sleeping place to some fortunate members of that vast community. Yet what is now this little centre of a suburban building estate was, in mediaeval England, a borough and township with many other features belonging to the civic life of the time.

There were two great general political events which gave particular prominence to its position, the settling of the great Warren family at the Conquest and the murder of Thomas a Becket, at Canterbury, at the instigation of Henry II. To the first we owe the erection of the Castle and Priory and the creation of the town as a Parliamentary borough, and to the second an association with mediaeval church life.

Almost the first written records of its existence are in the pages of Domesday Book, which, as we know, was compiled by direction of William the Conqueror, as soon as possible when the country had settled down after the Norman Conquest, and the fair possessions of the Anglo-Saxons had been divided among the stern and merciless soldiers who followed the king to victory. The description given there is: "The king holds in demesne Cherchefelle which had been held by Eddid (Edith) the queen (Dowager). It was then assessed at 37i hides (3750 acres) now for the king's work at 34 hides (3400). The arable land is... There are three carucates in demesne, and 67 villeins and 11 bordars with 26 carucates. There are 2 mills at 12 shillings wanting 2d. and 12 acres of meadow. The wood yields 140 swine for pannage and 43 for herbage (*i. e.,* 1400 fat swine and 430 lean ditto in all). It is now valued at £40, which is the amount it yields."

Though this information is brief it is to the point, and gives indisputable evidence of the state in which the town or hamlet was then placed, and indicates how the present castle stood with reference to the houses of the people. From research into the ways of the early Anglo-Saxon tribes it has been shown that their township was purely patriarchal; it consisted almost entirely of farmers' houses each in its own little croft surrounded by a stockade, outside of which lay the pastures, ploughlands and woods, all forming common land for the use of every freeman. So the early township of Cherchefelle had its existence and customs, but with an unusual feature in addition, it was held and probably inhabited by a Royal Dowager, and so the rude dwellings of those early farmers were grouped round the south slopes of a Saxon stronghold, often doubtless containing the person of such a loved sovereign. We can almost realize in our minds this early picture of liberty, the freemen living around the royal dwelling, not in servile subjection, but in manly independence and self-control, believing in their own wise laws and institutions, and, although surrounded by mile upon mile of deep and dark forest, in no constant fear of outlaws, because there was no tyrant to drive men to such extremes. How sudden and terrible was the news of the fall of Harold at Hastings, which fell upon this family, and how different was the scene soon after, when the stern and forbidding fortress of the Earls of Warren loomed on the raised ground of the Saxon palisade, with ditch and moat, drawbridge and portcullis, the tyrannical system of feudal power in full force, the happy freedom of the Saxon farmers lost, and instead the sad tale of 70 villeins or serfs chained to the soil, allowed only to live on the land of their forefathers, tending the swine in the manner so familiar in the pages of Ivanhoe, in the person of Gurth the swineherd.

The valley between the North Downs and the range of hills forming Redhill Common and Reigate Park, was known from time immemorial as Holmesdale, in connection with which there is a tradition that it was the only part of the kingdom that was not overrun by the Danes, the origin of the old couplet, "Vale of Holmesdale never wonne ne never shalle," adopted on the Borough Arms at the incorporation in 1863.

There have been many theories raised as to the two names by which the town has been called, and their derivations. I have already mentioned that the name in Saxon and Early Norman times was Cherchefelle, and, in about 1189, it is mentioned as Chrechesfeld in a grant of the Church to St. Mary Overy at Southwark. This name means Churchfield, signifying, it seems, that some church was then standing, though this is not mentioned in Domesday Book, but nothing more definite is known as to the words. From 1250 the name Rigge Gate or Reygate is used as in 1275, when one of Edward I's judges of assize was named John de Reygate. It was formerly supposed, as stated in Aubrey's *History of Surrey,* 1673, that the word Reygate stood for gate, meaning stoppingplacing or street, over rea, a Saxon word for water, but as there is no water or brook in the immediate vicinity, it was always somewhat difficult for antiquarians to see the application of the words.

It is now generally admitted that the name originated in quite a different manner, although we have still tw6 conflicting theories as to the exact idea. In the middle ages, the valley of the Weald, stretching from the North to the South Downs, was a vast forest, and roads were few and far between, and the ancient stone street reaching from near Chichester, through Ockley and Dorking to London was the main thoroughfare. Mr, Manning, the Surrey historian, writing about the end of the last century, thought that a branch road may have passed from Ockley, and on through Reigate to Croydon, etc., and so to travellers this town would be recognised one to another as Cherchfelle-on-Ridgegate, that is the passage of a road or "gate" over a ridge or hill, afterwards shortened by leaving out the first word and the latter corrupted to Riggegate, and so to Reygate. Another theory which leads to the same derivation inclines to the opinion that the old track known for centuries as the Pilgrims' Way, caused the term to be used. But, although used by the pilgrims, who twice a year crowded the long backs of the bushless downs clear of the impenetrable and boggy undergrowth of the valleys beneath, on their way to Canterbury, this roadway was much older, and was the great means of connection from the west of England to the east, and it is possible that, as Guildford meant to the traveller the fording of the stream of the Wey, so the next place to turn in was on the edge of the hill, and down the old way by the side of the pits familiar to us all.

The early history of Reigate is wrapt up entirely in feudal rights and customs. The town being close under the walls of a Norman Castle and lorded over by its mailed barons for so many centuries, all the institutions, local authorities, and customs emanated from that source. William, Earl of Warren, to whom the whole belonged, was a Baron of that name in Normandy, being a descendant from the same ancestor as William the Conqueror, and having married his daughter Gundred; the name Warren is derived from the town of Guarenna or Varenne in the county of Calais. He was one of the most intimate followers of the King, and displayed such prowess at the battle of Hastings, that his share in the conquered country was very large. There was no definite grant of Surrey to him, but a record shows that the town of Reigate was settled upon him with nearly the whole of Surrey, and he was raised to the dignity of the Earl of Surrey by William Rufus.

He erected the Castle upon a hard knob of sand forming a splendid natural camp, probably the site of a Saxon fort. The manor and estates passed through his descendants to Hamelin by marriage with a grand-daughter, afterwards to William, the 6th Earl. His son, John, the 7th Karl, came into the title when quite a lad, and died in 1304. John, the 8th Earl, died in 1347, and, leaving no male issue, the estates devolved on the family of FitzAlan, Earl of Arundel, by his marriage with Alice, sister to Earl Warren. He died in 1415, and, being childless, the estate came to his sister, Elizabeth, wife of the Mowbray, Duke of Norfolk. From him it was handed down through the family of Howard to Thomas, Duke of Norfolk, father of the poet Earl of Surrey. This duke was attainted of treason by Henry VIII., and the estates were confiscated by him. Edward VI. granted a portion of the estate and manor to Lord William Howard, of Effingham, the then owner of the Priory estate.

William, the 6th Earl of Warren, became involved in the troubles of the barons with King John. He was one of the few who at first sided with the King, but was obliged at length to oppose him, and it is quite possible that there is some truth in the tradition that Magna Charta was discussed in the cave under the castle. If not actually that, there is high probability that the Earl might concert political measures with other barons there.

In 1275, Edward I. appointed Justices of Assize to enquire into the titles of landowners, for the purpose of obtaining possession of many of their properties. John, the 7th Earl, was cited at Gloucester, and, on being asked " by what right he held his landes, he sodenly drawing forth an olde rusty sworde; by this instrument (sayde he) doe I holde my landes, and by the same I entende to defende them." The King, perceiving the hatred of the people to his schemes, and the temper of the man he had to deal with, abandoned his intentions. This stout act was one of the first attempts at showing the King a barrier to despotic power, and the commencement of the Barons siding with the people for constitutional rights. A fine historical picture of this remarkable scene was painted by an artist, Pine, and an engraving made, a print from which is contained in Watson's *Memoirs of the Earls of Warren.*

Probably there are few spots in the country where there exist the remains of a media?val fortress, and so little evidence is forthcoming by means of which one can speculate as to the shape and form that the buildings, which once reared their massive strength above the turf of to-day, took. That the Castle of Reigate was of a large size is indisputable, the amount of land enclosed within the fosse or dry ditch on the south and west sides, and the water moat on the north and east, point to this conclusion, apart from the fact of it having been erected by one of the foremost men of England. Mr. Reginald Palgrave, in his *Handbook to Reigate,* published in 1860, gives the opinion that it was not of a massive character, but used and intended as a residence. The Earl of Warren erected several more castles in the country, *viz.,* Castle Acre, Coningsburgh, and Lewes, and we know he chose the latter as his principal home. At the same time, the Reigate, or Holmesdale Castle, was several times resorted to by its owners for purposes of defence, as when John, the 7th Earl, after the murder of Baron de la Zouch, at Westminster, fled to the Castle for the sake of protection against the king, who summoned him to appear and answer for the crime. Edward, the Prince, himself appeared before the walls and commenced the attack, when Warren yielded and paid the heavy fine of 12000 marks, about equal to £24000 in modern value, besides offering a humble apology. He seems to have soon recovered his tone, for seven years after the castle was large and important enough for him to entertain Edward I. here in a style of great splendour. The succeeding and last Earl having fallen in love with Alice, wife of the Earl of Lancaster, sent a few knights to Canford, in Dorset, and brought her in triumph to Reygate Castle. The Earl of Lancaster obtained a divorce, and, in revenge, recovered damages against the Earl of Warren by burning his castle at Sandal, near Wakefield, and devastating his manors in the neighbourhood, a stirring incident of the wild and ungovernable spirits of the time. The only remaining occasion when Reigate was of service as a fortress was in the lifetime of the Earl of Arundel. This noble joined in resistence to Richard II., and the King attempted to surprise Fitz-Alan here, but so threatening were the walls that he retreated without venturing an attack. From this latter event, which happened about 1390, the only information we can get as to the state of the castle is a tale of ruins. A plan was taken in 1497—a century later, which refers to the castle as ruinous. The late Mr. Thornton, land-agent, etc., of Reigate, had possession of this plan, and worked up from it a view of what he conceived the castle to have been. In 1550, Lambarde tells us that even then only " the ruyns and rubbishe of an old castle, which some call Homesdale" were to be seen there. Camden describes it as " forlorne, for age ready to fall." In 1624, a survey was made which states " Sir Roger James holdeth from year to year the site of the Castle of Reigate with the warren and lodge there, called the Castle Warren, containing 17 acres 16 perches," and "the lordes of this house have a decayed castle with a very small house," &c.

During the civil wars the Parliament, fearing lest the mass of ruins should assist the Surrey men to revolt, directed the Derby House Comitee, 1648, to place it in such a condition that no use could be made of it. This order must have been carried out effectively, for in Aubrey's *History of Surrey,* 1673, *Magna Britannia,* 1695, and Salmon's *Antiquities of Surrey,* 1736, verylittle mention is made of walls standing above ground.

There is a book in the library of the parish church called *A Companion from London to Brighthelmston,* published about 1786, in which is inserted a capital engraving, dated 1793, showing a plan of the Castle at that time, and, in Watson's *Memoirs of the Earls of Warren,* published 1792, there is a somewhat similar plan, together with a view of the existing state of the walls, etc. From these imperfect records there does not seem much ground for speculation, but still there is sufficient evidence of the general arrangement of the buildings, moat, etc,, to give some idea of what formerly was erected by the Norman baron in the end of the 12th century.

On the two last maps I have referred to and on a similar map drawn in a manuscript volume entitled *List and State of Reygatc Burgages,* 1786, compiled by William Bryant, who was agent for Earl Hardwick, which Mr. Gilford kindly lent me for reference, and, in fact until the cutting of the tunnel into the town, the moat of which we see one end at the London Road entrance to the castle grounds, reached in a long curve round the northwest of the grounds up to the back of Dr. Walter's garden, there meeting the extremity of the southern ditch. Thus the area was protected from invasion, using natural means so far as these were available. In the maps and view still in existence, as in the old one now lost, it is clearly shown that, on the top of the steep bank of the dry ditch, there was built a series of round towers connected by a wall, a similar arrangement to the usual Norman castle, as at Portchester, etc. Both the towers and walls were pierced with narrow slips to enable defenders to shoot down the slopes without showing themselves, and the round towers were finished with battlements in the usual manner. There is no reason to doubt but that these towers, etc., were continued all round the inside of the moat, with stronger ones, and probably with gateways having the dropping portcullis, at the junction of the dry ditch with the moat, for an entrance. What there was inside this outer fortification no one knows; the usual building, of course, as in other early strongholds, would be a massy square keep, as at Guildford, the Tower of London, etc., but as this castle does not seem to have been built till the timegof William Rufus, probably it had no such keep, those in existence having been built at once after the Conquest. Whatever form the central building took, it must have been of considerable size to have been the place of entertainment for a King, and standing as it did upon the height above the town must have presented a grand and impregnable specimen of feudal power. The area to the west, in the true type of the Norman castle, was taken up with the Bailey yard for practising tournaments, etc., and surrounded with stables, barns, and other subordinate dwellings of the retainers who (in case of an invading host storming the outer moat and wall) would retire into the central building or castle proper. There is a MS. which states that the 1st Earl built a chapel within the Castle. So far as entrances into the castle existed, it seems most probable that the present entrance from the town was always such, for it is referred to as the entrance in the MS. volume by VV. Bryant, and (as at Guildford) would have a gateway with portcullis, close to the street, and thence a rising winding roadway, similar to the Castle of "Edinboro'," would pass under a gateway and tower at the present archway, and so into the precincts of the inner walls.

In most descriptions of the castle and caves the entrance to the latter at the bottom of the ditch is spoken of as an ancient entrance. Doubtless this opening would be used as a sally port in war, and a private entrance in times of peace. There is such an entrance to a castle described by Sir Walter Scott in *A Legend of Montrose,* leading from the foot of a castle and forming a water entrance, and one which could be easily defended from a storming party by a few resolute men. In the map of 1786 there is shown a small onestoried house on the northeast side of the Castle court. This was built about 1520, and is not part of the original structure. There is no record as to when this house was demolished, but it has not been standing for at least 60 years. The present archway was built in 1777 by Richard Barnes, an attorney of Reigate, who, at that time, was tenant of the grounds, and wrote an inscription in Latin, which, with a translation, was placed on the archway some seven years back.

The caves under the central court are the only remains which admit of more certain speculation. They consist of a large curved apartment, with pointed vaulted ceiling, and with a continuous seat round one end, and a similar small recess is pointed out as the guard chamber, but, as it is not shown on old maps, it is probably modern. There is a map and description of the Castle in the number of the *Gentleman's Magazine* for July, 1842, which shows another circular cave with domed ceiling and a low seat all round the sides, near the present entrance from below, and it is suggested that it was a dungeon, but no trace of this now exists.

The long vaulted chamber, known as the Barons' Cave, is sufficiently spacious for the purpose for which the tradition says it was used, and the imaginative antiquary may fear no contradiction if he should maintain that such a council chamber was the very fittest place of the whole fortress for the unclasping of

"A secret book," and reading— "Matter deep and dangerous,
As full of peril and adventurous spirit
As to o'erwalk a current roaring loud
On the unsteadfast footing of a spear."

Although something may be said in favour of the patriarchal protection and noble hospitalities dispensed under the feudal system, it was only adapted for the then half-civilised state of society. The barons we're so many powerful reguli, holding their dependents in servile vassalage, and often controlling even their king. The influence of religion itself was maintained by the most absurd fictions and gross superstitions, not by the written Word of God.

No longer do the hauberks of chain mail and the pot-shaped helms of warlike knights and barons, gleam by torchlight in the vaults of Reigate Castle. Their generations and their manners have passed away. The defences of some of their oldest castellated dwellings are reduced to steep earthern mounds; of their chambers, the lowest in the caverns of the earth, alone remain. Yet, erst, descending from the inner and the outer wards, the flanking towers or donjon keep, which once frowned over such water-worn and shapeles heaps, "The yeoman tall
The iron-studded gates unbarred,
Raised the portcullis' ponderous guard,
The lofty palisade unsparred,
And let the drawbridge fall."

The Castle and the Priory of Reigate have always been in near relationship both as to position and their various owners. The foundation of the latter was granted to a body of Friars who came into England with a bull or charter from Pope Innocent III., creating them a new order of Friars of the Holy Cross, and as a badge they were to carry crosses upon their staves. They were unable to settle for some years until William, the 6th Earl Warren, about 1240, gave them land out of his manor, and afterwards endowed them with other lands and houses in the town. These Friars were afterwards called '' Crouched '' or Cross-bearing Friars (though there is another account which says that they were Austin Canons). The body was considerably strengthened by the addition, in 1296, of two citizens of London who took upon themselves the religious habit. In 1315, John, the 8th Earl, the last of his name and ancient family, gave them many other lands, besides releasing them from all rents paid to his ancestors, and in Manning's *History of Surrey* there are copies of several bequests of land which are interesting, *viz.* , Geoffrey Wallensis granted 19 acres of pasture land abutting on the original Priory land, towards the highway, west (i.e., Bell Street), and the hill on the south. In 1329, William Clark, of Nutfield, granted 50 acres of land upon the lower slope of Reigate Hill, then, as now, called the Brokes. Alice Skinner, in 1392, granted 8 acres of meadow called " Parke Ponde Meade," lying north of lands called Seal Hill, the old name of the Park Hill; this forms part of the Priory lawn, and includes the present pond.

John Lymden was the last Prior, and, at the dissolution of monasteries and confiscation by Henry VIII., was granted a pension of £10 per annum for his life. At their resignation in 1539, it is supposed to satisfy the conscience of the king, the monks had to use a certain form of declaration whereby they confessed that the perfection of a Christian did not consist in unnecessary ceremonies, in wearing a grey habit, in being girt with a knotted cord, &c, and that being disposed to lead a better life, they submitted to the king as supreme head of the Church, and abandoned their monastery to him, &c. The revenue amounted then to about £68.

As a consequence of the many grants of the Earls of Warren, there belonged to the Priory a manor called the Manor of the Priory of Reigate, for which Courts-barori were held, and several houses in the town, besides farms in the parish and elsewhere, were held under the Manor. At the dissolution many tenants discovered that men of business made harder landlords than the old monks.

The site of the Priory, together with all its advowsons and rectories, its rights and privileges, was granted to William, Lord Howard, of Effingham, by a charter dated June 6th, 1541 (a copy of which is contained in Mr. Bryant's MS. book), by Henry VIII., in exchange for the Rectory of Tottenham, in Middlesex, and afterwards a moiety of the manorial rights, and the site of the castle were presented to him by Edward

VI., being called the Manor of Reigate, as I have before mentioned. It descended from him to his son, Lord Charles Howard, afterwards created Earl of Nottingham, and High Admiral of the English Navy against the Spanish Armada. He is buried in Reigate Church. This nobleman married twice, and settled the estates at his death on his second wife, Mary, daughter of James Stewart, Earl of Moray. She soon after married William, Lord Monson, who purchased the other half of the Manor of Reigate, and so the whole property came into one ownership. At the death of his wife, the Countess of Peterboro', grand-daughter of the first wife of Lord Howard, claimed the Priory estate as her right. In the unsettled times of the Civil war fortunes quickly changed; Lord Monson favoured the Puritans, and was actually concerned in the execution of Charles I. , and sooner than give up the house he mounted it with cannon and dared the Countess to come near. At the Restoration, Lord Monson was attainted of treason, the whole of his estates confiscated hy the Crown, and he was cast into the tower and died there. The Countess of Peterboro', on the other hand, in consideration of the services of her son, John Mordaunt, to the king, had the Prioryestate confirmed to her. John.Mordaunt became second Earl of Peterboro', and, at his death, in 1675, the estate fell to his son Charles, created Baron.Mordaunt of Reigate, and third Earl of Peterboro', a clever man, but reckless of his property. He became deeply involved in debt, and to pay off these the estate was sold in 1681 to Sir John Parsons, Lord Mayor of London. His son, Humphrey Parsons, succeeded to the property, and the estate was again sold at his death in 1766, to Richard Ireland, of Dorking. After his death it came to Arthur Jones, Esq., who sold it to George Mowbray, Esq., in 1801. Lord Somers bought it in 1808.

With regard to the buildings of the old Priory, there were manyportions existing until the purchase of the house by Richard Ireland in 1766, who has been referred to by many visitors to Reigate late in the 18th century, as having pulled down whole piles of monastic buildings. The Howard family made use of the Priory buildings for their residence and inserted the beautiful chimney piece in the Hall, which is the only relic of their occupation. Sir John Parsons pulled down a portion of the buildings of mediaeval and Elizabethan times, and erected a typical Georgian house, of which I have an engraving; he built the beautiful oak staircase and employed Verrio to decorate the walls and ceiling with classical paintings. Richard Ireland altered this house and introduced many external features in detail, which are still retained and are shown in another engraving dated 1785; these engravings are interesting as indicating the gradual evolution of the house during the 18th century.

At the forfeiture of the Manor of Reigate to the crown by the attainder of Lord Monson, the King, Charles II. , gave it to James, Duke of York, afterwards James II. By the Revolution (1688), the estate again passed to the crown in the person of William III., who granted the Manor, with the site of the Castle, and all rights and privileges over commons, to Sir J. Jekyll in trust for Lord Chancellor Somers. Lord Somers died unmarried in 1716, and the Manor descended to his sister, Lady Jekyll, and, on her death, in 1745, the property went to her nephew, Mr. James Cocks. After being held by various members of that family it descended to Mr. Charles Cocks, M.P. for Reigate, whose services were rewarded by the grant of the old title of Earl Somers, and, in addition, Viscount Hasnor, and, from his descendants, the estate passed to Lady Henry Somerset, the daughter of the last Earl.

The old Church, dedicated to St. Mary Magdalene, possesses many objects of interest; I have prepared a plan of it, showing the various dates of building, in different etchings. It undoubtedly was originally a complete Norman erection, but the only remains of that date are the nave, columns and arches of late Norman character. They are built of the local fire-stone, which will not bear exposure to the various changes of temperature, a cause undoubtedly of there being no remains of external Norman walling. Two chapels were erected in the "Decorated" period, as aisles to the chancel, and the remainder of the church was rebuilt in the " Perpendicular" style, but, as the work generally was again carried out in local firestone, the windows have been necessarily restored from time to time, and, unfortunately about 1840, a very well designed, but altogether inappropriate, window of Early English character was inserted at the east end. The tower was practically rebuilt in 1874 under the direction of Mr. Gilbert Scott. The screen across the east end of the church, dividing the chancel and side chapels, is a good specimen of "Perpendicular" woodwork, and the sedilia in the east extension of the chancel are exceptionally good. There are no monuments of Gothic date, but several of Jacobean design, erected to the families of Thurland and Bludder, and there is also a fine classic monument, dated 1730, to Richard Ladbroke of Frenches, unfortunately quite hidden by the organ. There is a brass plate over the door to the vestry, with an inscription in Latin, stating that the vestry was erected in the year 1513 by John Skinner, Gent., for the souls of Richard Knight, William Laker and others, and also for those of his own parents. Several members of the Howard family are interred in the chancel, the most distinguished being Charles Howard, Earl of Nottingham, "Lorde High Admyral and General of Queen Elizabeth's Navye Royal at Sea agaynst the Spaniards Invincible Navye"; he died in 1624. The last member of the Howard family was buried in 1811; in the same vault are buried Humphrey Parsons, Lord Mayor of London, who died in 1740, and Richard Ireland, another proprietor of the PrioryHouse, who died in 1780. The church possesses an unique old library, which will well repay a visit; there are some 2000 volumes, including early manuscript histories, and Bibles in Latin, and such books as Dudgale's *Monasticott Anglicanum,* Spencer's *Faery Queen,* Milton's *Paradise Lost,* all in early editions; also a copy of the Breeches Bible, a black letter Wycliffe's Bible, Lord Howard's Prayer-Book, and many other priceless literary treasures, too numerous to mention here. The library was formed by Andrew

Cranston in 1701, and the original catalogue and complete list of persons who have given hooks is preserved; the room is redolent of the early 18th century, and is such a charming relic, apart from its utility, that no one could wish it altered.

I have said very little up to now about the town of Reigate, and 1 propose to go back again to the first records of the existing town, and spend the last part of my time in talking about its old institutions and the changes that years have brought in its appearance.

Of the gradual evolution of the town from the time of Domesday (with its 70 villeins) to the 13th century, when mention is made of tradesmen and mechanics, such as John the blacksmith, &c, and when there is record of fairs and markets, it is difficult to speak apart from our general knowledge of the feudal system. Under that system, men who had done some service to the Baron, or made a name in arms, were created freemen, and property was given them, not on lease, but on quit rent, with the liability of being called upon to follow the Baron to strife in his own petty quarrels, or for weightier purposes. Several houses in the town are still held by copyhold, and many others were in various ways created freehold.

A great event, which occurred in 1170, viz., the murder of Thomas a Becket, at Canterbury Cathedral, affected the little town for many years, and was the cause, no doubt, of its early growth. It is difficult for us to comprehend the lively faith in Becket's intercessory powers which led so many thousands from remote counties to his shrine, but of one fact we are well assured, and that is the extent of the system of pilgrimage. Two days in each year were specially devoted to his memory, the feast of his martyrdom (December 29th) and the festival (July 7th). The July feast was the most frequented, chiefly at the recurrence of the jubilees. On the anniversary, in 1420, no less than 100000 persons were thus collected. As the old track on the hills, for all west countrymen and the Irish and foreigners who landed at Southampton, was so near Reigate, many thousands must have traversed our streets. What troops, clad in the picturesque costumes of many lands, flocked down from the hill all dusty and footsore into the town, when the long shadows were glancing eastward up the valley— Irish, Welsh, and Normans—some on horseback, some on foot, with music and song, and merrie tales!

The Red Cross inn, at the west end of the High Street, is the only remaining link to connect the present day with those old times, and has held that name from time immemorial, but there were then doubtless many other inns and hostels for rich and poor, and what is still more interesting, three chapels erected probably for the special use of these pious strangers. A chapel dedicated to St. Thomas a Becket stood in the place of the present Town Hall, but a little to the east of it. What a picturesque and romantic picture it must have presented, with Gothic windows, buttress and parapets, recalling scenes in old continental towns 1 Another chapel stood at the west end of High Street, and was dedicated to the Holy Cross, and it seems probable was connected in worship with the Priory; the third chapel, dedicated to St. Lawrence, is the only one of which remains are yet in existence; the walls are still in the chemist's shop in Bell Street. It can be seen from the front that the south wall is three feet in thickness, and, at the back, the grey stonework looks mediaeval; inside are some remains of carved corbels, etc. In 1530, the last jubilee took place, and, in 1538, the feast was abolished and all chapels to Becket's memory ordered to be pulled down or desecrated, so that the loss to Reigate commercially must have been serious. Aubrey says "The Chapel of St. Lawrence was converted into a mercer's shop, the other of the Holy Cross, into a barn. Thus are the houses of God converted into shops of merchandise and granaries for corn."

A great institution of former days (before the time of good roads, and when the opportunity of seeing new inventions and manufactures was seldom obtained) was the annual fair. John, Earl of Warren, mentions three in existence in 1279, on Wednesday in Whitsun week, on the eve and day of St. Lawrence, and the eve and day of the Exaltation of the Cross. The first was in existence within the memory of many present, and was held in Bell Street. The second, on September 14th, has been obsolete for a considerable time and the third, the sole survivor, has been held since 1796 on December 9th.

The cattle fair supplies a demand, but the small, so-called, pleasure fair, which is the real relic of the mediaeval fair, is doomed, and will soon be one of the memories of the past.

The reference to the fairs by John, the 7th Earl, was "at Court of Pleas held at Guilford, where he claimed as his right of inheritance, the castle, honour, and town of Reigate, with all the members thereof, and the liberties thereto belonging, also a free market to be holden on Saturday in every week, and three fairs annually, with the tolls and customs pertaining to the same. Also assize of bread and ale, pillory, and tumbril, with all judicial process attendant thereupon, infang-theof, custody of the prison and gallows, with court-baron and court-leet, and all animals called cumeling."

The Saturday market was probably not a success, for John, the 8th Earl Warren, obtained a charter dated July 24th, 1313, from Edward II. for a market to be held in every week on Tuesday with all liberties and free customs to such. This market, though it dropped out of existence for short intervals, is still held as a corn market, and was within the memory of many a busy scene. It is the only connecting link between Reigate and the mail-clad barons who lorded over it so many hundreds of years; and thus, "while the monasteries founded by the Warrens are dissolved, the castles that they reared have crumbled into dust, and the vast estates they acquired are dispersed, this old custom is continued for nearly 600 years by virtue of their charters."

The privilege, formerly possessed by Reigate of being represented in Parliament, is still older than the market. As long ago as 1272, Roger le Quarrener and Robert Sabel were returned to Edward I's Parliament, held at Westminster. A copy of the writ served on the Sheriff of Surrey by Edward II., is in the MS. volume of Mr. W. Bryant, " We do command you that you cause to be chosen from the county aforesaid from every borough two Burgesses of the most discreet sort and most able for business, and that you cause them to come to us without delay at the aforesaid day and place" (the Feast of Pentecost at York). The writ is dated from Fulham, April 13th, 1298. On the back is the Sheriffs return, "Borough of Reygate." John de Bocham who is assured to go by Stephen of Maldon and Thomas Atte Church, Philip Pechy is assured to go by Wm. Atte Denne and Wm. Blauncestyn.

The appearance of the town of Reigate and the general arrangement of the streets was much the same up till 20 years since, as it had been for centuries, with the exception of the west end. The list of notes on the elections I have referred to contains a complete account of every house in the old Parliamentary Borough, with, in many cases, the various uses the building has been put to from 1693 until 1786, and, in some instances, references to its use previous to the first date. From this most interesting and unique record, and references to properties in old deeds, &c, it is possible to work up some fairly accurate plans of the old town. The space at the west end of the High Street was, until about 1800, occupied with a central group of houses, known as the East Row Island; there were four houses, one formerly being an oat mill; this block was removed to allow space for coaches to turn at the Red Cross corner.

In the premises now occupied for the motor shop of Mr. Guy, stood, until the time of the Reformation, the old market-house. The tenement is called " The Owlde Market Place," in a deed dated January 30th, 1586. The only remains of the building are an underground apartment, the object and use of which it is difficult to determine. It is about 12 feet wide, and rectangular, on one side is the staircase downward, and, on the other side, a small windowonly 9 inches wide and nearly 2 feet high. The room is vaulted over with heavy stone ribs to a flat curve, and I have made a sketch of the place, which can be inspected, showing it to have had a typical mediaeval appearance, and, from the small window and massive construction, 1 should imagine that it was the prison or dungeon under the Market House or Guildhall, and should give the date at about 1200 to 1230. From this building, which stood opposite the old entrance to the town by way of Nutley Lane and from Dorking, was, and is, a narrow street, now called Upper West Street, but formerly Pudding Lane, and referred to as " the ancient lane leading from the Castle of Reigate." These facts tell us that that portion of the town was, in pre-Reformation times, the centre for trade and for vindicating the authority of law, for the open space now fronting Nutley Lane probably contained the pillory and gallows of John de Warren, and, in imagination, one can see many scenes of rude violence and bloodshed witnessed on this spot; there, many a brave Saxon outlaw, who, like Hereward, would never do homage to the Norman baron, having been at length captured in the forest, paid the last debt of nature by a fearful end. How often, too, this quiet yard has resounded with the shouts of the retainers of the castle and the leather-clad burgesses of the township in loud clamour and dispute as to how far their respective rights extended. It was on this spot that the traders first carried on their markets and fairs after the invaluable privilege of holding them had first been gained by their earls. These busy scenes have for three centuries passed away and been well-nigh forgotten.

After the disuse of the Castle as a residence, which terminated its actual life in the town, the market was removed to its present position; doubtless, also, the narrow limits of Slipshoe Street, West Street, and the lanes round the houses in the centre of the High Street, militated against the continued importance of the west part of the town. At the desecration of the old chapel of Thomas a Becket, in 1538, it was utilised as a market-house and assize court until 1703, when the present building was erected by Sir Joseph Jekyll, then owner of the Manor of Reigate. He also, in 1725, pulled down the old chapel, and, in its place, built a prison surmounted with a clock; in this prison were detained persons awaiting trial at the sessions. The Town Hall and prison in the original state are shown on an engraving of the 18th century. The clock was removed from the prison and placed upon the Town Hall in 1801, and the prison demolished, when the workmen came to the foundations of the old chapel. The prison, or cage, was rebuilt in what was called the Mint Yard, and bears the inscription, "This building was erected in the year 1811 upon ground belonging to the Right Honourable John Sommers, Lord Sommers."

Although there are few buildings in Reigate which give indications in themselves of ancient foundations, and the town in this way is not so attractive as many other Surrey towns, this is to he accounted for to a large extent by the use for building of the local stone which is not of a lasting character. Although many houses are old they have had to be re-faced, and many early ones were no doubt demolished in Georgian times in consequence of their instability. Another reason for the absence of large mediaeval erections is that the town never had a strong civic life, and, until the middle of the 18th century, was under feudal influence. The Town Hall itself was erected as a memorial court by the Lord of the Manor, and, therefore, differs from such buildings in other towns which were Guild Halls erected by the community. There is also ample evidence that the various houses and premises were held under feudal tenure, since some are still copyhold. Moreover, what local government existed before the charter of incorporation was obtained was under seignorial influence, the officials in charge of local affairs were elected at the annual Manorial Court, and were styled High Bailiff, Aletaster, Leather-tryer, &c. There was, until about fifty years ago, a building over the entrance to the Castle from High Street called the "Old Building," granted with the Manor of Reigate, and used by the person who hired the corn tolls of the market. Such of the houses as had been enfranchised were held by individual owners until the beginning of the 18th century, and were purchased gradually, until about 1750, by the families of Yorke, Earl of Hardwicke, and Somers Cocks. One of the latter held afterwards the title of Lord Somers, which had been extinct for a time. The control of Parliamentary elections had a deal of influence in the purchase of the houses; the two families were joined by marriage in the beginning of the 19th century. There are several Elizabethan and Jacobean fireplaces in some of the houses, and in Slipshoe Street are buildings of the 15th century, or perhaps earlier date. We have, unfortunately for the antiquarian, lost the best of the old houses this year in West Street. The one at the corner was especially interesting, having been originally a one-storey building with an open timber roof, and was referred to in manuscripts as the old Red Lyon Inn. It was of 15th century date, and so was used by the pilgrims to Canterbury, and was one of the last relics of them.

The only industry which existed, apart from retail trade and the market, was the manufacture of oatmeal. Some twenty mills were at work in the early 18th century, and two or three existed until the end of that century.

The feeling of dependence upon seignorial power contributed to a large extent to the deficiency of large mediaeval and later buildings and enterprises, and was emphasised, even up to the early part of the 19th century, in the carrying out of the cutting and tunnel from the old London Road to the Market Place, to shorten the turnpike road, an improvement undertaken by Lord Somers.

Old times have passed away, new times are pressing with a heavy hand upon our old town, and it is fast losing the quaintness of its mediaeval origin. I think that no time and trouble devoted to recording its old life and appearance are wasted and I hope to be able shortly to give all persons interested an opportunity of possessing copes of the old prints and other drawings that have been preserved and kindly lent to me. My friend, Mr. W. Brooks, has also made lantern slides of them, which will be shown at the Municipal Buildings tomorrow evening.

By S. W. KERSHAW, F.S.A.

For convenience, my subject may be divided into the history of Gatton in early times, and, secondly, when it figured as a Parliamentary borough, for which it has an unique interest.

The situation of Gatton opens up a large field of enquiry. Gatton, the " ton " or settlement by the gate or road (known as the Pilgrim's Way), which also gives its name to Reigate, the Rigegate, or road on the ridge. Aubrey, the Surrey historian, describes Gatton as the "town upon the gate " or road, which, running eastward from the Roman causeway, Stone Street, here crossing the North Downs, passed on towards Croydon. The late Mr. LevesonGower, the distinguished Surrey antiquary, maintained that the so-called Pilgrims' Way was an old British track, anterior to the Romans, but used by them, as was shown by the villas that lie along this track, at Abinger, Bletchingly, and Titsey. One discovered in 18(S5 is still in the grounds of Mr. Leveson-Gower's seat. there. The proximity of this road invests Gatton with varied and peculiar interest. Of the town, Aubrey, who wrote in 1719, says, " it is small and inconsiderate at present; was well known by the Romans; coins and other remains of antiquity have been formerly-discovered." The same writer affirms that "the town was formerly situate more to the west, towards the top of White Hill, that is, the hill above Reigate." Dr. Richard Pocock in his *Travels* in 1750, published by the Camden Society, wrote:—"At Gatton, said formerly to have been a town, Roman coins have been found."

The finding of coins reveals a possible theory that they mayhave been brought by the pilgrims travelling from Canterbury or Winchester, both of which places were of greater influence then than London, and mints existed in each of those cities. As soon as Kent was incorporated into Wessex, in which Surrey lay, the whole district assumed greater importance. The neighbouring town of Croydon is said to have had a mint, but I think, on very slight authority, though the name Mint walk existed in 1619, and is mentioned in Steinman's history of this place, 1834. The Archbishops of Canterbury, who lived for centuries at Croydon, had the right of coinage, but there is no evidence that they had a mint there. Again, during the Civil war, local mints were set up, in order to coin money as required for supplies. The war raged along the border land from Reigate to Farnham, and the defence works of these castles, with Sterborough in Surrey, were ordered to be demolished, so that the enemy should not attempt a surprise. Thus, either to Roman or later times, even to the Commonwealth, may be ascribed the finding of coins in this district, a fact supported by local history. It is not unlikely that some of these finds may have been trade tokens, which would show the commercial importance of Surrey in past days.

To Reigate Castle, sometimes called Lord Monson's "Castle of Holmsdale," much interest attaches, especially as Lord Monson was a former owner of Gatton House and its beautiful surroundings. Lambarde, the Kentish writer, described Reigate Castle in the time of Queen Elizabeth as " ruins and rubbish," but even now in its present condition this statement seems unworthy of so careful a chronicler. The old and famous rhyme-"The Vale of Holmsdale, Never conquered, never shall." is associated with the " Men of Kent," who more especially dwelt in this long valley, which stretched from Dorking.to Sevenoaks. In remote times, Holmsdale formed part of the great southern forest of " Anderida," described in Mr. Green's *making of England,* "as a wedge of forest and scrub that filled the hollow between the North and South Downs, that stretched in an unbroken mass for 120 miles, from Hampshire to the valley of the Medway."

Gatton, on high ground, must have been an important outpost in early days, as associated with British history, and one name, among others, that of " Battlebridge Farm," would seem to indicate the scene of some conflict.

This district also recalls the first Danish invasion; the Danes advanced into Surrey by Sutton, and their course was not stayed till they met defeat at Ockley from the Saxons, an incident which forms a learned paper in the *Surrey Archceological Journal.*

Parliamentary Gatton.

The chief and unique fame of this place consists in its having returned two members to Parliament since the 29th year of Henry VI. (1451). This privilege continued without interruption until the Reform Bill of 1832 abolished the right. The elections took place in a small building known as the Town Hall, a little temple among the trees behind the Mansion house. A somewhat similar example of a meeting-place for election and other purposes, exists at Aldborough, in Suffolk, where the Moat, or Mote, Hall, a halftimbered building of the 16th century, was the rendezvous of the Borough Corporation. Aldborough, like Gatton, returned two members until disfranchised in 1832. It will be remembered that Aldborough formed the subject of the "Borough" of the poet Crabbe, a native of this old East Anglian town. We turn again to Salmon in his *Antiquities* and find he says— "Gatton is an old borough town, something hard to be paralleled, in that three places which send members to Parliament are so near together, as Reigate, Gatton, Bletchingly."

Gatton's constituency consisted for some years of one person, the Lord of the Manor for the time being. It was this privilege which generally made the property fetch a high price in the market. The first two names returned for Gatton were Thomas Bentham and Hugo Hulls, in 1451. In 1542, Sir Richard Copley, knight, who describes himself as a burgess and inhabitant of the borough and town of Gatton, returns, that he hath been "freely elected.' In Queen Mary's reign, another of the family, Thomas Copley, was made member. In 1585, Sir Thomas Bishop, secretary under Sir Francis Walsingham, succeeded to the honour, being nominated by the Crown, and the same year he was sheriff for Surrey and Sussex, as the two counties had only one sheriff at that time. The Gatton estate of the Bishop family was confiscated by Oliver Cromwell because they defended Arundel Castle against the Parliamentary army in 1643.

Other members for Gatton, of fame in various ways, were John Puckering, in 1586, Sergeant-at-Law, and afterwards Speaker; he was employed in unravelling the Babington and other plots in the cause of Mary Stuart; he was a favourite with Queen Elizabeth, whom he entertained at his villa at Kew, in 1591.

In 1603, Sir Thomas Gresham, of Titsey, was member of Parliament; he is remembered as founder of the Royal Exchange, and of the " Lectures," continued to the present day, in a learned and popular way, at Gresham College. Sir Thomas was the intimate friend of Cecil, Lord Burleigh; as ambassador to the Netherlands and political adviser, he won great honour. He took interest in the coinage, and established the first English paper mills at Osterly, in 1565. The late Mr. Leveson-Gower, F.S.A., was a descendant of Sir Thomas Gresham, whose portrait, by Antonio More, forms one of the treasures of the family mansion at Titsey Place. Those who desire further information respecting this family will find it in the *Genealogy of the Gresham Family,* compiled and published in 1883, with illustrations.

In 1621, Sir Edward Bowyer represented Gatton—his family was much connected with the manor of Camberwell. Succeeding members were Sir Charles Howard, of Lingfield, and, in 1685, Sir John Thompson, who was a constant speaker in the House of Commons. Sir Mark Wood was the last owner of the borough, and shortly before the Reform Bill of 1832, his descendants sold it to Lord Monson.

As the burgesses never mustered more than 30 votes, the Lord of the Manor directed the election; this, in the 16th century, was no strange proceeding, but when it occurred in 1832 it was thought the election farce had reached the climax of absurdity. Another instance of what may he called a " mock election " was at Garrett, near Wandsworth, when the Mayor was elected at every new Parliament, and this once popular scene of confusion gave rise to Foote's amusing comedy, "The Mayor of Garrett." The last election was in 1796. The annals of Gatton, however, are unique in the country, as regards Parliamentary lore, and one may well treasure the scanty intelligence that has come down to us through local historians and other sources.

As near Gatton, Reigate may be mentioned as having returned two members, the Reform Bill left it with one, and, in 1867, it was disfranchised, the last member of Parliament having been Mr. Leveson-Gower, F.S.A., of Titsey, near Limpsfield.

Gatton's history has figured in some early documents and papers, and in the Loseley MSS., belonging to Mr. More MolyneuK, of Loseley Park, near Guildford, of special Surrey interest, Gatton is frequently mentioned. An account of the manor, nomination of burgesses, and other matters, are given, and one famous personage, Sir Thomas Cawarden, of Bletchingly, cannot be unnoticed. Sir Thomas was "Master of the Revels," and it is of interest to note that, under his auspices, a school of actors had been formed, indirectly connected with the old theatres in Bankside. He was also keeper of Nonsuch Palace and Park in the reigns of Henry VIII., Edward VI., and' Elizabeth, and royal favour was extended, so that he was permitted to have armed retainers in his feudal-like castle at Bletchingly. On his death, his manuscripts passed to his friend and executor, Sir William More of Loseley, thus adding to the great value of that unique Surrey collection. No more interesting or valuable study of local matters in Tudor and Stuart periods could be obtained than by an examination of the historical abstracts of these documents, in vol. vii. of the *Historical MS. Commission Report,* 1879.

Bletchingly can hardly be unnoticed here, for some likeness in Parliamentary procedure to Gatton. The first return of members was made in 1294-5, and continued till the year 1832. Before 1733 the elections took place in a large house, and afterwards at the White Hart Inn; in later times the number of voters who attended was only 8 or 10. Many irregular proceedings occurred at these elections, of which record is made by some local and other writers. The instance of Gatton, however, has even a more unique character in Surrey annals, and, though the story of its constituency has long passed, it may afford reflection to the student of town or village communities, and thus add a page to a somewhat obscure phase of County government and custom

By A. E. TONGE.

Although a considerable amount of work has been done from time to time in making drawings by hand of the eggs of our moths and butterflies as seen under the microscope, it has been, on the whole, of an intermittent character, and, apparently, no work is obtainable illustrating a sufficient number of our British species drawn to scale, to be of use as a means of identification, excepting in the case of a few individual species or groups. It would therefore appear that such a collection of illustrations would be of considerable utility to students of entomological science, and this I am making an effort to get together by the aid of photomicrography.

1 have dabbled in the so-called "black art" for some years, but, until less than three years ago, it never occured to me that I might utilise my slight knowledge for the advancement of another and more prominent of my hobbies, namely, entomology; but having experienced great difficulty in identifying some wild ova which I had found, my attention was drawn to the lack of information obtainable on the subject.

I constructed several cameras for use with the only microscope I possessed, and found out as far as I could their special defects and shortcomings, finally arriving at the apparatus shown upon the screen which, if homemade, and consequently somewhat rough and ready, in construction, has enabled me to obtain something at least approaching the results at which I aimed.

As you will see, the apparatus consists of a very long extension camera, racking out at the back, and connected with the draw-tube of the microscope in such a manner as to exclude extraneous light while permitting freedom in focussing. The whole is mounted on a substantial baseboard in such a manner as to ensure proper alignment. Focussing is effected by means of a brass rod passing under the entire length of the camera, and connected with the fine adjustment of the microscope by an indiarubber ring. I use objectives of various foci, according to the magnification desired, the one shown on the screen is a photographic lens of 5" equivalent focus, adapted to the body-tube of the microscope, and necessitating the use of a special slide carrier behind the microscope substage, but I invariably work at an exactly ascertained magnification of 10, 20, or 40 diameters, and, if possible, mount at least one ovum on its side, so that measurements from two points of view at least can be taken from the resultant prints. All my photographs are taken by artificial light to ensure equality of illumination, and 1 make notes with each exposure of the apparent coloration of the eggs, and, if available, of the length of time which has elapsed since they were laid.

Lepidopterous ova are, as no doubt many of us are aware, very beautiful objects under tbe microscope. They are separately contained in an outer envelope, which assumes a large variety of shapes, and is, in many instances, sculptured with horizontal, perpendicular, or polygonal ridges, and may be hard and brittle, or soft and leathery; sometimes opaque, but more often semi-transparent, and permitting the coloration of the larva inside to be seen through the shell as the time for hatching approaches. When first laid they are usually white or very pale in colour, and many of them subsequently assume most beautiful colours and markings as the embryos mature. If no change in the colour takes place after the eggs are laid they may almost invariably be assumed to be infertile, and will gradually shrivel up instead of hatching.

At one point in the shell there is always a minute more or less circular rosette, usually situated in the centre of a slight depression, and formed by the various ribs approaching each other and splitting up into fine network surrounding the micropyle, which is presumably the aperture in the shell through which fertilisation is affected. The accompanying slide will give some idea of the appearance of the micropyle when very greatly magnified.

Roughly we may divide lepidopterous ova into two groups, which can be described as upright and flat. The former, which includes the eggs of most of the *Noctiiides, Notodontides, Arctiides,* and all the butterflies, consists of those which are so laid that the micropyle is at the end or side opposite to, and parallel with, the surface on which the ovum is laid; while the latter, comprising amongst other groups those of the *Geometrides, Sphingides,* and the bulk of the " micros," have the micropyle at a point at right angles to the base or surface of adhesion. I have endeavoured, in the next two slides which will be thrown upon the screen, to illustrate my somewhat involved description of this marked difference between the two groups of ova.

In a natural state some ova are laid singly, as for instance, *Euchlo'e cardamines* and *Centra vinula,* others in clusters, either evenly spread over a flat surface, or clustered closely on the top of each other like those of *Aglais urticae* or *Triphaena pronuba.* Some, again, are dropped at random amongst the grass or roots upon which the larva; will feed, such as *Melanargia galathea* and *Hepialus lupulinus* . Some are laid in a neat ring round a twig of the foodplant, as those of *Malacosoma neustria* and *Hybcmia leitcophaearia*, the latter species even covering the eggs with a light thatch of hairs detached from the body of the parent insect, while the former embeds the eggs in a hard cement-like substance, which makes it impossible to detach them without injury. *Crocallis elingitaria*, a very common insect, of which the ova may easily be found in a wild state on blackthorn and other shrubs, lays a very curiously brick-shaped egg in regular rows along the twig. In fact, the various methods of deposition are almost infinite, and ova may be found inside dead stems of the foodplant, on leaves, twigs, and flowers, both on and in the cocoon formed by the parent when in the larval or caterpillar stage. I have myself found a batch of *Macrothylacia rubi* ova on oak paling, and in at least one instance, for which I can vouch, ova have been laid on the painted iron frame of a street lamp.

The eggs are naturally difficult to find owing to their small size and to the apparent care taken by the parent to make them as inconspicuous as possible amongst their natural surroundings, so that the usual method of getting ova is to capture a wild female imago of the species and confine her in a suitable cage with a sprig of the foodplant. If placed in a shady place, and fed with moistened sugar or syrup meanwhile, most moths will lay. Small glass tumblers, with a flat piece of glass laid over the top, are very handy to keep the moths in, as they do not appear to require much air, and I have in this way kept many species alive for a month or more, in most cases successfully obtaining the ova I desired. I usually line the tumbler with paper so that the eggs will be more easy to remove, as if laid upon the glass it is often impossible to get them off unbroken. Butterflies being of a sun-loving nature are rather more difficult to deal with, but may often be induced to layby potting a living plant on which the larva can feed and either covering this with a muslin sleeve or a glass cylinder with muslin over the top. The insect can then be put in this and placed in the sunshine and the sugar syrup introduced on a piece of sponge, or a lump of sugar tied up in muslin and dipped in water can be hung up inside for her to feed on.

It is very often necessary to send ova by post, either for purposes of exchange or to assist entomological friends, so that a word as to the best method of packing may not be out of place. For a small number of ova, especially when loose, nothing can beat a short length of quill, in an ordinary envelope, but for large batches or a number of different species the best way is undoubtedly to screw up each one in a small piece of tissue paper, with the name of the species written on it, and to pack them all in a tin box with cotton wool. Do not place your postage stamp on the outside of the parcel but attach a tie-on label, so that, when the stamp is defaced in the post, there will not be any danger of the tin being smashed in and the eggs crushed.

Mr. Tonge, then exhibited about 80 slides of lepidopterous ova from photographs obtained by the methods described, to illustrate his subject, some of which were remarkably well done.

By (Miss) E. R. SAUNDERS, F.L.S.

The rediscovery of the work of Mendel, and of the conclusions to which he was led, have now become widely known, nevertheless a brief description of some of his experiments, and of the theortical conclusions which follow from them, will perhaps interest some who may have had neither time nor opportunity to study earlier and fuller accounts.

Gregor Mendel was born in 1822; at the age of twenty-one he entered an Augustinian monastery at Briinn, and later was ordained priest. He subsequently spent some time in the study of the natural sciences at Vienna, eventually returning to Briinn and carrying out his now well-known experiments on cross-breeding in the cloister garden. The results appeared in the *Proceedings of the Natural History Society of Briinn*, in 1865, but, although this publication was then being circulated among many of the learned societies of Europe, this important contribution to the study of heredity attracted no notice at the time. Only after the lapse of more than 30 years has attention been drawn to Mendel's work, in the writings of three Continental botanists, first by de Vries, and almost simultaneously by Correns and Tschermak.

Mendel's interest in hybridisation was aroused by observation" of the crosses commonly made with ornamental plants. He noted that such crosses constantly led to the production of a certain limited number of forms, which could be relied upon to reappear when the cross was repeated, and he was led to conclude that certain definite, ascertainable laws must determine the production of these hybrids, and that cross-breeding experiments conducted upon suitable lines should reveal these laws.

Mendel's aim was to reduce the problem to its simplest form, to select for crossing types, which, so far as could be seen, differed from one another in respect of a single character only, and thus to investigate separately the inheritance of each particular character. His choice fell upon the eating pea, *Pisiim sativum*. This species furnished several pairs of varieties which fulfilled the necessary conditions. Of the several differentiating characters selected for investigation, two will suffice for illustration, *viz.*, seed-shape (whether round or wrinkled), and seed-colour (whether yellow or green). Preliminary trials showed that in each case the characters A translation of Mendel's paper appeared in the *Jour. R. Hort. Soc,* 1901, xxvi., and later in Mendel's *Principles of Heredity,* by W. Bateson. *Mendelism,* by R. C. Punnett (1905), contains an exposition of the Mendelian theory, illustrated with instances of various characters already shown to have a Mendelian inheritance.

were constant. A full account of the experimental results is here impossible, and is indeed unnecessary, since a translation of the original paper has already appeared. Briefly stated, the results were as follows: The peas obtained from a cross between two varieties differing in respect of the colour character were all similar, and all yellow; moreover they were indistinguishable from the yellow peas of the pure-bred parent. Similarly, when the cross was between round and wrinkled, the first generation of peas consisted of all round, peas that were indistinguishable from the pure-bred round peas. Of the two contrary characters constituting a pair, the one which alone appeared in the first (Fi) generation was termed *dominant,* the other which disappeared in the first generation, to appear, as we shall see presently, in the second generation, *recessive.* In the two cases quoted above roundness and yellow colour proved to be dominant; wrinkledness and green colour, recessive. If these F peas were sown, plants were obtained, each of which, if allowed to fertilise itself, produced a *mixed* lot of peas constituting the second or F2 generation. Some showed the dominant, some the recessive character. Further the two forms occurred in the definite proportion of 3 dominant (D)to 1 recessive (R), *e.g.,* 3 round to 1 wrinkled, or 3 yellow to 1 green, as the case might be. In another experiment the plants raised from the Fi peas were crossed back with the original recessive type, self-fertilisation being rendered impossible by the removal of the anthers. Here also constant results were obtained. The F2 peas were again mixed, the proportion in this case being 1 D to 1 R. The experiments were continued and several later generations were raised. It was found that the *recessive* F2 peas obtained from self-fertilisation of the crossbreds produced plants which bore only recessive peas, *i.e.,* from the F2 green peas and wrinkled peas, plants bearing only green peas and wrinkled peas were obtained in the several succeeding generations during which the experiments were continued. The *dominant F%* peas proved to be of two kinds. Some like their recessive sisters, henceforth bred true, and produced only plants bearing dominant peas. These, which may be distinguished *as pure* dominants, formed about one-third of the whole number of dominants. The remaining two-thirds behaved as *impure* dominants (dominant cross-breds), and yielded plants with both sorts of peas, the proportion being again roughly 3 D to 1 R. This difference in the behaviour of the F2 dominants is indicated by writing the ratio thus—1 DD: 2 DR: 1 RR, instead of simply 3D: 1 R.

The subjoined scheme graphically represents the results just described:—

When the experiment was of a more complicated form, and varieties differing in respect of two (or more) characters were crossed together, a similar statistical regularity was observed. For example, in the cross round-yellow X wrinkled-green, the Fi peas were round and yellow, for, as we have seen, roundness and yellow colour are dominant over wrinkledness and green colour. Each such round-yellow Fi pea produced a plant bearing a mixture of *four* sorts of peas, in the proportion (roughly) of 9 round and yellow: 3 round and green: 3 wrinkled and yellow: 1 wrinkled and green. The two characters, shape and colour, are inherited independently, and, in such a way, that new combinations arise. Taking the colour character alone into consideration, we have, as in the simpler experiment, the ratio 3 yellow to 1 green. The same proportion holds good in regard to shape, the whole number of round is to the whole of wrinkled as 3 to I. But among peas of similar shape some are yellow, others green, and conversely, peas of similar colour may be either round or wrinkled. Of the four sorts of peas only the wrinkled-green (*i.e.,* only those which exhibit both recessive characters) breed true as a class. In each of the three other categories some individuals breed true, others behave as crossbreds, the proportion in each case being definite.::

Results of a similar kind were obtained in the case of each of the other characters selected for investigation. How were they to be accounted for? What was the meaning of this regular occurrence of constant forms among the offspring of crossbreds? Such constant forms must surely be due to unions between egg-cells and pollen grains of like character. Thus Mendel arrived at the conclusion that the reproductive cells in the ovules and pollen of hybrids must be of different kinds, of as many different kinds, in fact, as there are constant forms among the offspring. This conception of the existence of differentiation among the germ cells--and herein lies Mendel's great discovery--offers us an explanation, as simple as it is sufficient, of such statistical results as those given above, and furnishes us with an underlying principle which has already proved of wide applicability, and has enabled us to unravel results of far greater complexity than those which occurred in Mendel's own experiments. Moreover, we now have the clue enabling us to interpret many of the facts recorded in the earlier literature on hybridisation which had previously seemed hopelessly confused and unintelligible.

We will now examine such a simple case as one of those given above, in the light of the new theory. The theory assumes that the germ cells produced by a crossbred may, in respect of a given For turther details the reader is referred to one of the fuller accounts mentioned on p. 57.

character, be pure, transmitting one form of that character and one only; that,e.g., a germ cell of *Pisutn* will carry, in the case of colour, *either* yellowness *or* greenness, in the case of shape, *either* the round form *or* the wrinkled. Thus, in the case of a cross between yellow and green the resulting crossbred plants will produce:— *(a)* pollen germ cells carrying green colour (c) egg cells carrying green colour *(b)* ,,,,,, ,, yellow,, *(it)* ,,,,,, yellow ,,

If now we further assume, in the absence of evidence to the contrary, that the different kinds of germ cells are produced in equal numbers, and that, on the whole, the several possible kinds of unions occur with equal frequency, then we should have the following proportion of unions among the germ cells of Fi crossbred.

1 between green and green (when *a* meets *c)* 1 ,, yellow,, yellow (,, *b* ,, /) 2,, yellow,, green (,, ;,, *d* and when *b* meets c)

Of these four kinds of unions, the first

(a x c) alone will yield green peas; from the other three the resulting peas will be yellow, for we have seen from the result in F1, that, when yellow meets green, yellow is dominant. Upon the assumptions mentioned above we should therefore expect that peas obtained in *F2* will be mixed in the proportion of 3 yellow to 1 green, that all the green and-J of the yellows will breed true, and that the remaining yellows will behave as crossbreds, and yield again 3 yellow to 1 green in F3. Or, stating these facts in more general terms, since similar results were obtained in the case of each pair of characters studied, we may express the ratio in F2 as 3 D to 1 R or, more precisely, as 1 DD: 2 DR: 1 RR. We have, in fact, exact agreement between the results demanded by the theory, and those obtained by actual experiment, and for these and similar cases the principle of the differentiation or segregation of germ cells may therefore be regarded as fully established.

But we may go further than this. Five years have now elapsed since the rediscovery of Mendel's work; in the interval, facts have been rapidly accumulating, and we have now a large body of evidence drawn from experiments on animals as well as plants, on which to support and extend the new theory. Much of this evidence is derived from cases, similar to those studied by Mendel, where the differentiating characters are related to one another as dominant to recessive. But this kind of relationship between contrary characters is by no means universal, as everyone who has had any experience as a breeder is well aware. Mendel's choice of such cases for experiment was a fortunate accident, for here the results of crossbreeding are seen in their simplest form. In other cases the results are evidently much more complex, as, *e.g.*, where the offspring in F, though uniform, does not exhibit the differentiating character of either parent, and where the F generation shows a whole series of forms unlike either parents or grandparents. Such cases require careful analysis which may involve a continuance of the experiment through several generations. But the fundamental principle of the purity of the germ cells stands no less firmly because of such complications, for, as I hope to prove immediately, it is not necessarily bound up either with the phenomenon of dominance in Fj, or with the occurrence of the Mendelian ratios in F2. Instances of the occurrence of such disturbing phenomena as those mentioned above are well seen in the case of Garden Stocks. The wild hoary Stock *(Matthiola incana)* is covered on the stem, leaves, calyx, and ovary with a dense felt of hairs which gave the plant a grey appearance. In the so-called "Wall-flower-leaved " varieties the whole plant is glabrous", and of a bright green. A cross between *incana* and a glabrous Ten-week strain gives a simple Mendelian result as regards the surface character. Fi is all hoary (D) and Fa is mixed hoary and glabrous in the proportion of 3 D: 1 R or 1 D: 1 R according as Fi is self-fertilised, or is crossed back with the original recessive (glabrous) strain. As regards flower colour the result is more complex. When the white variety of *incana* is crossed with glabrous Ten-Week stocks of various colours as, *e.g.*, red, cream, or even white, the resulting crossbreds are in each *c&sepurplc*, like the *type* form of *incana*. Here then we meet with the phenomenon of reversion, where, in other cases, we find dominance. In view of this result in Fi the simplest expectation with regard to F2 is a mixed offspring of three colours *(viz.*, those of the two parents and the reversionary purple of the Fi crossbreds), the three colours occurring among both the hoary and the glabrous individuals, in each case in the proportion of one of each of the parental colours to two of the purple. In the case of the cross, *whhe-incana* x *redglabrous*, this is precisely what occurs; *F-* is mixed red, and purple, and white, and the three colours occur among both hoary and glabrous plants in the proportion (roughly) of 1 red: 2 purple: 1 white. When the cross is between *xhte-incana* x *cream-glabrous*, the F2 result shows a further complication. Here we not only have the expected white and cream and purple, but, in addition, a wholly new and unexpected colour, *vis.*, red. In this experiment two causes contribute to render the result in F2 somewhat complex, *viz.*, reversion and "resolution "; the latter term is applied to those cases where new forms appear to arise from the splitting up of a compound character, in this instance, colour. Such forms not infrequently breed true at once. When resolution occurs it is In some specimens a single hair or tuft of hairs may be found at the leaf apex.

obvious that the results cannot be expressed by the simple ratio 1 DD: 2 DR: 1 RR. Still more interesting, perhaps, are the results which follow from the cross, glabrous-cream x glabrouswhite. In this case the Fi plants are *all hoary and purple flowered*; F2 shows a mixture of hoary and glabrous, the glabrous individuals being all white or cream, while the hoary plants all have the cell sap of some shade of red or purple. Here we have to take into account—(1) Reversion in the character which differentiates the parents (*i.e.*, flower-colour); (2) reversion in an apparently unrelated character in which both parents resemble one another (surface character); (3) resolution; (4) association or linking together of sap colour with hoariness, and of the glabrous character with white or cream. The occurrence of these phenomena in no wise affects the validity of the fundamental principle of germ purity. The segregation of the germ cells occurs independently of reversion, resolution, and the like. A ready proof of the truth of this statement is afforded by those cases where by appropriate matings, we can at will obtain from the same individual either the apparently non-Mendelian result, or a simple Mendelian ratio. For example, a DR crossbred from *white-incana* x glabrous-red will yield a simple Mendelian ratio in Fj when self-fertilised, or when crossed backwith the original recessive strain, *viz.*, 3 hoary to 1 glabrous in the former case and 1 hoary to 1 glabrous in the latter. But, if the *same* crossbred is mated with a glabrous strain which is cream or white in colour, *the F-± generation will be all hoary*, though here, as before, a glabrous strain has twice been introduced into the pedigree. Such results show us that the phenomenon of dominance, and the occurrence of the Mendelian ratios are characteristic only of certain simple cases, they do not constitute an essential part of the principle of germ purity. When they occur the proof is obtained at once, in their absence further experiments alone can show whether we are dealing with a genuine exception or not.

In illustrating the complex nature of the phenomena which may appear in cross-breeding, I have taken my examples exclusively from the results obtained with a single genus of plants. It must not be supposed, however, that these are isolated cases, peculiar to stocks. Already numerous instances of a similar kind are known to us among both plants and animals, and the experiments now being carried out by many workers in this field are steadily adding to the number. Experiments of this kind are within the reach of most; the breeding of plants can be carried out on a small plot of ground and with little outlay. Time and labour, accuracy and patience and perseverance, these are essential; bestowing these, each may, if he wish, contribute his share to the elucidation of the problems of heredity.

By R. H. WELCHMAN, B.A., and C. E. SALMON, F.L.S.

This account of the Flora of the Reigate District, naturally resolves itself into a list of species and localities, which it is hoped by comparison with the floras of other districts may be of interest to the Congress; at the same time it cannot fail to draw attention to the large number of rare and interesting plants which are found in the neighbourhood. As, however, the enumeration of species, more or less well-known to all, at least by hearsay, tacked on to the names of places unknown to strangers to Reigate, can convey little meaning, it seems necessary to give a slight sketch of the surrounding country, pointing out those localities, which will be mentioned later on in connection with different plants, and mentioning the soil, absence or presence of water, and other incidentals. The term district is somewhat elastic, and we have thought it best to tajie as a limit the ground that can be covered by an average walker in a reasonable time, and, though here again considerable variety of opinion may occur, we think a seven mile radius will serve the purpose.

The three divisions of chalk, sand, and clay run parallel east and west in the order named. Of these, the chalk escarpment of the North Downs is botanically by far the most characteristic of the neighbourhood, though the greensand outcrop furnishes us with a larger number of highly interesting plants. The clay district of the Weald is, in comparison with the other two, poor in the variety and interest of its flora.

The charming line of the North Downs, as seen from Reigate, is by now familiar to you all. Above Dorking, on the west side of the Mole, which here bends sharply to the north and saws its way through the Downs, lies Ranmore Common, with beechwood glades and open spaces, the happy hunting-ground of botanist and entomologist. North of it lie Fetchan Downs, with Norbury Park sloping down eastwards to the Mole opposite Mickleham. Facing Ranmore, at the east of the valley, is Boxhill, north of which lie the Gallops, above Mickleham, one of the most delightful and secluded bits of down in England. Going eastward along the escarpment comes the buttress, known as Betchworth Clump, from an outstanding cluster of beeches, which form a landmark for miles around; the Pebblecombe, where the road winds pass-like up and over the downs; then Buckland and Juniper Hills, and finally with Colley Hill and the Pilgrim's Way we find ourselves at Reigate. Back in the angle formed by the Gallops, Boxhill and Betchworth lies Headley Heath with Headley Lane running up from the Mole; while east of it, fronted by the Buckland Hills, lies Walton Heath, a splendid open stretch, recently given over to the Golfing fraternity. North of Walton village lie Walton and Epsom Downs, to the northeast are Burgh Heath and Banstead Down, from the top of Colley Hill, behind the fort near " the Beeches," to the suspension bridge on the Reigate Road, bounding on the south the Margery estate, while the part between the Reigate Road and Wray Lane is Reigate Hill proper. North of Redhill and Gatton the chalk recedes, but east, again, of Merstham there is a splendid stretch of hills above Bletchingly and Godstone, forming the southern ridge of the Caterham valley. On the top between the Epsom and Croydon main roads lies the Chipstead district. All these districts are devoid of chalk streams or springs. We shall have frequent occasion to mention Colley Hill and Reigate Hill (up which the main road runs), and would therefore draw particular attention to them. We are rightly proud that, within a few minutes' walk from the centre of the town, such a large number of interesting plants are to be found, and, though many are decreasing in frequency from one cause or another, we trust that, as the land, by a generous gift, is now public property, the public themselves will aid in their protection and preservation.

We now give a list of those plants which are characteristic of the chalk hills—one or two of them are practically found on the chalk only, but for the majority no monopoly is claimed—but merely that they grow much more luxuriantly and freely on the chalk than elsewhere, and therefore we may venture to style them typical chalk plants: *Clematis vitalba, Arabis hirsuta, Reseda Ititea, R. luteola, Helianthemum chamaecistus, Viola liirta, V. calcarea, Polygala calcarea, Hypericum hirsutum, Liiium cartharticum, Geranium columbinum, Eiionymus europaeus, Trifolium medium, Hippocrepis comosa, Anthyllis vulneraria, Ouobrychis viciaefolia, Spiraea filipendula, Poterium sanguisorba, Rosa rubiginosa, Pyrus aria, Bryonia dioica, Pimpiuella saxifraga, P. major, Peucedanum sativum, Daiicns carota, Viburnum, lantana, Galium erectum, Asperula cynanchica, Scabiosa columbaria, Cardans nutans, Carliua vulgaris, Centaurea scabiosa, Campanula glomerata, Blackstonia perfoliata, Echium vulgare, Euphrasia kerneri, Origanum vulgare, Thymus chamaedrys, Daphne laureola, Juniperus communis, Taxus baccata, Epipactis latifolia, Cephalanthera pollens, Habeuaria conopsea, Ophrys musci/era, O. apifera, Avefta pubescens, Bromus credits, Brachypodium pinnatutn, B. sylvaticum.*

The greensand district comes a good second to the chalk in botanical interest. The secondary range of hills, which dies away east of Leith Hill, rises again in Park hill, Reigate, and continues along Cockshott Hill, Redhill Common, and Redstone Park to Nutfleld. The Holmesdale runs between these hills and the North Downs. In the sand belt is also included the lowlying Reigate Heath, with its three slight eminences, Skimmington Castle in the south, Trumpet Hill in the west, and Windmill Hill in the centre. It also possesses a *Sphagnum* bog along its southern side, the only one close by, and another, though less interesting morass, in the northwest corner. As the district is poor generally in damp situations, we have to thank the Heath for many species of bog, as well as sand-loving plants, which otherwise would not be included in the flora of the immediate neighbourhood, in fact more than twenty species would have to be expunged from our list, were Reigate Heath not part of the district. Park Hill, Cockshott Hill, and Redhill Common also furnish us with several species, and Redstone Hill, west of Redhill, was formerly well-known to botanists. Gatton Park, which lies between Reigate and Redhill and the North Downs, is partly sand and partly chalk, and thanks to its lakes we can add several species to our rather scanty list of aquatic plants. Its woods are also very interesting to the botanist. The following is a list of plants which may be called typical'of the sand and are absent from the chalk: *Centuiiculus minimus, Teesdalia nudicaulis, Cerastium quaternellum, Radiola linoides, Trifolium striatum, T. subterraneum, T. filiforme, Potentilla argentea* (all the above are found on Reigate Heath), *Saxifraga grauulata* (Park Lane).

South of the Greensand zone we strike the clay of the Weald, through which the Mole runs westward from east of Redhill to Betchworth, and flows under Sidlow, Flanchford, and Rice bridges in the order named. Though there is nothing particularly characteristic of its flora, a few waterside plants, such as *Lythrutn salicaria* and *Biitomiis umbellatus* are found along its banks, but, not to our knowledge, elsewhere in the neigbourhood. For the various outlying uncultivated spaces which may be of interest, or which we shall have occasion to mention, Holmwood Common, on the clay at the eastern foot of Leith Hill, south of Dorking, is the most important locality, while Earlswood Common (sand-clay), just south of Reigate, Petridge Common, on the Horley Road, and Burstow Common, southeast of Horley, are all more or less interesting to the botanist. Nutfleld Marsh (clay), abutting on the south side of the London and Brighton main line, close to Redhill station, will also be mentioned occasionally. The district may be called fairly well wooded, beech and yew on chalk, and oaks on the Weald.

It will be seen from the foregoing rough description of the neighbourhood that it abounds in dry, chalky, and sandy situations, that marshy districts are scarce, and trickling hill streams, with their accompanying bogs, are altogether absent, consequently the botanist can form a very fair estimate of the kind of list we are about to read, a list, with which pains have been taken to make it as complete as possible, but, at the same time, we crave indulgence, if, in selecting the list of plants which we consider rare or interesting, we have included any which may seem to be too common, and further, we hope we have omitted none which might have been of interest.

The published floras to which reference will be made, are Luxford's *Reigate flora*, 1838; Brewer's *Flora of Reigate*, 1856; the most recent, and as far as we know, the only floras of the immediate neighbourhood, Brewer's *Flora of Surrey*, 1863, and Dunn's *Flora of S.W. Surrey*, 1893. We give some old records of plants which have not to our knowledge been found here in the last few years because we do not like to omit them, and to state definitely that any species which has once been found in the district, no longer exists there. In a few cases, with regard to the exact situation of rare plants, we have not taken the Congress entirely into our confidence; it is not because we have any distrust of the Congress itself, but because, since the Transactions of the Congress are more or less public property, we fear the presence of these plants may become known to professional plant-hunters or unscrupulous collectors, and the species may disappear from the district. We have already suffered in this respect, a notable example being *Ajuga chamaepitys,* which has been nearly all grubbed up from its favourite spot on the hill, and consequently few plants left to seed. (*T.Jialictrum flavum,* between Sidlow and Flanchford Bridges;) *Myosurus minimus,* Trumpet Hill and South Park; *Ranunculus tripartitus,* Holmwood Common; *(R. hirsutus,* near Reigate Heath;) *R. parviflorus,* Redhill Common, near Cottage Hospital; *Helleborus viridis,* near Mickleham and near Ranmore Common; *H.foetidus,* Woodmansterneand Headley Lane; *Aquilegia vulgaris,* near Reigate Hill undoubtedly wild situation; *(Castalia speciosa,* in old brickfields on Horley road reported;) *Nckeria clavicnlata,* around Reigate Heath; *Fumaria vaillantii,* Colley Hill; *Nasturtium sylvestre,* ditch near St. Mark's Church; *Cardaminc amara,* marsh near Redhill Station and Reigate Heath; *(Iberis amara,* east end of Reigate Hill;) *Viola palustris,* Reigate Heath; *V. calcaria,* Buckland and Betchworth Hills; *V. sylvestris,* Wray Lane, etc.; *(Dianthus armeria,* Sandy Lane, Redstone Hill, and elsewhere,) but not recently recorded; *Saponaria officinalis,* Linkfield Lane; *Sagina nodosa,* Chipstead; *(Arenaria tenuifolia,* Banstead Downs;) *Cerastium pumilum,* Walton Down; *C. arvense,* Walton Down and Smitham Bottom; *Hpyericum dubium,* on Buckland Hill; *(H. montanum,* Reigate and Buckland Hills; *Linumangustifolium,* Reigate Heath;) *Geranium pratense,* foot of Reigate Hill, probably extinct;

G. pyrenaicum, railway bank at Merstham, also found at Betchworth; *Althaea hirsuta,* near Reigate. A notice, by C. E. Salmon, of the discovery of this species in Surrey, in the *Journal of Botany,* describes the situation as a rough chalky field amongst rather wooded country. In all probability the field had been, many years back, in cultivation, but so long ago, that now it is almost similar in aspect to many untouched portions of the Downs. The spot is far from houses. Other supposed native stations are Kent and Somerset, possibly also Gloucester and Herts. *Genista tinctoria,* near Charlwood; *Medicagoarabica,* near Nutfleld Marsh; *Trifolitttn fragiferum,* Wray Common; *Lotus tenuis,* Reigate Hill; *Astragalus glycyphyllos,* near Merstham; *Vicia gemella,* near Reigate Heath and at Outwood: *Lathyrus nissolia,* foot of Colley Hill and Outwood by roadside; *L. sylvestris,* Merstham; *Primus insititia,.* Reigate Hill and Betchworth; *(P. domestica,* frequent in woods and hedges about Reigate and in hedges round Reigate Heath;) *P. avium,* Reigate Hill; *(P.cerasns,* Reigate Hill;) *Rubi,* a variety of *Rubi* may be found on and round Reigate Heath; *(Potentilla palustris,* bogs near Reigate Heath;) *Alchemilla vulgaris,* near Dorking; *(Pyrus communis,* on the right hand of the road from Nuffield to Horley, just inside a gate on the descent of a short hill—a fine old tree—perfectly wild, record by J. S. Mill in Brewer;) *Pyrus torminalis,* Parkgate, south of Leigh; *Pyrus germanica,* Redstone Hill; *Clirysospleiiiuin alternifolium,* swampy copse near Reigate Heath; *Hippuris vulgaris,* Bury Hill, Dorking; *Myriophyllum verticillatum,* Gatton lake, and at Bury Hill, Dorking. *Epilobium roseum,* Outfield Marsh; *(liitpleurum falcatum,* Reigate Heath;) *Bupleurum rotundifolium,* Buckland Hill; *GEnanthc fistulosa,* Merstham; CE. *silaifolia,* near Sidlow; *Authriscus vulgaris,* Reigate Heath; *(Caucalis daucoides,* Buckland Hill;) *Smymium olusatrum,* Cockshott Hill; *(Sanibuctis ebulus,* Gatton;) *Galium uliginosum,* Reigate Heath; *(G. tricorne,* Betchworth;) *G. sylvestre,* top of Colley Hill

(Dipsacus pilosus, near Betchworth; *Gnaphalium sylvaticum,* Redstone Hill; *Artemisia absinthium,* Chipstead;) *Petasitcs officinalis,* Merstham; *Carduus nutans x crispus,* Colley Hill; *Cnicus pratensis,* near Reigate Heath; *(Hypochaeris glabra,* Reigate Heath; *Pulicaria vulgaris,* EarlswTood Common:) *Jasione montana,* Redhill Common; *(Phyteumaorbiciilare,* between Dorking and Ranmore;) *Campanula latifolia,* Bagden Valley, Dorking (Dunn;) *C.rabunculus,* Railway bank at Betchworth; (C. *rapuncidoides,* near Leath Hill; *Wahlenbergia hedcracca,* Reigate Heath, probably extinct;) *Hypopitys monotropa,* Reigate Hill; *(Hottonia palustris.* Burstow Common;) *Erythraea pulchella,* Buckland Hill; *(Gctitiaua germanica,* between Banstead Downs and Chipstead;) *Metiyanthes trifoliata,* Reigate Heath; *Lithospermum officinale,* Walton Down; *Cnoglossum germanicum,* near Ashstead; *Cuscitfa europaea,* Rice

Bridge; *Atropa belladonna,* Colley Hill and Headley Heath; *(Hyoscyamus niger,* about Redhill and Reigate;) *Orobanche major.* South Park; *O. elatior,* Boxhill; *Lathraea squamaria,* Chipstead; *Verbascum lychniiis,* near Smitham Bottom; *V. blattaria,* roadside near Kingswood; *Limoscila aquatica,* Holmwood; *Euphrasia rostkoviana,* Reigate Heath; *Salvia pratensis,* near Reigate; *S. verbenaca.* Castle Mound; *Mentha rotundifolia,* near stream by Hooley House, Redhill;) *M. rubra,* Nutfield Marsh; *M. gentilis,* Dawes Green; *(Calatnintha parviflora,* Redhill;) *C. officinalis,* Bagden Valley, Dorking; *Ncpeta cataria,* roadside near Reigate Heath; *(Marrubium vulgare,* Reigate Heath;) *Teucrium botrys,* Boxhill; *Ajuga chamaepitys,* sparsely scattered along chalk hills; *(Littorella juncea,* Reigate Heath; *Cheuopodium polyspermum,* in cornfields near Redhill; *C. Rubrum,* near Nutfield J. S. Mill; *C. urbicum,* opposite west entrance to Reigate Park;) *C.ficifolium,* near Woodmansterne; *(Rumex pulcher,* about Reigate; *R. hydrolapathiim,* Broome Park, Betchworth; *R. acutus,* roadsides in Wray Common;) *Polygonum dume-*

torum, near Reigate Heath; *P. bistorta,* near Reigate Heath; *(Daphne mezereou,* Betchworth Hill;) *Viscnm album,* abundant on poplar at Burford bridge; *Thesium humifusum,* Ranmore Common; *(Euphorbia platyphyllos* Buckland Hill;) *Mercurialis annua,* Betchworth Castle; *Buxus sempervirens,* probably more abundant on Boxhill than anywhere else in the kingdom.

The following is a short summary from an article on "Box in Britain," J. B., 1901. Graf zu Solius Lanbrach suggested that the box and yew trees of Boxhill might probably be the remains of a native forest which originally clothed the North Downs, and that it is probably the only thing of its kind in the world. A receipt for box trees cut down, dated 1608, is mentioned in Manning and Bray's *History of Surrey,* and it seems certain that the wood must have been there in 1500, and improbable that it was planted at such an early date as prior to 1500. The writer goes on to point out that box is a palaeotropical form which has outlived the ice age, and once more penetrated to the northwest. If anywhere in Britain, it is certainly a native in this locality.

Typha angustifolia, pond near Merstham; *Sparganium neglectum,* near Mole; *Acorus calamus,* Bury Hill, Dorking, probably planted; *(Potamogeton obtusifolius,* New Pond, Earlswood; *P friesii,* Gatton Lake; *P. zosteraefolius,* Gatton Lake, J. S. Mill;) *Zannichellia palustris,* Holmwood; *Butomus urnbellatus,* near Mole, near Brockham; *Sagittaria sagittifolia,* Mole, near Boxhill; *Damasonium stellatum,* Walton-on-Hill, Earlswood Common; *Hydrocharis tnorsus-ranae,* in pond, now filled in, on High Trees estate up to 1902; *Epipactis palustris,* pond near Merstham; *E. violacea,* Reigate Hill; *Cephalanthera ensifolia,* near Headley Lane; *Listera ovata,* abundant on the chalk hills; *Xcottia nidusavis,* abundant on the chalk hills; *Spirauthes autumualis,* Reigate Hill; *Orchis morio,* Sidlow Bridge; (O. *militaris,* between Boxhill and Mickleham, record from a *Journal of Botany,* 1874 Dunn;) *O. ustulata,* Reigate Hill; *O. latifolia,* near Reigate

Heath; *O. hircina*—"This species is stated in the *English Flora* to have been found on Boxhill. If this is correct it is probably now extinct there, as it has not been met with for many years."—Brewer's *Surrey Flora*, 1863; *O. simia*---" A single plant of this beautiful orchis was found by Miss Porter, in a copse near the Lower Lodge of Gatton Park, at the beginning of June, 1853, shortly after which the spot was well searched in the hope of finding more of it, but without success. The gardener stated that he had seen it in the same spot on two or three former occasions."—Brewer's *Surrey Flora*, 1863; (O. *purpurea*, Buckland Hill; *Habcnaria viridis*, Buckland Hill;) *Aceras anthropopliora*, plentiful on the chalk hills Especially abundant this yearj; *Herininium monorchis*, Headley Lane; *Ophrys aranifera*, near Bletchingley; (O. *arachnites*, very sparingly on Buckland Hill;) *Iris foetidissima*, Reigate Hill and Clifton's Lane; *Narcissus pseudonarcissus*, Holmwood; *Galanthus nivalis*, abundant by the Mole near Brockham; *Paris quadrifolia*, Nutwood, Gatton Park; *Convallaria magalis*, near Kingswood; *Ruscus aculeatus*, near Buckland Hill; *Tulipa sylvestris*, still in orchard near Buckland; *Scilla antumnalis*, Coulsdon; *Allium ursinum*, Buckland; (*Colchicnm autuinnale*, formerly on the Wray Park Estate, probably now extinct;) *Luzula maxima*, near Headley; *L.forsteri*, Redstone Hill; *Rhynchospora alba*, Reigate Heath; (*Eriophorum latifolium*, near Reigate Heath;) *Carcxpulicaris*, Reigate Heath; *C. carta*, Reigate Heath; *C. axillaris*, Gatton Lake and Clifton's Lane; *C. boeutiinghauseniana*, near Reigate Heath; (*C. distichia*, near Blackboro' Windmill, probably extinct; *C. teratiuscula*, ditch near Wiggy, Redhill, probably extinct; *C. acuta*, below Woodhatch Hill;) *C. strigosa*, Colley Copse and Gatton Park; *C. pseudocyperus*, Gatton Park; *C. rostrata*, Reigate Heath; *Homalocenchrus oryzoides*, Brockham; (*Alopecnrus fulvus*, formerly in ditch pond in cattle fields, where the Municipal buildings now stand;) *Hordeutu sylvaticum*, near Ranmore Common; (*Lolium arvense*, near Redhill;) *Celarach ofticinarum*, Headley Lane; (*Osmunda regalis*, Reigate Heath;) *Botrychiuin lunaria*, Reigate Heath; *Pilularia globulifera*, Earlswood Common; *Equisetumsylvaticum*, Holmwood. We have given several old records from Brewer, thinking they might be interesting to those who are engaged in the melancholy task of noting the disappearance of species from known localities. In some cases as *Hydrocharis* on the High Trees Estate and many of the Wray ParH records, the causes of disappearance are plainly due to building operations. Other cases are less obvious, *Dianthus armeria*, for example, is given for several spots close to Reigate, which have not been disturbed, but to our knowledge the plant has not been recently found there. Let us not, however, abandon hope. We will now conclude the paper with a short list of aliens which are more or less established in the neighbourhood, omitting those species which have appeared one year and disappeared the next, because, though interesting in themselves, they cannot be included as adding to the botanical interest of the district.

Alien List. *Papaver somniferum*, cornfields about Headley, well established; *Barbarea intermedia*, Reigate Hill, Betchworth and Brockham; *Claytonia perfoliata*, well established at Bury Hill, Dorking, and is also found near footpath west of Reigate Heath; *Erigeron canadense*, established in many places about Reigate, as on Cronks Hill, Pym's sand-pits, etc.; *Ornitliogalum nutans*, still exists in a lane near Wray Common, where it has grown for 50 years; *Liliutn martagon*, copse at Headley, where it is thoroughly naturalised, known there 50 years ago.

Notk. Parenthesis ( ) denote records from published Floras, the occurrences not having been recently verified.

Report On Wild Plant Protection.

By G. S. BOULGER, F.L.S., F.G.S., F.R.H.S.

(Presented June 8th, 1905.)

As resolved at last year's Congress, a circular was sent out by our Hon. Secretary last autumn. It ran as follows:—" The Council are desirous of eliciting information as to the danger of extermination of wild flowering plants and ferns, and as to any means—other than educational—of checking the same. Will you, therefore, kindly bring the matter before your Society at an early date, and inform the Council whether, in the opinion of your Society (1) any particular species or groups are, in your district, in present danger of extermination; and (2) if so, from what cause; and (3) whether your Society is of opinion that legislative or other action can, and should, be taken to check such extermination?"

With regard to the wording of these queries, for which I am responsible, 1 find from one reply that it is necessary to point out once more that the phrase, " other than educational," was merely introduced because everyone interested in the matter is agreed as to the desirability of educational measures for the object in view.

The results of this circular have not been very satisfactory. About a dozen replies have been received, two of which are mere promises to write again, while a third suggests inquiry as to the action of the Devon County Council, as to which everyone working at this subject is naturally fully informed. Two societies are opposed to legislation, two recommend it, and two more think it would be useless, or useless if not enforced. Eastbourne reports no species known to be in danger; Horsham thinks *Lycopodiiaii clavatum* alone rather scarcer; and North London thinks no species in present danger, though the inevitable growth of the Metropolis lessens the area of wild plants. On the other hand, Northfleet reports that dealers have made some orchids rarer. Mr. Step, on behalf of the South London Society, considers the danger within the twenty-mile radius of London a real one, and Mr. Griffin, in a long and interesting report on behalf of the Catford Society, reports the loss of eight species of orchids out of ten from Darwin's orchis bank at Down, and of fourteen species of ferns from the Ightham district during the memory of Mr. Benjamin Harrison, the loss of the annual *Blackstonia perfoliata* from one locality, due to the constant plucking of its blossoms, and the increasing scarcity of *Vtola odorata, Cynoglossitm, Vcrbasctim, Lychuitis*, etc., from similar causes. Mr. Griffin also speaks of the damage done to the marshland flora by the smoke and coal-dust of cement-works, and recommends the extension eastward of the operation of the Metropolitan smoke-abatement legislation. In common with several other reports, he refers to the depredations of hawkers, and especially to the wholesale rooting up of briar-rose stocks and orchids for sale to nurserymen. With reference to this, his society recommends legislation against the sale of scheduled wild plants for profit. Several societies recommend appeals through the public press, especially during the tourist season, the issue of such cards as the " Don't" circular of the Selbourne Society, and of leaflets. The Eastbourne Society suggests that this Union should address a circular to societies and schools; the St. Leonard's College Society suggests addressing landowners, and Mr. Step recommends that the clergy and teachers should be asked to urge some consideration for others upon school-children before their departure for school treats.

Though somewhat outside the subject of the circular, I feel bound to report a suggestion by Mr. Griffin a propos of the numerous aliens constantly added to our flora, that this Union, or its affiliated societies, should do something to assist financially in the preparation of a descriptive alien flora.

To this summary of suggestions from the Societies, I may add that I submitted a tolerably complete statement of the question to the Royal Horticultural Society, in August last, which is to appear *in extenso* in the next issue of their Journal. In that lecture I ventured to suggest that every local Natural History Society, or Field Club, should enumerate among its objects "the discouragement of the practice of removing rare plants from the localities of which they are characteristic, and of exterminating rare birds, fish, or other animals"; and that in the agenda for every meeting of such societies should appear the questions, " Has any Member to report that any plants, animals or objects of interest are in danger"?

As to legislation there is no doubt considerable difference of opinion as to the success or failure of the Wild Birds' Protection Acts, and still more misconception as to what sort of legislation is suggested in the case of plants. I have ventured a rough draft of a Bill, which I will lay on the table, and for which I alone am responsible. Naturally I quite agree with Mr. Step that this—or any other—legislation will be absolutely useless unless it is enforced.

By Professor W. M. FLINDERS PETRIE, D.C.L., LL.I)., F.R.S.

Written history is a very small part of the history of mankind. Material history, as contained in monuments and other prehistoric remains, takes us farther back. Material history takes us back at least to the times of the Neolithic peoples in Egypt.

The fall of the great civilization of the Roman Empire is to many people a mystery. It seems to them, so far, unexampled.

But, if we look back and go over the great epic of man, we find that no civilization has lasted much more than 1,000 years, and that none has survived 2,000 years. We find, moreover, that civilizations recur as regularly as the seasons. And the important fact is that material history forms by far the greater part of the record.

Egypt offers a good field for examination because it affords the greatest amount of knowledge with the least amount of work. Egypt, therefore, is a convenient ground for studying the history of man. Other countries are not so good; Palestine for instance, is barren in comparison with Egypt.

Until recently people laboured under two great fallacies: (1) That Egypt had nothing to do with the rest of the world. (2) That we could never get back beyond the highest civilization of Egypt, in the Pyramid times.

The work of the last quarter of a century has overturned these two fallacies.

We find that Egypt was unite as closely bound up with the Mediterranean as were the rest of the peoples round the Mediterranean.

We can trace the whole rise of its civilization from the time when man in Egypt was clad in goat-skins, made a small amount of pottery, and used but little metal. How he came to make use of metal we do not know, as the smelting of copper was not the earliest nor the easiest of the arts.

We know, too, that in Egypt before 6,000 B.C. man rose to the production of works of fine art, that he made beautiful vases, and that his commerce extended round the Mediterranean.

The prehistoric ships were the largest galleys known: they were larger than the greatest of the Roman galleys, or those of the Venetians in the Middle Ages.

We find the Egyptians using such vessels 6,000 years B.C. to carry on their trade all round the Mediterranean.

Then later we find their civilization decaying and cheapening in all its products.

This fresh view we have acquired in the excavations of the last ten or twelve years. This long period of prehistoric civilization was followed by a thousand years of the early monarchy and then comes a later period of Egyptian History, that of the Pyramid builders. We have thus a clear view of the rise of their civilization, in place of the blank which has existed in our knowledge hitherto. In later times we find that Egypt was in contact with neighbouring Semitic races, the oldest monument mentioning the name of Israel dates from about 1,200 u.c. This was found at Thebes, and mentions them as being in Palestine itself, where these people were contemporary with allied races in Egypt.

Merenptah, King of Egypt, is stated to have warred against the Israelites in Palestine, and despoiled them.

Beyond that, other points have been cleared up; the wars between Jewish Kings and Egypt prove to be correct and fall into their places quite naturally.

During the past winter for 3i months we have been centring our work in Sinai.

The geological features of Sinai are at

the base sea-worn granite islands, over which has been deposited a mass of carboniferous sandstone, 600 or 800 feet thick. Above this occurs a ferruginous band, beneath which the turquoise was mined; over that there has been a lava flow, and over the basalt a sandstone of Tertiary age has been deposited.

The sandstone has been weathered out into gorges 600 to 800 feet in depth.

The whole front has been cut away by a valley, down to the granite floor, so that its sides present a cliff 700 feet high and at the lower part a long slope of talus.

The Egyptians visited Sinai in early times for turquoise, they probably traded for it at first, just as they did with Persia in prehistoric times for lapis lazuli.

During the first Egyptian dynasty they took possession of Sinai, the king is represented on the rock-cut inscriptions as smiting down the Bedawi Chief of Sinai.

From 4,500 B.C. to 1,100 n.c. Egypt was lord of Sinai.

In the Bible, Musr is the kingdom of Egypt, and also includes Sinai; there is no need to suppose any other country to be intended.

The oldest remains in Sinai are the Bethels or places of offering upon the high hills.

We know from the history of Jacob how stone pillars were set up to mark a sacred spot.

The local goddess, identified by the Egyptians as Hat-hor, was the goddess of turquoise, and around her shrine were built stone shelters to break the wind; in these shelters the turquoise hunters who came to the shrine of the goddess to pray for success, used to sleep. The stone pillars are set up in these shelters.

These pilgrims when they received their dream set up a pillar in commemoration thereof.

At Sarabit el Khadim there is also a large temple occupying the summit of the hill. In front of the primitive cave is an area of burnt offerings, 100 feet long by 50 wide, pointing indubitably to Semitic worship.

At first the buildings consisted of small fore-courts in front of the cave. In 1500 B.C. the Egyptians extended the temple and built chamber after chamber before the fore-courts; they did not follow Egyptian plans, but built according to Semitic ideas. The ashes were preserved by laying the temple foundations upon them. The capitals of the pillars are ornamented with heads of the goddess Hat-hor; and in the courts there are basins for ablutions (as Mohammedans have) surrounded by four pillars, but dating from 2,000 years before the time of Mohammed.

Although the temple related to Semitic worship, it was carried out by Egyptians, thus helping in the preservation of Semitic work. The later additions to the temple consist of a large number of chambers banked over with earth and stones, as subterranean sleeping places for pilgrims.

Hitherto we have had nothing but literary evidence concerning Semitic worship, now we have material evidence. Twenty or more steles were erected in the Temple to recount the expeditions of the Egyptians in the Twelfth and Eigteenth Dynasties. Some of those pillars commemorate the finding of turquoise, with the help of Hat-hor.

On the face is inscribed the King's name, and on one side the name of the leader of the Expedition; there is also a record of the many minor officials. Mention is made of two or three artists, 500 soldiers, 300 peasants (as labourers), 30 inspectors, and two or three sculptors. There is one record also of 500 donkeys for carrying the provisions and water-supply from the coast; such a number of men would require about this amount of transport, as the conditions of the country were probably much the same then as now.

We found also the head of a black-statuette of Queen Thy, wife of Amenhotep III. It is of fine workmanship and shews great delicacy of carving, on a small scale, the face being about an inch and a half high.

With regard to the whole series of civilizations round the Mediterranean we had no sounding line as to time, until the correlation of these civilizations with that of Egypt, which has put back the beginnings of culture to a much earlier age. There are still uncertainties before 1500 B.C., but the outlines are fairly clear. When I first went to Egypt it was doubted whether there was any reliable Greek evidence before the Ptolemaic period. But we have gone since then steadily backward.

As a first step, the connexion of Egypt with the Greek Classic world was put back to nearly 700 B.c. The forts at Naukratis and Daphnae certainly take us to the middle of the 7th Century B.C.

The next discovery was that of hundreds of fragments of early Greek Pottery which were found in an Egyptian town of 1400 B.C. These fragments represented Pottery from the./Egean Islands. This proved that trade had been going on with the Greeks all through this period.

Then the palace-rubbish from Tell-el-Amarna—1,000 fragments of Greek vases, absolutely dated on the Egyptian side—formed the next important step. Every piece was copied by me in watercolours and the volume of these copies presented to the British School at Athens where it forms a standard text book. The next stage was that of finding pottery of a style yet unknown in the mounds of Kahun. Everything proved that this pottery belonged to about 2,500 B.C. But there was no connection of it found in Greece till some years later, when it was first observed at Kamares in Crete. Since then it has been found in all deposits of that age, and is now well known on the Cretan side. The ornament is laid on in a free and flowing manner, and in several colours. Thus the middle period of Cretan civilization has been linked with the Twelfth Dynasty of Egypt.

The pottery from a certain class of Royal Tombs of the time of the First Dynasty is also unmistakably Greek. It has all the characteristics of Greek pottery, but of an earlier period than any yet found in the i'Egean itself, as it dates from nearly 5000 B.C. The actual clay of which it was made was non-Egyptian, and the forms of the vases, and their pattern belong essentially to the Ægean. Amongst the black ware of this group of pottery at Abydos a pointed amphora was found. A fragment of this ware I took with me to Crete, and saw a similar specimen at Knossos. I produced my fragment from Abydos and found them absolutely identical. The time of the First Dynasty in Egypt and the late Neolithic civilization in Crete seem to correspond in many respects, and are synchronous in date.

The still earlier connection of Egypt in prehistoric times with Europe is shewn by the figures of large ships, and the pottery which was carried from one land to another in the course of this early commerce.

Obituary.

Prof. G. B. HOWES, D.Sc, LL.D., F.R.S.

At the request of the Council the following obituary notice of Prof. G. B. Howes, D.Sc, LL.D., F.R.S., was drawn up by Mr. F. W. Rudler, I.S.O., President of the Association at the time of Prof. Howes' decease.

Biological Science in Britain has suffered a grave loss by the premature death of Professor Howes, a loss which will be personally felt, with much keenness, by those who were present at our Brighton Congress in 1900, when he presided over the meeting with such conspicuous ability and geniality as to win golden opinions from all quarters. George Bond Howes was born on September 7th, 1853, and consequently at the time of his death he was but 51 years of age. Yet into that limited range of life he had managed to compress an exceptional amount of scientific work. This was the more remarkable inasmuch as he had not enjoyed in early life any opportunity of systematic study in natural science. Descended from a Huguenot family, and educated at a private school, he struck into a scientific career at a great disadvantage, and success was achieved only by remarkable perseverance, associated with natural gifts of a high order.

At the age of 22 he obtained the position of an assistant in the biological laboratory which Professor Huxley had recently established at South Kensington. There he was occupied at first in executing a large series of anatomical drawings, but he soon advanced far beyond the position of a scientific draughtsman; for the professor, with characteristic penetration, was not slow in detecting the talent of his young assistant, especially in the careful preparation of dissections. At that time, the late Mr. T. Jeffery Parker was the Demonstrator in Biology, and on his acceptance of a professorship in New Zealand, Mr. Howes was appointed to succeed him. This was in 1880. Five years later, on Huxley's partial retirement through declining health, Howes was made Assistant-Professor of Zoology, and in 1895, on the death of Professor Huxley, who had been Honorary Professor of Biology at the Royal College of Science, a chair of Zoology and a chair of Botany were established, the former being given to Mr. Howes.

Honours were not slow in falling upon the new professor. In recognition of his valuable original work, chiefly in Vertebrate Morphology, he was elected into the Royal Society in 1897. From the Victoria University he received the Degree of D.Sc, and from St. Andrews that of LL.D. Professor Howes was a Vice-president of the Zoological Society, an ex-president of the Malacological Society and Zoological Secretary of the Linnean Society. In 1902 he presided over the Zoological Section of the British Association, and on that occasion delivered so elaborate an address that no fewer than 186 references to the authorities quoted are given in an appendix.

This address may be accepted as an illustration of the thoroughness which characterised all his work; and there can be no doubt that his lavish expenditure of intellectual energy in the discharge of duty, led to his ultimate decline. In command of.a great wealth of knowledge, and gifted with a singularly retentive memory, possessed, too, of remarkable geniality and modesty of manner, Professor Howes was extremely popular at scientific meetings; but his engagements left him too little rest, and led to a continuous state of mental strain. To succeed Professor Huxley was indeed a great honour, but it also involved a grave responsibility. It is not too much to say that, though some unfortunate accidents contributed to his fatal illness, he really fell a victim to his scientific zeal. After a lingering illness, with occasional revival, Professor Howes passed away early in February, leaving behind him the memory of a model teacher—admired by all who could appreciate his scientific work, and beloved by those who had the privilege of his friendship.

F. W. R.

LIFE MEMBERS. *(Instituted at the Canterbury Congress, 1902.)*

Adkin, Robert, F.E.S. 4, Lingard's Road, Lewisham, S.E.

Adkin, Mrs. R.

Bennett, F. J., F.G.S. "The Acacias," West Mailing, Kent

Mullen, Rev. R. Ashington, F.L.S., F.G.S. Pryford Vicarage, Woking Surrey (Hon. Sec.) (D)

Mullen, Mrs. Ashington. Pyrford Vicarage, Woking, Surrey

CoomaraswSmy, Anandy K., F.L.S., F.G.S. Walden, Worplesdon, Guildford

Foran, C. Elm Grove, Southsea

Gray, H. Norman, P.A.S.I. Newlyn House, 131, Earlham Grove, Forest Gate, E.

Howorth, Sir H. H., K.C.I.E., F.R.S. , F.G.S. (V.P.). 30, Collingham Place. Earl's Court, S.W. (Ex-Pres.)

Meeson, F. 98, Sutherland Avenue, Maida Vale

Merrifield, F., F.E.S. (V.P.). 24, Vernon Terrace, Mrighton

Neatc, P. J., J.P. Watt's Avenue, Rochester

Rudler, F. W., I.S.O., F.G.S., &c. 18, St. George's Road, Kilburn. N.W. (Ex-Pres.)

Sargant, Miss E., F.L.S. Quarry Hill, Rcigatc

Stebbing, Rev. T. R. R., F.R.S., F.L.S. (V.P.). Ephraim Lodge, The Common, Tunbridge Wells. (Ex-Pres.)

Stebbing, Mrs. T. R. R., F.L.S. Ephraim Lodge, The Common, Tunbridge Wells

Stebbing, Miss Grace. Catton, Southborough

Stirling, Sir J., Hart., F.R.S., Finch-

cocks, Goudhurst, Kent
Turner, Miss E. L., F. L. S. Langton Green, Tunbridge Wells
Vardon, Rev. S. A., M.A. Langton Green, Tunbridge Wells
Walker, A. O., F.L.S. Ulcombe Place, Maidstone
Walmisley, A. T., M.I.C.E. Atherstone, Castle Avenue, Dover
Whitaker, W., F.R.S., F.G.S. (V.P.). 3, Campden Road Croydon (Ex-Pres.)

DELEGATES.

Abbott, G.. M.R.C.S., F.G.S., Tunbridge Wells N.H.S. 33, Upper Grosvenor Road, Tunbridge Wells
Adkin, R., F.E.S.. South Lond. Entom. 4, Lingard's Road, Lewisham, S.E.
Adeney. Mrs., Tunbridge Wells Nat. Hist. Soc. Tunbridge Wells
Baldock, J. H., F.C.S., Photo Survey, &c, Surrey, Croydon N.H. St. Leonard's Road, Croydon
Bedford, E. J., Eastbourne N.H. "Anderida," Gorringe Road, Eastbourne
Blackburne, F.., Hastings and St. Leonard's
Campbell-Bayard, F., Croydon N.H. Cotswold, Wallington
Chamberlain, Joseph, Tunbridge Wells, Amat. Phot. Soc. 28, Cambridge Street, Tunbridge Wells
Chapman, T. A., M.D. City of Lond. Entom., Holmesdale N.H. Club, Betula, Reigate
Comber, A. T., Horsham Museum. 3, Worcester Terrace, Reigate
Dodd, Mrs. C. T., Tunbridge Wells Nat. Hist. Soc. Tunbridge Wells
Elgar, H. J., Maidstone and Mid Kent. The Museum, Maidstone
Edwards, S., F.L.S., West Kent N.H. 15, St. German's Pl., Blackheath, S.E.
Eele, Jas. W., Haslemere N.H. Shottermill, Haslemere
Fleay, F. G., M.A., Balham Antiq. & N.H. 27, Dafforne Road, S.W.
Frost, R. C, Woolwich Antiq. 11, St. John's Road, Plumstead
Gray, H. Norman, P.A.S.I., City of Lond. Coll. Sci. 131, Earlham Grove, Forest Gate, E
Griffin, W. H., Catford and District N. H. 6, Rutland Park, Perry Hill, S.E.
Haldane-Harrison, T., Photo Survey, &c, Surrey
Halliday, J., Hastings and St. Leonard's. Linden House, Hastings
Halliday, Mrs. J., Hastings and St. Leonard's. Linden House, Hastings
Hembry, F. W., F.L.S., Sidcup Lit. and Sci. Langford, Sidcup
Inge, E. Gane, Haslemere N.H. High Street, Haslemere
Inniss, F. J., New Brompton N.H. 1, Adelaide Road, New Brompton
Large, Eustace, Catford and District. Catford
Mathews, Paul, M.A., Rochester Phil. 32, South Avenue, Rochester
McDakin, Capt. Gordon, R.E., East Kent N.H. 12, Pencester Road, Dover
McDakin, Mrs. G., Dover Sciences,,,,
Meeson, F., Woking Field Club. 2, Marchmont Gardens, Richmond, Surrey
Merrifield, F., F.E.S., Brighton and Hove N.H. 24, Vernon Terrace, Brighton
Nicholson, C. P., F.L.S., North London N.H. 22, Crouch Hill Road, Crouch End, N.
Otter, J. L., Selbourne. 20, Hanover Sq., W.
Payne, E. S., Hampstead Sci. 45, Roslyn Hill, Hampstead
Robarts, N. F., F.G.S., Croydon N.H. 23, Oliver Grove, S. Norwood
Smith-Pearce, Rev. T. N. H., M.A., Epsom College. Epsom College
Sparks, H., Eastbourne N.H. Arundel House, Eastbourne
Spitta, E. J., M.D., Brighton and Hove N.H. 41, Ventnor Villas, Hove
Swanton, E. W., F. MYC. SOC, Haslemere N.H. Brockton, Haslemere
Sweetland, Miss A. K., Southborough N.H. 3, Park Road, Southborough
Symons, W. T., Northfleet and District N.H. 12, Springhead Road, Northfleet
Stebbing, Mrs., Tunbridge Wells Nat. Hist. Soc. Ephraim Lodge, Tunbridge Wells.

MEMBERS AND ASSOCIATES FOR 1905,

Abbott, G., M.R.C.S., F.G.S. 33, Upper Grosvenor Road, Tunbridge Wells (Hon. Treas.)
Adeney, Miss. Cranford, Reigate (A)
Adkin' B. W. Trenoweth, Hope Park, Bromley (M)
Allen, Mrs. S. "Hesperia," Doods Park Road, Reigate (M,
Allchin, J. H. The Museum, Maidstone (M)
Allfrey, Miss. Friston, Reigate (A)
Allfrey, Miss M.,, ,, (A)
Ashby, J. W. Wyresdale, Redhill (M)
Ashby, Mrs. J. W.,,,, (A)
Ashby, J. Sterry. Brendon, Redhill (A)
Baker, Rev. E. J., M.A. Nutley Lane Parsonage, Reigate (M). (Hon. Local Treas.)
Baker, F. H. 95, Belsize Road, Hampstead, N.W. (M)
Bannerman, W. Bruce, F.L.S., F.G.S., F.S.A. The Lindens, Sydenham Road, Croydon (M)
Barlow, Mrs. The Dial Ground, Reigate (M)
Baxter, Miss. Heatherwood, Reigate (M)
Beach, Mrs. 11, Parkhill Road, Haverstock Hill, N.W. (M)
Bennett, A., F.L.S. 143, High Street, Croydon. (Referee)
Bennett, Miss S. 16, Oakhill Road, Reigate (A)
Bird, C, B.A., F.G.S., Mathematical School Rochester (M)
Bloomfield, Rev. E. N., M.A., F.E.S. Guestling Rectory, Sussex (V.-P.)
Bostock, Miss. Nuffield Priory, Redhill (A)
Boulenger, G. A., F.R.S. 8, Courtfield Road, S.W. (V.-P.)
Boulenger, Mrs. G. A.,,,,,,"," (A)
Boulger, Prof. G. S. F.L.S., F.G.S. 11, Onslow Road, Richmond (V.-P)
Bowyer, Miss A. Brookside, Redhill (M)
Box, J. W. 25, Henry Street, Chatham (M)
Brackett, A. W., F.S.I. 51, Queen's Road, Tunbridge Wells (M)
Bristow, E. High Street, Reigate (A)
Britton, C. E. 25, Victoria House, South Lambeth (A)
Buckley, Miss. Great Doods, Reigate (A)
Buckley, Miss J. ,,,, ,, (A)
Buckley, Miss W. E. Great Doods, Reigate (A)
Bullen, iMiss. Doods Road,, (A)
Burr, Malcolm, B.A., F.Z.S., F.E.S. 23, Blomfield Court, Maida Vale.W.

(M)

Burt, Mrs. Nahor Rani, Alders Road, Reigate (A)

Butler, G. W., B.A., F.Z.S., F.G.S. Candahar, York Road, Reigate (M)

Butler, Mrs. G. W.

Campion, Miss A. E. Mayfield, Redhill (M)

Campion, Miss M. E.,,,, (A)

Chapman, Miss. Betula, Reigate (A)

Chapman, Miss L. M.,,,, (M)

Clayton, Miss. Side Elms, Reigate (A)

Cole, Miss E. 53, London Road, Canterbury (M)

Collier, H. S. Furze Hill House, 'Reigate (A)

Cooper, Miss E. Greenhaycs, Reigate (A)

Cooper, Mrs. J. H. Dashwood, Gravesend (A)

Cornell, George. 12 & 14, Gabriel's Hill, Maidstone (M)

Crosfield, A. J., J.P. Carr End, Reigate (M)

Crostield, Mrs. A. J.

Crosfield, J. B. Undercroft. Raglan Road, Reigate (M)

Crosfield, Miss.

Crosfield, H. Leith Cottage, Yorke Road, Reigate (A)

Crafer, Mrs. 102, Beaconsfield Villas, Preston Park, Brighton (M)

Cudworth, Mrs. P. Whitfield, Hlackboro' Road, Reigate (M)

Davison, C. Glenfeulen, Reigate (A)

Davison, Mrs. ,,,,,

Davies, G. C. 25, High Street. Reigate (A)

Day, W. H. 42, Earl Street, Maidstone (M)

Dickens, Miss. Melsonby, Deering's Road, Reigate (A)

Dobson, Mrs. Way Close, Reigate (A)

Donisthorpe, H. St. J. K., F.Z.S., F.E. S. 58, Kensington Mansions, S.W. (Referee)

Edmonds, F. B. 6, Clements Inn, Strand, W.C. (M)

Eumorphopoulos, Mrs. Sunny Hank, Meadville, Redhill (A)

Every, Mrs. The Cedars, Maison Dieu Road, Dover (M)

Enoc'k, F., F.L.S., F.E.S. 13, Tufnell Park Road, N. (M)

Fielding, Rev. C. H., M.A. West Mailing (M)

Fletcher, K. R. Buckland, Betchworth (A)

Fletcher, Mrs.,,,,

Fremlin, R. J. Heathfield, Maidstone (M)

Frisby, G. E. 9, Fengatcs Road, Redhill.(Hon. Local Sec, 1905). (Referee)

Frisby, Mrs..,,, (A)

Frost, Mrs. R. C. 11, St. John's Road, Plumstead (A)

Gibb, Miss L. M., c/o. Geo. Butler, Esq. Candahar, Reigate (A)

Grant, Mrs. Hale Edge, Nutfield (A)

Green, Laurence. F.C.S. 3, St. Michael's Road, Maidstone (M)

Green, Mrs. 3, St. Michael's Road Maidstone (M)

Griffiths, Dr. Rosemead, Horley (A)

Griffiths, Miss.,,,,

Grinling, G. H. 6, Russell Place, Woolwich (M)

Groves, Jas. 55, Jeffery's Road, Clapham Rise, S.W. (Referee)

Gruner, Miss Joan F. Oakhill, Hindhead (M)

Halsted, Mrs. T. Oak Lodge, Reigatc (A)

Hanncn, Hon. H. The Hall, West Farleigh (M)

Harrison, Miss M. Little Santon, Merstham (A)

Harrison Miss. North View, Somers Road, Reigate (A)

Heisch, Mrs. 13, Somers Road, Reigate (A)

Hewetson, J., M.D. Holmficld, Reigate (A)

Hollams, Mrs. Dean Park, Tunbridgc Wells (A)

Holmes, E. M., F.L.S. Ruthven, Sevcnoaks (V.-P.)

Home, Miss. Kent Cottage, Warwick Road, Redhill M)

Hughes, F. Fairleigh, Reigate (M)

Hughes, Mrs. F.,,,, (A)

Hughes, H. F. ,, ,, (M)

Hutchinson, Jonathan, F.R.S. Inval, Haslemcre (V.-P)

Hutchinson, R. R. 28, Prince's Street Tunbridgc, Wells (M)

Hyde, A. 23, High Street, Reigatc (A)

Innis, F. J. 1, Adelaide Road, New Brompton (M)

Jenner, J. H. A., F.E.S. 209, School Hill, Lewes (M)

Johnson, Mrs. Beulah, Chart Lane, Reigatc (A)

Johnston, W. Oak Bank, Redhill (M)

Johnston, Miss A.,,,, ,,

Johnston, Miss A. J.,, ,,

Jones, E. Dukinfield, F.Z.S. Castro. Blandford Road, Reigate (M)

Kearton, R., F.Z.S. Ardingly, Catcrham Valley (M)

Keeble, F. H. The Manor House, TatsKcId (A)

Kelsey, Dr. Park Gate, Reigate (A)

Kelsey, Mrs.,,,, ,,

Kensett, Miss. 106, New Street. Horsham (M)

Kerr, R. 2, Sibthorpe Terrace, London Road. Mitcham (A)

Kerr, Miss.,,,,,,,,,

Klaasscn, Miss A. C. Abcrfcldy, Campdcn Road, Croydon (M)

Klaasscn, Miss C. F.,,,,,,,,,

Laker, G. 24, Gloucester Road, Redhill (A)

Laker, Miss,,,,,, (A)

Lathom. Mrs. Thornfield, Reigatc (A)

Lemon, Mrs. F. Hillcrest, Redhill (A)

Littler, Frank M. Launccston. Tasmania (A)

Livctt, Rev. G. M., B.A. Watcringbury Vicarage (M)

Lowne, B. T. Bromley Road, Catford, S.E. (M)

Marriage T. S. Bell Street, Reigatc (A)

Marriage, Mrs.,,,, (A)

Marten, C. J. 3, Mason's Avenue. Croydon (M)

Martin, E. A., F.G.S. 23 Campbell Road, Croydon (Photo Sec.)

Mason. T. A. Temple Court, Reigatc (A)

Masters, Mrs. 2, Bower Street, Maidstone (M)

Merriman, G. L. Tremadoc, Smoke Lane, Reigatc (M)

Mitchener, T. H. Evcrton, Reigate (M)

Mitton, W. Hurstpicrpoint, Sussex (Referee)

Monro, W. 138, Britton Street, Gillingham (Ml

Mordan, P. Sandfels, Park Lane,

Reigate (A)
  Mordan, Mrs.,,,,,, (A)
  Morgan, J. The Museum, Grange Park, South Norwood (M)
  Morris, J. 17, Throgmorton Avenue, E.C. (M)
  Mould, Mrs.-Cotswold, Evesham Road, Reigate (A)
  Mould, Miss A..,,,,,, (M)
  Mould, Miss E. ,,,,,, (A)
  Mould, Miss L., ,,,, (A)
  Newman, T. P. Hazelhurst, Haslemcre, Surrey (M)
  Newmarch, Major-Gen., R. E. 6, Norfolk Terrace, Brighton (M)
  Nicholson, E. S., F.E.S. 22, Crouch Hill Road, Crouch End, N. (M)
  Nicholson, Miss G. High School for Girls, Reigate (A)
  Nicholson, W. E., F.E.S. Lewes (M)
  Nottidge, A. J. Yardley Lodge, Tonbridge (M)
  Pannell, C. East Street, Haslemcre (M)
  Parkinson, S. T. South-Eastern Agricultural College, Wye (M)
  Payne, E. S. 45, Roslyn Hill, Hampstead, N.W. (M)
  Payne, M. R. Elmhurst, Evesham Road, Reigate (M)
  Payne, Miss F. ,,,,,, (A)
  Phillips, Miss C. Southcote, Smoke Lane, Reigate (A)
  Pierson, H. 57, Castle Hill, Hertford (A)
  Pilleau, Miss G. Torridon, Reigate (A)
  Pollen, F. Hungerford. Farley, Reigate (A)
  Pollen, Mrs. F. H.,,,, (A)
  Pollock, Sir F., Bart., LL.D. Hindhcad Copse, Shottermill (V.-P.)
  Powell, Mrs. J. Ivanhoe, Reigate (M)
  Ra'gg, R. S., B.A. The Grammar School, Reigate (M)
  Ragg, Mrs. R. S.,,,, (M)
  Randall, Mr. West View, Reigate Hill, Reigate (A)
  Reid, Capt. Savile G., R.E. The Elms, Yalding (M)
  Roberts, Miss. 27, Pearfield Road, Perry Vale, S.E. (M)
  Rogers, Rev. W. Moyle. Grosvenor Road, Bournemouth (Referee)
  Rowlstone, Miss. 78, Darnlcv Road, Gravesend (A)
  Russell, the Hon. F. G. Rollo' F.R. Met. Soc. Dunrozcl, Haslemere (M)
  Salmon, S. Clevelands, Reigate (M)
  Salmon, Miss..,,, (M)
  Salmon, C. E., F.L.S.., (M)
  Sambrook, J. U. Rosckestal, Reigate (A)
  Sambrook. Mrs.,, ,, (A)
  Sargant, Mrs. Quarry Hill, Reigate (M)
  Sargant, E. B.,,,, (A)
  Sargant, Miss.,, (A)
  Saunders, Sibert. Springfield House, Whitstable (M)
  Saunders, Miss E. R., F.L.S. Newnham College, Cambridge (A)
  Semark, Miss. 46, Ashford Road, Maidstone
  Simmonds, Rev. A., M.A. St. Mark's Vicarage, Reigate (A)
  Simpson, Mrs. G. Wray Park, Reigate (M)
  Smith-Pearce, Rev. T. N. H., M.A. Epsom College (M)
  Soames. Rev. H. A., F.L.S. The Hawthorns. Otford (M)
  Stallworthy, Rev. G. B. The Manse, Hindhead (M)
  Starling, E. A., M.D. Chillingworth House, Tunbridge Wells (M)
  Steadman. O. F. Dover Lodge, Northflect (M)
  Stebbing, W. P. D., F.G.S., F.S.A. Playfair Mansions, Queen's Club Gardens, S.W. (M) Stebbing, Mrs. W. P. I). Playfair Mansions, Queen's Club Gardens, S.W. (A) Stone, H. S. , M.D. Bccchwood, Reigate (A) Stone, Mrs.,,,, (A)
  Storr, Rayner. Hindhcad (M)
  Taylor, G. R. S., Clear's Corner, Reigate (H)
  Taylor, Mrs. G. R. S. ,, ,, (M) (Hon. Local Sec, 1905)
  Taylor, W. F.,,,, (A)
  Taylor, Mrs. W. F.,,,, (A)
  Taylor, Miss M. C. Margery Hall, Reigate (M)
  Taylor, Miss A.,, (A)
  Taylor, Miss C. c/o G. W. Butler, Esq., Candahar, Reigate (M)
  Taylor, Rev. J., M.A. 31, Marine Parade, Dover (M)
  Tapsfield, Miss C. 27, High Street, Maidstone (M)
  Tremayne, L. J. 29, Cockspur Street, London, S.W. (M)
  Treutler, W. J. M.D., F.L.S. Goldstone Villas, Hove (V.-P.)
  Treutler, Mrs.,,,,,, (M)
  Thornton, Rev. J., M.A. The Meadows, Betchworth (A)
  Tonge, A. E. Aincroft, Reigate (M)
  Tonge, Mrs. ,, ,, (A)
  Triggs, Rev. A. Corona, Reigate (A)
  Trollope, W. T., L.D.S. Camden Park, Tunbridge (M)
  Trollope. Miss. Eastnor Cottage, Reigate (A)
  Turner, H. J. 98, Drakefcll Road, New Cross, S.E. (M)
  Turner, J. W., B.A., B.Sc. Lindfield Lodge, Folkestone (M)
  Tutt, J. W., F.E.S. 119, Wcstcombe Hill, Blackhcath, S.E. (Editor)
  Venner, A. vv. Crosslev, Redstone Hill (A)
  Vowell, A. S., C.E. Stockton Villas, South Park, Reigate (M)
  Wallis, B. Undercroft, Reigate (A)
  Walters, J., M.D. Church Street, Reigate (M)
  Walters, Mrs. J.,,., (A)
  Ward, H. Snowden. Hadlow, Kent (M)
  Ward, Mrs. Doods, Doods Road, Reigate (A)
  Warren, H. 3, Lansdowne Terrace, Maxton, Dover (M)
  Webb, Sydney. 9, Waterloo Crescent, Dover (M)
  Webb, Mrs. S. ,,,, (A)
  Welchman, R. H., B.A. Caragh, Reigate (M)
  West, W. 15, Horton Place, Bradford (Referee)
  Whitehead, Miss. Whitefield. Blackboro' Road, Reigate (A)
  Whitley, Miss E. 18, Westbourne Terrace Road, Hyde Park (A)
  Williams, J. A. Wheelside, Hindhead (M)
  Willey, Dr. Somerstield, London Road, Reigate (A)
  Willey, Mrs. ,,,,,, (A)
  Window, Miss Harriet. Howbery, Haslemcre (M)
  Yeo, A. W. Oakhampstead, Reigate (A) TO THE

S.E. NATURALIST VOL. X., 1905.
Editor— J. W. TUTT, F.B.S., 119, Westcombe Hill, Blackheath, S.E. NOTES ON AN ADDRESS
Professor W. M. FLINDERS PETRIE.
D.C.L., LXD., F.R.F., FRIDAY, JUNE 8, 1905,
In the Public Hall, Re.gatu, at the Tenth Annual Congress of the South Eastern Union of Scientific Societies.
Sir H. H. HOWORTH, K.C.I.E., F.R.S., Etc.,
In the Chair.
WOKING: S E. STEER, Printer, Bookbinder & Stationer, Chobham Road.
NOTES ON AN ADDRESS BY
Professor W. M. FLINDERS PETRIE, D.c.l, L.ld., F.r.s. WRITTEN history is a very small part of the-history of mankind. Material history, as contained in monuments and other prehistoric remains, takes us farther back. Material history takes us back at least to the times of the Neolithic peoples in Egypt.

The fall of the great civilization of the Roman Empire is to many people a mystery. It seems to them, so far, unexampled.

But, if we look back and go over the great epic of man, we find that no civilization has lasted much more than 1,000 years, and that none has survived 2,000 years. We find, moreover, that civilizations recur as regularly as the seasons. And the important fact is that material history forms by far the greater part of the record.

Egypt offers a good field for examination because it affords the greatest amount of knowledge with the least amount of work. Egypt, therefore, is a convenient ground for studying the history of man. Other countries are not so good; Palestine, for instance, is barren in comparison with Egypt.

Until recently people laboured under two great fallacies: (1) That Egypt had nothing to do with the rest of the world. (2) That we could never get back beyond the highest civilization of Egypt, in the Pyramid times.

The work of the last quarter of a century has overturned these two fallacies.

We find that Egypt was quite as closely bound up with the Mediterranean as were the rest of the peoples round the Mediterranean.

We can trace the whole rise of its civilization from the time when man in Egypt was clad in goat-skins, made a small amount of pottery, and used but little metal. How he came to make use of metal we do not know, as the smelting of copper was not the earliest nor the easiest of the arts.

We know, too, that in Egypt before 6,000 B.c man rose to the production of works of fine art, that he made beautiful vases, and that his commerce extended round the Mediterranean.

The prehistoric ships were the largest galleys known: they were larger than the greatest of the Roman galleys, or those of the Venetians in the Middle Ages.

We find tne Egytians using such vessels 6,000 years B.C. to carry on their trade all round the Mediterranean.

Then later we find their civilization decaying and cheapening in all its products.

This fresh view we have acquired in the excavations of the last ten or twelve years. This long period of prehistoric civilization was followed by a thousand years of the early monarchy and then comes a later period of Egyptian History, that of the Pyramid builders. We have thus a clear view of the rise of their civilization, in place of the blank which has existed in our knowledge hitherto. In later times we find that Egypt was in contact with neighbouring Semitic races, the oldest minunient mentioning the name of Israel dates from about 1,200 B.C. This was found at Thebes, and mentions them as being in Palestine itself, where these people were contemporary with allied races in Egypt.

Merenptah, King of Egypt, is stated to have warred against the Israelites in Palestine, and despoiled them.

Beyond that, other points have been cleared up; the wars between Jewish Kings and Egypt prove to be correct and fall into their places quite naturally.

During the past winter for 3 months we have been centring our work, in Sinai.

The geological features of Sinai are at the base sea-worn granite islands, over which has been deposited a mass of carboniferous sandstone, 600 or 800 feet thick. Above this occurs a ferruginous band, beneath which the turquoise was mined; over that there has been a lava flow, and over the basalt a sandstone of Tertiary age has been deposited.

The sandstone has been weathered out into gorges 600 to 800 feet in depth.

The whole front has been cut away by a valley, down to the granite floor, so that its sides present a cliff 700 feet high and at the lower part a long slope of talus.

The Egyptians visited Sinai in early times for turquoise, they probably traded for it at first, just as they did with Persia in prehistoric times for lapis lazuli.

During the first Egyptian dynasty they took possession of Sinai, the king is represented on the rock-cut inscriptions as smiting down the Bedawi Chief of Sinai.

From 4,500 B.C. to 1,100 B.c. Egypt was lord of Sinai.

In the Bible, Musr is the kingdom of Egypt, and also includes Sinai; there is no need to suppose any other country to be intended.

The oldest remains in Sinai are the Bethels or places of offering upon the high hills.

We know from the history of Jacob how stone pillars were set up to mark a sacred spot.

The local goddess, identified by the Egyptians as Hat-hor, was the goddess of turquoise, and around her shrine were built stone shelters to break the wind; in these shelters the turquoise hunters who came to the shrine of the goddess to pray for success, used to sleep. The stone pillars are set up in these shelters.

These pilgrims when they received their dream set up a pillar in commemoration thereof.

At Sarabit el Khadim there is also a large temple occupying the summit of the hill. In front of the primitive cave is an area of burnt offerings, 100 feet long by 50 wide, pointing indubitably to Semitic worship.

At first the buildings consisted of

small fore-courts in front of the cave. In 1,500 B.C. the Egyptians extended the temple and built chamber after chamber before the fore-courts; they did not follow Egyptian plans, but built according to Semitic ideas. The ashes were preserved by laying the temple foundations upon them. The capitals of the pillars are ornamented with heads of the goddess Hat-hor; and in the courts thsre are basins for ablutions (as Mohammedans have) surrounded by four pillars, but dating from 2,000 years before the time of Mohammed.

Although the temple related to Semitic worship, it was carried out by Egyptians, thus helping in the preservation of Semitic work. The later additions to the temple consist of a large number of chambers banked over with earth and stones, as subterranean sleeping places for pilgrims.

Hitherto we have had nothing but literary evidence concerning Semitic worship, now we have material evidence. Twenty or more steles were erected in the Temple to recount the expeditions of the Egyptians in the Twelfth and Eighteenth Dynasties. Some of those pillars commemorate the finding of torquoise, with the help of Hat-hor.

On the face is inscribed the King's name, and on one side the name of the leader of the Expedition; there is also a record of the many minor officials. Mention is made of two or three artists, 500 soldiers, 300 peasants (as labourers), 30 inspectors, and two or three sculptors. There is one record also of 500 donkeys for carrying the provisions and water-supply from the coast; such a number of men would require about this amount of transport, as the conditions of the country were probably much the same then as now.

We found also the head of a black-statuette of Queen Thy, wife of Amenhotep III. It is of fine workmanship and shews great delicacy of carving, on a small scale, the face being about an inch and a half high.

With regard to the whole series of civilizations round the.Mediterranean we had no sounding line as to time, until the correlation of these civilizations with that of Egypt, which has put back the beginnings of culture to a much earlier age. There are still uncertainties before 1,500 B,c, but the outlines are fairly clear. When I first went to Egypt it was doubted whether there was any reliable Greek evidence before the Ptolemaic period. But we have gone since then steadily backward.

As a first step, the connexion of Egypt with the Greek Classic world was put back to nearly 700 u.c. The forts at Naukratis and Daphnae certainly take us to the middle of the 7th Century B.c.

The next discovery was that of hundreds of fragments of early Greek Pottery which were found in an Egyptian town of 1,400 B.C. These fragments represented Pottery from the /Egean Islands. This proved that trade had been going on with the Greeks all through this period.
8

Then the palace-rubbish from Tell-el-Amai na—1,000 fragments of Greek vases, absolutely dated on the Egyptian side—formed the next important step. Every piece was copied by me in water-colours and the volume of these copies presented to the British School at Athens where it forms a standard text book. The next stage was that of finding pottery of a style yet unknown in the mounds of Kahun. Everything proved that this pottery belonged to about 2,500 B.C. But there was no connection of it found in Greece till some years later, when it was first observed at Kamares in Crete. Since then it has been found in all deposits of that age, and is now well known on the Cretan side. The ornament is laid on in a free and flowing manner, and in several colours. Thus the middle period of Cretan civilization has been linked with the Twelfth Dynasty of Egypt.

The pottery from a certain class of Royal Tombs of the time of the First Dynasty is also unmistakably Greek. It has all the characteristics of Greek pottery, but of an earlier period than any yet found in the./Egean itself, as it dates from nearly 5000 B.c. The actual clay of which it was made was non-Egyptian, and the forms of the vases, and their pattern belong essentially to the./Egean. Amongst the black ware of this group of pottery at Abydos a pointed amphora was found. A fragment of this ware I took with me to Crete, and saw a similar specimen at Knossos. I produced my fragment from Abydos and found them absolutely identical. The time of the First Dynasty in Egypt and the late Neolithic civilization in Crete seem to correspond in many respects, and are synchronous in date.

The still earlier connection of Egypt in prehistoric times with Europe is shewn by the figures of large ships, and the pottery which was carried from one land to another in the course of this early commerce.

ADDENDA ET CORRIGENDA.

Page 4, line 1!)—For " Kiihlmann " read " Kihlmann."

Page 8—Transpose *(loc. cit.* p. 266) to place after Schimper.

Page 10, lines 39-40—Delete " by the same treatment."

Page 13, line 11—Commence new paragraph at "It is well known," and run on.

Page 15, line 14—For "Most" read "most."

Page 17, line 17—After " which " add " however."

Page 75—Add to list " Mrs. R. J. Fremlin, Maidstone."

Plate IV., opposite page 56—Note: The dark shades represent depressions, the light ones elevations. The space enclosed is about two acres.

THE BKINU THE TRANSACTIONS OK THE SOUTH-EASTERN UNION OF SCIENTIFIC SOCIETIES ALSO THE PROCEEDINGS AT THE ELEVENTH ANNUAL CONGRESS HELD AT EASTBOURNE, JUNE 6th, 7th, 8th, and 9th, 1906. The objects of the Union are to systematise Scientific Work among the different Societies composing it, to give greater impetns to Scientific research, and, in general, to promote the study and advancement of Science by Co-operation. In view of these objects, School Natural History Societies receive special consideration, and are admitted on payment of a nominal fee.

Authors are entirely responsible for the facts and opinions contained in their papers.

Readers of Papers are requested to send to the Editor, J. W. Tctt, Bayleigh Villa, Westcombe Hill, S.E., a list of any *errata* they may detect in the present volume.

The Congress for 1007 will be held at Woolwich, on

June 12th, 13th, 14th, and 15th, under the presidency of

Professor SILVANUS PHILLIPS THOMPSON, D.Sc,

F.R.S., F.R.A.S.

The Editor would be glad to exchange Transaction with other Unions and Natural History Societies. All communications relating thereto should he addressed to J. W. Tutt, Westcombe Hill, S.R.

Objects of the South-Eastern Union of Scientific Societies
    Date and Place of Congress for 1907
    Places where Meetings have been held and names of Past
Presidents
    Officers and Council for 1906-1907
    Eastbourne Local Committee...
    Rules
    Bye Laws
    Original work done by Members of Affiliated Societies, in—
        Archaeology
        Astronomy
        Botany
        Entomology
        Geology...
        Miscellaneous
        Microscopy
        Photography
        Physics Zoology
Papers Published by Affiliated Societies
List of Affiliated Societies
Referees
    Botanical Research Committee
Help wanted for Scientific purposes...
List of Lecturers
Tenth Annual Report...
    Balance Sheet, 1906
    Report of Autumn Meeting...
Report of the Photographic Secretary
    S.E.U.S.'S. Lantern Slides
    Photographic Survey and Record of Surrey...

Report of Delegates to British Association...
Museum Notes...
    Proceedings of the Congress, 1906...
Presidential Address......
    Nature near Eastbourne
    Comparison between the Sussex and the British Lists of Birds
    The Flora of the Eastbourne District
    Sea Erosion and Coast Protection
    The Geology of the Upper Ravensbourne Valley, with notes on the Flora..
.

Pevensey and its Lords
Michelham—A Sussex Priory
Notes on the Flora of Eastbourne as observed during the
Congress
    Nature Study...
    Life Members...
    Delegates
    Members and Associates for 1906
    Those marked are Officers and Council of the Eastbourne Natural History Society.

RULES. *As revised at the 4th Annual Congress, held at Rochester, May 27th, 1899, with subsequent amendments.* i. Objects.—The objects of the Union shall be to systematise work among the various Societies composing it, to give a greater impetus to research, and to promote the interests of the Societies by co-operation. 2. Management.—The affairs of the Union shall be managed by a Council and a General Committee. 3. The Council shall consist of a President, Vice-Presidents, General Secretary, Treasurer, Editor of the Transactions, and seven other persons, three to form a quorum; all to be elected annually, and none except the Vice-Presidents, Secretaries, Treasurer, and Editor, to be eligible in the same position for more than two years in immediate succession. The filling of casual vacancies to be at the discretion of the Council itself. 4. The General Committee shall consist of the Council, Past-Presidents, and the Delegates. 5. Affiliation.—All Scientific Societies in Hampshire, Kent, London, Middlesex, Surrey, and Sussex, shall be eligible to join the Union, provided that the Society claiming to join comprises at least 10 members. 6. Congress.—A Congress for the furtherance of the general work of the Union, and for the reading and discussion of papers, shall be held annually in June, at such place and at such date as may be decided on by the General Committee at the preceding Congress, or, failing such decision, by the Council. 7. Delegates.—A minimum Annual Subscription of 5s., *payable in advance at leant a fortniyht before the Congress,* shall entitle a Society to affiliation and a voting ticket for one Delegate at the Annual Congress. Societies with more than 50 members, exclusive of honorary members, shall if they so desire, be entitled to voting tickets for additional Delegates in the proportion of one for every additional 50 members, and one for the number (not less than 10) in excess of every multiple of 50, on payment of 6s. for each ticket. 7a.—School Natural History Societies may affiliate for a subscription of 2s. 6d. (rule added June, 1900). 8. Members.—Members of Affiliated Societies shall be admitted to the Congress on payment of 2s. 6d. 9. Associates.—Persons unattached to any Affiliated Society may, at the discretion of the Council, be admitted to the Congress on payment of 3s. 6d. 9a. Life-Members.—Members, Associates, and other persons, at the discretion of the Council, may compound for the Annual Subscription by a single payment of £2 2s. for Life-Membership. (Rule added June, 1902.) 10. General Meetings.—The meetings at the Congress shall be for the reception of reports of work and the reading and discussion of papers. 11. The General Committee shall, at some time during the Annual Congress, receive a Statement of Accounts, appoint an Auditor, elect the Council for the ensuing year (by ballot if demanded), appoint such Sub-Committees as may be required, decide on the next place of meeting, and, when necessary, revise the Rules. 12. Executive.—All other affairs of the Union throughout the year shall be managed by the Council. 13. Transactions.—Such Transactions of the Union as may be published shall be issued free to all Affiliated Societies, Members, and Associates. 14.

Local Receptions.—Each Society or Town inviting a visit of the Union shall appoint a Local Committee and Local Secretary to assist the General Secretary in drawing up the Programme of the Congress, which shall be arranged at least a month before the said Congress. 15. Expenses. — The expenses of printing and general management shall be paid out of the funds of the Union; those of providing rooms for the meetings of the Congress by the Society or Town issuing the invitation. 6. Changes in the Rules may be proposed and discussed at any meeting of the General Committee, but cannot be passed until the following year, unless they have been submitted to the General Secretary at least three months in advance so that he may report the proposals to the Affiliated Societies before the Congress. BYE LAWS. i.—The Union shall have the right, at its discretion, of printing *in extmso* in its Transactions all papers read at the Annual Meeting. The copyright of a paper read before any meeting of the Association, and the illustrations of the same which have been provided at his expense, shall remain the property of the author; but he shall not be at liberty to print it or allow it to be printed elsewhere, either in *ej-tenso* or in abstract amounting to as much as one-half of the length of the paper, before the first of November next after the paper is read. 2.-The author of any paper printed in the Transactions shall be entitled to receive 25 separate copies of it gratis, and to have any further number printed at his own expense by private arrangement with the printers to the Association. 3. —If proofs of papers to be published in the Transactions be sent to authors for correction, and are retained by them beyond four days for each sheet of proof, to be reckoned from the day marked thereon by the printers, but not including the time needful for transmission by post, such proof shall be assumed to require no further correction. 4.—Should the extra charges for small type, and types other than those known as Roman or Italic, and for the author's corrections of the press, in any paper published in the Transactions, amount to a greater sum than in the proportion of ten shillings per sheet, such excess shall be borne by the author himself. 5.—-A time limit of 25 minutes is prescribed for each paper, with 5 minutes for each speaker, and the discussion of the subject is to be closed at the end of one hour (March 81st, 1900). 6.—That the whole paper of each author should be sent in four weeks before the meeting. 7.—That each paper be accompanied by an abstract suitable for newspaper reporters and available for them before each meeting. 8.—That the discussion be open to all newspapers to report, but only the statement written by each speaker be used for the Proceedings (September 28th, 1904). ORIGINAL WORK DONE, OR BEING DONE, BY MEMBERS OF AFFILIATED SOCIETIES. (Furnished by their Secretaries.)

Brighton and Hove N.
H.800.
Eastbourne N.H. Soc.
Brighton and Hove..
City of London Coll.
N.H.Soc.
　Dover Sciences Soc...
　Eastbourne N.H. Soc.
　HolmesdaleN.H.Club.
　Hastings N.H. Soc...
Northflect N.H. Soc.
North London N.H.
Soc.
South London En torn.
and N.H. Soc. Bournemouth and
　New Forest (jointly)
　N.H. Soc.
Eclipses of the Sun
　The Spectroscope and its application to Astronomical Investigations.
BOTANY.
Some problems in plant fertilisation..
The preservation of our wild plants
Journ. B, Hort. Soc, vol. xxix., Pt. 4).
Report of noteworthy additions and localities to the flora'of the district.
The galls of the British oakB
　The Swiss flora
　A botanical demonstration
　Some features of interest in the East bourne flora.
The anatomy of the scutellum of *Zea mats Annul of Botany*, vol. xix., 1905, p. 115)

Homology of the grass embryo
Silent dubia. Herb., in Britain *(Journ. Bot.,* 1906, p. 127.
Holostrum umbellatum, L., in Surrey (ibid., 1905, p. 189).
Notes on the flora of Sussex (ibid., 1906,
pp. 8 and 47).
PUtntago lanceolata var. *npliaerottachya*. Kohl (ibid., 1906, p. 126).
Botany of Suffolk (for Victorian County Hist.).
LichenB of the Hastings District
The essentia conditions of Plant Life..
　Local Lists of Plants
　The Life of a Moss
　Mendel's Law of Heredity
　A list of Fungi of the New Forest
　R. J. Ryle, M.D.
F. Ash, B.Sc.
　H. Edmonds, B.Sc.
Prof. Boulger, F.L.S.,
F.G.8.
　Rev. J. Taylor, M.A., B.D.
　E. Connold, F.E.S.
　A. H. Maude, J.P.
N. S. Whitney, M.B.
Miss B. Milner, B.A.
　Miss E. Sargant, F.L.S.,
and Miss A. Robertson.
　Miss Ethel Sargant, F.L.S.
C. E. Salmon, F.L.S.
Dr. H. Stanley.
A. J. Rayner.
Research Committee.
L. B. Hall, F.L.S.
D. J. Scourfleld, F.R.M.8.
J. F. Rayner.
ENTOMOLOGY.
Revision of Aluoitidie (Pteropholidre)..
T. A. Chapman, M.D.,
　F.E.S.,ondA.W.Bacot,
F.E.8.
　The British species of the genus Peri-
L. B. Prout, F.E.8.
zonia Emmeletia).
The *Witumnptfra hcutaia* group.. . u
Polyommatu corydon, vars. et abs. C. P. Pickett, F.E.S.
　Collecting Lepidoptera A. N. Battley.
　Aculeate Hymenoptera of the Reigate
G. E. Frisby.
　District.
　Local Lists of Lepidoptera Research

Committee.
The Coleophorid Moths H. J. Turner, F.E.S.
Paliearctio Lepidoptera T. A.Chapman, M.D., and
J. W. Tutt, F.E.S.
Revision of the British Plume Moths. . T. A. Chapman, M.D., A.
Sich, F.E.S. and J. W.
Tutt, F.E.S.
Biological Revision of British Butter- J. W. Tutt. F.E.S., and flies. T. A. Chapman. M.D. GEOLOGY.
The chalk area of N.E. 8urrey *(Proc.* G. W. Young, F.G.S.
(ieol. Aitor., vol. xix., pt. iv., 1905)
Some Surrey Wells W.Whitaker,B.A. ,F.R.8.,
F.G.S., &c.
Intermittent streams of East Kent.. C. Buckingham.
Some Notes on Deneholes W. H. Steadman.
Quaternary Man of the Lower Thames M. H. Heys.
Valley.
Origin of Concretionary structure of G. Abbott, F.G.S.
Magnesian Limestone and other Rocks
The recent exposure of Fossiliferous J. L. Toucar, B.Sc.
Beds at Well Hall, Kent.
Mitchum gravels A. J. Hogg.
MISCELLANEOUS.
' On Natural History Museums .. F.W. Rudler.I.S.O.,F.G.S.
The Philosophy of Dress A. W. Williams, M.D.
Beach combing E. Counold, F.E.S.
Portugal and its Natural History.. G. F. Chambers, F.R.A.tS.
Nature Study J. C. Wright.
"Museum Gazette and Journal of Field J. Hutchinson, F.R.S.,
Study" D.Sc.&E.W.Swanton.
Ancient Egypt Rev. Mr. Fletcher.
Natural Perfumes from Animals and J. O. Braithwaite and
Plants. 8. W. Bradley.
Footpath preservation Protection Committee.
Chapters on paper-making, vol. II... Clayton Beadle.
A glance at the music of the Period,

H. H. Riches, Mus. Boo.
1650-1750. MICROSCOPY.
Practical Hints on Microscopic Manip- F. N. Clark,
ulation
PHOTOGRAPHY.
Photomicrography E. J. Spitta, M.D.
Lantern slides of the ova of the British Alfred E. Tonge, K.E.S.,
Lepidoptera. and F. N. Clark.
Photography of Wild Bird life.... Miss E. L. Turner, F.L.S.
Photographic slides of Alpine Scenery.. W. H. Gover.
Notes on Photographic sketching.. W. R. Stretton.
PHYSICS.
Atoms, Radium, and Life G. W.Burman,M.A.,B.8c
Clouds, Wind, and Weather.. J. J. Hollway.
Description and explanation of the R. C. Cann Lippincott.
Production of Colour by Colourless Media.
Eastbourne N.H. Soo.
Haslemere N.H. Soo. Hastings and St.Leonards N.H. Soc.
Chemistry in Daily Life
Demonntrations of the Spectroscope. .
How to draw a Straight Line
ZOOLOGY.
Notes on some British Wild Bees
Cuckoo Questions
British non-Marine Mollusca.
A plea for the further recognition of subspecies in Ornithology *(Zoologist,,* Feb. 1906).
Nestling Plumages of British Birds
Aranete of the Hastings District..
Notes on *Pleistocene* and Recent Shells from Crete (Map) *(Proc. Malac. Soc,* vol. vi.. no. 6, 1905).
Notes on land and freshwater shells from the Alhambra Ditch, Granada; and from Carmona, Seville; and on land, freshwater, and marine shells from Holocene Deposits at and near Carmona (oi. cit., ail loo.).
On some Land and Freshwater Mollusca
from Sumatra, Part I., Plate *Proc. Malac. Soc,* vol. vii., No. 1, 1906).
Neo-Lamarckianism, Part i

The British Woodlice, being a Monograph of the Terrestrial Isopod Crustacea of the British Isles.
Isopoda from the Gulf of Manaar, with 12 plates; in Herdman's *Pearl Fixh eries of Ceylon,* published by the Royal Society, 1905.
"Crustaceans" in Victoria History of the counties of England *(tontinued).*
A new Costa-Rican Amphipod, with plate: in the *Proceedings of the United States National Museum,* 1906.
Amphipoda Gammaridea, with 127 figures; in *Das Tierreich,* published by the Royal Prussian Academy of Sciences, 1906.
K.B.—The object of this List is not only to form a record of work done, but also to assist workers to communicate with each other—see last year's list. Mere lists of Lectures given before Societies are not wanted for insertion, but only particulars of *original* work, whether published or not.
Research Committee.
Wilfred Mark Webb, F.L.S., and Charles
sUlem.
Rev. T. R. R. Stebbing, M.A., F.R.S., &c.
Recent Local Records, Lepidoptera and Flora.
Additions to Nat. Hist. Dept., Geology, Mineralogy, Zoology Botany, Entomology, also Antiquities.
Bird song.
Our Roads from Pre-Historic Times to the Present Day.
Why I gave up Photography.
Physical Degeneration.
Slaugham Church, Sussex.
American Notes.
Longevity in Empires.
Insects and Flowers.
The British Species of *Primula.*
A few Notes on Birds.
Life in Ponds and Ditches.
Some Aspects of Geology.
Pictures from Pagan Times.
Description and explanation of the socalled Solar Spectrum.
Some Italian Hill Towns.
The Lady's Slipper Orchis and other plants of the Caldecott Herbarium.
British Vegetable Galls.

Some NoteB on the Cinque Ports, especially Hastings and its Members.
Notes on *Notolophut gonottigma*.
The Origin of Life.
Correlation between Structure said Environment in Plant Life.
The Great Pyramid, when, why, and by whom constructed; a new theory.
A simple Calculating Machine.
A few notes on the *Lrmnaeer.*
Botanical Report (1) Dates of Blooming of Garden Flowers (1906), and (2) List of wild plants observed near Epsom (1906) with dates of flowering of many species.
Entomological captures at Oxshott and Ran more.
List of Birds near Boxhill and Leatherhead 1905.
Photography.
Meteorological Report and Barometrical
Variations, 1905. Echinoderms. The genus *Eurymv (Colia)*, with special reference to *E. eurytheme.* Random Notes on the Entomology of the
Lowlands of Oahu. Report: Seal Chart Field Meeting (Map.),, Reigate Field Meeting., Clandon Field Meeting.
Chislehurst Field Meeting. ,, Oxshott, Fungus Foray. The life of Phineas Pette, Woolwich Shipwright. Woolwich Bibliography.
; Signifies that this Society is, or has been,-Delegates, whose names are printed in italics, were detained, and unable represented on the Council of the S.E.U.S-S. to be present. REFEREES.
BOTANICAL. (Additions to this list would be welcomed).
The following gentlemen have kindly consented to name & limited number of specimens for our Members and Associates
(A stamped directed envelope should always be sent, or no replies need be expected).
Cryptogams (not microscopic).—Thomas Howse, F.L.S., 26, Friar Stile Road, Richmond.
Freshwater Algae.—W. West, 15, Horton Lane, Bradford. Specimens of species of *Zygnema, Spirogyra,* and *Mougeotia* should be fruiting. They are best sent in small tubes in water. Habitat must be always stated. Permanent reedy ponds and ditches yield the best results, especially those where *UtricuUiria* occurs. Plants like *Utricularia,* leaves and peduncles of *Nymphaea, Nuphar,* etc., might be sent in tin boxes; and Mr. West will examine these for minute forms. Gelatinous or slimy coverings of damp, shady, or trickling rocks should also be sent in small tins.
Marine Algae (excluding diatoms and desmids).—E. M. Holmes, F.L.S., Ruthven, Sevenoaks.
Fresh Algte should be rolled separately in old muslin or calico, so that one plant does not touch another; tben pack so as to be free from pressure in tins or boxes.
Mosses.—Vacant.
Phanerogams.—A. Bennett, F.L.S., 148, High Street, Croydon, and Rev. E. N. Bi.oomfield, Guestling, Sussex.
In some orders like *Vrucifer/ie, Cyperaceae, Umbelliferae,* the fruit is almost a necessity.
W. H. Griffin, 6, Rutland Park, Perry Hill, S.E.
Cyperaceae.—A. Bennett, F.L.S., 148, High Street, Croydon.
Hieracii.—Rev. W. R. Lynton, Shirley Vicarage, Derby.
ZOOLOGICAL.
Aculeate Hymenoptera.—G. E. Frisby, 9, Fengates Road, Redhill.
Diptera.—Rev. E. N. Rloomfield, Guestling, Sussex.
Tenthredinidae.—Ditto.
He will also be glad to hear of any " finds" in either the Fauna or Flora of his part of Sussex.
Coleoptera.—H. St. J. K. Donisthorpe, F.Z.S., F.E.S., 58, Kensington Mansions, South Kensington, S.W.
Lepidoptcra. J. W. Tutt, F.E.S., Rayleigh Villa, Westcombe Hill, S.E.
Orthoptera.—M. Burr, F.Z.S., F.E.S., Royal Societies' Club, S.W.
Galls.—E. Connold, F.E.S., Hon. See. of Hastings Nat. Hist. Soc., 7, Magdalen Terrace, St. Leonard's.
Hydroida (Calyptoblastea).—Rev. H. A. Soames, F.L.S., Hawthorns, Otford, Sevenoaks.
BOTANICAL RESEARCH COMMITTEE.
This will be added to from time to time so that it may embrace 2 or 8 representatives from each county in the district.
Prof. G. S. Boulger, F.L.S., 11, Onslow Road, Richmond, S.W.
W. H. Beeby, F.L.S., Hildasay, Thames Ditton.
Jas. Groves, F.L.S., 58, Jeffreys Road, Clapham Rise, S.W.
E. C. Horreix, F.L.S., 58, Copleston Road, Denmark Hill, S.E.
T. Howse, F.L.S., Guy's Cliffe, Richmond, S.W.
Rev. E. N. Bloomfield, F.E.S., Guestling Rectory.
W. E. Nicholson, F.E.S., Lewes.
E. M. Holmes, F.L.S., *Chairman,* Ruthven, Sevenoaks.
HELP WANTED FOR SCIENTIFIC PURPOSES BY MEMBERS.
Mr. A. E. Tonge, "Aincroft," Eggs of British Lepidoptera for photoSchool Hill. Reigate. graphing.
Mr. J. W. Tutt, 119, Westcombe Data concerning " habits" (in all stages), Hill, S.E. "distribution," and "time of appearance" of British "plume" moths and British butterflies.
Mr. Wilfred Mark Webb, "Od-Mr. Webb has been asked by the Ray stock," Hanwell, W. Society to prepare a Monograph on
British Centipedes and Millipedes. He would be glad to receive specimens, and would furnish correspondents with tubes of spirit and an illustrated pamphlet of instructions.
Where no fee is mentioned it may be assumed that none is expected, but the travelling expenses must be paid and accommodation for the night provided if required. For Lantern Lectures the Society will please provide Lantern and operator. In future issues of this list names of ladies or gentlemen *recommended* by Affiliated Societies will be inserted. It is expected that many of them will be able to give at least the name of one capable Lecturer willing to repeat his or her lecture before other societies in this Union.
Lf.ctubes. Surtects.
Mr. J. H. Allchin, The Museum, 1. A Glance at the Early History of Kent Maidstone. Private address, and the life of William Caxton.

Chillington House. 2. Some Kentish Celebrities—Wm. Harvey,

Fee, 2 guineas, and travelling Wm. Shipley, Wm. Alexander, Wm. expenses. Woollett. Both with lantern illustra tions.! rrof. G. S. Boulger, F.L.S., F.G.S., Ed. of *Nature Notes,* 11, Onslow Road, Richmond, S.W.
Fee on application.
1. Insectivorous Plants. 2. Seed-dispersal. 3. Geological Photography lantern slides. all with

Mr. A. W. Brackett, F.S.I., 51, 1. A simple calculating machine.
Queen's Road, Tunbridge Wells. 2. Phenomena outside our apprehension. 3. Ether.

Your Council has again the pleasure of announcing a successful year's working. Their pleasure is diminished, however, by the resignation of Dr. Abbott, whose health necessitates prolonged absences from England. He feels that a treasurer should be able to keep in closer touch with the woik than he can in future hope to do.

The Council, therefore, has no choice but to accept his resignation, and to express its deep sense of indebtedness to him as the founder of the Union, its secretary for nine years, and, since 1904, its treasurer, and as one who has, ever since its commencement, given time and thought ungrudgingly to its welfare. The Council accepts his resignation with profound regret, and recommends his election by the Delegates as a Vice-President.

Mr. Robert Adkin, F.E.S., being willing to accept office, is recommended as a worthy successor to Dr. Abbott.

The Tenth Congress, at Reigate, was very successful, both in numbers and general interest, in spite of the fact that three days were wet, and the excursions arranged had to be abandoned on the last two.

The Mayor's reception was crowded, and the Holmesdale Natural History Club and Local Committee achieved the success which they?,o richly deserved.

There is the same number of Life Members as last year, 28.

The Photographic Secretary has forwarded an encouraging report, and suggestions for increasing the usefulness of the Lantern Slide Loan Department. The orchid slides have travelled, among other places, to Beverley, Yorkshire.

The number of Societies affiliated to the Union is 48, a net increase of G. Seven Societies, hailing from Sussex, Bournemouth, New Forest, Greenwich, Lewes, Morley College, S.W., and Reigate Grammar School, have sought affiliation during the past year, and one (Fulham Field Club) has ceased to exist.

The following have withdrawn from the Union:—Sir F. Pollock, Bart., V.-P., and Mr. Sibert Saunders. Your Council regrets to record the death of Mr. John Morris, for many years a Member. They also thank Mr. E. A. Martin, F.G.S., for his services as lecturer, which his want of leisure compels him to give up. There is no doubt the list of lectures could be increased with mutual advantage to the affiliated societies. Your Council feels that the time has come, after ten years' working, to readjust the finances of the Union, as the expenses have now outrun the resources. They do not propose to alter the terms on which Societies are affiliated, but as they are faced year by year with annual deficits of greater or less extent, this year, for instance, the small bank reserve has been Three new Life Members have since joined.
again drawn upon to the extent of £5, it is evident that the subscription has been fixed too low. Calculating the cost of the *SoiithEastern Naturalist* for the last two years your Council finds that the volume has been issued to 2s. 6d. subscribers at a loss of about 3d. per Member. In consequence of this constantly recurring loss, the money required for postages, stationery, expenses of Council Meetings, etc., has to be met out of reserves. Your Council recommends, therefore, the raising of the annual subscriptions of new Members to 3s. 6d., and of new Associates to 5s., or whatever sums are decided, or to readjust the classification of Members, if deemed advisable by the Delegates. By Rule 16 this increase cannot become operative before twelve months' time, when it must be passed by the General Committee at next year's Congress.

Publications have been received from tho Horsham Museum Society, Maidstone Museum, Dover Sciences Society, Hastings and St. Leonard's N.H. Soc, City of London College Sci. Soc, Brighton and Hove N.H. Soc, North London N.H. Soc, Nottingham Naturalists, Rochester Nat. Club, Fulham Public Libraries, Manchester Museum, Tiffin's Boys' School, Bournemouth and District Soc. of Nat. Sci., Woolwich District Antiquarian, Horniman Museum, 1904 and 1905, Epsom College, Eastbourne N.H. Soc, Photographic Survey and Record of Surrey, South London Entom. Soc, Haslemere Museum, also the Report of the Iron Ore Deposits in Sydvaranger of Norway, and last, but not least, the Report of the Corr. Societies Committee and Conference of Delegates of the British Association. Messrs. E. A. Martin and H. Norman Gray contribute separate reports of their departments.

At the Council Meeting held on April 6th last, a sub-Committee, consisting of Mr. F. W. Rudler, I.S.O., F.G.S., Dr. William Martin, and Mr. H. Norman Gray was appointed to investigate the ways of the protection of Treasure Trove, and to send out circulars to schoolmasters, postmasters, and others likely to promote that end.

Your Council have received with regret the resignation of the Rev. W. Moyle Rogers, of Bournemouth, and thank him for having acted as Referee for the *Ritbi.* BALANCE SHEET, 1905.
RECEIPTS. EXPENDITURE.
Balance brought forward from 1904 account
Reigate Congress.—Proceeds of tickets sold by Holmesdale club
Delegates1 Subscriptions, 1905..
Members' Subscriptions, 1905
Associates' Subscriptions, 1905..
Bales of S.E. Naturalist
Bank Interest on Deposit account
Withdrawn from Birkbeck Bank
   Printing Congress Programmes, Tickets, &c
   Printing S.E.Naturalist, 350copies
Supplement to S.E. Naturalist
Expenses *re* Autumn Meeting
Expenses *rr* Museum at Congress

Postages and parcels
Balance
X s. d.
Examined bv Touchers and receipts, 23rd May, 1906. W. T. TROLLOPE.

REPORT OF THE AUTUMN MEETING.

The first Autumn Meeting of the South-Eastern Union of Scientific Societies took place on Saturday afternoon, December 9th. 1905, at the Guildhall (London) Library and Museum, etc.

The visitors, in number about 140, after assembling in the main porch, entered one of the Committee rooms, where they were received by Mr. Pitman (Chairman of the Library Committee), who extended to them, in the name of his Committee, a most hearty welcome, to which the Rev. R. Ashington Bullen responded, expressing the visitors' sincere thanks for the graceful manner in which they had been received, as well as for the opportunity for visiting such a historical place.

Mr. Welch, the Librarian, next delivered a brief, but most interesting historical account of the Library and Museum, together with the buildings forming the same, and referred to a number of the books which had been specially got out and placed on the tables for inspection, very many of them being of great value.

The visitors were then conducted by Mr. Welch through the Museum—the most important features and articles being pointed out *en route*—into the crypt, then into the Guildhall itself, from thence to the Aldermen's Council Chamber, and finishing up in the large Council Chamber, where Mr. Welch's conductorship terminated.

Mr. W. Whitaker, past president, here took the opportunity of moving a most hearty and sincere vote of thanks to the Library Committee, and other authorities of the Guildhall, for their kind permission to make this visit, and also for their kindness in allowing them to have the use of a room in which some of them would shortly be able to take tea, and last, but not least, to Mr. Welch, for the delightful manner in which he had shown them round and described the various points of interest, and also for the eminently satisfactory way he had arranged for, and conducted, such a large party.

Mr. F. W. Rudler seconded the vote of thanks, and alluded to the enormous amount of work the arrangement of such a Library and Museum, and the compilation of the necessary literature in connection therewith, must have entailed, but which was all so excellently done by Mr. Welch.

The vote of thanks having been put and passed with acclamation, Mr. Welch, in replying, intimated the pleasure he had felt in receiving and conducting round such a party, and his readiness, if at any time any of the visitors wanted any further information, to do all in his power to assist them.

The visitors then retired, about 84 remaining to have tea in a room very kindly placed at their disposal by the authorities.

The following places were represented at the meeting:— Farnboro', Tunbridge Wells, St. Leonard's, Maidstone, Catford, Redhill, Wallington, Westerham, Enfield, Sidcup, Sevenoaks, Eastbourne, Reigate, and Brighton.—H. Norman Gray, Hon. Sec. Autumn Meeting.

REPORT OF THE PHOTOGRAPHIC SECRETARY —MR. EDWARD A. MARTIN, F.G.S.

The duty which has fallen to me to perform has been chiefly the custody of the sets of Union Lantern Slides.

There has been an increased use made of the various sets by affiliated societies, and there can be no doubt that they have proved extremely useful, especially to those societies which have a limited membership and consequently a small income.

Any Society having a set to spare would do well to present it to the Union, or lend it for an indefinite period.

Sets illustrating the common wild flowers (with dissections), trees, and birds and their nests, etc., would, I am convinced, be of great utility.

S.E.U.S.S. LANTERN SLIDES.

Tbe following sets of Lantern Slides are available for use by affiliated Societies on application eitber to the General or Photographic Secretary: 1.—*Some British Orchids* (50 slides) contributed by Mr. S. Horeley, M.I.C.E., with explanatory lecture. 2. — *The Oault and Lower Greenland* (about 80 slides), with lecture. 3.—*The fVralden Formation* (about 50 slides), with explanatory notes. i.—*Ice Flowers and Crystals* (small set), with explanatory notes, by O. Abbott.

No charge is made except for carriage both ways. The orchid slides are a new and interesting set, dealing with the general and detailed structure of many British species and their adaptation to insect-fertilisation.

The Society for the Protection of Birds, 826, Holborn, W.C., also lend *to their subscribers* very beautiful Lantern Slides relating to Birds, which are well worth the attention of Secretaries and Lecturers.

The notice of Secretaries is particularly called to the suggestions made in the Photographic Secretary's Report that the Union should solicit loans or gifts of *small* sets of slides (a dozen or so) illustrating *any* particular scientific phenomenon or limited branch of scientific work. Such sets, with full explanatory notes, to occupy about balf-an-hour for exhibition, would doubtless be much appreciated for the purpose of soirees or other occasions in which time is necessarily limited, while two sets might furnish material for an ordinary evening meeting. Many members of our Societies possess sets of this character which they have prepared for their own use, and which they would be willing to lend for the use of the affiliated Societies. Secretaries are hereby asked to furnish the Photographic Secretary as soon as possible with the names and addresses of any of their members who, in their opinion, might be induced to cooperate.

Contributions are still solicited towards the following larger sets that are in course of formation: 1.—*Pre-historic man in S.£. England*. 2.—*English Wild Flowers,* with special reference to forms of capsules and their dehiscence. 3.—*Photomicrographs*. 4.—*Coast Erosion in S.E. England*.

Contributions of lantern slides may be sent to the Photographic Secretary,

Mr. E. A. Martin, F.G.S., 58, Whitworth Road, South Norwood, S.E., who will be glad to give any information as to this branch of the work of the Union.

SURVEY AND RECORD OF SURREY. Photographic Survey And Record Of Surrey.—As delegate from the Croydon branch of the Survey, I have been asked to bring before the meeting an account of the work already accomplished. Up to the present time 2,107 prints have been received, being an average of 650 prints a year since the Survey was first founded, a number which is very satisfactory, and exceeds anticipation. A careful selection of prints has been sent to various exhibitions, and 100 are to be seen in the present exhibition at Eastbourne, but unless the greatest care is exercised, damage to the collection is sure to arise in time, and it is suggested that frames be used in future to obviate this, and, as the *mounts* used are all of the same size, this should not involve much difficulty or expense. With regard to the number of prints sent in to the six sections, into which the Survey was divided during last year, Architecture beads the list with 870 prints largely due to the efforts of Dr. Hodson, the Hon. Secretary of this section, who had made Architecture a special study. Next comes Natural History with 88 additions, some excellent examples of Lepidoptera and Mollusca being contributed by Mr. H. D. Gower. Art and Literature number 47, and contain many interesting examples. Anthropological prints number 42, and comprise most interesting objects from the Horniraan Museum, Old Bermondsey Abbey, British Camp at Carshalton-on-the Hill, and pit-dwellings at Worms Heath. It is disappointing to find so few prints in the Geological and Topography and Passing Events sections, because one would have thought subjects in both these sections would have been numerous. With regard to the processes employed to produce the prints, there can be no doubt that, if permanence is desired, and for Survey work this is all important, carbon or platinum are absolutely essential, bromides are looked on with suspicion unless very carefully made, *i. e.*, with thorough fixing and washing; while silver prints it would be better to entirely exclude. It is suggested that those who cannot do carbon or platinum prints might hand their negatives to those who can and will. At the same time it may be pointed out that the platinum prints are the easiest, quickest, and cheapest (in the end) of any. Another *very* important point is that more than half the value of a print depends on correct data, and particulars being given, many prints being sent in with scarcely any detail concerning them. Lantern slides are not nearly so numerous as could be wished, and it is hoped that a good collection of these will be made to lend to exhibitions and meetings, for the double purpose of being able to show Survey work to an audience, and for making the Survey itself better known and appreciated. Finally, it is suggested that workers should rather specialize in their work, in any particular section, than send, as it wore, a miscellaneous selection of prints in all sections, the collection thus having a greater useful and educational value than by a more or less disjointed system.—J. H. Baldock, F.C.S., F.R.P.S. TO THE BRITISH ASSOCIATION FOR THE ADVANCEMENT OF SCIENCE.

The meetings were held in London, at the rooms of theLinnean Society, on October 80th and 81st, 1905, under the chairmanship of Dr. A. Smith Woodward, F.R.S. Your delegate attended both meetings.

The Corresponding Societies' Committee reported (through Mr. F. W. Rudlor, Secretary) that there will, in future, be two classes of Local Societies eligible for relationship with the British Association.

(1) Affiliated Societies, *i.e.,* those that undertake local scientific investiga tion and publish the results. (2) Associated Societies formed for the purpose of encouraging the study of science, which have been in existence at least 3 years and number not fewer than 50 members. Each society of class 1 or 2 will have the right to appoint a delegate to attend the Annual Conference, who must be a member or associate of the British Association; they will have the same rights, except that only delegates from class 1 are eligible for membership of the General Committee.

Dr. Smith Woodward's address dealt with the work of local Scientific Societies, which form so prominent a feature of intellectual life in Great Britain and Ireland.

He deprecated the way in which some societies allow their excursions to become luxurious picnics, in which natural history is allowed to occupy a quite subordinate part of the programme.

He thought that the best work done by the smaller societies was that of instruction in the current progress of science and its presentation in such a form as to rouse interest in scientific pursuits.

He said that it became more and more expedient for societies to restrict their meetings to general discourses and demonstrations; nearly all original papers being so technical that they must be studied closely, in print, before they can be appreciated, and so the most active societies make a special feature of exhibits, and take the majority of papers as read.

For the guidance of more advanced students there is a distinct need of books to treat of ignorance rather than of knowledge, and until they arc forthcoming the societies cannot do more important service than that of supplying the deficiency.

Another duty the local societies might perform would be the selection of books on natural science for many of the smaller public libraries. Rarely now can the ordinary student find in them the most important books, even on the natural history of the district in which the library exists.

He deprecated the all-pervading mania for "tit-bits," so characteristic of this restless age, which has even penetrated some of our societies and reduced the amount of systematic and persistent plodding necessary for many important lines of research.

An animated discussion followed, too long to condense for this notice.

Dr. William Martin read a paper on "The Law of Treasure Trove, especially in relation to local scientific societies. " After discussing the law as it stands he said that "an union of antiquarian and allied societies was desirable for the purpose of submitting to His Majesty's Commissioners of the Treasury one or more resolutions which relate to the amendment of the law of treasure trove, and of the administration of that law."

Necessary steps (he thought) should be taken to bring to the knowledge of the public through the post office:— (1) The importance of the preservation of treasure trove.
(2) The remuneration which is offered to finders; and, if considered expedient, (3) The desirability for the remuneration offered to finders being put on a statutory basis; and (4) An extension of the law of treasure trove so as to cover other relics of antiquity, thereby bringing the law into harmony with what is believed to be the law of Scotland.

Mr. W. M. Colles spoke on the " Law of Copyright as affecting the Proceedings of Scientific Societies." He considered that:— (1) As regarded scientific papers published in the general sense, by being placed for sale, the ordinary rules of copyright law applied.
(2) Papers only privately printed were in the same position as unpublished MSS. , and the copyright, if unassigned, remained in the authors by common law.
(3) It was questionable whether the byelaws of societies (declaring the copyright of accepted papers to be vested in them) were sufficient to give those societies a good title to the copyright of such papers, unless they have been actually assigned by legal deed.

Mr. Harold Hardy dealt with the position of the author of a paper which might be read at a meeting of a scientific society.

So long as the paper was unread or unpublished the author was entitled to copyright in his literary composition. Or, if the audience consisted merely of members of a society, or a limited number of persons invited and admitted by ticket, the author was entitled to copyright.

If, however, the general public were admitted to the reading of a paper, the author, to protect his copyright, must either (1) have printed his paper before oral delivery, and registered it as a book at Stationers' Hall, or (2) must have given notice to two magistrates living within five miles of the place where the paper was read, which would give him protection for 28 years.

He thought this latter proceeding should be abolished, and that the law should protect the oral delivery of lectures, since there was at present no protection for the right of publishing a lecture by oral delivery as apart from the copyright.

On Tuesday, October 81st, Professor Boulger introduced the subject of "The preservation of our native plants." Among the causes of extermination he placed:— (1) Natural causes, *e.g.,* encroachments of the sea, the increasing density of population with its absorption of land in building operations.
(2) Avoidable causes, excesses of children, tourists, botanists, and the work of trade-collectors.

Among protective measures he mentioned concealment and enclosure of the localities of rare plants, the cultivation of wild forms, and legislation.—R. Ashinoton Hullen. MUSEUM NOTES. Edited By The HONORARY CURATOR.

It is pleasurable to report no diminution of interest in the Congress Museum, the exhibits being quite as varied and interesting as in previous years. It has been suggested that full descriptions should always accompany the specimens exhibited. Hitherto, the notes for the museum report have been obtained by writing to the respective exhibitors after the Congress is over.

The Curator cannot bo expected to do all the labelling, he is not a cyclopaedia of general information in natural history. In future, will exhibitors kindly send or place with their specimens, such detailed particulars as they wish to appear in the *Transactions'* These should be given to the Curator on the Saturday morning. If this idea is carried out, much needless correspondence will be avoided, and assistance given in the speedy issue of the *Transactions.*

In connection with the exhibition of photographs by the Photographic SurveysofKent, Surrey, and Sussex—which always forms one of the most interesting sections of the Museum—it has been suggested that the Surveys provide their own screens. Temporary screens, such as have been kindly provided every year by the local committee, are never satisfactory, and moreover, are expensive in the long run. If the Survey Committees could agree as to a standard screen, and if packing cases to enable such screens to be packed quickly and sent by rail without any fear of damage were also provided, much saving of labour and time would be effected. It is well-nigh an impossibility, properly to affix the prints to the screens without putting pins through the mounts, but if each Survey kept a special set of all its photographs for exhibition purposes, the question of wear-and-tear would not be a serious one. The providing of suitable screens would probably contribute to the very desirable result of reducing the number of prints sent for exhibition. It may be hinted that—as far as the S.E.U.S.S. is concerned—70 should be the maximum number allowed for each Survey, and that care be taken only to exhibit photographs of special interest in the different sections. Photographs of such ubiquitous molluscs as *Clausilia laminata* and *Limnaea stagnalis,* that cosmopolitan fungus *Polyparut squamosum,* and such a frequent plant as the bog asphodel *(Narthecium omfragiim)* might well have been omitted from the selection sent to Eastbourne.

Mb. G. Abbott, F.G.S., Tunbridge Wells.—(1) Galls on *Quercus pubexams* produced by *Cynips ylutinosa* from Ligure, Italy. (2) Pod-like gall, bent into S shape, measuring 18J inches, from Ligure, Italy. (3) Fossil Sponges (?) found loose in balls of flint in cliffs at Cuckmere Haven—some of them are made up of several pieces which dovetail into each other, yet no two specimens seem to be alike. (4) Beekite form of *Chalcedony* from Champigny, near Paris. (5) Specimens of *Mitnda phalloides,* Chev., which had grown on dead pine needles and cones: Broadwater,

Tunbridge Wells. (6) Structures from Magnesian limestone, Sunderland—one variety, hitherto undescribed (ball and band structure), consisting of alternate bands of solid limestone and layers of balls; a peculiar result of the segregation process in a bed of balls. (7) Sharks' Teeth, from Miocene, Italy. (7) Photograph of *Chert paramoudra,* measuring about four by five feet, on shore at Cuckmere Haven, taken in 1905, and still remaining.

Mrs. C. Beach, Haverstock Hill.—A collection of British mosses from various localities. It included the rare *Pterigynandrum tiliforme* from Creagna-Callich, Perth; *Bryum pseudo-triquetrum* from Glen Helen, Isle of Man; *Tetraplodon nmioides,* taken from the head of a dead sheep on a mountain in mid-Perth; the uncommon *Antitrichia curtipendula* from trees near Princetown, Dartmoor; *Hypniiui wilsoni* from Skipwith Common, Yorkshire. The following species from the neighbourhood of Eastbourne: *Neckera cumplanata, Brachytherium rutabulum, Hypnuin ctipressiforme,* and *Barbula uni/uilata.* Also *Fanosia hygrometrica* from Mortola, Italy; the rare *f,eptodon smithii* from trunk of an old olive-tree at Bordighera; *Pterogonium yracile* and *Eurynchium speciosum* from Hyeres, France. Mr. F. J. Bennett, F.G.S.—(1) Relics from a kitchen midden within Pevensey Castle, including oyster and cockle shells, scale of ray, bones of rabbits and birds, flint flakes and fragments of pottery. (2) " Starch," or prismatic fracture (?) flint implements from a Lower Terrace, Thames Gravel, near Gravesend. The variety of flint giving this peculiar fracture has always been regarded as of non-human origin, *i.e.,* as a natural fracture. The flints exhibited were a core and some flat-side flakes with no bulb or conchoidal fracture. Some of the long, thin, flakes had one serrated edge, and one small flake was distinctly a saw. The interesting point was that most of the worked flints in that pit seemed to give this starch fracture, though it also yields very early, and much patinated, palwolithic flakes and implements. The exhibitor also examined the unworked flints in that pit, many of which would not yield any bulb or conchoidal fracture, but broke up into flatsided and cubical pieces. He concludes that the fracture is governed by the *nature* of the flint, and in that sense all fractures, whether by man or nature are *natural fractures, (it)* Nearly 100 photographs of Kentish gravestones taken for the exhibitor by Mr. E. W. Filkens, of Gravesend, all from churchyards within a radius of 10 miles from West Mailing. The churchyard at West Mailing, contains all the types exhibited, and the best examples of them. They were classified into the following types:— *1a)* Plain early stones (1623), head and shoulder pattern.

b) Early faces (1687), very few of these exist.

(c) Skull, Doll's head and face types.

(f) Cherub (1740-1810), early and late types.

le) Profiles (1717-17BO), early and late types.

(/) Allegorical, including n variety of examples taken from scriptural subjects, etc.

For particulars respecting the gravestones in churchyards in the district around West Mailing, the reader is referred to Mr. Bennett's "History of West Mailing Church," and to Museum Notes, *Transactions S.E.U.S.S.,* 1905.

Rev. E. N. Bloomfield, M.A., F.E.S. —(1) A case containing the following Diptera: *I'abatms bovinus, Asiliis crabroniforniis, liowbyliiitt dispar, Calleprobola s/wciosa, Crinrhina ranunculi (ru/icaiula), fiericomyia bnrealis,* "sings" while at rest; *(lastrophiluseaui* (? 2) larva lives in the stomach of the horse, *(Kstrus oris,* larva in frontal sinuses of the sheep, *Cephenomyia rnfibarbis,* larva in throat of red deer; *F.chinomyia yrossa* larva, a parasite in large caterpillars. (2) A type collection of British Diptera, including 72 species,arranged by Mr. W. Harwood, of Colchester, including the male and female St. Mark's Fly *(liibio inarci),* the Humble-bee Fly *(Rombylius major),* which hovers at flowers in the spring; two varieties of the *Volucella bombylans,* these flies are parasitic on Humble Bees, and very much like them; *Conops ceriiformis, Physocrphala rufipes,* and *Sicus ferruyineus,* these species parasitic on wasps, etc., living within them; the Flesh Fly *(Sarcopltaya carnaria),* which produces its young alive, the scarce Blue Bottle or Blow Fly *(Mesembrina meridiana),* and the Green Bottle Fly *(Scatophaya stercoraria).* This collection has been very kindly presented to the Haslemere Museum by Mr. Bloomfield.

The Eastbourne Natural History Society exhibited the remains of a Roman font found in May, 1906, during the demolition of a part of the wall surrounding the Saffrons, in Corapton Place Road, Eastbourne. It was presented to the Eastbourne Museum by His Grace the Duke of Devonshire, in June, 1906.

Dr. Edward Gilbert exhibited an interesting series of pressed seaweeds and some of the rarer local plants, including *Rubi.*

Mr. W. H. Griffin (catford And District Natural History Society) Exhibited: (1) Pressed and mounted specimens of all the plants mentioned in his paper on "The Geology of the Upper Ravensbourne Valley," printed pp. *50etseq.* (2) The flint implements, and teeth and bones of extinct Pleistocene mammals mentioned in the same paper. (3) A human skull of Negroid type which was dug up in a garden in north London many years ago. Unfortunately, the member of his Society who brought him the skull for examination, could not furnish any particulars as to the exact locality, the nature of the soil, and the depth from which the skull was obtained. A descriptive newspaper article written by Mr. Griffin was placed with the skull in the temporary museum. From this it appears:— 1. That the skull appertained to an adult, although it is very small.

2. The cranial index is 70.37, indicating a dolichocephalic race. 3. The index of the nasal aperture is 50, or hyper-leptorhine. 4. The base of the aperture is markedly simian. 5. The brow ridges are strongly developed, although the frontal develop ment is fair. 6. The jaw is very prognathous. 7. The extreme circumfer-

ence of the skull is 19j inches; that of the skulls of seven adult English females, taken at random, was 21j inches. 8. The teeth are not worn down by the mastication of uncooked food as they frequently are with adult savages. 9. The appearance of the skull indicated a long period of interment.

Mr. Griffin's suggestion was that the skull appertained to a person of mixed African and European race, pointing to the offspring of a union between a Roman soldier and a negress during the Roman occupation of Britain.

Mr. E. M. Holmes, F.L.S., contributed a large collection of rare British plants.

Mr. Jonathan Hutchinson, L.L.D., F.R.S., D.Sc.—(1) A series of swing frames containing illustrations of the common plants of the seashore, and of birds frequenting our coasts. A special corner of the Haslemere Museum is set apart for illustrations of seaside objects and actual specimens. It is suggested that a seaside section should find an important place in all museums situated on our coasts; at the present time no special attempt has been made in this direction in any of the seaside museums with which we are acquainted. (2) The following objects were sent from his Educational Museum at Haslemere: *(a)* Existing and fossil shells of *Trigonia,* and *(b)* the jaws and egg-cases of the Port Jackson shark.

The large and characteristic *Trigonia* fossils are shells of animals allied to cockles, or still more closely to arcs. They first appear in Oolitic strata, and are the " horse-heads " of the Portland quarrymen. They occur in some places in such abundance that rocks of several feet in thickness are almost entirely made up of them. Curiously enough, none have been found in Tertiary strata, though half-a-dozen living species still exist on the Australian shores. Fossil *Trif/oniae* are, more or less, triangular, and are usually ornamented with ribs and warts. The shells are very thin, and more often than not, they are entirely absent in the fossils, only the "cast" remaining. In the modern species, the interior of the shell is beautifully pearly, and a noticeable feature is the posterior inclination of the umbones, which is of very unusual occurrence. The animal is provided with a large and powerful foot, which it uses both for crawling and leaping. The byssus, or fine silky threads by which all other members of this group of molluscs (the mussels, &c.) attach themselves to the sea-bottom and to rocks, is wanting in the genus *Tritjonia*. These important fossils are very common, and are valuable as characterising certain strata. The cast of the interior of the shell when formed in Portland Oolite or Oxford Clay forms a very beautiful object, being very sharply defined, and looking as if taken in biscuit paste. The apparent three-cornered shape is caused by the great width of the shell posterior to the hinge. The following Trigonias were exhibited:—

IT. *cuxtnla* from the Great Oolite. Minchinhampton.

Casts of-*T. alaeformit* from the Lower Greensand.

I *T. irregularis* from the Oxford Clay.

For convenience of comparison recent shells of *Triyoniae (T. larmarckii* and *T. pectinata)* and cockles were placed with them.

The remarkable egg-cases of the Port Jackson shark *(Ctstracion phili/ipi)* are as common during August and Septeinl»er, on Australian beaches around Port Jackson as are the egg-cases of the Dog-fish on English shores. They are found in moderately shallow water, wedged, crown outward, between rocks, and cannot be removed except by actually unscrewing them, or by turning them round and withdrawing the small end first. In an allied species *((' . yaleatits)* from the same locality the egg-cases have not such well-developed spirals. The spirals are becoming rudimentary through disuse, the egg-cases—instead of being wedged between rocks and held by the spirals—are fastened to seaweeds by long tendrils in the same way as those of the common Dogfish Shark. The Cestracionts are of special interest because of their great antiquity, one of the earliest fossil genera—*Otodiis*—occurring in Carboniferous strata. They were abundant in the chalk, where the genus *Cestracion* is associated with the ancient mollusc *Trit/onia*. Curiously enough, they still occur together in Port Jackson harbour. There are scarcely any representatives of either genus in Tertiary strata, and both present striking illustrations of decadent families, now very reduced in genera and species. The *Cestraciontidae* are known as the Pavement-Toothed Sharks because of their peculiar dentition, the teeth which cover the margins and inner surface of the jaws being pavement-like. Another peculiarity of these sharks is the presence of a strong spine on the front edge of each of the two dorsal fins. The lower jaw of a Port Jackson Shark was exhibited with the egg-cases, and, for purposes of comparison, a skull of a typical shark.

(8) The so-called il Vegetable Caterpillar," a species of fungus, *('ordiceps (Torrubia) robertsii,* growing upon the subterranean larva of a New Zealand moth *(Hepiahis virescens)* allied to our Ghost Swift Moth. After growing in the interior of the living caterpillar, the fungus sprouts out as a tall stem from its head, and produces, above the ground, a club-shaped fructifying part, the entire plant being four or five inches long. Sometimes two fungi spring from one caterpillar. Many other insects are affected by *Cordiceps. C.entomorrhiza* has been foundonthelarvreof a British speciesof Swift Moth. (4) A case containing several species of common British Pondsnails, all belonging to the genus *Limnaea*. The differences, apart from size, are slight variations in the spire and aperture. The family likeness is unmistakable, and it becomes a question whether they are all entitled to specific rank. Very interesting phylogenetic evidence has been brought forward which points to *L. peregra* (by far the commonest of all the species) as the ancestral form from which all the others have been derived. The progeny arising from an union between a 3 *L. statjnalis* and a £ *L. auricularia* were carefully watched until full-grown, when they could not in any way be distinguished from typical *L. peregra!* (5) Skulls of the Musk Deer *(Moschus moschiferus),*

male and female. In the male the tusks project considerably below the mouth, a feature which does not exist in any other member of the deer family. In the absence of antlers these curious little deer agree with the Chinese Water Deer. (6) Skulls of rabbits and rats showing abnormal growth of the incisors. All were obtained at Haslemere. Such malformations are not uncommon. With these was exhibited the skull of a large Indian porcupine, in which the teeth were normal. (7) Skulls of poisonous and non-poisonous reptiles. *(a) Python moltirus,* showing the rows of sharp-pointed teeth, neither grooved nor perforated, and of almost equal size. *(b)* The poisonous *Echidna eleyans,* showing the pair of long poison fangs in the upper jaw. These are channelled, and are connected by a duct with the poison-gland at their extremity. (8) Examples of *Arasehnia prorsalerana* to illustrate seasonal dimorphism. *Lecana,* the spring brood, is yellowish-red, marked with black and blue; *prorsa,* the summer brood, is black, with a yellowish-white band across the wings. *Prorsa* passes the winter in the pupal state, and emerges in spring as *levana,* the eggs of which produce *prorsa.*

Mr. B. T. Low.nk (catford And District Natural History Society) exhibited—(1) Five species of living pseudo-scorpions, eis.: *Cheiridium museoriim, Chernes cimicoides, Chelifer latreillii, Obisium muscorum, Chithoniu rayi.* (1) About 80 sheets of dried plants taken from his herbarium to illustrate the life-history of the species, all comprising as nearly as possible:—*(a)* The seed. *(b)* Several stages of the young plant to show development of cotyledons and first leaves, (r) Complete specimens at the flowering period, *(d)* Specimens at the fruiting period, completing the cycle which brings us to the seed again, *(e)* Separate leaves, flowers, fruits, buds, etc., to show perfect outline, upper and under surface, etc. The following is a list of the species exhibited:— *Lepidium ruderale, Hilda rubra, Hypericum perforatum, H. hirsutum, llejr aquifolium, Medicayo falcata, Anthyllis ruhieraria, Lathyrus aphaca, L. pratenxix, Primus spinosa, Hubus fruticosus, Seandt'jr pecten-reneris, Lonicera.cylosteiim, Gnaplialiuni sylraticum, Trayopoyon pratenxix, Orobanche miliar, l.iuaria rixcida, L'alamintha officinalis, Xejieta cataria, Chenopodium polyspermuni, Parietaria officinalis, Castauea satira, Alnus ylutinosa. Iris pseudacorux, Tulipa sylrestris, Sayittaria sayittitolia,* and several sheets of seedlings arranged according to their natural orders for comparing habit of growth, shape, etc., one with another.

Miss Milnkk, B.A., exhibited specimens of rare local plants.

Mr. J. W. Rodda exhibited photographs of the rock tables in Upper Greensand, formerly to be seen beneath Beachy Head, but now buried in gravel.

Mr. E. W. Swanton, Brockton, Haslemere.—(1) Two examples of the scalariform monstrosity of *Ileli.r ax/ierxa* found in the creel of a "wall-fish" collector, near Bruton, Somerset (see *Museum Gazette).* It is noteworthy that monstrosities of our land and freshwater shells occur more abundantly in some districts than in others. Trochiform *II. Iiortensis* and *//. arbustoruin,* as well as scalariform *H. caperata,* have been taken from hedge-banks near to the spot in which these scalarid *//. aspersa* were found. Prof. T. D. A. Cockerell and other conchologists have noted "that certain variations occur in all allied species under certain conditions." The conditions favouring the elongation of the spire in the abovementioned cases are unknown. (2) *Helix F.uparypha) pisana,* specimens given to Mr. F. N. Townsend, F.L.S., in 1850, by Mr. Hawker, a well-known conchologist, who stated he had found theiu on the Sussex coast. So far as we are aware, this species does not occur in Sussex at the present day. The only British localities known to us are sandhills on the Cornish coast, south Wales, and southeast Ireland. It is a Lusitanian species. (8) Illustrations of new and rare British fungi, from the *Transactions of the British Mycoloyical Society for 1905,* including *Sparassis laminosa,* found by Mr. Douglas Taylor in Woolmer Forest, during the Society's visit there in September last. This fine species, hitherto unobserved in this country, differs from its ally, the local *Sparassis crispa,* in the more laminar patent branches, and in becoming putrescent soon after being gathered. (4) *Polystictm tnontagnei,* found by Mr. E. W. Swanton near Grayswood. Easily distinguished from *P. perennis* in the cyathiform and zoneless pileus, and the large entire pores. (5) *Polyparm benzoinus* (near to *P. rerinosus).*—A sweet-scented and sub-imbricate species found by the Rev. W. L. W. Eyre on a fallen cedar tree, Swarraton, Hants, in January, 1905. (6) *Pleurottis decoru,* remarkable in the yellow pileus and stem covered with linear bistre fibrils, that become black with age. Found by Mr. Angus Grant, near Drumnadrochit,. Inverness, in September, 1905. (7) A series of pigmy flints from Blackdown, near Haslemere. With the point directed away from the observer, all showed secondary chippings (very numerous and minute) on the left side only. For other implements found on this site, see *Transactions S.E.U.S.S.*, 1904, p. 55 (with illustration). (8) Photographs of an extensive series of vessels of Late Keltic times, found by Mr. E. W. Swanton, in November, 1905, near the site where those were obtained which are figured and described in the *Transactions S.E.U.S.S.,* 1904, p.56. Three almost perfect cinerary urns, with several perfect accessory vessels placed around them. The ware is of five or six qualities. Two pieces may be considered as imported Samian ware. The accessories contained nothing more than earth, but from the cineraries were taken, in addition to calcined human bone, several rudely chipped flints, and from one of them a fragment of a bronze fibula. Many broken cineraries were found in this " family circle " of interments. The base of one had been repaired by the insertion of a plug of lead. There may be seen in the History Room at the Haslemere Museum, a Samian patera, found in London, which has been repaired in precisely the same manner, and we are told that a similar one is in the York Museum. The date of the Haslemere Urnfield may now be placed approximately at B.C. 50.

Mr. W. Plomer Youncj, F.R.M.S.—(1) Geological photomicrographs, including: Cornish serpentine, from the Lizard; Pumice stone, from Lipari; Blue lias, from Lyme Regis; Oolitic limestone, from Bristol; Hornblende rock, from, Scotland; Pikrite, Porphyritic granite, from Westmoreland; Dacite, from Kremnitz, Hungary; Schorle rock (Black Tourmalin), from Land's End; Chalcedony, from Cornwall; Iridescent agate, from Sussex; Olivine, Basalt cum Labradorite. (2) Photographs of Norman and Gothic Doorways: Bakewell Church, west doorway, Norman; St. Nicholas, Godstone, west doorway, Norman; St. Mary's, Axminster, east doorway, Norman; St. Andrew's Cathedral, Rochester, west doorway, Norman; St. Katherine, Merstham, west doorway, Early English; St. Hilda's Abbey, Whitby, great west door, Early English; St. Mary's, Axminster, west doorway, Decorated; St. Mary's, Oxted, west doorway, Decorated; St. Mary's, Axminster, north doorway, Decorated; St. Peter and Paul, Godalming, south doorway, Perpendicular; St. Mary's, Oxted, south porch, Perpendicular; Farnham Church, west doorway, Perpendicular.

Kent Photographic Record (sir David Salomons, President).— The Kent Photographic Record and Survey had on view a large number of prints of various subjects, botanical, geological, archaeological, etc. Among them may be specially mentioned the photographs of orchids, in situ, by Mr. Elgar, of the Maidstone Museum, remains of Boxley Abbey, by the Rev. Gardner Waterman, views of vanishing Rochester by members of the local Naturalists' Club, etc. The Survey still feels the want of members, who will devote themselves to a systematic line of research, and attention is specially directed to the recording of views of buildings, which are destroyed to make room for street widenings, etc. The Hon. Sec. is endeavouring to obtain a set of negatives of the historic houses in Rochester, with views of the Roman Wall and other antiquarian relics.

The Photographic Survey Of Sussex. —This Survey contributed to the Exhibition at Eastbourne 187 photographs of various sizes ranging up to 15x11ins. The subjects represented were almost entirely confined to those of Ecclesiastical and Domestic Architecture, the Survey not possessing sections for other scientific purposes. An interesting series of studies of churches and their details, such as doorways, capitals, chancel arches, fonts, etc., by Mr. G. C. Druce, showed excellent work and great judgment in selection. The neighbourhood of Eastbourne afforded to Messrs. Sparks and Kelsey nice studies at Ratton and Willingdon, and to Colonel Owen good photographs at Wostham. Valuable photographs of old engravings of Brighton subjects by Mr. Haines, and of old sketches (lately the property of the Rev. W. D. Parish) by Mr. Roods, and a copy of a charter to Cuckfield. contributed by Canon Cooper, were of much interest. Especial notice should be made of valuable work by Mrs. Padwick of photographs of Church Plate, the only contribution of this subject in the exhibition. A series of views, out of a large number, of Bodiam Castle, by Mr. Sands, F.S.A., were all worth close study. Old stone-slated farmhouses in the west of Sussex were well rendered by Mr. Robertson. Mr. Salmon's views of Shoreham Church are useful, and that of the " Old Pad Inn," at Lancing, lately destroyed by fire, is an instance of the great value of a County Survey. The Honorary Secretary to the Sussex Survey contributed various examples of work, amongst them one of the interesting wall-paintings lately uncovered at Trotton Church, this being the only specimen in the exhibition.

PROCEEDINGS OF THE CONGRESS OF THE SOUTH-EASTERN UNION OF SCIENTIFIC SOCIETIES, 1906.

The Eleventh Annual Congress of the Union commenced at Eastbourne on June 6th, by a Geological, Botanical, and Entomological excursion to Beachy Head, at which the retiring President, Professor W. M. Flinders Petrie, was present. Mr. W. H. Griffin kindly contributes notes of the Botanical Section (see page 68). Mr. R. Adkin reports that the entomological results are not sufficiently interesting to record.

In the evening a meeting was held, at which there was a good attendance. Professor W. M. Flinders Petrie, LL.D., D.C.L., F.R.S., the retiring president, extended a hearty welcome to Mr. Francis Darwin. F.R.S., the President-Elect, saying that Darwin's name was a household word, and one of the best known in the world of science. He was sure they had much pleasure in welcoming, as their President, one who was such a fine illustration of his own father's doctrines, which were to determine the wherefore of things. The Professor remarked that it was with the greatest satisfaction and pleasure that he vacated the chair for so worthy a successor (applause).

The President's Address.

Mr. Darwin, who was enthusiastically received, then proceeded with the annual address *(jiustm* pp. 1-17).

At the conclusion of Mr. Darwin's address, the Rev. T. R. R. Stebbing, F.R.S., said that he had been asked to move a resolution, the substance of which could easily be predicted. In introducing their new President, his predecessor, Professor Flinders Petrie, had eloquently enlarged both on the scientific work done by Mr. Francis Darwin himself, and on the striking examples of hereditary genius which his family afforded. In the latter respect there was one matter of detail deserving of special prominence. Charles Darwin wrott a most interesting life of Erasmus Darwin, and Francis Darwin followed suit with a life of Charles Darwin, which was one of the most delightful biographies in the English language. The earlier of these works showed us what charmingly ingenious young people Erasmus and his sister must have been, the schoolboy maintaining that roast goose was compatible with the strictest Lenten diet, because all flesh is grass, and vegetables may be eaten at any part of the year. The President, without going quite so far as this youthful logician, had touched on very interesting similarities of behaviour between animals and plants, and had earned their gratitude for his philosophic and scientific address. The terms of the resolution were that the thanks of the Congress be given to the President for his address, and that he be requested to allow it to be printed in the *Transactions of the South Eastern Union*. This was seconded by Dr. Jonathan Hutchinson, F.R.S., and carried amid much cheering.

Thursday. *June 7th.*—The day's proceedings commenced with a meeting of delegates and others forming the General Committee, which was presided over by the President, Mr. Darwin.

The Secretary, the Rev. R. Ashington Bullen, read the tenth annual Report (see page xx).

The Reports of the Photographic Secretary, Mr. Edward A. Martin (see page xxiii), the " Autumn Meeting" Secretary, Mr. H. Norman Gray (see page xxii), and of the Treasure Trove SubCommittee followed. It was understood that the expenses of the Treasure Trove Sub-Committee should not fall upon the General Fund of the I'nion, but that special help should be solicited from persons interested in the subject.

Dr. Abbott read the balance-sheet, which showed that an amount of 16s. Id. had been brought forward from last year. The receipts totalled £'55 His. 6d. , and a balance of £2 odd would be carried forward. Money to meet the year's expenses had been drawn from the small Rank reserve.

A discussion took place as to whether it would be advisable to increase the subscription of the members from 2s. 6d. to 8s. 6d., and the latter sum was eventually agreed to.

Dr. Abbott, having intimated his intention of retiring from the treasurership, Mr. Robert Adkin was appointed to the office, which he kindly accepted.

Dr. Abbott consented to act as auditor.

The ex-President, Professor Y. M. Flinders Petrie, was unanimously chosen as a Vice-President.

Mr. J. H. Baldock (Photographic Survey of Surrey) made some general remarks upon the possessions of the Survey, a number of which were on view in the room (see page xxv).

"Nature Near Eastbourne."—Mr. J. H. A. Jenner, F.E.S. (of Lewes), read his paper, entitled "Nature near Eastbourne" (printed in full on pp. 18 *et seq.*).

The President characterised the paper as a most interesting summary of the beauties of a part of England with which most of the members of the Congress were unfamiliar, and invited discussion thereon.

Mr. E. A. Martin (Croydon) desired to have further information in regard to the making of dew-ponds. Mr. Jenner had referred to a family at Alfriston who were accustomed to make them. Did they make them with a foundation of thatch and puddled clay, or merely with concrete, as he understood was the method of making them on the Downs in the neighbourhood of Eastbourne? If the latter, they probably scarcely merit the title of dew-ponds. In regard to the threatened extinction of certain species of butterflies and moths in the district, Mr. Martin advocated the strengthening of the stock of such species by artificial propagation and liberation when extinction was threatened. There would thus be no falsification of local entomology, and these ornaments of the countryside would not be lost. He hoped that, with the study and capture of occasional insects, the undue destruction of more than were absolutely necessary for study would be discountenanced.

Mr. Wilfred Mark Webb (Selborne) offered the suggestion that, in the London Parks, butterflies were taken by birds, and that these, and not altogether the schoolboy, were responsible for the destruction of beautiful insects. He had very much enjoyed listening to Mr. Jenner's interesting paper, and with regard to what some of the other speakers had maintained, he thought that too many objections hail been made to collecting. After all, it was only the beautiful things which people made a fuss about, and that quite rightly, for everyone could enjoy them, but nobody objected to ugly things being destroyed, and attention might be given to these. He was glad to hear from Mr. Martin that the Selborne Society was encouraging the study of nature.

Dr. Treutler (Brighton) said: "I do not think the appearance of the swallowtail butterfly in Sussex, is always due to the insect's escape from confinement. I have twice seen it in remote and secluded parts of the country—near Fletching, and near Hurstwood. While the collection and study of insects and butterflies are to be encouraged, I strongly protest against their wholesale slaughter, and arrangement in cases after the manner of a gaudy Turkish carpet.

Mr. Henry Davey (Brighton) remarked that Eastbourne was to be congratulated on having near it Abbott's Wood, so rich a locality for lepidoptera. The decrease in moths and butterflies was not due wholly to captures, but to man's interference in other ways, *e.g.,* the use of many rural places as building sites, etc. Butterflies like a *waste* space, they are plentiful enough in the Alps of Central Europe.

Mr. J. L. Otter (Selborne) asked Mr. Jenner if he had given his attention to the artificially levelled surfaces, of apparently ancient origin, which frequently occur on the South Downs; and, if so, whether he accepted the suggestion that these plateaux date from Neolithic times, and were formed as stations, from which the defence of Hocks and herds against the attacks of wolves could be effectively made?

Mr. Jenner, in reply, said that he was not in agreement with all the statements of Messrs. Hubbard. Dew-ponds were like an inverted watch-glass, and the bottom was covered with clay. Oxen trampled the latter into the ground, and if a dew-pond bottom became perforated the pond became dry-. The family he referred to as dew-pond makers, was named Pettitt, and was not a wandering family, but lived at Alfriston. The extinction of lepidoptera was partly due to school children, but more, perhaps, to the extensive establishment of pheasantries, which was a most serious thing if established in a wood frequented by rare insects. Draining laud also led to extinction of rare insects, by the destruction of the plants on which their larv fed. Much, he thought, might be said against the re-introduction of insects and plants. If a species died out, much as they regretted it, they should note the fact. But artificially introduced species could not be called members of the native fauna or flora. As to the existence of the swallow-tail in the Fens, he considered that it was on its margin of distribution, and that the Fens were the last land severed from the continent. Mr. Whitaker, F.R.S., F.G.S., does not accept this as a geological fact, and expressed his dissent to Mr. Jenner privately. With regard to the theory of the artificial plateaux on the Downs being for the protection of flocks, etc., from wolves, he thought there might be something in it, but that the great probability was that Neolithic Man had more to fear from predatory Man.

"Flora Of Easthourne District."—Mr. Bulman, M.A. (President of the Eastbourne Natural History Society), read a paper on the "Flora of the Eastbourne district," which was rendered additionally interesting and instructive by specimens of some of the flowers. The paper had been written by the late Dr. Whitney and Miss B. Milner, B.A. (see page 30).

Mr. Bulman prefaced the reading of the paper by expressing, in feeling words, the loss which the Eastbourne Natural History Society bad sustained in the sudden death of Dr. Whitney.

The President expressed the thanks of the Union to the authors.

Dr. Treucler (Brighton) said: "*Sibtlwrpia euiojiaea* is more widely distributed in the county than was at first thought. It occurs plentifully in Heathfield Park. I have found it also near Pitt Down, Uckfield. *Statire auriculae folia*, Vahl., occurs on chalk cliffs, east of the C'uckmere. *Bupleurum aristatum* was collected by Mr. Roper and myself on turf not far from Meads, 20 years ago, but I fear the station is now lost. *Phyteuma spicatum* occurs sparingly in other parts of the county besides Abbott's Wood; I have found it at Cross-in-Hand, at lsfield, and at Possingworth."

Dr. Gilbert (Tunbridge Wells) remarked that *I'liyteuma spicatum* had been discovered in Devonshire and Cornwall. *Sibthorpia eurofiaea* occurs in the southwest counties, and is abundant at Heathfield, in Sussex.

Mr. A. H. Maude (Eastbourne) said: "I have found *Sibtfwrjria europaea* growing abundantly near Hurstraonceux, though, in the New Victoria County Guide, it is said not to grow so far to the east of Sussex. *Phyteuma spicatum* is a very common plant in Switzerland, and it is curious how its floral spike lengthens out from two inches to six or eight inches, before it has finished flowering. *Centaurea calcitrapa* is abundant on the Downs in certain parts, with its curious dichotomous branches and stiff thorns. *Habenaria* (or *Caeloylossum*) *ciride* varies very much in size according to soil and situation, here, on the Downs, being only a few inches, but, in Switzerland, reaching nearly a foot in size in choice situations, with leaves one-and-half inches broad, and a very marked chocolate tip. *Senecio campestris* is another plant abundant on the Downs, and I have never found it away from them."

Afternoon Excursion.—Delightful weather favoured an excursion to Alfriston and the Cuckmere Valley in the afternoon, about 65 members making the journey. Alfriston Church, "The Cathedral of the South Downs," and the Clergy House, were visited by one party, and others paid a visit to the small church at Lullington. A walk was then made across the fields to Littlington, the return journey to Alfriston being made by way of the river. Afterwards tea was taken at the historic Star Inn. The drive home, by way of Polegate and Willingdon, was most pleasurable.

The Educational Value Of Museums. —This was the subject of the discourse given in the evening by Dr. Jonathan Hutchinson, F.R.S. (founder of the Haslemere Educational Museum). In introducing Dr. Hutchinson, the President remarked how much they appreciated the honour done to the Union by a man of the eminence of Dr. Hutchinson, who, although he was a most busy man, was yet good enough to make time to come amongst them and give them that lecture.

Dr. Hutchinson said his idea of a museum was something totally different from the old-fashioned provincial museum, and the elaborate scientific museum which was now in all their large cities. He was thinking of an educational museum, not an elaborate or expensive building, but one with plenty of space and light and good protection from the weather. Such a museum was an essential part of education (applause). They ought to get rid, to a considerable extent, of the attempt to train the mind by resorting to mere text books, and, instead, bring the student into contact with the thing—from the book to the thing, and from the thing back to the book. A museum should be subject to change. It should not be allowed to get old, but there should be fresh specimens introduced, and periodic exhibitions of special things. They had a very nice museum in Eastbourne. It contained some very interesting and exceedingly valuable specimens, but to his mind it was already crowded. What he thought was needed was a place ten times as large, and it might be had very easily, as there was a nice plot of ground at the rear of the premises which might be covered in at very little expense. In time, Eastbourne would be vigorous enough not to be content with sheds, but to build a place three or four storeys high, and have a model educational museum. Eastbourne had a moveable population of scholars attending schools, amounting to upwards of 8,000. He would suggest an institution to which they could resort and gain knowledge without fatigue to the memory by becoming familiar with the things themselves. Mere book learning fatigued the memory, as the student had to try to remember what he read. It was a constant effort to retain that knowledge. But memory which concerned things he had seen was quite without that effort. Knowledge gained from actual familiarity with the thing itself was productive of increased knowledge. It was germinal in itself; developed the mental powers and incited a person to further research. He thought an elementary conception in regard to space and time—the undying past and the future that was to come—was most important. He did not favour the teaching of history according to a particular reign or period, but considered it was well to begin with giving a wide conception of events. The object of education was not to store the mind with innumerable items of education, to be retained with difficulty, but to enable the mind to generate knowledge by its own knowledge of facts. He advocated a development of knowledge amongst the whole community for the purpose of bringing about broader ideas, and a general rising to the higher state of knowledge, and a higher conception of duty.

Dr. Treutler said: "We are greatly indebted to you for your most interesting and deeply suggestive address, in which you have travelled over a wide range of topics, anyone of which would alone furnish food for deep reflection. The education derived from museum-teaching, such as you indicate, would produce most important results. It would tend to correct man's estimate of himself; it would teach him his place in Nature, as only one among a host of sentient, and often highly intelligent, beings; it would improve man's relation to animals, his conduct and bearing towards them would lead him to treat them with kindness and respect. I emphatically share and endorse your remarks on the subject of death. The study and observation of nature, will teach man to look on (loath, not as the King of Terrors, but as a welcome rest shared by all living things, so that he will learn to dread his grave as little as his bed; while his ideas and hopes of a future existence will be placed on a sounder and more rational basis."

Mr. W. T. Vincent (Woolwich Antiquarian Society) said that most of those present were acquainted with Dr. Hutchinson's work at Haslemere, but everyone ought to know that the learned lecturer had established at Haslemere a splendid museum, upon the model which he had just described, and that visitors thereto would be gladly welcomed.

Mr. Abbott expressed his great appreciation of Dr. Hutchinson's address, and heartily thanked him for it. He suggested that it would be a very suitable address to be given before the affiliated Societies during the coming winter session. In this way, the members would learn the value of museums, and the necessity for their establishment on educational lines.

Friday. *Jum 8th.*—The morning session was taken up with a paper by Mr. E. A. Martin, F.G.S., on "Sea Erosion and Coast Defence" (see page 35), and one by Mr. W. Ruskin Butterfield, M.B.O.U., on the " Birds of Sussex compared with the list of the Ornithologists' Union " (see page 27).

The President said they were all much obliged to Mr. Martin for his very lucid statements. He asked some of the delegates present for comments.

Dr. West said he knew the coast near Bournemouth and Christchurch Bay, and there the erosion was entirely due to the currents, and not to the actual battering of the sea. The coast wore away quite three feet a year in parts, and the sea simply swept away what was brought down. In '79 the river at Christchurch Bay was going at a tremendous pace, and ate the cliffs away, and tons fell every few months. He tried to get along the shore on one occasion, but he had to turn back, as it was highly dangerous—as much as his life was worth. The cause of the tremendous waste at Barton Cliffs was because the bottom bed of the cliffs was a bed of lignite, and Dr. West described how the undercliffs were formed. The same thing took place in Bournemouth Bay, though to a somewhat less extent. Besides that there was the actual wearing of the cliffs by the rain, which was very great indeed. Beautiful "alpine chains " were to be found with little streams and waterfalls!

Mr. Whitaker, after some humorous observations, said he did not agree with Mr. Martin about prophets being inaccurate. He thought it was perfectly easy to prophesy with a feeling that it would be fulfilled. He prophesied at Folkestone that he believed that a piece of cliff would go if not looked after, and it went within a year. That was very gratifying to him. He considered that prophecies were accurate generally. Mr. Whitaker told the Congress that, when at Dunwich on a geological survey, he measured the distance from a point at the top of the cliff to the edge of a ruined church, and on going to his hotel he picked up an old book. He there found the same measure taken about 100 years ago, and the. loss was trifling. As mentioned by Mr. Martin, in reference to Shoreham's action in protecting their harbour at the expense of Brighton, it was money thrown away—to save your land at your neighbour's expense. Concerning this subject the proper course to be taken was for the Government to institute a proper inquiry, and this was what it was going to do, and a Royal Commission had been appointed for the purpose. How utterly impossible it was for the State to do anything for the long stretches. They would spend £100 to save £10. If they were to stop the erosion of the sea, where on earth were the geologists to get their sections from? They wanted erosion in places, or they would lose facts and theories. He hoped they would have some more local information, and get the local societies to take careful records of what was going on. That was a very easy thing. One was provided with maps by the Government; they were dated, etc., and all one had to do was to make rough measurements to see what had been lost since the Government measured. He thought it ought to be one of the selfimposed duties of societies to take measurements of the coast. If the discussion only resulted in that it would be a very good result.

Mr. R. W. Brown, having made a short remark about provision in Government leases against erosion, Mr. Martin replied, saying it was an almost inexhaustible subject. He was very much obliged to Mr. Whitaker for his observations. As regarded prophecies, they were, he admitted, sometimes correct, but they were ofttimes incorrect. He agreed with Mr. Whitaker that local societies ought to keep records of erosions. They ought, also, to photograph the sections and let him have them for lantern slides. They could then be lent round to the various societies.

This concluded the discussion.

With regard to the paper on the "Birds of Sussex" the President said the paper was the result of a great deal of extremely careful work. He would like to ask Mr. Butterfield how it was that the bustard was chased by greyhounds—and captured? How could the dog catch a bustard, which was such a strong flying bird. He knew the frost and cold weather made their wings useless, but he could not understand how the birds could be caught as a regular practice.

Mr. T. Parkin spoke of the power of the bustard to conceal itself, and told of one of his experiences. As regarded the President's question, he might say that the birds would allow the greyhounds to get very close indeed, and, in addition to this, the bustard depended a great deal, almost entirely, upon its legs. Mr. Parkin went on to speak of herons, saying that they were very numerous at Windmill Place, where they were unmolested. There was no question that the heron had leen on the increase. In a wood near Rye there were two or three nests, and the old birds did everything in their power to protect the young.

The Rev. T. R. R. Stebbing told of a heron which came to a place in Tunbridge Wells and stole goldfish.

Mr. Bradford called attention to the fact that the museum at Brighton had an excellent collection of Sussex birds; he thought the members would find it useful.

Dr. Treutler asked if there had been a recent visit of the roller? It did occur in 1863. Then, again, the white thrush was found on a doorstep at Hove a few days previously. It had occurred on several occasions before, but he did not know whether it had been recorded from the rest of Sussex.

Mr. Whitaker remarked that, in his opinion, the visitors should not be taken into account. He thought there ought to be a list of the birds that nested and lived in Sussex, and that it should not include amateur casuals. When there was an increase in birds the people claimed some of the credit. Birds should be well treated and encouraged to come again.

Mr. Hewlett spoke of the peregrine falcon, and said he had been told that a so-called sportsman had shot one of them at Beachy Head during the breeding-season. He thought it was a shocking thing, and something ought to be done in the matter.

The President asked if the offender could not be punished, and was answered that he was liable to be summoned.

Mr. Hewlett said it ought to go forth to the so-called sportsman that his action had met with the greatest disapproval of the-Congress (applause).

Mr. Butterfield, in reply, said it had been his purpose to tell what birds visited the county in recent years, since the days of Mr. Borrer. He believed that Sussex occupied the first place, or a better place than Norfolk, even as far as resident birds were concerned. Whether as to species or individuals he believed Sussex came out at the top.

The President, having thanked Mr. Butterfield, the discussion closed.

Excursion To Michei.ham Priory.—The excursion in the afternoon was to Michelham Priory, Folkington, the journey being made in brakes. Through the kindness of Mr. J. E. A. Gwynne, F.S.A. , F.Z.S., the party were conducted over the building. The moat was inspected, and the company were shown over the tower. The Rev. E. E. Crake, M.A., Rector of Jevington, read an interesting paper on the history of the Priory (see pp. 64 *et xeq.*), and at the conclusion he was accorded the hearty thanks of the Union, through the President, who also thanked Mr. Gwynne for his kindness. The group of delegates and members were photographed amid the mediaeval surroundings (see frontispiece).

The evening was the occasion of the reception by the Mayor of Eastbourne (Councillor S. N. Fox), and Mrs. Fox. The reception was held in the spacious Assembly Room of the Town Hall, the apartment being gaily adorned with bunting, and the chairs and small tables being most comfortably arranged in conversazione fashion. The Mayor and Mayoress stood at the entrance, and extended a welcome to the guests as they arrived, this ceremony lasting over half-an-hour. The time passed very pleasantly, for the Municipal Orchestra, under the direction of Mr. Theo. Ward, played some delightful selections. Another source of entertainment was the number of microscopes on view in the room, exhibited by Messrs. G. W. Bulman (President of the Eastbourne Natural History Society), Miss Milner and Mr. G. D. Plomer, and microscopic objects were shown with the oxy-hydrogen projection lantern by Mr. Y. Sparrow. The Council Chamber served as an apartment for the supply of light refreshments.

An Interesting Stone.—The guests, who included a number of the members of the Town Council, having assembled, the Mayor remarked that, on behalf of the townspeople, he desired to welcome the Congress to Eastbourne. He congratulated the Union upon securing such an illustrious man for their President as Mr. Darwin, a son of a genius. The Mayor mentioned that Eastbourne held memories of several scientific experts, principal among them being those of Thomas Huxley (applause). He trusted that the Union had found its visit to Eastbourne a fruitful one—fruitful in scientific research, and also in pleasure in connection with the excursions; and he hoped that they had found the district a fertile field for discovery (applause).

Mr. Darwin thanked the Mayor for his kind words of welcome, and also for the hospitality extended to the members during the Congress. The Union were very much in earnest in spite, or, rather in consequence, of being a comparatively new organisation. The members came to Eastbourne to enjoy themselves, and, therefore, were very thankful to the Mayor for providing such a pleasant evening (applause). Eastbourne would benefit a little by the holding of the Congress in the town. An extremely interesting stone had been discovered by a visitor taking part in the Congress, and he was glad to be able to say that through that gentleman's activity in the matter, the Duke of Devonshire had been approached, and had very kindly consented to give it to the Eastbourne Museum (applause).

The stone is a Roman font, and was found last month during the demolition of a part of the wall surrounding the Saffrons, in Compton Place Road. It was conveyed to the office of the Duke of Devonshire's agent, Mr. J. P. Cockerell, and remained there until yesterday, when Mr. Preston, a member of the Congress, was told about it by Mr. Howard. Mr. Preston examined the relic, and through his activity, and the kindness of the Duke, it will be placed in the local Museum. The Eastbourne Society will have the stone photographed, and a copy of the photograph will be sent to the Society of Antiquaries, who will presumably estimate its value.

The concluding portion of the evening was devoted to a lecturette by Mr. E. J. Bedford, on "Bird Architecture," which was illustrated by some splendid lantern-slides, and the admirable manner in which the lecturer described various birds and their nests was greatly appreciated by those present.

Saturday. *June 9th.*—Delegates' Meeting.—Saturday's proceedings opened in the morning with a meeting, at which the President, Mr. Francis Darwin, F.R.S., presided. The general Hon. Secretary, the Rev. R. A. Bullen, first of all gave a report, as representative of the South-Eastern Union, of the Conference of Delegates to the British Association for the Advancement of Science (see p. xxvi).

It was then notified that the retiring members of the Council were MissSargant, F.L.S., Mr. W. H. Griffin, the Rev. W. Hudson, M.A., F.S.A., and Mr. W. Mark Webb, F.L.S. To fill these vacancies, Mr. Connold, Mr. Meeson, Mr. Roods, and Mr. Plomer Young were nominated. Mr. Connold, however, kindly offered to withdraw in favour of Mr. Thomas Parkin, M.A., J.P., the Hastings and St. Leonard's President, who, with the remaining three nominees, was elected.

Dr. Treutler, M.D., F.L.S., proposed a vote of thanks to the late President (Professor W. M. Flinders Petrie, LL.D., D.C.L., F.R.S.), and also to the retiring members of the Council, remarking on the debt of gratitude they owed to those gentlemen. Anyone who knew what it was to manage the affairs of the Union would appreciate the amount of work that was done when they looked round and saw the extreme perfection of the arrangements of the Union. It was like navigating a ship—when everything went well the officers and crew deserved the credit, and he thought they would agree that the ship of that Union had been splendidly navigated from haven to haven (applause).

Mr. F. W. Rudler, I.S.O., F.G.S., seconded, and the vote was carried with acclamation.

The President next proposed a vote of thanks to the Mayor and Mayoress of Eastbourne, the local Committee, the local Society, the Hospitality Committee, and the exhibitors and lecturers at the Mayor's reception. He spoke chiefly of their indebtedness to the local Natural History Society, for it was owing to their kind invitation that they were holding that Congress at Eastbourne. In connection witli this, he mentioned especially the names of Messrs. Bedford, Hollway, and Sparkes.

Mr. Mark Webb seconded, and, the vote being carried, Mr. Bedford and Mr. Sparkes replied, the former remarking that if their excursions had been more in the shape of picnics, it was as he wished, as he wanted them to see the whole locality, and they could explore on other occasions.

In proposing the President for 1907, Mr. Stebbing said: "Mr. President, Ladies and Gentlemen, during the last two or three days, many of us, perhaps, for the first time in our lives, or certainly, with more seriousness than ever before, have been led to ponder on the virtues and curious effects of *Periodicity.* To it, no doubt, must be attributed the circumstance that those who control these matters have charged me to bring forward the resolution I now hold in my hand. They have bidden me do the same

kind of thing once or twice before, so that now, merely under the influence of periodicity, they tell me to do it again. But, just as this mysterious power leads us, the older we grow, the more frequently to tell the same well-known stories, so in proposing a periodic resolution, I can scarcely escape from dwelling upon topics which the occasion itself suggests. The delicacy of this annual task consists in this, that it has to be carried out in the presence of the President recently come into office, and while we are basking in the sunshine, so to speak, of his genial administration, he finds us already worrying about his successor. There is also a recurrent difficulty in the very renown of the men who of late have condescended to preside over our meetings. To obtain others of like calibre to follow them is no longer easy. In the pleasant discourse with which you enlivened the Mayor's party last night, you, Mr. President, compared, or rather contrasted, the great congress of the British Association with our small one, ruling, as we were all delighted to hear, that the advantage was much on our side. Encouraged by this comparison, I shall permit myself to recall some incidents connected with the meeting of the sister Association some years ago at Bristol. On that occasion, the President, Sir William Crookes. created a kind of scare by explaining that between 80 and 40 years hence, all the wheat-growing area of the globe would be exhausted, and that consequently, all of us who depend on wheaten bread as the staff of life were in danger of becoming extinct. There was, however, one chance for us, namely, to obtain from the atmosphere a fertilizing product with which it is abundantly supplied, though we do not know how to extract it. On the following Sunday, in what is known as the Association sermon, the Bishop of Bristol, with quaint humour remarked, that miracles in theology need no longer be a stumbling-block, when men of science were inciting us to feed on burnt sky! Well, Sir, this might very naturally seem an improbable source of nutrition. Nevertheless, from a lecture delivered not long ago, it appears that some ingenious

Norwegians have devised machinery for making the requisite extract of air. The abundant water power of Norway makes it already profitable to work the invention in that country. The lecturer was able, not only to explain the working of the apparatus and show pictures of it, but also to exhibit the solid products yielded under suitable process by the impalpable atmosphere. The lecturer was Professor Silvanus Phillips Thompson, D.Sc.,F.R.8., Principal and Professor of Physics in the City and Guilds of London Technical College. Those who have heard him lecturing on this or other scientific subjects, will, I believe, agree that we could scarcely have a more suitable man to preside over us at such a centre as Woolwich, and I therefore beg to move that he be President-elect for the Congress in 1907.

Mr. W. Whitaker, F.R.S., seconded, and the proposition was carried unanimously.

Mr. W. T. Vincent (President of the Woolwich Antiquarian Society) invited the South-Eastern Union of Scientific Societies to hold its next Annual Congress at Woolwich, and said he was the bearer of assurances from the Mayor and Borough Council, and the Governors of the Woolwich Polytechnic, that the members would be heartily welcomed and afforded all possible facilities for their meetings and pursuits. It would be the first of the Union's Congresses in the Metropolis, and any shortcomings on the part of Woolwich would be compensated by having all London to draw upon.

Mr. R. C. Frost (Treasurer of the Woolwich Antiquarian Society) supported the invitation, saying that, after his experiences of these Congresses, he was afraid that Woolwich would have some difficulty in deserving such an honour, but might be relied upon to do its best.

Dr. Abbott, as a member of the Woolwich Society, commended its organisation and activities, and thought a Congress at Woolwich would be successful.

The President formally accepted the invitation, amid cheers.

Questions in regard to the finances then came forward for discussion, and eventually Mr. Vincent moved that the consideration of the question of sending free copies of the "Transactions" to delegates be referred to the Council to consider and report at the next general meeting. Dr. Abbott seconded, and the motion was agreed to.

Dr. Abbott also proposed that the Council consider whether any alteration of the rules was desirable, and Mr. Mark Webb seconding, this was also carried.

Lastly, a vote of thanks to the Rev. R. Ashington Bullen for his many services to the South-Eastern Union was proposed and seconded. This evoked such an outburst of enthusiasm that the Hon. Sec, could with difficulty control his emotion. His reply took the form of an amusing story, the moral of which was to deprecate so exciting an experience as that which the kindness of the delegates had brought upon him. This concluded the business before the Delegates' meeting.

Geology And Nature Study.—Business being concluded, an excellent paper was read by Mr. W. H. Griffin (Hon. Secretary, Catford Natural History Society) on " The Geology of the Upper Ravensbourne Valley," with notes on the flora (see p. 50). The lecturer impressed upon the younger section of his audience the importance of practical work in the field, observation instead of mere cram, and the constant testing and sifting of matter given in text-books. He warned his hearers strongly against the so-called popular variety of nature notes to be found in some periodicals, and finally advised them first to read Huxley's "Physiography," and then thoroughly examine the course of any stream in the locality where they reside.

At the conclusion there was a short discussion.

Afterwards Mr. Mark Webb gave a paper on "Nature Study," illustrated by lantern slides. He remarked that our life had become more and more artificial, and the education given to children was found for this reason to be open to several grave objections. Speaking generally, it created no interest, it was unnatural, repressed activity, and did not satisfy inherent longings and instincts, such as those for exploration or travel, hunting, and the like. As a matter of fact, it had been pointed out that many of the most celebrated of men had been those, like Thoreau, Robert Louis Stevenson, Walt Whitman, and Darwin, who broke away from the trammels of school, played truant, and educated themselves in certain important directions. The education of the schools of which the feature had been the imparting of a second-hand information had been based on the assumption that the mind of the child was like that of an adult, and the result had been the loss of power on the part of pupils of acquiring knowledge for themselves and of acting under newly-risen circumstances. Aims Of Nature Study. —The aims of nature study were to obviate the objections to ordinary education which had been described. In its highest form it should be carried on out of doors, and information given must be reduced to a minimum. Bearing in mind also that we were dealing with a part of the general education of young children who should "face existence whole," we should, while insisting on accurate observations and deduction, avoid the introduction of logical connections which would turn our work into what was called elementary science, which should come at a later stage. The brain of the child stored up this and that piece of experience that interested it, and later on each fragment was assigned to its proper position with regard to the others. He did not, however, suggest that observations on one particular object should not be continued, for the value and interest of them was thereby increased. There were many difficulties in the way of the teachers, and he would not minimise them in the slightest. He could not help thinking, however, that in many cases the chief one was jumping the chasm that separated "thinking that nothing can be done " and "seeing whether something could not be attempted." The materials for nature study, it was pointed out, were endless, and the choice depended on local conditions. The whole object was not to afford information, but, by the gratification of properly directed curiosity, to develop mental powers.

Mr. Griffin and Mr. Webb were each thanked for their papers.

The excursion on Saturday afternoon was to Pevensey. The Ancient Norman Church of West Ham was first visited, under the guidance of the Vicar (Rev. Howard Hopley); afterwards, Archdeacon Sutton, M.A., conducted the members over Pevensey Church. The party subsequently visited the Old Mint, where the Pevensey coins used to be struck, and Rev. E. E. Crake, M.A., kindly read a paper on "Pevensey and its Lords" (see p. 60), on the greensward of the Castle. The various guides were thanked by one of our VicePresidents, Mr. Whitaker. At suitable times, and after tea, the members returned to Eastbourne by motor-train, and the Congress of 1906 concluded.

TRANSACTIONS OK THE SOUTH EASTERN UNION OF SCIENTIFIC SOCIETIES. *1906.* PRESIDENTIAL ADDRESS. By FRANCIS DA11W1N, M.A., F.R.K., Hon. Fellow of Christ's College.

The pleasantest part of my duty tonight, is to tbank you for the honour you have done me in choosing me for your President. I am afraid that I know but little in detail of the work of the Union, but I warmly sympathise with the objects for which it was founded, and with the aims and activities of its constituent Societies. I hope, therefore, that you will take the will for the deed, and be lenient to my shortcomings. I must also ask your consideration in regard to my Address to-night. In choosing Periodicity as my subject, I hoped that I might perhaps interest our biological members, but it is not always easy to be interesting on a somewhat difficult question. I shall, at any rate, do my best not to be unreasonably lengthy, and if I prove to be tedious to some, I can do no more than apologise.

One of the most striking features of living things is their periodic or rhythmic character. Life itself may be described as a rhythm, made up of alternate destruction and reconstruction. Protoplasm, "the physical basis of life," is alternately falling to pieces by a degradation into simpler compounds, and rebuilding itself from the food materials supplied.

Simpler instances are found in the process of reproduction. A seed germinates, and, as the plant grows to maturity, it passes through an ordered series of changes, which finally terminate in the production of a seed. Or, to take a better example, the buds of a deciduous tree are at rest in the winter; in the spring, they shoot forth by the elongation of the dwarf branch of which the bud consists. The branch grows for a certain time, and then elongation ceases, and fresh buds develop, which will rest till next spring, so that the growth of a branch is made up of alternate periods of growth and rest. Here we have got to the regions of Natural History. We have here a comparatively simple rhythm of phenomena which has been investigated by innumerable people who love to watch the changes in the country side, as well as by many more professionally scientific workers who have given their lives to what is known as the science of phrenology. One of these workers was the Rev. L. Jenyns, afterwards known as Blomefield, who kept, for many years, near Cambridge, a minute diary of the time of flowering of plants, of the leafing of trees, the song of birds, the appearance of various insects. These were published in his *Observation in Natural History,* in 1848, and recently reprinted, in 1903, by the Cambridge Press. I may, perhaps, be allowed to quote occasionally from my introductory remarks to this work.

Let us return to the leafing of trees, and, by looking at Jenyns' calendar, see what is the character of the periodicity which is presented by trees in our climate. The following table gives the earliest and latest dates on which a variety of trees appeared in leaf:

The first thing that strikes one is the great variability in the dates. It is clear, whatever may be the nature of the machinery that brings about the periodicity of leafage, that it is not a machine that works regularly. In saying this, I am imagining that we do not know what is really so familiar, namely, that in cold years the leaves come out later. The figures tell something else than that the machinery is irregular. If we take the number of days between the earliest and latest dates, we see that the number diminishes from elder 79, honeysuckle 62, to ash 20, oak 22, plane 28, vine 17, and, as we know that cold weather is rarer in April than in February, we might suspect that the variations in date of leafage are dependent on temperature. But an absolute statement of this sort conveys a very partial and incorrect idea of the whole problem. Before showing how this is so, it is worth while to refer to the manner in which primitive people have treated the sequence of natural phenomena probably from time immemorial. Thus, as Jenyns says, "the middle of March may be, in the long run, the most suitable time for sowing various kinds of grain," but the husbandman may easily go wrong in this or other operations if he sticks to a fixed date. But if he knows that the conditions necessary for his purpose are also necessary for the flowering of some familiar herb, he will be safer in waiting for it to flower than in going by date. He will act on the assumption that all plants are equally delayed by a cold year, or brought on quickly by a favourable one. How far this assumption is accurate I do not know, but that it was anciently acted on is clear. Stillingfleet quotes from Aristophanes that "the crane points out the time of sowing," and "the kite when it is time to shear your sheep." Again, an old Swedish proverb has it that "when you see the white wagtail you may turn your sheep into the fields, and when you see the wheatear you may sow your grain." The poet, Gray, acted on this principle, for he writes to his friends not to expect him at Cambridge till "the codlin hedge at Pembroke was out in bloom."

The only English proverbs that I have come across are these:— "When the sloe tree is as white as a sheet, you must sow your barley, be it dry or wet" (Forster's *Perennial 'alendar,* 1824). Miss Jekyll, in *Old West Surrey,* speaking of the wryneck, quotes "When we hears that, we very soon thinks about rining (barking) the oaks."

It is interesting to note a case where two phenomena, occurring in one country, occur also together in a more rigorous climate, being delayed together in the way just referred to. Wood-sorrel was known in medieval Latin as *Panix rnnilli,* and as cuckoo's sorrel by the Saxons. In Stillingfleet's *Calendar* (1755) it is said to flower on April 16th, and the cuckoo's note begins on April 17th. According to a Swedish calendarist the cuckoo sings first on May 12th, and the wood-sorrel flowers May 18th. There is probably some luck about this coincidence of date, interesting as it is.

Cases such as these have induced phfenologists to try to explain the periodic phenomena as strictly regulated by the temperature. Perhaps the most striking case in their favour is the behaviour of Arctic plants. In the Polar regions *(e. ij.,* Nova Zembla) the Bummer consists of two months. July and August, during which the mean temperature is about 5C.: In such conditions, cases like the following occur:—At Pitlekaj, the last nine days of June gave a mean temperature below 0C, while the mean for the first nine days of July was between +4 and +6; and on July 10th all the four species of willow were in full flower, the dwarf birch, *Sedum paluxtre, Polygonum, (assio/ie, Diapensia,* were also out, and within a week the whole vegetation was in flower!. Thus there is a great rush of all sorts of flowers as soon as the temperature is above zero, and not that dropping fire of flowering, which with us begins with the mezereon and ends with ivy in the autumn. Another instance shows the rapidity of the reply to warmth. On May 28th,,' the temperature at Ssagastyr was above 0C. for the first time for 250 days. On May 29th, the first flowers were seen, *riz., ('lin/xoxplenium* and *Draba.* In September the cold comes on and shuts down vegetation like a lid, giving the whole country the look of a southern region caught in an early frost. It is no wonder that Arctic plants are dwarfed, so that, for instance, a species of *Artemisia,*

which is from two to four feet in height in Norway or Sweden, only grows to four or five inches in Polar regions. Yet some of them live long, and Kiihlmann describes a juniper on the tree-limit of north Lapland with 544 annual rings, averaging rJ5 in. (02mm.) in widthj. It is not surprising that, under such conditions, temperature should seem to give an absolute ruler.

The methods of phrenologists have been various. The general idea has been to take the average temperature of each day and add them together, for two different stations. Thus, for instance, they would find that, in England, the catkins of the hazel open after having experienced a certain number of days of moderate temperature; while, in Sweden, the same thing occurs, only after a greater number of days of lower temperature. The hope of the phrenologist is to find that a total, got by multiplying the number of days by the average temperature, will be the same for both places. A classical example of this attempt is that of Boussingault. who took the time necessary to ripen barley in different places:

Alsace
Kingston, U.S.
Cumbal
Sta Fe de Bogota..
Estbonia
Upsala
Katisbon
Bavaria

Here five out of eight totals coine fairly together, '.«., those between 1780 and 1798, the remaining three are very irregular.

The following is a better case, namely, the comparison of the dates of flowering of the lilac, birch, and *Primus padus* in Germany and Sweden (Upsala):

This case is interesting because the totals are not calculated from the mean shade temperatures, but the maximum sun temperature on each day. This is an instance of straits to which pheenologists have been driven in their efforts to find some law which will fit all facts. In this way they have tried the squares of the temperatures, and they have dealt in various ways with the time element, sometimes making January 1st their

starting point, or taking the day on which the temperature first rises to the freezingpoint. It has also been suggested that December 21st, the winter solstice, would be a satisfactory initial date.

What seems to me the best of all attempts at phienologic law, is that of Linsser. He compares the times of flowering at St. Petersburg and Brussels. But instead of comparing single species, he divides the whole flora into six groups, according to their date of flowering, and takes the mean period of flowering for each group at Brussels and St. Petersburg. For the initial date he takes at Brussels, January 16th, when the winter temperature began definitely to rise, and for St. Petersburg, April 8th, when the thermometer rose to zero:

In this table, March 16th is the average date of flowering of group 1 at Brussels;; 51 means that it was 51 days later at St. Petersburg. The last two columns give the sum of the temperatures above zero, received by the two groups. At Brussels, the group has received twice as much warmth as St. Petersburg. This does not look hopeful; but here comes in Linsser's ingenious point, *riz.,* his comparison of the heat actually received with that of the whole year. These figures are for Brussels, 8687, for St. Petersburg, 2253. Now on March 16th, group 1 had received at Brussels 184=W = 0-05 of the whole available heat, and 51 days later, when group 1 flowered at St. Petersburg, it had received rfj¥ − 0,04 of the total heat. The following table gives corresponding figures for all six groups: 1. 2. j 3. 4. 5.. 6.

Brussels....-05 I-09-15-21 -28 '-40
St. Petersburg.. 04 10 19-37 31 42

The merit of this method is that it does not treat plants as so many thermometers, each requiring so much heat to drive it to a certain point, but as organisms having different innate qualities. It assumed a hereditary knowledge on the part of the plant, that the right time to Mower is when it has received a certain percentage of the available heat. In other words, since a plant flowering at any time of year is guided, not by the total heat received, but by the percentage, it applies an inherited experience of the total available heat. This law is, as 1 have said, probably the best, but, on general grounds, as 1 shall now show, no law based on temperature alone can be universally applicable. Take for instance, such an obvious instance as the desert flora of Egypt. Here the factor is water supply, not temperature. Or to take the extreme northern parts of Europe, where, even if the winter were as warm as the summer, the distribution of light would regulate the activity of all green plants. But even when such, extreme cases are omitted from consideration many formidabledifficulties remain.

Thus, the fact that an increased temperature acts differently on different functions of the plant is fatal to any general law of temperature. Thus, a high temperature which is extremely favourable to the growth of leaves, may be unfavourable, or even fatal, to the development of flowers and fruit, and, therefore, the same phonological law cannot apply to both. The best known. instance is that of European fruit-trees in certain parts of the, tropics, where they grow luxuriantly, and yet produce no fruit.

Again, it is hard to see how the same law can apply to both annuals and perennials. If a plant is to flower, it must have a reserve or store of food material in the form of starch, etc. A perennial has a store-house in its bulb or rhi/.ome—a store made last year, but an annual plant has to manufacture its reserve material before it can flower.

An instance occurs in the growth of buds. At the end of summer, a tree, such as a horse-chestnut, or a beech, has developed the buds which are to remain dormant through the winter and!row into branches the following year. Why do they not grow in the summer during which they are formed? It cannot be want of heat, as the temperature is more favourable in late summer than at their normal time of growth. Moreover, they are actually capable of growth in summer under certain circumstances. Thus, if the leaves are all removed from a branch, the buds develop. We have, therefore, parts of a plant capable of growth and exposed to a favourable temperature, and yet not growing. We can only explain it by saying that the plant, and not the temperature, is master of the situation. That it is, in fact, guided by internal rather than external conditions. The bud has to go through certain invisible changes during its winter-rest, before it is ready for its normal growth. These invisible changes are part of the plant's automatic rhythmic capacity, which enables it to be independent, to a large extent, of external changes.

Askenasy's well-known work *Botanisclie Xeituny,* 1877) on the flower-buds of *Prunux avium,* throws light on this question. He took the weight of 100 buds, at regular periods throughout the year, and thus got the following figures; the column o gives the weight in grammes— ii. *a.*

July 1st........ 1 January 1st...... 4
August 1st...... 2 February 1st 14
September 1st.. 3 March 1st...... 6
October 1st 4 April 2nd *ii*
November 1st.. 4 April Hth.. 43
December 1st.... 4

There are thus two periods of growth, and a resting period between. The first period of growth is while the buds are developing, from early summer to October. Then a period of rest, from October to the end of January, in which no increase of weight takes place. Then a rapid increase in weight, ending in the opening of the buds early in April. So much it is necessary to know, but the really interesting point is the effect of forcing the tree at various times of the year. Thus, branches removed at the end of October, and put in water in a warm room, showed absolutely no tendency to break into flower. In December they could be forced, and as time went on the buds showed themselves more and more amenable to the influence of temperature, proving that the invisible process of preparing for the spring was steadily proceeding. The following table gives the number of days necessary to make the cherry branches flower at different dates:

December 14th 27 I March 11th loJ
January 10th 18 March 23rd 8
February 2nd 17 April 3rd 5
March 2nd 12

This result shows how greatly the effect of heat depends on the invisible changes in the plant. A fact which is sometimes summarised, as I have previously pointed out, by saying that periodicity depends on the plant, not on the conditions of life. This is, of course, an exaggeration—like so many proverbs.

Some of the best instances of the plant being the master in this matter of periodicity are to be found in tropical vegetation, and, indeed, among the very phenomena by which it has been attempted to prove the opposite, *i.e.,* the paramount importance of the conditions. Thus, it has been said that in an uniform climate the trees are in flower, fruit, and leaf throughout the year. In other words, that the plants show no periodicity because the conditions are not periodic. This, according to Schimper, *l'rianzen Geoijraphie,* p. 262, is far from the truth. The trees undergo regular periods of rest and activity; but these periods are not attached to any special periods of the year, but occur at different times in different individuals. Thus he describes (p. 2G1), at Singapore, trees of *Pohireana reijia* in full leaf, growing side by side with specimens which have shed the whole of their foliage. There is, also, the curious case observed at Buitenzorg of *Urnxtigma,* which drops its leaves about every two months, a process which certainly is not connected with periodic changes in climate. This individuality in periodicity is sometimes found, not only in different trees, but in different parts of the same trees. The mango shows this well, the young leaves are red, and the full-grown ones dark green, so that a leafing bough is clearly visible as a red patch against the dark green of the rest of the tree. Schimper studied European fruit-trees growing at Tijbodas *(loc.cit.,* p. 26G), in Java, perhaps the most uniform climate in the world. Each branch had its own cycle of periodicity, so that he found spring, summer, autumn, and winter branches on one tree. The phrenologist would say that this proves his case, since the behaviour of the tree «« *a whole* is quite different in the variable climate of Europe. But if he were asked how he accounted for the periodicity of each single branch, he would be puzzled to do so by an appeal to climatic differences. The truth is that changes occurring in the plant must be considered as the primary thing, while the rhythm of growth depending on these may either be independent of the external conditions, or it may become attached to, or associated with, periodic climatic changes. It is probable that the rhythm of growth may be controlled by, or associated with, all kinds of circumstances. Thus, I have no doubt that the tendency in many trees to flower before the leaves appear is an adaptation to the means of fertilisation. I am thinking of such cases as the hazel, where the pollen is distributed by wind. It is obvious that the pollen will find easier access to the stigma before the female flowers are hidden by the leaves. Thus it has come about that the flowers develop at & lower temperature than the leaves. It is this sort of fact which has led to the statement to which I have already referred, that the optimum temperature for flowering is lower than that for leafing. But this seems to me a wrong way of putting the thing. It is not that the act of flowering essentially, or fundamentally, requires a lower temperature, but that since it is to the advantage of the plant to flower before its leaves appear, the act of flowering has become associated, or correlated, with a low temperature. The plant wants to flower early because this favours fertilisation, and it is guided in recognising the early part of the year by the lowness of the temperature. Thus its reaction to low temperature is a secondary, not & primary, phenomenon.

I believe that the same act of association has taken place in the willow; but here the advantage of early flowering, is that the tree is rendered more conspicuous, and thus more attractive to insects. Other cases of association may be found among European plants grown in tropical climates. Thus it has been found that each plant, instead of resting during the cool part of the year, does so during the dry season, thus one controlling circumstance has been substituted for another.

In all these cases, as I have already said, we distinguish a quality of periodicity residing in the plant, which shows itself in a tendency to repeat the great features of vegetation at regular intervals. We distinguish this, I say, from the tendency in the rhythm to attach itself to periodic conditions of climate, or other external circumstances. When I come to the rhythm observable in the movements of plants, it will be easy to show you the difference between what is purely internal and what depends on external change. For the phenomena we are considering this is not so easily done, but I may at least give one instance taken from Sehiibeler's book on the vegetation of Norway. Schiibeler obtained seedE of a variety of maize from Stuttgart, and sowed them in Norway, on May 26th, 1852. In that year, the crop was ripe on September 22nd, *i. e.*, after 120 days. He continued to cultivate the variety, and found that it ripened each year in a shorter period,:so that by 18.r)7, only 90 days had elapsed before it was ready for harvesting. In order to check his results, he had obtained more seeds from Germany, and sowed them alongside of the descendants of the 1852 plants, and found that they required 122 days to ripen; this proves that the shortening of the period of the acclimatised plants was not due to any peculinrity of the season in 1857, since the imported seeds still (as in 1852) took 122 days to come to maturity. Here we have evidence of the periodicity of the maize being a hereditary quality. For in spite of the change of climate, it contained its German rhythm in Norway, and only gradually lost it.

I now propose to show you that the same general laws of periodicity hold good in a different class of phenomena, namely in those which are associated with a daily, not a yearly, period of time. I shall begin with what is known as the sleep of flowers, *i.e.*, the power which certain flowers have of closing in the night-time. There is, I think, uo doubt that this is an adaptation for the protection of pollen from rain, and possibly from cold. I have said that the flowers close at night, but their movement is not strictly regulated by the diurnal alternation of darkness and light. A bed of crocus, with every flower closed, is a familiar sight on a cold wet day. Thus, though they are always shut at night, they are not necessarily open in the day. The phenomena are complicated by the fact that they are partly due to changes of temperature, partly to changes in illumination. If, as I have suggested, the closure is primarily a protection against rain, it is easy to understand that when the sun is overcast and the rain threatens, some species would be more sensitive to the fall in temperature, others to the diminished light, just as a blind man would make his weather forecast from the sudden chill preceding a shower, whereas a normal man, who happened to be indoors, would only be conscious of the darkened sky. I should like, once for all, to say that I do not attribute consciousness to plants, but that I do attribute to them a sensitiveness which does not differ in kind from that of the lower animals. So that, just as I should say that a worm perceives the difference between light and darkness, I should not hesitate to use the word perception in regard to plants. In this sense, it may be said that a crocus perceives a difference of 1C, or, in other words, that it can be made to open slightly by raising the temperature of the air 1. I have never observed this degree of sensitiveness, but I have seen a tulip open and shut when thp temperature varied 2C.

From our point of view the most interesting experiments are those which show the different effect of heat at different times of the day. Thus, in the morning, a snowdropt can be made to open by raising the temperature, whereas, in the evening, the same increase of temperature has practically no effect. The experiment was performed on a plant placed in the dark, so that the change in environment was strictly limited to temperature by the same treatment. This case is precisely similar to Askenasy's experiments on cherry buds, where attempts to force them fail in the autumn, and only succeed when the natural period of growth is approaching.

"Die Ptianzenwelt Norwegens," 1873, *Ahjemeiner Theil,* p. 80. Since the above passage was written I have read Wille's paper in the *Iiiolog. Ceiitraiblatt,* 1905, where he shows that no great reliance can be placed on Schiibeler's results. + Pteffer, *Phytiolugitche Untertuchunyen,* p. 194.

In both cases the results are to be explained by assuming that the inherent periodicity of the plant has become attached to the cycle of the plant's existence, so that a certain act of the plant, *e.g.,* opening of the flower or growth of the bud, has become attached to, or correlated with, a certain season, *i.e.,* morning in one case, spring in the other. Tbe important point is that the plant is stimulated to open not exclusively by the natural accompaniments of morning, *i.e.,* light and heat, but also by the arrival of a certain stage in the cycle of internal changes proceeding in the plant. This may be further illustrated by the following experiment. If a patch of daisies be darkened in the evening by covering them with an inverted box, they will close in the usual manner, but they will open next morning in spite of the absence of the accustomed stimulus of light. The act of opening in the morning has become associated not merely with the return of light, but also with a certain stage in the nutrition of the plant, just as a man may know that it is dinner-time by his sensations, and not solely by the external stimulus of the dinner bell.

The same class of phenomena may be better studied in the sleep of leaves. The utility of the sleep of leaves is not quite clear. The

Fio. 1 shows a, species of *Ca»ia* in the sleeping and wakinK position.

characteristic change, occurring as night comes on, is that the leaf ceases to be more or less horizontal, and takes up a roughly vertical position. This may be either effected by a dropping, or rising, or twisting into the nocturnal position.

In any case the result follows that the leaf is less exposed to radiation at night, and is, therefore, slightly warmer than if it had remained horizontal. In evidence of this difference of temperature, the leaves of sleeping plants may lx found dry in the morning, when the horizontal leaves of other plants are covered with dew. We have not time to go into the possible value to the plant of the sleep movement, but must consider it from the standpoint of periodicity. What I have described in the case of the daisy is especially wellmarked with sleeping leaves, *i.e.,* if darkened in the afternoon they will be found to have taken up the diurnal position next morning, although still kept in darkness. What is more remarkable, they may be kept for several days in the dark, and continue to show some periodicity, though the movements become slighter and more irregular, and finally the leaves come to rest in a position half-way between those characteristic of day and night. The plant has been compared witha pendulum,keptswinging by a force applied at regular intervals. Such a pendulum will continue to swing with diminishing amplitude when the rhythmic force has ceased to be applied. I am told that this analogy is not strictly admissible, but to the nonmechanical mind it is certainly a help.

I have lately made out an interesting case of habit in a sleeping plant—the scarlet-runner *(Phaseolits).* Like other plants, it adapts itself to one-sided illumination by placing its leaves obliquely, so that they are at right angles to the line of illumination, and get the full advantage of the light. In fig. 2, the arrow represents the direction of the incident light, and it will be seen that the leaves are at right angles to it. The two right-hand plants are supposed to be in a dark cupboard. The figures do not, of course, represent *Phaseolns* leaves, but are merely diagrammatic.

If a *Phaseolus,* which has assumed this oblique position, is allowed to go to sleep at night as usual, and is then placed in a dark cupboard, it will, next morning, as already described for other sleeping plants, assume the diurnal position in spite of the continuous darkness. What is especially interesting is that it does not return to its normal day position, *i.e.,* with horizontal leaves, but takes up the oblique position already shown.

This is a remarkable case of what looks like a reminiscence of its former position. It is interesting, psychologically, since it might almost be described as an instance of a plant taking advantage of its individual experience. Personally, I think it unnecessary to make this comparison. Another experiment, published by Miss Pertz and myself," some years ago, has some value in this connection. It is well known that if a strongly growing vertical shoot is placed horizontally it bends geotropically till it is once more vertical.

We took a shoot of this sort and fixed it to a rotating horizontal spindle, which was controlled by clock-work, so that it made a sudden half-turn every half-hour. During the first half-hour the plant will be stimulated to curve upwards. During the next period it will again be stimulated to do the same thing, but, owing to the spindle's rotation through 180, the new upward curvature will be in the opposite direction to the result of the first stimulus. Thus the plant is subjected to alternate reverse stimuli, occurring at intervals of half-an-hour, with the result that the shoot curves up and down in a half-hourly rhythm.

In fig. 8 the rhythm is not yet regular, but it shows some curious points. The strong horizontal lines give the times at which the spindle of the machine made a half-turn, the fainter *AnnaU of Botany,* 1892.

lines divide the spaces into three periods of ten minutes each. The letters U and D show the direction in which the plant is curving. Thus, in lowest space, the line (marked with an arrow-head) is directed obliquely towards D, signifying that the plant curved downwards for 25 minutes; then came the half-turn of the spindle. The plant continued its curvature unaltered, as is evident from the line running on into the third period in its original direction. But now the curvature is upward, owing to the rotation of the 3pindle. In the middle of the third period the curve is reversed, *i.e.,* the plant curves down, and continues to do so for about 25 minutes, until, in the middle of the fourth period it again begins to bend down. Here the plant had not completely learned the half-hourly rhythm, though in its last swing in the fifth and sixth periods, its movement in one direction lasts 2H minutes, which is fairly accurate.

In Figure 4 the shoot is swinging to and fro in almost exactly a half-hourly rhythm. It is a striking experience to sit watching a plant in this condition, and to be able to prophecy that at a certain time it will reverse its curvature, and to see one'3 prophecy fulfilled. Here the alternating conditions of life have built up a rhythmic condition of a definite period. But the most interesting fact is obtained when the clock-work escapement is thrown out of gear, so that the spindle remains stationary. It might be expected that the rhythm of the plant would cease, but this is not the case. In fig. 4 it will be seen that in the three upper periods the letters D and U do not change sides, all the D's being on the left, and the U's on the right. This is the graphic method of representing that the spindle no longer rotates. It will be seen that during the last hour-and-half the shoot reversed its curve twice, at intervals of almost exactly half-an-hour.

To me the interest of this fact is that it proves the existence of purely internal periodicity. To take an example from human habits: Most people are accustomed to wake at a certain time each morning without being called, and this is no doubt an internal rhythm. But we cannot be certain of this, because there are a number of changes in our environment which might give a signal to our bodies to wake.

But in the case of our geotropic plants, we are certain that there are no external changes occurring, at intervals of half-anhour, which could stimulate the curving shoot after the clock had stopped. There must, therefore, be a time-measuring capacity in the plant, and this internal chronometer I imagine to be connected with the nutritive changes going on. I am convinced that our artificial rhythm is part of the same great class of phenomena as those I described in the earlier part of this address, where I insisted that annual periodic phenomena can only be understood by reference to an internal or automatic periodicity, which is capable of becoming adherent to, or correlated with, a rhythmic environment.

One of the most beautiful instances of automatic movement" occurs in *Arerrlwa bilimbi,* a shrub belonging to the *O. ralvlae,* a leaf
Fig. 5.
of which is shown in fig. 6. When the plant is kept at a temperature of about 25C. the leaflets are found to be in constant move The movement was first described by Mr. Lynch, of the Cambridge Botanic Garden, see *TAnn. Snc. Journal,* xvi., 1877, 231. ment, each acting independently, falling suddenly through 20 degrees, and rising slowly to its original position.

Fig. 6 gives a graphic representation of the dropping of an *Averrhoa* leaf as it goes to sleep. The numbers at the sides give the angles with the vertical made by the leaf under observation. It will be seen the fall to the vertical position, *i.e.,* to the zero line, is made up by a series of rhythmic movements, in each of which the fall is greater than the rise. Similar, but much slower.
movements may be seen in such plants as clover and the scarlet runner.

The rhythmic movements described in the " Movement of Plants" are of the same type as those of *Arcrrhoa,* although differing superficially from them. The typical case is that of a twining plant. Everyone is familiar with the appearance of the terminal shoot of a hop or a scarlet runner; it is not vertical, but hangs over to one side. It is in constant movement, thus it might, for instance, be pointing N., and if examined a little later, would be found pointing N.-W., then W., then S.-W., and so on, till it had swung round to its original position.

In the " Power of Movement in Plants" it was shown that circumnutation on a much smaller scale exists throughout the vegetable kingdom. There can, indeed, be no doubt that the revolving of twining plants is nothing more than an amplification of the rhythmic manner of growth common to plants in general. My father's point of view, which has not been generally accepted, was that the curvatures of plants upwards, downwards, to the light, or in response to other stimuli, are merely developments of circumnutation; so that circumnutation was considered as the basis from which other movements have been evolved. It is, perhaps, permissible to say that, as the evolution of a new species depends on a basis of variability in structure, so a given curvature may be held to depend on the rhythmical variations in position, which we call circumnutation.

Circumnutation is, as it were, the raw material out of which movement in response to stimuli has been developed. This is the widest, and in one sense the most important application of our knowledge of rhythm—a quality which, as I have tried to show, dominates the most diverse phenomena in the life of plants, and with this example, I will bring my address to an end.

Nature Near Eastbourne.
By J. H. A. JENNER, F.E.8.

The town of Eastbourne possesses exceptional advantages for the study of all branches of natural history. Among these advantages may be mentioned its southern latitude, varied soil, and geological formations, its mild and equable climate, and, like other Sussex watering-places, its large amount of sunshine—and we may add to these the remains of the works of prehistoric man, which are so numerous.

The district which has been worked by the Eastbourne Natural History Society, is, roughly speaking, bounded by Seaford on the west, St. Leonard's on the east, and Cross-in-Hand, Heathfield, and Dallington on tlu north; but, for the purpose of this paper. I shall take a district round Eastbourne, which is within an easy day's journey. This will practically include, as it were, an epitome of the characteristics of nearly the whole of Sussex.

To begin with the geological formations:—We are on the southern margin of the wealden formation, with its numerous Woods on the weald clay. We have also, close to us, representatives of nearly the entire cretaceous formation. *vh.,* the gault, the upper greensand, the lower chalk, and the upper chalk. In addition to these we have a remnant of the tertiaries in the capping of Woolwich beds at Seaford and Newhaven. We have also the more recent post-tertiary beds, near the Wish, with remains of rhinoceros, hippopotamus, and elephant, specimens of which may be seen in the Museum. This formation is analogous to the wellknown elephant bed of Brighton, and was evidently produced in some way by the destruction of the chalk hills, long subsequent to their elevation from the bed of the sea. Then again, we have what are called submarine forests, the relics of a former land surface destroyed by inroads of the sea, and now to be seen at lowtide at various places on the coast between Eastbourne and Rye. Next we have the very curious collection of shingles, called the Crumbles-produced by the continuous action of the sea currents from west to east, which have carried eastward, and at times quite blocked, the mouths of all the Sussex rivers, and, in this instance, have blocked in that rich expanse of alluvial land called Pevensey Level. Beachy Head has evidently, in this instance, acted as a natural groin for this accumulation of shingle.

The gault, the lowest member of the chalk formation near Eastbourne, was formerly exposed on the front, but is now walled up. It exists as a narrow band, skirting the downs between Eastbourne and Berwick, and there is a somewhat larger exposure near Glynde.

The upper greensand also occurred on the front, but is now obscured by the parades, though it is still in evidence about the town, and its hard, grey-green stone is frequently used for walls.

The series of the various layers of the lower and upper chalk proper, is well seen between Eastbourne and Seaford — and especially in the fine cliff of Heachy Head and the Seven Sisters—forming one of the most beautiful sea views in England.

The South Downs are entirely composed of chalk. Each formation has its special characteristics. The chalk is remarkable for the beautiful rounded contour of the Downs—with their dry valleys, or combes, now entirely without streams, though undoubtedly formed by water action—also for its peculiar flora and fauna. The hills near Eastbourne rise to between 500 feet and 600 feet. The weald consists of flat clay land, mostly productive of wood, while its harder members produce the lofty hills around Crowborough, and the cliffs of Hastings.

Standing, as we do, on the edge of the English Channel, we are tempted to ask—whence the channel? This, though comparatively recent geologically, is relatively very ancient—and had, of course, a momentous effect on the natural history of these islands. It is evident that the chalk was once continuous with that on the shore of France, as the weald formation was probably continuous with that of Belgium. It is to this fact that we owe the possession of many plants which came in in this way, and are now left stranded on our island.

There is one small river, the Cuckmere, in the district. This little river, rising in the weald by two small streams, one near Waldron, the other near Heathfield Park, flows southwards, and enters the sea at Cuckmere Haven, between Eastbourne and Seaford. It is an admirable example of the way in which the Sussex rivers were formed by the raising of the anticlinal of the weald, which formed rifts in the chalk formation, through which the Sussex rivers flow. In its upper reaches it has no special characteristics, but as it nears the sea below Alfriston, its narrow valley, with its estuarine cliffs and its shingle bank pushing its mouth continually eastward, close against the sou cliff, are worthy of attentive observation.

Including man in the scheme of "Nature near Eastbourne," we may pass on to the traces of prehistoric man which we have in the neighbourhood. Of the Palaeolithic Age, which a writer has described as " almost below our horizon," there are few traces in Sussex, although undoubted implements have recently been found in the west of Sussex, in the river gravels of the Rother, and a few have been picked up near Friston. It is possible that the river terraces of the Cuckmere, of which there are a few, might yield examples, but they are not easily explored, being mostly grass land.

Of the Neolithic Age, remains are most abundant. There is hardly a field on the Downs where flakes, scrapers, and other flint manufactures may not be found, showing how extensively the Downs were populated in those days—but many spots are much richer than others—evidently camping grounds, and splendid celta and other weapons have been found near East Dean, Exceat, Birling, Seaford, etc. There is a beautiful knife in the Museum from Southbourne, which in workmanship equals some of the best Danish examples.

We have also, of perhaps rather later age, the Kitchen Midden, at Newhaven, and the very interesting Rock Shelters in the cliffs at Hastings (specimens of remains from the latter may be seen in the Hastings Museum).

We next come to a very distinct and interesting series of antiquities, the Hill Camps. These have recently had much attention given to them. They are probably rather later than the true Neolithic Age, as we find, at Cissbury, the Neolithic shafts for obtaining flint have no connection whatever with the very fine entrenchments. There are good examples at Birling and Seafonl, and others on various parts of the downs-one of the best defined being on Mount Caburn, between Eastbourne and Lewes. The question to be considered is—What was their object? Were they places for observation, or living places, or places of refuge? Also, why have such prominent heights as Firle Beacon, or the hills above Willingdon, no traces'? Again, if places of residence, how did the inhabitants obtain water? To answer the last question, a recently published book has pleaded the antiquity of the dew pond or sheep pond, which is still a familiar object on our hills, and has the peculiarity of filling with water from condensation from dews and fogs. They are still used, and a family at Alfriston has had the reputation of being especially proficient in making them for the last three or four generations.

These earthworks, Messrs. Hubbard say, were made by men they call earthworkers, in contradistinction to the stoneworkers, who were later, and constructed such monuments as Stonehenge, but the principal reason for earthworks in this neighbourhood appears to have been the absence of stone. It is possible that they were constructed from the Neolithic Age through the Bronze Age. Manyother marks of early life may be seen on the Downs, such as the socalled Neolithic areas of early civilisation, and barrows and tumuli — places of early sepulture—which are very numerous. Also, we have the very remarkable figure on the side of the Downs called the Wilmington Giant, concerning which all sorts of explanations are given, but I think it is undoubtedly ancient, and probably coeval with the earthworkers. For a long time our figure was almost unnoticed.
and its form could only be seen in certain lights, but it has now been outlined with white bricks, and forms a conspicuous object from a distance. There is a somewhat similar figure at Cerne Abbas. Numerous relics of the Bronze Age have been found in the neighbourhood. Some years ago many bronze celts were found near the submarine forest at Pevensey, and other specimens are in the Archaeological Museum at Lewes.

In a paper I read a short time ago at Brighton, I said that it was doubtful whether Sussex had any prehistoric stone monuments, and suggested that some Sarsen stones might possibly be such. Since then, Mr. Toms, of the Brighton Museum, has given the matter some attention, and it seems probable that some of these stones may have been portions of so-called Druidical Circles—we have such names as Standean, Stanmer, the Goldstone, also Amberstone, near Hailsham, and the idea of Great upon Little, near West Hoathly, having been a Druidical Deity is well known. These Sarsen stones, which are certainly relics of the lower beds of the tertiaries, do not appear to be numerous mar Eastbourne. Of the early Iron Age specimens are not numerous, and are never very certainly ascertained, as a rusty iron object, apart from other confirmation, may be possibly of any age.

Passing on, we come to the Boman occupation of Britain. Of this there are numerous evidences near, and in, Eastbourne; as early as 1717 a Boman pavement was discovered on the site of Cavendish Place, and minor discoveries have continued to be made since. A considerable Bomano-British cemetery existed at Seaford on ground now used as golf links. Hut the finest Boman relic is undoubtedly the fortress of Pevensey Castle, one of the most perfect Boman constructions in the country. Here we have evidences of what I have mentioned before, that is, the continuous and successive occupation of certain suitable sites. Just as the Neolithic or Bronze Age camps were used in succession by Bomans and Saxons, so here we see a Neolithic site with remains of flint implements, and to my mind something like a crannoge, occupied subsequently as a Boman fortress, and still later by an early English castle.

Pursuing our course downwards, Saxon interments have been discovered near Eastbourne, and Saxon coins very frequently near Alfriston, Wilmington, and other places. Later, we have admirable examples of Norman work at Hastings, Lewes, and Pevensey Castles, besides in many churches in the neighbourhood. Mediaeval work is abundant, but mention may perhaps be made of Wilmington and Michlcham Priories, the Clergy House and Star Inn at Alfriston, also the Market Cross at the last-named place.

Eastbourne is a town of modern growth, though Borne is mentioned in Domesday. Its great growth has been certainly since 1800. Very few houses stood on the Front then—it is said there were only seven in 1760. This modern progress has, of course, rather tended to drive nature away from it, but still not to any serious extent. The richness of the surrounding neighbourhood will be proved by the large lists of birds, plants, and insects recorded in the recently published *Victoria County History of Sussex.*

Commencing with the plants, as all living nature is dependent upon, and starts from, them, I could say much, but a paper dealing more directly with the subject is to follow. The varied soil, of course, has its resultant effects on the fauna and flora. Anyone travelling by rail cannot fail to observe the change in vegetation as he passes from one formation to another.

The woods in the cretaceous portion of the district are few and limited in extent, but on the wealden beds they are still rather extensive—they are all probably relies of the great forest of Andred, which is said formerly to have covered the whole of the weald, and probably extended, with few interruptions, westward to the New Forest, the insect faunas of the two districts possessing great resemblance. This forest was considered almost impenetrable, but it probably had its open spaces and backways, as it is not uncommon to find flint implements in some parts of it.

On the top of the Dowrns one notices at once the shortness of the turf and the dwarfing of the general herbage—it is only where rainwash has accumulated soil on the slopes and in the valleys that one finds any luxuriant growth. It is probable that the tops of the Downs have always been treeless, both from shallowness of soil and effect of wind. The slightest observation will show how markedly the trend of all our trees in this district is away from the southwest, the direction of our prevailing wind. The Downs afford an opportunity for the investigation of many problems in natural history, *e.y.,* the struggle for existence among plants; the way in which one plant stamps out its opponents, while another avoids competition by raising itself above the common herd; the re-peopling, as it were, of bare patches caused, for instance, by beacon fires, or the cutting of turf, is also an interesting study; the arrival of new species in various ways, and their eventual extinction by the inured veterans of the soil, may sometimes be witnessed. Then, again, how do the isolated dew ponds get their sometimes numerous population? During strong winds the distribution of seeds over wide areas is facilitated on these hills, and one may frequently see beech leaves miles away from the trees which produced them. The cryptogams of the district have been fairly well worked out. and more especially the mosses, by Messrs. Dixon and Jameson. Lists of algre and lichens may be found in the *Transactions of the Eastbourne Natural History Society*. The marsh ditches of Pevensey Levels produce many Desmids and *Diatowaceae*.

But we must pass on to the animal kingdom, and, as a paper on birds is to follow, I will only say that Knox records that Beachy Head in his time was the haunt of, among others, the raven, the chough, and the peregrine falcon; these have perhaps disappeared, but, with the preservation act in force, may re-establish themselves, as many of the smaller birds are doing. The most famous Eastbourne bird is, perhaps, the wheatear, of which thousands used to be captured by the shepherds and sold as the English ortolan.

I am not able to say anything of the fishes, or of the marine mollusca, but I have no doubt plenty of material for study may be found. Of the land and freshwater mollusca—of which there are about 106 Sussex species—a large proportion may be found near Eastbourne. I contributed a list of the Eastbourne species to the *Transactions of the Eastbourne Society* some years ago, and I fancy very few additions have been made since. I may, perhaps, mention a few of interest. Of *Helicella barbara*, which is common in the southwest of England, I discovered a small colony near Mill Gap, some years ago, the only Sussex locality—the spot is now entirely built over—but if not an intentional introduction, the species should occur elsewhere in the neighbourhood. *Helicella itala* and *H. I'iri/ata*, especially the latter, occur sometimes on ihe turf in incredible quantities, so that it is almost impossible to walk without treading upon them. *Heliciyo-na lapicida*, usually a limestone species, has occurred under faggots near Abbot's Wood, and *Helix nemoralis* may be found with numerous fine varieties, especially in the beech hangers of the Downs. The marsh ditches yield most of the species of *Limnaea* and *I'lanorbu*. One may note the absence of *Helix pomatia*, the Eoman snail, which is said to occur in places of Roman occupation —also of //. *obvoluta*, a shell which is found sparingly on the north sides of the Downs in West Sussex. I think it might possibly be found in this district.

The insect world is probably as well represented around Eastbourne as in any part of England, indeed the lists recently published in the *Victoria County History of Sussex* are as large as, or larger than, those (or any other county. As is natuial in the extreme south of a country, many species occur that are on the extreme limit of their distribution, and, in some years, the number is added to by various immigrants.

We will take the Lepidoptera first, and begin with the butterflies. It is astonishing what a stumbling block butterflies are to the man in the street, and even to the press. Only a short time ago I heard a man say " I didn't know there were so many butterflies—I know a white one, a brown one, and—after a long pause—there's a yellow one!" And just lately a London paper reported that "a butterfly, one of the scarlet variety," had been caught in the Twopenny Tube. We have about 50 species recorded from our district out of the small and select company of 68 recorded from the whole of Britain. Of course, our insular list is very small compared with that of Europe. For the purpose of my paper I have divided the list as follows:—

General.... 16
Casual visitors.... 6
Chalk species...... 6
Woods and lanes.... 16
Extinct, or nearing extinction 5
Doubtful 3

Commencing with the casual visitors we have *Pontia daplidice, Aryynnis lathonia*, and *Euvanessa antinpa*, all of which have occasionally appeared here, and *'olias hyale* and *C. edusa*, also *Pyrameit rardui*, which sometimes appear in great numbers, apparently immigrants. I used to think all our species were always with us, merely varying in rarity from year to year. If we consider the causes for or against insect abundance, it seems possible that it is so. If we reckon the factors for or against insect abundance as 100, and they probably are as many—when say 75 are for, and 25 against, we get the insect in perhaps its normal abundance, and it is easily seen how potent the variation in these numbers would be. What these factors are we do not always know, some, perhaps, are inscrutable. It seems, however, that many species would cease to exist were they not recruited from abroad. The chalk species—*Polyommattu eorydun, P. bellargus, P. astrarehe*, and *Cupido minima*, may all be found, not uncommonly, on the Downs, also *Aryynnis aylaia* and *Hipparchia semele*, which last has an interesting habit of mimicry, settling on discoloured chalk stones, where it is discovered with difficulty. From the amount of wild thyme on our hills it is possible that *Lycaena avion* may have formerly occurred, if so, it has long been extinct. Among those of the woods and lanes several of great, interest occur: *Apntura ins* sometimes sails over the oaks at Abbot's Wood, where also *Limenitis sibyllo* is rarely seen. *Melitaea athalia*, an exceedingly local butterfly, is still seen at Abbot's Wood, but not in the numbers that it once was; *Melanaryia yalathea* is sometimes seen in numbers, and occurs in one or two spots on the Downs; *Thecla w-album* is rare, and *Xeweobius litrina* is very local. None of the general species need special comment, except *Emjonia polychlaros*, which seems to be getting commoner again. The whites, *Pieris brasxicae* and *I'. rapae*, would at times be serious pests were they not kept in check by a small parasitic ichneumon. As to those extinct, or nearing extinction,

Ijeucophasia sinapix, formerly common, has not been seen for years, and the same may be said of *Aporia rrataeyi. Melitaea aiirinia, Polyyonia c-albmn,* and *Nomiadcs semianjus* are hardly ever seen, and some others seem to be lessening in numbers and gradually disappearing—why is this? Probably there are many reasons, perhaps some are climatic, but the most potent are the changes in character of the land, drainage, and the encroachment of cultivation and bricks and mortar on the country. If the last be the reason we can only console ourselves, as was done when a rare plant was threatened with extinction:—

"Better is the straight onward furrow of progress than a thousand wild lobelias."

The doubtful records are *Papilio machaon* and *P. podaliritu,* almost certainly escapes from someone's breeding-cages, and *Melitaea cinxia,* most likely an error, though it occurred on the coast at Folkestone, and still occurs on that of the Isle of Wight in similar places to Cow Gap.

It would be impossible for me to go through the list of moths in this short paper. Of the hawk moths we have *Manduca atropos, Agriux convolculi, Celerio gallii, I'hryxux livornica,* and *Hippotion celerio,* all of which five species are probably like some of the butterflies we mentioned, foreign immigrants. The electric light has a great attraction for them. Two of the Anthrocerids may be mentioned, viz., *Rliagades globidariae* and *Adscita yeryon,* which exist in small local colonies on the Downs. They never seem to spread far from headquarters, though their foodplant is widely spread. The first-named is evidently a species on the extreme margin, although the latter goes much farther north, of its distribution. *Fuichelia jacobaeae,* formerly rare about here, but common in the forest region, has lately become common nearer the coast. Again, why this change? Are the newcomers immigrants? Among the Noctuids, a class of insects most frequently taken by "sugaring" trees, are *Moma urion, Aeronycta alni, A. auricoma* (extinct?); *Ccrastis erythrocephala,* of very rare and uncertain occurrence, and the exceedingly rare *Opliindes lunarn* and *Catephia alchymhta.* All the species of the genus *Taeniocampa* have been taken, mostly at sallow bloom; and on the Downs such rare and local insects as *Agrotis cinerea, Aporophyla australis,* and *Luperina cespitix* find their headquarters. Among the Geometrid moths, *(tnophos obxenraria,* on the chalk, assumes a light tint, being black in peaty districts, a fine instance of protective resemblance, another phase of which occurs in *Bryophila ylandifera,* which, settling on the greensand walls, is almost invisible. *Aeidalia immorata,* discovered in 1887, has been found in no other locality in England. Of the Pyralides—*Ayrotera nemoralix* was suddenly common at Abbot's Wood in the "seventies," but is now either very rare or extinct. The *Tortrieina* and *Tineina* have been well worked, but I must refer you to the very full and comprehensive lists in the *Victoria County History.*

Of the other orders of insects, the orthoptera call for little notice, except that, I suppose, the introduced pest, *Blatta orientalist,* can be found by anyone in want of it! Time will not permit to say much of the hymenoptera, except that many species of humblebees abound, and at this time of year (early June) *Andrenidae,* with their parasites, the Nomadas, are much in evidence. The list of coleoptera is a large one, and very many fine species occur. Of the diptera we have the largest list of any county in England; many more species of this neglected order are certain to occur.

A visit to the woods, say Abbot's Wood, at this time of the year is always instructive and pleasant. We may find among plants— *Phyteuma spicatum* and the local *Ranunculus intermedins,* and in the clearings a profusion of flowers—one year it is *Viola tricolor,* another *Aquileyia vuli/aris,* another *Ai/raphis nutans,* but always the primrose. Attention to these flowers will produce many of the butterflies I have mentioned, including *Melitaea atlialia,* and the large clearwing moth, *Hmaris fuciformis,* accompanied by a large humble bee, which somewhat mimics its flight. Then the nests of the wood ant, *Formica rufa,* will perhaps produce the rare ladybird beetle, *Coeciiulla labilis,* which has the curious habit of living in the nests. The hawthorn, now in blossom, will deserve a visit. It has been said it is the only flower that attracts both men and flies, it will certainly attract plenty of flies; shallow pools will produce *A;iahi, Ih/tisci,* and other water-beetles; sitting on the treetrunks, and mimicking the lichens, the careful eye will detect many species of moths; beating the undergrowth and "whacking" branches will dislodge others. Then, as dusk comes on, many other insects will come out to be captured by net or sugar.

I think I have said enough to show that Eastbourne is a remarkably fine centre for nature study, and although the increase of the town has necessarily driven away many things, yet we must congratulate the authorities on having perpetuated, and also on having, as it were, led back into the town many natural beauties. As to nature study—it should be study, not destruction—a red admiral on a bramble flower, with all its wondrous colouring and its mysterious living organism, is a far more splendid object than a specimen in a cabinet, though for purposes of study specimens are absolutely necessary. I must plead guilty to having acquired many specimens, yet when one thinks of it the destruction of any life becomes rather repulsive, and one realises that "a living dog is better than a dead lion." Nature study needs great attention to minutite and detail, the most trivial fact may be of use. Is it worth while? Is it worth the trouble? I say yes!" It is better to be great and attend to little things than to be little and only thinking of great things." Nature study opens a fresh world to its attentive follower. To nature students even "The wilderness and the solitary place shall be glad for them, and the desert shall rejoice and blossom as the rose."

Comparison Between The Sussex And The British Lists Of

Birds.

By W. RUSKIN BUTTEHFIELD.

There can, I think, be little doubt that the total number of species and sub-species of birds known to have occurred in Sussex is greater than that of any other county in the British Islands. For many years the county of Norfolk held the position which I here claim for Sussex; at the present time, however, the Norfolk avifauna comprises 815 forms (as I am informed by Mr. J. H. Gurney), against 328 for Sussex. In estimating the total number of British birds, I have followed, for convenience, Mr. Howard Saunders' *Illustrated Manual,* 2nd ed., 1899, wherein 884 forms are admitted, while 21 have since been added, *viz.* :—Mediterranean Shearwater, *I'uffinus kuhliikuhlii* (Boie); Pevensey, Sussex, February 21st, 1906 *(Bulletin of the British Ornithologists' ('lub,* xvi., p. 71; April 2nd, 1906). Larger Snow Goose, *Chen nivalis* (Forst.); Behuullet, Co. Mayo, September, 1886 *(Bull. B. (). ('lub,* x., p. xv; November 30th, 1899). Baer's Pochard, *Nyroca baeri,* Badde; Tring, Herts., November 5th, 1901 *(Bull. B. (. Club,* xii., p. 25; November 80, 1901). Pacific Eider, *Sumateria v-niijrum,* G. R. Gray; Graemsay, Orkney, December 14tb, 1904 *(Bull. B. (). Club,* xv., p. 32, January 26th, 1905). Baird's Sandpiper, *Jhtrropyiiia bairdi* (Coues); Rye Harbour, Sussex, October 11th, 1900 *(Bull. B. O. Club,* xi., p. 27; November 28th, 1900). Black-winged Pratincole, *Glareola welanoptera,* Nordm.; Jury Gap, Kent, June 1st, 1908 *(Bull. I'. 0. club,* xiii., p. 78; June 30th, 1903). Greyheaded Wagtail, *Motaeilla tiara borealis,* Sundevall; Halifax, Yorks., spring of 1901 *(Bull. B. (). Club,* xiii., p. G8; May 80th, 1903). Black-headed Wagtail, *M. tiava melavocephala,* Lichtenstein; Willingdon, Sussex, May 13th, 1903 *(Bull. B. 0. Club,* xiii., p. 69; May 30th, 1908). Sykes' Wagtail, *M. ftara beema* (Sykes); Rottingdean, Sussex, April 20th. 1898 *(Zoologist,* 1902, p. 232; c/'. *Die Voijel der palaarktisehen Fauna.* Heft iii., p. 290). Dusky Thrush, *Turdus fuscatus,* Pallas (=7'. *dubius,* Bechst.); Gunthorpe, Notts. , October 18th, 1905 *(Bull. B. . ('lub,* xvi., p. 45; January 81st, 1906). Western Black-eared Chat, *Saxicola sta/iazina caterinae,* Whitaker; Polegate, Sussex, May 28th, 1902 *(Bull. B. O. Club,* xii., p. 78; June 30th, 1902). Eastern Blackeared Chat,.S'. *xtapazina staiazina,* Linn6 (=*S. aurita,* Temm.); Pett, Sussex, September 9th, 1905 *(Bull. B. 0. Club,* xvi., p. 22; November 1st, 1905). Eastern Stonechat, *I'ratincula mama* (Pallas); Cley, Norfolk, September 2nd, 1904 *(Bull. B.). Club,* xvi., p. 10; November 1st, 1905). White-spotted Bluethroat, *Cyanecula le.ucocyana,* C. L. Brehm; Dungeness, Kent, October 6th, 1902 *(Bull. B. 0. Club,* xiii., p. 14; October 81st, 1902). Cetti's Warbler, *Cettia eetti* (Marm.); Battle, Sussex, May 12th, 1904 *(Bull. B. O. Club,* xiv., p. 84; May 80th, 1904). Nubian Shrike, *Lanius nubicus,* Licbtenstein; Woodchurch, Kent, July 11th, 1905 *(Bull. B. 0. Club,* xvi., p. 22; November 1st, 1905). Willow Tit, *I'arus atricapiUut kleinsckmidti,* Hellmayr; Finchley, 1897 *(Ornithol. Jahrbuch,* xi., p. 212; *cf. Zool.,* 1898, p. 116). Snow Finch, *Montifrinqilla nivalis,* Linne; Rye Harbour, Sussex, February 22nd, 1905 *(Bull. B. O. Club,* xv., p. 58; March 24th, 1905). Greater Redpoll, *Carduelis flammea rostrata,* Coues; Isle of Barra, Scotland *(Ann. Scott. Nat. Hist..* 1901, p. 131). Meadow Bunting, *Emberiza da,* Linne; Shorehain, Sussex, October, 1902 *(Bull. B.). Club,* xiii., p. 88; January 80th, 1903). Yellowbreasted Bunting, *Emberiza aureola,* Pallas; Cley, Norfolk, September 21st. 1905 *(Bull. B. . Club,* xvi., p. 10; November 1st, 1905).

The following birds, which have been obtained in the county, are not included in Borrer's *Birds of Sussex,* all of them, with the exception of the Golden Eagle, having been procured since the publication of that work in 1891:—Mediterranean Shearwater, Little Dusky Shearwater (Bexhill, December 28tb, 1900—*Bull. B. (). Club,* xi., p. 45); Bulwer's Petrel (Beachy Head, February 3rd, 1903—*Bull. B. O. Club,* xiii., p. 51; and St. Leonard's, February 4th, 1904—*Bull. B. O. Club,* xiv., p. 49); Golden Eagle *(cf.* Saunders' *Manual,* 2nd ed., p. 827); Goshawk; Baird's Sandpiper; Solitary Sandpiper (Rye, August 7th, 1904—*Bull. B. O. Club,* xv., p. 12); Black-winged Pratincole (Rye, June 18th, 1908—*Bull. B. O. Club,* xiv., p. 17); Collared Pratincole; Whiskered Tern; Black-headed Wagtail; Grey-headed Wagtail (Willingdon, May 13th, 1903—/i»M. *B. O. Club,* xiii., p. 69); Sykes" Wagtail; Western Black-eared Chat; Eastern Black-eared Chat; Whitespotted Blue-throat; Icterine Warbler (Burwash, April 80th, 1897— *Bull. B. O. Club,* vi., p. li); Melodious Warbler; Marsh Warbler; Cetti's Warbler; Willow Tit (St. Leonard's, several, 1900— *Bull. B. (). Club,* xi., p. 27); Wall Creeper (Winchelsea—*Bull. B. O. Club,* vi., p. viii; and Hastings, December 26th, 1905—*Bull. B. O. Club,* xvi., p. 44); Snow Finch; Two-barred Crossbill (Westfield, February 28rd,' 1899—*Bull. B. O. ('lub,* viii., p. lix); Meadow Bunting.

The subjoined table shows how the Sussex list of birds compares with the British list. Species like the Crane and the Spoonbill that formerly bred in the British Islands, or in the county of Sussex, are not included in the third and fourth columns. In the order and value of the families I have followed the *Cambridge Natural History,* vol. ix., 'Birds. ' I may add that the total number of Palrearctic birds given in Mr. H. E. Dresser's *Manual* is 1219, and that the number recorded in the English translation of Gatke's *Heligoland* is 896.

'Including Blue-headed Wagtail and Grey Wagtail (Dr. N. F. Tioehurst).

t Including Marsh Warbler *(cf. Hutory of Harting,* lff77, p. 276). : Including Willow Tit (nest found near St. Leonard's by Mr. M. J. Niooll).

J Including Lesser Redpoll and Crossbill.

The Flora Of The Eastbourne District.

By NEVILLE S. WHITNEY, M.B. (Lond.), M.R.C.S., and Miss B. MILNER, B.A. (Lond.).

Botanists are fortunate with regard to the Mora of the Eastbourne, or Cuckmere, district, because it is both comparatively rich and very varied. It may be said to consist of four different classes of plants:— 1. Those growing upon the downs. 2. Those found upon the marshes. 3. The maritime species. 4. The rural plants of the fields, meadows, lanes, etc.

The county of Sussex, containing 1461 square miles, has 959 species in all indigenous to it; the Eastbourne district, containing 160 square miles, had, at the time of the publication of the *Flora of Eastbourne,* by the late Mr. F. C. S. Roper, 780 species, to which have been added, since the publication of the flora, 25 species, raising the total to H05 species. These figures apply to phanerogams only. The Eastbourne district is also a good hunting-ground for the cryptogamic botanist, the Rev. H. G. Jameson, of Eastbourne, having himself collected over 200 species of mosses in the district. The uncommon little adder's tongue fern, *Ophioiflonxum vnlijatum,* was formerly abundant in the marshes east of Lewes Road.

First, with regard to the rare plants, of which the district contains several. Most of the facts concerning these are recorded in the aforementioned flora compiled by Mr. Roper, a work wrhich received commendation from that eminent authority in topographical botany, Mr. Watson. The most interesting of these is the *I'hyteiima tpicatum,* or white spiked rampion, which is only known as a Sussex plant, and only found in this district. It was mentioned in 1633 by Gerarde, and in 1610 by Parkinson. Two centuries later (1825) it was rediscovered on IIadlow Down, near Mavlield, and the next year was recorded at Cross-in-Hand, in the Eastbourne district. Mr. Roper records it in Abbott's Wood, and last year (1905) specimens were gathered in a wood at Horeham Road. They were tall, fine specimens with rather slender spikes of yellowish, greenish, white flowers. Thus it has been established in the district for 80 years, and may fairly rank as a Sussex plant, though there is a probability that it originally escaped from the monastic gardens of Warbleton or Michleham priories, which, however, are at least a mile away from the places where it has been found.

Of the *nine* rare plants found in the Eastbourne district, the next three belong to the natural order *Scrophulariaceae. Sibthorpia eitropaea,* or the Cornish money-wort, has been reported from Cornwall, Devon, and Somerset, but nowhere else in England except East Sussex. It occurs also in South Wales, the Channel Isles, and the south of Ireland. It has been said to occur very sparingly at Waldron, within this district, since 1837, and about 35 years ago was said to be extinct, but specimens were again found in 1875 in the original locality, and again, as late as 1898, by the Rev. W. R. Andrews, of Eastbourne. It is remarkable how so essentially western a plant should be located in this eastern district. The genus named after W. H. Sibthorp, Professor of Botany at Oxford, is a small one confined to this one European species, two from central America, and two from the coast of Africa. It is a slender trailing plant with small pinkish or yellowish axillary flowers. *Bartsia ciscosa* has a restricted range, being found only upon the western side of Great Britain, except for the one locality in the Eastbourne district, a lane near Bexhill Common, whence it was reported in 1835, and it was seen growing in the same locality in 1872, so that it had been there nearly 40 years. It is not uncommon in the southwest of England, from Dorset to Devon and Cornwall, growing plentifully in the bogs near Land's End. It has handsome yellow flowers. *Srro/hularia ehrharti* was classed by Dr. Hooker as a sub-species of *S. nodosa,* and by Bentham as a variety of *5. aquatica.* It was reported, both in Smith's *Old English Botany* and Dr. Syme's New Edition, as having been met with at Wilmington, within this district, where it was found by Mr. Jenner and Mr. Reeves. Mr. Roper himself never found it. and considered that it might have disappeared through improved drainage, but Mr. Reeves found it again in 1886, and another specimen was found by Mr. Jameson about 10 years ago.

The next two of our rarities belong to the *Uuibelliferae,* viz., *Bupleurum aristatum* (Narrow-leaved hare's ear) and *beseli libanotis.* The former was first met with by the Rev. E. A. Holmes in 18G0, on the downs between Beachy Head and Eastbourne. It was never found by Mr. Roper, who considered that its insignificant size and inconspicuous appearance account for the fact that it is so seldom seen. A fairly large patch was found, both in 1903 and 1904, by Miss Milner, near Beachy Head, and from this patch the specimen shown was gathered. It flowers about the middle of June, and a search for it last week in the locality where it was previously found was unsuccessful. Besides East Sussex, the plant is only found in Torquay and the Channel Islands. *Seseli libanotis* has never been recorded in this county, except in the Eastbourne district, and only occurs in Herts and Cambridgeshire.

Among our rarer plants there are also two species belonging to the *Compositae, Hz., Crept foetida* and *Centaurea calcitrapa.* The former was first detected in this district by Mr. Roper, growing upon the shingle beach between Pevensey and Bexhill. It occurs in only nine counties in England. *Centaurea calcitrapa* is confined to the southern half of the kingdom, its northern limit being Norfolk, Cambridge, and Glamorgan, in Wales. It occurs locally on the downs near the sea, where the specimen shown was gathered, and has been reported from only 18 counties in England. The stiff spreading spines of the involucral bracts, which produce the resemblance to the old device for laming horses, from which its name "calcitrapa" is derived, render this species a particularly easy one for even a novice to identify. The last of our rarer species is *Cerat(iihi/Uinii aquaticiim* sub-sp. *submersion,* found by Mr. Roper in ponds to the east of this town. Of these nine rare plants, three belong to Mr. Watson's Atlantic type, viz., *Sibthnrpia europaea, Bartsia riscosa,* and *Bupleurum aristatum:* three belong to his Germanic type, viz., *Seseli libanotis, Crepis fnetida,* and *I'entaurea calcitrapa;* two to his English type, viz., *Scrophularia ehrharti* and *Ceratophyllum aquaticiim:* and one to his "local" type, viz., *Phyteu-*

ma spicatum.

We have also two plants within our district of considerable interest, to which attention should be drawn in any notice of the local flora, viz., *Crambe niaritima* and *Raiiunculus lingua. Crambe maritima,* or the wild sea-kale, is noted by Sir William Hooker as "not very general" in its distribution, and by Sir Joseph Hooker as " rare." It is locally abundant within the Eastbourne district, growing upon the chalk cliffs, and upon the shingles east of the town. The plant has a sturdy cabbage-like appearance, with large, sinuate, waved, and glaucous leaves and rounded heads of white flowers, arranged in branched corymbose racemes, and possessing a pleasant scent. The four Ions stamens have their filaments forked at their upper ends.

Marsh Plants:—*Ranunculus linijua,* or the greater spearwort, is not a very generally distributed species of its genus, but within the Eastbourne district it is to be found " locally abundant " upon the marshes. Besides the interest attaching to the fact of its being among the few species of *Ranunculus* possessing undivided leaves, and of its bearing considerably larger flowers than any other of the British species, there is an additional interest for us here, that it was Mr. Roper who was the first botanist to observe and describe the submerged reniform leaves borne by this plant, altogether unlike the usual lanceolate leaves which alone had previously been described in British floras. Mr. Roper made the discovery about 1888, and having been assured by Sir Thistleton Dyer that the observation of the existence of these leaves was altogether new, he read a paper upon the subject before the Linnean Society in London. We know that some of the varieties of *R. aquatilu* possess two kinds of leaves, floating and submerged. In the ditches in the marshes is also to be found our only insectivorous plant— *Utricularia vulgaris,* the bladderwort whose handsome yellow blooms are very conspicuous in the ditches bordering the road to Pevensey. One of our supplementary lists also reports the finding/ on one occasion, of *Utricularia minor.* The same dykes also give us the interesting frog-bit *(Hydrocharis morxus-ranae),* which is very variable in regard to position, owing to its method of vegetative reproduction by means of winter-buds, which often travel some distance from the place at which they sank. The showy flowers of the water violet *(Uottonia palustris),* the white water lily, and the rose-coloured umbels of the flowering rush, also make this marsh flora a very attractive section.

Downs Flora:—Just at present the hollows of the downs and the slopes near the sea are nearing their best; the ground is yellow with the blossoms of *Lotus corniculatus* (bird's foot trefoil) and the sweet-scented *Hippocrepis* (horse-shoe vetch), the *Orchis ustulata* and *Habenaria viridis* (frog orchis) are plentiful in the short grass, great spikes of the "bee" orchis are just opening, but the "spider" orchis is nearly over. *Rosa spinosissima* is plentiful, and blue columbines are found in the hollows. Tall spikes of the vivid blue "viper's" bugloss *(Echium rulyare)* are found on the cliff slopes, and, a little later, the pink of the rest-harrow *Ononis arrensis)* and the white of the dropwort *(Spiraea filipendula)* will predominate. Three other orchids—*Orchis pyramidalis, Habenaria conopsea,* and *Uephalanthera pallens*—are also found on the downs, or in downs plantations, this month. The downs also possess several semiparasitic plants, the most uncommon of which is *Thesium linophyllum,* a plant only found in the chalky lands of the southern counties. Its procumbent stems and minute flowers are to be found in the hollows near the sea.

Maritime Species.—The somewhat dreary-looking waste at the East of the town known as the Crumbles and the salt marsh at the mouth of the Cuckmere, near Exceat Bridge, afford a most fascinating group of plants. These are the succulent-leaved Halophytes: — *Saiicornia herbacea,* glass-wort; *Crithmum tiiaritinntm,* samphire; *Snaeda maritima; Glau.c maritima,* sea milkwort. A group of typical salt-marsh plants are:—*Statice limonium,* sea-lavender; *Armeria maritimum,* sea-thrift; *b'rankenia laevis,* seaheath; *Aster tripolium,* sea-aster.

Near Exceat grow:—*Bupleurum temtissimum,* slender hare's-ear; *I.iimm angustifolium,* narrow-leaved flax; *Althaea officinalis,* marshmallow. On the Crumbles is found an interesting little group of clovers:—*Trifolium striatum, T. subterranean, T. maritimum, T. resnpinatnm:* other rather interesting plants on these shingles are— *Huscut aculeatus,* butcher's broom; *Saponaria officinalis,* soapwort; and *Glauciitui luteuin,* yellow sea-poppy.

The large natural orders are represented as follows: — *Hannnculaceae,* 24 species; *Leijuminosae,* 40 species; the little grass-pea, *Lathyrus nissolia,* plentiful near Wilmington, being worthy of mention; *Rosaceae,* 28 species; *Caryophyllaceae,* 26species; *('ruciferae,* 31 species; *Uinbelliferae,* 86 species; *Compositae,* 63 species; *Scrophulariaceae,* 33 species, including three of our rarities, and 12 of the 17 species of *Veronica: Orchidaceae,* 17 species, the most uncommon being *Nenttia niditsaris,* the socalled bird's-nest orchis, and *Ophrys museifera,* the fly orchis; the *Cyperaceae* are represented by 82 species; of the *Graiuineae* we have 27 genera, and, altogether, 54 species. During the last few years the following, not included in Mr. Roper's flora, have been observed:— *Ribes yrossularia* and *Iris foetidixxima,* by Miss Milner; *Medicago sylrestris,* by Air. Hilton, of Brighton; *Campanula rot unit i folia,* by Dr. McQueen, near Battle, and by Dr. Whitney, upon the downs, near Paradise; *Rosa rubiyinosa,* near East Dean, upon the downs and near Holywell; *Rhamniu eatharticns,* near Folkington; *Ophrys museifera,* upon the downs near Wannock; *Habenaria bifolia,* between Heathfield and Crossin I I.iiki. These last four species were found by Dr. Whitney. A concluding point of interest is the distribution of the plants over the district in relation to the geological formations, or the nature of the soil upon which they are found, the plants being distributed according to

the classification adopted in Mr. Roper's flora, in eight divisions:— 1. Chalk. .... 165 species.
2. Greensand and Weald Clay.... 55,, 3. Hnstings Beds 52 4. Common to above three...... 31,, 5. Marsh and water plants. ... 104,, 6. Shingles and maritime...... 60,, 7. Common to last two...... 20,, 8. Common to all.... 13,,

It must not be imagined from this that the species are strictly confined to the particular deposit or situation above stated. It is simply a tabular statement of the localities in which the plants have been usually observed.

Since the above paper was written, *Drosera rntundifolia,* the round-leaved sundew, has been definitely added to the flora of the Eastbourne district; and *Narthtcium ossifrayum,* the bog asphodel, has been reported.

Sea Erosion And Coast Protection.
By EDWARD A. MAETIN. F.G.S.

During the last few years the combined action which has been taken by certain local authorities who are interested in the matter of coast protection, has called public attention to the loss which some of our shores are suffering, and have suffered for centuries, by the denuding action of the sea. The movement has had in view possible assistance from the National Exchequer in the direction of meeting the cost which has fallen on such authorities, and of seeking relief from the necessity of having to pay for what is held by them to be a matter of national interest.

In this connection the well-known case of Attorney General *v.* Tomline, is cited, in which the former obtained an injunction against the lord of the manor, restraining him from removing shingle from the sea-shore at Bawdsey, a few miles north of Felixstowe, and it was then held that it was the duty of the Crown to protect the realm from erosion by the sea, by maintaining such natural barriers as existed against it; and it is also pointed out that the removal of shingle from Spurn Head was prohibited by the Board of Trade.

Foremost amongst those parts of the coast which have suffered from inroads of the sea, we must place the whole line of coast stretching from near Plamborough Head, in Yorkshire, to the mouth of the Thain&s, consisting, for the most part, of cliffs of more or less incoherent material, such as glacial drift, alluvial clay, London clay, or the easily disintegrated strata of the lower London tertiaries. Next in importance, so far as loss by denudation is concerned, we have the line of coast stretching from the North Foreland to Folkestone, consisting of chalk, with superimposed beds of eocene age at Pegwell Bay; and between Eastbourne and Brighton, also consisting of chalk cliffs, but with important beds thereon of eocene age, as at Seaford and Newhaven, and with post-glacial beds on the eastern side of Brighton. Westward, thence, the trend of the chalk hills inland has exposed the tertiaries of West Sussex to the sea, and the evidence of the Isle of Wight as an island is, on the face of it, evidence of loss of land by erosion of the eocene and oligocene beds which were, at one time, continuous across the Solent and Spithead.

On our western coasts history records less waste by sea erosion, in consequence of the harder nature of the strata exposed, those parts of Lancashire and Cumberland where the glacial beds come down to the sea being the more noticeable as eroded coasts.

Looking at the progress of erosion as evidenced to us in the historical period, we find that, generally speaking, the younger the age of the rocks exposed the greater has been the loss by denudation. Situated as is Great Britain, on the edge of a continent, the island has been peculiarly prone to the influence of land movements, and the great variety and number of its geological formations show that such movements have had greater results in its case than in the case of the adjoining mainland. On its final rise from the last great subsidence which it underwent, it rose with the strata of which it is formed tilted in such a way as to leave them now with a general dip to the east, although it is not satisfactorily shown that this dip may not have been given to the strata at some earlier period, prior to the great subsidence of the glacial age. The result of the dip is that the western coasts consist in the main of palaeozoic rocks, and as one travels across England from west to east one encounters rocks of newer age, until, on the eastern coasts, one finds the most recent of all.

If one regards contour alone as a guide to the denudation undergone, one might well imagine that the broken coast line on the west of England was good evidence that here the action of the sea had been very great. That a tremendous amount of landcarving, and, therefore, probably sea-coast erosion, took place here before historical, and possibly before human, times, is shown by the deep river-valleys which have been. sounded, cut deep into hard palaeozoic rocks in pre-glacial times, and subsequently filled up by glacial or alluvial clay, but it is not improbable that the coastcarving was aided by land subsidence, which gradually brought the formidable paheozoic cliffs more and more within the power of marine denudation. Similarly we can scarcely attribute the loss of the greater part of Lyonesse to erosion by the sea. Leaving aside all legendary tales connected with this area, we find that Strabo speaks of the Scillies as consisting of not more than ten islands, whereas there are now about 150 small islands. Marine denudation would scarcely account for this great change within historical times, and I think we may safely judge this change to have taken place as a result of actual subsidence, since it has split up the larger islands into many smaller ones, by the inflowing of the sea over the low-lying valleys which now separate them, this being more than sufficient to account for any loss of land which may have occurred through the complete disappearance of any of the islands which may, in bis time, have been near sea-level.

For evidences of loss of laud on the western coasts by the incoming of the sea during historical times, we are, therefore, compelled to look to subsidence as the cause, except in those places where the line of coast exposed is made up of glacial or alluvial clays and drift.

Of sea-erosion on our eastern and south-eastern coasts, as attributable to. subsidence, I have been unable to trace any evidence. The subject of my paper is scarcely one which will justify me in tracing the possible origin of the North Sea, and the consequent bringing to birth of the cliffs of boulder-clay of our Holderness and East Anglian coasts. And yet we must be careful to remember that this coast denudation, which so much concerns our maritime local authorities, is but a continuation of what has gone before, and, although the coast may now be stationary, so far as vertical movement is concerned, the only difference is that now this very condition results in far less rapid erosion than was formerly the case. If we imagine that, after the last great rise of the land, the area of what is now the North Sea was reduced by various sub-aerial agencies to a level not very greatly above that of the sea at high water, and these agencies could do no more, but rather less, then the sea would rapidly be enabled, in times of storm and flood, to carry on the process of denudation which had already set in. Then, when a subsidence of the land again took place, and brought our submerged forests below the level of the sea, the North Sen became submerged, and marine agencies commenced the attack on our eastern shores. The process of cliff-formation then immediately commenced by the denudation of the land, which, by the fact of its lying low, was then within reach of high water. The present erosion of our eastern clay cliffs is thus the continuation of what commenced long before history records, and it follows, too, that the very existence of cliffs at all is evidence of great denudation. The process of eating into the land would be checked, but not stopped, as the land assumed the stationary condition in which we now find it. Erosion then proceeded until the present day was reached, when, although for the most part the coast line has been worn back into a series of cliffs, there are yet parts where the coast is almost at sea level, or where the land at the back of a low line of sea-cliffs lies at, or below, the level of the sea.

Probably erosion has nowhere been so great as on the Norfolk and Suffolk coasts, and, as an instance, the disappearance of the thriving town and seaport of Dunwich may be mentioned. The history of this place is very complete, and, in Domesday Book, we are informed that tracts of land which had been taxed in Edward the Confessor's time had since been devoured by the sea. In Gardner's history of the borough, published in 1754, we are introduced to successive losses which were sustained; the monastery, many churches, the old port, 400 houses at one time, the town hall, gaol, etc., all disappearing in turn. Ray says that "ancieDt writings make mention of a wood a mile and a half to the east of Dunwich, the site of which must at present be so far within the sea." The old tradition mny here be mentioned that "the tailors sat in their shops at Dunwich and saw the ships in Yarmouth Bay," and from this we may deduce the fact that Dunwich must then have been at Ray, *Phys. Theol. Discourses.* the most easterly part of the East Anglian coast, since between the two there is now the bulge of the land at Lowestoft.

Without attempting to specify the many instances of erosion, which local histories can easily supply, one should, perhaps, call attention to the Isle of Sheppey, and thereby point a moral for the benefit of geologists. The London clay cliffs, which are constantly being eaten away here, have furnished specimens for many a museum. In 1827, Lyell wrote that the island was about six miles long by four in breadth. "The church at Minster, now near the coast, is said to have been in the middle of the island 50 years ago, and it has been calculated," says Lyell, " that, at the present rate of destruction, the whole isle will be annihilated in about half a century.";: Needless to say, the island is still there, and it has not yet been annihilated, although nearly 80 years have since elapsed. The rate of destruction has clearly decreased, and it is wrell in such cases, we may now judge, not to prophecy in geological nintters.

The rate at which the coast is eroded varies considerably according to the nature of the material of the cliffs. At two spots, even in close proximity to one another, the rates may vary. In some places the erosion proceeds by substantial falls of cliff, between which falls there is no apparent alteration in the face of the cliff; in these cases it is, however, possible to arrive at an average annual rate. To the line of coast between Bridlington and Kilnsea, a distance of 40 miles, a rate of erosion is assigned by Mr. W. H. Wheeler of 6f feet per annum.! Smeaton put the loss at 80 feet per annum.J Mr. Shelford put it at 6J feet.§ Other estimates have been made for limited areas on the same stretch of coast, and these vary, not only between themselves, but for the same stretch in successive periods. Thus, since the date of Domesday, in 1080, Kilnsea has lost 709 acres, or 7 feet a year. In 67 years, however, from 1888, the rate was CJ feet, and from 1838 to 1847 the average loss was no less than 15 feet per annum. These particulars were recorded by Mr. 11. Pickwell, together with the rates of erosion affecting other localities on this const, as a result of a close examination of various historical records.%

The following table may be useful as showing various recorded estimates of rates of erosion:—

Locality. Katk. Authority.

Owthorne

SliBrrinKliam

Thanet(Bedlam Farm) Lyell, *Principle of Geology.* t The " Sea Coast," page 2. J Building of the Eddystone Lighthouse. § Min. ot I'roc. Inst. C. E., vol. xxviii., p. 472. 1 R. Piokwell, Assoc. Inst. C.E. in " Proo. Inst. C.E.," vol. li., p. 193.

Locality. Rate. Authority.

Reoulvers to North 2ft. a year.. Lyell: *Principle of*

Foreland (11 miles) *Geology.*

Easington.... 14yds. a year, from 1771 to R. Piokwell.
1852.

do. 3'3yds. a year, from 1852 to do.
1870.

Dimlington.... l-8yds. a year, from 1771

to do. 1852.
do. 6lyds. a year, from 1852 to do. 1876.
Holmpton.... 5yds. a year, from 1788 to do. 1802.
do. 2yds. a year, from 1802 to do. 1833.
do. 3yds. a year, from 1852 to do. 1876.
Withernsea (old vil-1-lyds. a year, from 1794 to do. lage). 1852. do. 6yds. a year, from 1852 to do. 176. Withernsea (opposite '5yds. a year, from 1794 to do. poorhouses). 1852. do. 56yd8. a year, from 1852 to do. 1876.
Owthorne.. l-2yds. a year, from 1786 to do. 1822.
do. 2"3yds. a year, from 1822 to do. 1852.
do. 3yds. a vear, from 1852 to do. 1868.
Waxholme.... 26yds. a year, from 1844 to do. 1852.
do. l"7yds. a year, from 1852 to do. 1876. St. Margaret's Bay.. 10 ft. a year. ..... Wheeler: *The Seu Coait.*
Holderness coast (36 2yds., or 30 acres, a year.. Phillips: *Geulogy ofthe* miles). *Yorkshire Coast.*
Holderness coast (part 5ft. to 10ft. a year, from 1852 Capt. Kenny, R.E.: of). to 1889. *Hep. Brit. Attn.,* 1895.
Holderness coast (12 132ft. in 40 years.... Capt. Salversen, R.E. miles). Westward Hoi 30ft. a year, from 1875 to H. G. Spearing: *Q.J.* 1884. *G. S.* , 1884.
Warden, Sheerness.. upwards of 220 acres in 220 Col. Le Messurier, R.E.: yearB. *Reji. Ilrit. Assn.,* 1S85.
Eastbourne (Splash nearly a furlong, from 1717 F. W. Bourdillon, M.A.:
Point). to 1841. *Hep. Brit. Assn.,* 1888.
Eastbourne (Round 3 acres in ten years, to 1857 *Guide tn Kasthourne,* House). 1857.
Eastbourne (east of 238ft. in 49 years. ... J. B. Redman: *Rep.* circular redoubt). *Brit. Assn.,* 1888.

Bulverhythe (Hast-about 10ft. in 12 months.. Col. E. C. Kim, R.E.: ings). *Hep. Ilrit. Assn.,* 1885.
Worthing.... 70ft. in 12 months.... Dixon: *Geol. of Sussex.*
Lyme Regis.... 3ft. a year in soft mud, lft. G. Roberts: *Hist, and* in harder rocks. *Avtiq. of Lyme Regis,* 1824.
Locality.
Sidmoutb
Bridport (East Cliff)..
Bridportj West Cliff)..
Hampshire Coast
Between Selsey Bill and Chichester Harbour.
Between Littlehampton and Bofmor.
Coast of Durham between T'. ne and Wear.
Coast between Dun wich andCovehithe: i. Easton Bavents ii. EastonHighCliff iii. Covehitlie Cliff
Hoylake to Birken head.
Lancashire coast (Rossall Point).

Leaving aside the considerations of falls of cliff, due entirely to want of natural drainage and other aerial agencies, cliff erosion arises from waves hreaking on the sea-shore, and thus removing fallen detritus: this, when raised on the shoulders of the tidal wave, pounds the cliffs like a battering rain, portions broken off being in turn utilised as a means of further assault. The force which it is possible for a wave thus to expend is governed by its amplitude, and this can be but small on a shelving strand. So where there is a natural beach stretching between the cliffs and the sea, this forms the best possible protection to the cliffs, and efforts only are necessary to induce the retention of such a beach, where there is a likelihood of it being carried away by natural means or artificially for the making of concrete. The breaking of waves on such a beach during u storm may be disastrous to its existence, unless steps have been taken by means of artificial breakwaters, groynes, etc., to reduce the force with which they are finally thrown on to the beach. The wind waves are merely the distant echoes of greater waves elsewhere. They originate in the pressure of the wind upon the surface of the open sea. Immediately an irregularity of pressure is set up, oscillatory movements radiate from the place of origin in gradually increasing circles, until they either die out as equilibrium is restored, or they reach a shore. They are purely undulations, and there is no permanent forward movement of the water until they reach an obstacle to their movement. If this be a shore line, the effect of the shoaling is seen in the decreased length of each wave; the depth of the water becomes insufficient to allow of the formation of the complete undulation, and this, combined with friction at the bed of the sea, metamorphoses the undulatory movement into one of actual translation, especially at the free and nnentangled crest of the wave. Then the shoaling having arrested the lower portion of the waves, the upper part pitches forward, and the energy imparted into the wave depression far out at sea by the wind, that is, by an irregularity in the pressure of the atmosphere, is finally expended in the work of the wave as it dashes with tremendous power upon the strand.

Wind waves *per se* would, however, in many places, have but little effect in degrading the cliffs, if they were not elevated into a position of being able to assault the cliffs, by riding on the surface of the tidal wave. On our southern coasts the term "tidal wave" seems almost a misnomer, seeing that no actual wave exists in the popular sense of the word, but rather a gradual rise in the surface of the water takes place, filling up bays and covering up low-lying strands by an almost imperceptible creeping movement. So, as the tidal inpour fills up the English Channel, the wind waves are raised into position on the shoulders of the former, to assault the cliffs at the back of the strand. In gales and storms masses of rock are dislodged from the cliffs. These in turn are used by the waves to further the bombardment of the cliff, and, in consequence of the continual attrition to which they are thus exposed, they become ground down in time to small fragments, and finally to sand or claysilt. At the hands of the sea the beach becomes sorted out, the finer sand sinking through boulders and shingle to form the bottom layer of a beach, then conies the fine shingle, and above it the larger pebbles next, and, over all, the boulders which have, as yet, received least disintegration, or which constitute a "Full" brought up by the last spring tide. The alluvium brought down by a river, or torn from a stripped cliff, having a very small specific gravity, is carried farthest out to sea, to settle where the power of suspension is feeble, owing to the river current having lost its ordinary velocity, or by the opposition of the under-water effect of windwaves. The tidal current, when flowing or ebbing strongly, will, however, overcome both, and is all-powerful in transporting the finely-divided material brought witbin its reach by the reflex movement following the breaking of waves upon a shore, or by river action.

Waves of great percussive force are only possible when the rise of the tide gives the depth that is necessary to allow of their complete formation. Heavy gales blowing on shore at the time of spring tides afford the opportunity for the greatest denuding effect of waves, the height of the tide being increased by such a gale. Under exceptional circumstances the height of an ordinary high tide may be increased by this means to 5 feet, or even 7 feet. It "Effect of Wind Pressure on the Tides," by W. H. Wheeler, M.I.C.E., *Brit. Asm.*, Ipswich, 1895.

is a vexed question as to the extent to which the tidal currents are responsible for the stripping of beaches. At spring tides, with onshore winds, it is quite possible that such currents may have considerable velocity, even close to the shore, at low water, and, whereas, a few feet below an ordinary wind-wave of some magnitude, a diver can ply his vocation without discomfort, he cannot hold his ground when the tide is running full. An experiment, which was made in Lake Ontario, showed that a box, anchored at a depth of 6 feet, during a storm became filled with sand; whilst another, anchored at 20 feet, remained completely empty. Perhaps, however, had such an experiment been made on the south coast at the time that the tide was running up the Channel, the result might have been different. In times of storm, as everyone knows, much seaweed is stripped from the lied of the sea, and tossed on shore, frequently with the blocks of chalk, on which they grew, attached to them. In this case, we seem to have a combination of the effects of ordinary wind-waves changing their undulatory movement into a translator one, owing to the shallowing of the bed, together with those of strong tidal currents. To the latter I think must be attributed the major portion of the tearing up of the seaweed, weakened as it is about the roots, in many cases, by the work of boring molluscs, such as *I'holas tlartylus, Teredo navalis,* etc., although there is something yet to be learned, I am convinced, concerning the depths to which the trunshitory motion of storm wind-waves can be felt.

The breaking of waves has an enormous effect on a beach on which they break, and is caused by the more rapid progression of the upper half of a wave oscillation than the lower half, owing to the latter being subjected to the friction of the shallowing sea-bed. So the former collapses forward, with a force commensurate with the amount of momentum which the undulation has retained during its passage from the source of origin in the open sea. The downward pounding action scatters the shingle on which it falls both backward and forward, and the undertow drags back into the sea the bench which has been loosened by the fall of the breaking upper half of the wave.

The kinetic energy developed by a wave having a height of a foot from trough to crest, has been estimated by Mr. YV. H. Wheeler as sufficient to raise 165lbs. of pebbles a foot high. With 15 waves to the minute, and estimating the average dimensions of pebbles to be two inches in diameter, energy would be evolved sufficient to raise 2,87(5,000 pebbles in a single tide, or a total weight of 266-4 tons of stone a foot high.

Owing to the fluid nature of moving water, it is not possible to accurately estimate the force of the impact with which waves would strike a cliff. Various estimates and experiments have been made, and these vary considerably. One, by Mr. Wheeler, gave 4-27 tons rW. H. Wheeler, C.E.,"7p7 *lirit. Atm.*, 1898. to the foot super, or, in other words, a wave 10 feet high would be capable of raising 21-85 tons of material to a height of 1 foot. A mean of a number of experiments by Mr. Thomas Stevenson gave 1-86 tons to the square foot; others by Mr. Frank Latham, at Penzance, gave 18cvt. to 20 cwt. per square foot; whilst others at Cherbourg, during the construction of the breakwater, gave from 600 lbs. to 800 lbs. to the square foot.

The process of sea-erosion is, therefore, as follows: The tides lift the waves to a height at which they can become weapons of offence against the cliffs; the undertow acts the part of the scavenger in dragging seawards the material eroded from the cliffs, and the tidal current takes up the disintegrated material and transports it elsewhere.

The flow of shingle along our southern coasts is in the direction of the prevailing wind, namely, from the southwest. Except in isolated bays, where there is a return scour in a north-westerly direction, such as, for instance, between Axminster and Bridport, and between Newhaven and Seaford, the shingle moves uniformly in an easterly direction. This seems to show that the tide falling out of the Channel is unable to denude the beach, which has accumulated with the aid of the incoming tide during south-west gales. Wherever groynes have been erected on this coast the western sides have been, as a rule, the first to be filled with btach. The supply is frcm the west. The scour on the eastern sides of high groynes is often sufficient to actually deprive the shore of beach, which may have been there originally. As the tide comes up Channel, its centre is in advance of its edges. Where it touches the shallows it is retarded, and the result is seen in what I may call the "strike" of the waves, which, at such times, on our shores, runs from south-east to north-west. When the tide bus turned, the opposite effect may sometimes be seen, but the groynes then protect the accumulated gravel on their western sides, so that little effect from this course is felt. The shingle which accumulates on the eastern sides on rarer occasions may be that gathered in deeper water, and carried in a westerly direction by an outgoing tidal current.

The chief source of moving beach is, as I have said, from the west. Whence is it derived? This is a geological question, and I fear engineers are much at fault on this head. The supply is practically unlimited. From time immemorial it has been driven up from the west, and all efforts at coast-protection have been devoted to the invention of the best means of arresting some portion of the supply, in order to create a substantial strand of beach, this being acknowledged to be the best protection which can be afforded to a line of cliffs. It has been customary to attribute the flint shingle which is strewn on our coasts to the erosion of the chalk cliffs which is now going on. Cease the erosion, and the shingle becomes a finite quantity. From this view I dissent. The erosion of chalk cliffs, I contend, is now altogether inadequate to explain this constant supply of shingle, which is brought up from, and strewn along the coast at, places which are westward of any existing chalk cliffs. Mr. Jukes Browne has recently conducted, in association with Mr. William Hill, a series of exhaustive experiments, having in view the object of ascertaining how far chalk would be capable, without extraneous help, of forming that confusing super-cretaceous deposit known as " clay-with-flints." That part of his work which interests us here, is the conclusion arrived at, that the solution of 100 feet of *micraxter-anr/uinum* chalk would yield a bed of flints only 7 feet thick.'" This, it may be pointed out, is exactly the same conclusion which was arrived at by M. de Lamblardie, in his work on the coasts of Normandy, in which he showed that about 7 per cent, of the eroded cliffs consisted of flints. The question at issue is whether the constant eastward-flowing supply has other source of origin than that supplied by the falls of chalk, and in some places tertiary, cliffs which is proceeding at the present day. That the supply is ever-flowing up Channel has been shown over and over again. Each town which has in turn pushed groynes out into the sea has arrested a part of the supply, and each has in turn starved the coast-line for some miles to the east of it until its own groynes have become filled on their windward sides. The same supply has been a fruitful source of expense to various harbour authorities, in keeping open a channel to the sea.

Seventy years ago the outfall of Shoreham Harbour, or rather of the river Adur, was at Aldrington, a long spit of beach parallel to the coast having been formed, whilst the river pursued its way four miles along the coast before entering the sea. A new outfall was subsequently made near Shoreham, and a pier, which was constructed in 1.S74 on the west of the outfall, stopped the flow of shingle into the river channel, until now there is a bank of shingle over a quarter of a mile wide. But Brighton suffered through this appropriation of the shingle, as well as by the protective measures which were taken to preserve the spit of land lying between the "Basin" and the sea at Aldrington, on which, in 1871, the Brighton and Hove Gasworks had been built. In the course of time the requirements of these works were met, and the supply flowed on again. Hove has since trapped as much as she needs; Brighton has pushed her esplanade further out into the sea (the Aquarium is built on reclaimed foreshore), and new works, in the shape of walls and groynes, now extend as far as Black Rock, Kemp Town. Here they come to an end, and the sea sweeps round on to the unprotected cliffs, which, for a distance of about half a mile, consist, In a pnper read before the *Geol. Soe. of Lond.,* May, 1906. unfortunately, only in part of chalk, the upper three-fourths being of postglacial rubble, which is acted upon very rapidly by atmospheric agencies. The scour of the waves at high tide loosens the chalk platform on which the rubble-drift rests, and marine and sub-aerial denudation combine to act irresistibly upon the defenceless cliff.

Again, at Newhaven, the sea-wall and other works on the west side of the harbour entrance have resulted in the accumulation of a great amount of shingle, and have prevented the beach so arrested from filling up the harbour channel. Here again the river Ouse had, in former years, been turned to the east by the westwardflowing shingle, until it made its entrance into the sea at Seaford. Those who know that town now scarcely realise this important fact. Another instance of the kind is found in the west pier of the harbour at Folkestone, which was built in 1856 to prevent the easterly drift of the shingle across the mouth of the harbour, the bar of which was dry at low tide. In 17 years the beach advanced on the westward side 120 feet for a distance of 500 feet. The erection of this pier and subsequent extensions has resulted in a wide stretch of level beach, on which pleasure gardens, etc., have been constructed, the whole area having been reclaimed from the sea.

There is practically no cessation in the supply of the shingle, and I contend that the source is not so much to be found in the destruction of the cliffs which is now going on, but rather in that which went on a long time ago, although in recent geological times, when England was still joined to the Continent by an isthmus, from where are now Folkestone and Dover, to the coast of France. That this was the physical configuration of that part of our country at the close of the Glacial Period is satisfactorily borne out by many lines of evidence, and Mr. Jukes Browne's works may be consulted with advantage on this subject". The degradation of the chalk cliffs which is now going on is but slight as compared with that which must have taken place during the destruction of the whole of the chalk area which formerly connected England with France. The final severing of the connecting isthmus may have been accomplished by marine action, but it must have been degraded throughout almost the whole length of the English Channel by sub-aerial agencies, while yet the Channel was non-existent, and while our southward flowing rivers probably met those from France in mid-channel, and formed a complete river system, having its outfall in the distant Atlantic Ocean. With what result? We find far inland in our southern counties valuable beds of flints, all that has remained of chalk that has disappeared; in fact, we find them farther north where there is now no chalk at all. Can we imagine that similar beds of flints were not formed A. J. Jukes Browne: *The Building of the Britiih Islet.* by the river system of the dry English Channel? The denuded chalk flints would have been conveyed in a south-westerly and westerly direction from our existing shores, and would, on the final and gradual subsidence of the bed of the Channel to its present depth, have formed beds of more or less rolled shingle. This shingle is now being returned eastward by tidal action, and in this we have practically an inexhaustible supply. Present day erosion is but slight us compared with that which took place when finally the sea flowed up the old river-excavated channels, and, when once the Dover Straits were forced, the erosion by the sea, reinforced by the through-tidal current, would receive a great impetus. To these far away times I attribute the shingle beds of the English Channel, which are now supplying the westward sides of groynes and piers.

In order to protect coasts from erosion, we must imitate nature as far as possible, and discover what are the means she has employed in those places where sea-erosion is comparatively unknown. The chief protection is a sloping shingle beach. The principle of groyning is to intercept the flow of gravel, after having ascertained the direction whence it comes. Mere wind-waves have little effect more than a few feet below them. Tidal influence is all-powerful. No effects are felt close to the shores, and the energies of engineers are devoted towards robbing the tidal current of the burden which it carries as it passes by.

Groynes which have been pushed out from the shores have intercepted a portion of the gravel which is being brought up Channel, but the shingle seems in no way to have been diminished, and, provided that suitable structures are reared, the shingle is arrested. In certain places, owing to exceptional circumstances, the deposit of one tide is frequently cleared away by the next, so that a secondary problem in such cases presents itself, that of retaining a deposit when it has once been formed. Thirty years ago the favourite form of groyne was one of great height, formed of wood, and stretching nearly to low tide mark. In such cases, enormous quantities of gravel were arrested, but it was all heaped on one side, and for many years these groynes, as, for instance, those in front of the Brighton Aquarium, exhibited a precipitous fac« to the east, with a drop of as much as 20 feet. This form has now almost entirely been discarded in favour of the low type of groyne, formed of concrete, so that, as the windward side becomes filled up, the gravel falls over on to the other side, and so fills that side up much more rapidly than with the higher type of groyne. The tendency is, in these cases, for the groynes to become completely hidden in a few years by an equable deposition on both sides.

The acquisition in this way of a shelving beach brings about that natural state of affairs which is acknowledged to be the best kind of protection which a cliff or a sea-wall can have. The attempt to attract a beach in front of the sea-wall at Hove by means of groynes set diagonally to the southeast, instead of at right angles to the shore, was a failure, and the use of sloping groynes is not now recommended, unless a terminal portion be added at right angles to the shore.

Acting on the principle that the shallowing of a shore tends to cause the tidal current to drop its burden, 20,000 tons of shingle, dredged from the mouth of Shoreham Harbour, where it was very undesirable, were emptied from hopper barges into the sea in front of the Hove lawns, and this artificial accumulation arrested much of the westward flowing gravel, and, besides acting as the nucleus around which shingle has since accumulated, much of it has in the course of time been washed up on shore, and no fears are now felt as to the safety of the sea-wall. This artificial dumping of material might, I submit, be utilised more often in the future. The enormous waste of Cornish and Welsh quarries might thus find suitable dumping-ground, and, besides becoming gathering-ground for passing gravel, would break the force of the wave-rollers, and these would pass harmlessly on to the shore.

We frequently hear references made to the iron-bound rocks of Cornwall, or the west of Ireland, and possibly some of the rocks which face the Atlantic owe the power which tbey possess of withstanding the eroding action of the sea, to the fact that they are literally ironbound, that is, bound together by the quantity of iron oxide in the rocks. Large portions of our gravel-beaches are uncovered, save by an occasional spring tide, and when such a tide happens, it acts frequently with disastrous effect. In such instances, can we not imitate nature again, and, by artificially treating the beach with some tons of scrap iron, bring about sufficient quantity of a solution of peroxide of iron, and form of the beach a conglomerate such as we find in some of our tertiary pebble-beds? The conglomerate at Seaford is an instance in point, and that contained in the Oldhaven pebble-beds around London is sometimes similarly consolidated. I make this suggestion as one worthy the attention of our engineers, since to do what nature does on our western coasts we must imitate the way in which nature herself does it. Our knowledge as to the movements of gravel far out in the Channel is very slight. But the moving banks of gravel which strike the shores, and which groynes are intended to arrest, are possibly, and I may say probably, plentiful in their distribution farther out in the bed of the sea. To intercept shingle, which the longest of groynes do not reach, a new form of groyne has been invented by Mr. R. G. Allanson-Winn, although perhaps the word "groyne" is not the term to apply to it.

He suggests "a system of submarine groynes of heavy chain cables, running out to sea at right angles to See paper by R. G. Allanson-Winn, B.A., read before the United Service Institution. March 5th, 1906. the coast line, and having their lower or sea-ends fastened to a lateral chain parallel to the shore in deep water, the last-named chain being held in position by two or more anchors beyond the limits of the series of groynes." Woven into the mesh of such chain-cables, or bound to it by hoop iron, would be faggots of bushes, brushwood, etc., the object being to create an obstruction, which would arrest the power of suspension of the moving water, and cause it to drop its burden. It is directly opposed against that portion of the flow of gravel which is impinging near to the coast, but which requires some assistance to draw it coastward, and to cause it to form a gravel strand. It is pointed out that frequent falls of cliff do not shallow the sea in front of it. The material is carried away, and deep-sea erosion is as much responsible as shore-erosion in denuding the coast. The undermining of the invisible lower portion of the shore beyond low water-mark, must be prevented, as well as that part which is visible. The artificial deposition of ballast, such as that to which I have already referred, may effect this protection. Some form of artificial induration of the rocks, and the forming of a conglomerate of existing banks of gravel, might assist it. Mr. Allanson-Winn's chain-cable groynes have the merit of relative cheapness to recommend them. They touch an area to which engineers have given little or no attention. They have been successful where introduced by the inventor on the coast north of Bray Harbour. The point on which one must differ from him is that it must be acknowledged that ordinary means have no appreciable effect below the surface, and that the gravel which he would wish to arrest is conveyed by tidal currents. But it matters little which form of wave is the means of conveyance; the tidal wave must be regarded as the deepsea means by which the shingle is brought to the places where the influence of the wind-waves begins to be felt, and which raises the latter, in the absence of a beach, into the position that enables it to attack the cliffs at high tide.

The all-important question which has of late been raised is as to who shall bear the expense of protecting the coast. Local authorities have been at great expense to protect the mile or two of coast with which they are more particularly concerned, and, although the outcry is that the cost has fallen very hardly on 3mall municipalities, the chief practical question is the protection of lines of coast but sparsely populated, rural sea-frontages as they may be called, in contradistinction to urban sea-frontages. It must be borne in mind, too, that the works which have been reared by urban authorities have had much to do in the way of increasing the erosion of rural sea-frontages, and instance after instance can be cited in which some parts have been starved of their natural protection by the selfish disregard paid to them by urban authorities. It no doubt pays the latter to protect their frontages, and I submit that any expenditure which the State might undertake in this direction should be devoted for the most pare to the long stretches of coast away from the towns. When these are in private hands, it is, I think, a question for discussion whether such coast to be protected should be acquired by the State, as a preliminary to the expenditure of public money for its protection.

As to the actual cost of protection, this varies considerably at different parts of the coast, and of course where, as in the case of Hove in 1884, a wide strip of foreshore, in this case 80 yards, is reclaimed from the sea, the works necessary may reach a very high figure. At Hove the protective works along a frontage of 720 yards, including groynes, granite-faced sea-wall, and the artificial dumping already referred to, cost upwards of £60,000, or £85 per lineal yard. On the other hand, it has been estimated that a long strip of frontage of wasting cliffs not requiring a sea-wall, may, by a series of wooden groynes, be protected at a cost of £1 10s. per lineal yard of coast. Taking this as the cost of protecting a frontage of 1200 yards of coast, the interest on the capital involved at 5 per cent, would work out at £90 to £100 per annum. But if along the same frontage the loss by erosion is 2 yards per annum, the total loss of land will be from of an acre to 1 acre per annum, equal to £85 to £50 per annum. Thus the cost of protection would work out at double the value of the land protected.

I think, however, that Mr. Pickwell's estimate of the average value of land is too high, and that £25 an acre would be a fairer estimate nowadays, that is, about 1 £d. a square yard. If we reckon that for every lineal yard of coast washed away each year, two square yards are lost, the total loss would be 2d. per yard of coast. An average of various defence works which have been built works out at £19 per yard of coast. Interest on this amount and provision for sinking fund may be reckoned at 5 per cent., i.e., 19s. a year.

With an expenditure of 19s. a year, it would be possible to save land of the value of 2d. Small wonder, then, that private landowners shrink from protecting their coast-line from sea erosion.

With these remarks I bring my paper to a close. There is much more, of course, to be said. But the subject is one which must come to the fore more and more in the future, and the taxpayer will no doubt have a word to say, ere the public exchequer is drawn upon to protect land in which he has no direct pecuniary interest.

"B. Pickwell. Associate Inst. C.E.: "Encroachments of the Sea," etc., *ilin. Proc. Inst. Civil Engineer,* vol. li.

The Geology Of The Upper Ravensbourne Valley, With Notes On The Flora. By W. H. GRIFFIN (Hon. Sec. Gatford and District Natural History Society).

Beyond the outer margin of the South-Eastern Metropolitan Postal District, there is an elevated tract of Lower Eocene pebblegravel which stretches out from Beckenham and Bromley towards the west and south, and is fringed by the Addington Hills, and by West Wickham, Hayes, and Keston Commons. From 60 to 100 feet below the Commons there is an extensive area of arable, meadow, and wood, lands, which is coloured on the Geological Drift-Map as chalk, and " clay with flints."

For some years I had known that flint implements were frequently found in the last-mentioned area. By traversing it in all directions, for several successive seasons, in search of plants, I was led to regard it as "a dry and thirsty land where no water is." There are no brooks; whilst ditches and ponds are rarely seen. It is so arid that water for pastured cattle is conveyed there by artificial means, and, when trespassing through the woods, I have frequently seen nondescript vessels, in which the careful gamekeeper has placed water for his cherished pheasants. The abundance of flint implements found from time to time in the locality indicated that it had been a favourite resort of prehistoric man.

To obtain a sufficient supply of water, even in the humid climate of Britain, sometimes taxes the resources of modern civilisation. The question now is, from whence shall we bring the water. Our ancestors met the difficulty by going to where the water was. Therefore, it is certain that, in the Stone Age, there was water in this now arid district. It might have been a river. If so, which river?

The Darenth was out of the question. The present surface elevations made the Cray improbable. The poor, little, now impoverished, Ravensbourne seemed to be likely, and I determined to ascertain whether any connection had previously existed between the bed of the present small brook, and certain far-removed gravelpits which I had seen in my botanical rambles. Those pits are:— 1. A large pit still being quarried near Hayes Railway Station.
2. A small disused pit close to the Water Works pumpingstation, on the road between the Hayes pit and Addington. 8. A larger disused pit in the opposite direction on the road below the west side of Hayes Common.

The valley-gravels indicated on sheet 4 of the one-inch DriftMap of the London District linked up these pits, and showed that two rivers formerly met at the Hayes pit, whence the united stream passed by way of Bromley, Catford, Lewisham, and Deptford into the Thames.

It is possible that some of my audience are not familiar with the topography of the South-Eastern Metropolitan suburbs. I may, however, assume that they know there was formerly a Royal Dockyard at Deptford, where Czar Peter came for instruction in the art of shipbuilding. The Royal mechanic, it will be remembered, lodged for a time with gentle John Evelyn, at Sayes Court, near by,. and ruined the good diarist's valued holly-hedge by driving wheelbarrows through it. The Dockyard was placed at the junction of Deptford Creek with the Thames. The Creek is, in fact, the lower reach of the Ravensbourne, and it was there Admiral Francis Drake's little Pelican was moored after his famous voyage round the world, and Queen Elizabeth went on board and knighted him.

The name of the river is said to be derived from the fact that when Caesar's legions occupied the camp at Holwood, Keston, they were at a loss for water, and a shrewd soldier followed a raven which came to the camp to pick up trifles, and found it drinking at the spring on Keston Common, which is called "Caesar's Pool." That spring is still one of the sources of the existing stream. It is a pity to pluck the heart out of an old legend, but a regard for accuracy compels me to state:— 1. The Holwood Camp was not a Roman Camp, but a neolithic defensive enclosure, or an early British *oppidum*. It is, nevertheless, designated " Caesar's Camp" on the Ordnance Map.
2. Although the water from the spring finds its way through the Keston ponds, and by ditches to the bed of the old river at Hayes Ford, a little to the south of Bromley, it was never more than a very minor tributary, never in fact of sufficient force and volume to lay down a gravel. It is enough, therefore, to point out its course on the map and leave it (see footnote on p. 52).

Having linked up the gravels at the three pits mentioned, and satisfied myself that they appertained to the Ravensbourne, I proceeded to investigate the country to the southward of the third pit, *viz.,* that on the road which lies below the west side of Hayes Common, and I ultimately connected the gravel with a deep dry valley in the chalk which runs parallel with the Westerham Road from Leves Green, Keston, to Biggin Hill, Cudham.

In the Drift-Map the bottom of this valley is coloured green, indicating chalk. The surface is clay with flints, but, by exploring the meadows in the valley-bottom I found a small disused gravelpit which showed that the "clay with flints" here is a very superficial hill-wash, overlying the old river gravel. Above this point, spurs from the North Downs about Tatsfield, Surrey, run into the valley, and we may say that this branch of the former river commenced there.

Of the other branch, which joined the former at the large Hayes pit, little need be said. It originated on the high ground about Farley and Chelsham, Surrey, and passed through Addington village towards Hayes. The only exposed gravel I could find was in the small disused pit near the water-works, previously mentioned. This gravel consists in the main of Lower Eocene pebbles. The angular flints mixed with the latter are small, and are not comparable in quantity and dimensions with those found in the gravel of the other branch. I may add that the arable land, to a height of some 80 or 40 feet above the last-mentioned small pit, is thickly strewn with angular flints, not of the same prismatic character as those which are contained in the "clay with flints" of the geological maps, but similar to those found in a river-gravel. I will now invite you to follow with me on the map", the course of the existing stream, and its extinct branches, from Catford Bridge upwards.

A little above Catford Bridge the Ravensbourne is joined by another small stream called the Pool. Like its sister, the Pool had its origin on the Surrey North Downs, and is now the impoverished representative of a noble ancestry. The present surface configuration suggests that the two streams once fell into a lake, which extended from Lower Sydenham to Ladywell, and received also the waters of several streams then running down from the hills about Sydenham and Forest Hill. The high ground which now forms the "divide" between the Ravensbourne and Pool valleys, in the neighbourhood of Beckenham Hill railway station, consists of London Clay, but it is all "derived" clay. I have examined the hedgebanks and ditches in the locality, and, wherever a section is exposed, have always found angular and water-worn flints embedded in the clay.

Returning to the Ravensbourne proper, it will be seen by the Drift-Map that, at Langley Farm, on the east side of the Bromley Road, a little above Catford Bridge, a width of more than half a mile of gravel is indicated. Langley Farm is now in process of development as a building estate. By examining excavations made for sewers I have ascertained that there is here, next the surface, about four feet of derived clay, overlying about eight feet of rivergravel, which rests upon undisturbed London Clay. From the lowest stratum of gravel, and at about ten feet below the surface, Mr. Hurden, a member of the Catford Society, obtained the shaft of a femur of *Bos taurus*, and tibia? of *Bo longifrom*, which are numbered 1, 2, and 8 in the first photograph (pi. i., figs. 1, 2, 8). I invite your special attention to the thick end of the femur. You will observe that the condyle was removed by sawing. It is in the same condition in which it was The map referred to was a hand-drawn enlargement on a scale of six inches to the mile of the one inch Geological Drift-Map, but it is not adapted for photographic reproduction. The points mentioned may be readily found on the Drift-Map, or any modern map of Kent.

Opposite p. 32. Fragments Of Bone And Horn From The Gravel Of The Ravensbourne Valley. *The South Eastern Naturalist.* 1906. found, and the very neat 3aw-cut presents a difficult problem. It is difficult to conceive how the bone could have got into the position in which it was found, except at the period when the gravel was being deposited. We know that mammalian bones, smashed for the purpose of enabling our oftimes hungry ancestors to extract the marrow, have frequently been found in similar positions. I produce a small flint saw, two inches long, which came also from Ravensbourne gravel five miles higher up the valley, and I have seen other neolithic saws much longer, but none of them could, I think, have been driven through a bone of this diameter. Possibly, it might have been done with such a tool by sawing round the bone to a depth of a quarter of an inch. If so, the man of the Stone Aq;e who did it was a clever workman.

In the second week of May last, my friend, Mr. Hurden, also obtained two huge bones, which were dug up in Whitefoot Lane, a country bye-road leading from the east side of Bromley Road, at Southend village. They were in nearly an upright position, the tops being only a few inches below the surface. Fragments of brick and tiles were found in the gravel with them. Both bones were sawn through longitudinally, and contained remains of much corroded nails in the flat saw-cuts. One of them is, I think, part of a *femur,* and the other part of a humerus of *Klephas primigeniui* (the mammoth). The respective sizes are 2ft 11ins by 10ins at the widest portions, and 2ft. 9ins. by 8ins. In this case the circumstances suggest that the bones were originally disinterred from the river-gravel in comparatis'ely recent times, and utilised by some thrifty agriculturist as posts to support a manger.

The fragments of horn and bones (numbered 4 to 8 in the photograph) (pi. i., figs. 4-8) came from the gravel at Langley Farm, and probably also appertained to *Box taurus* or *11. lonyifrons*. It may be mentioned that the first-named, now extinct, ox, is supposed to have been the *Urus* mentioned by Cipsar as frequenting the forests of Germany, and to be also the ancestor of the Chillingham wild-cattle. *Bos longifrom* is believed to have been the ancestor of our smaller breeds of domesticated cattle.

In passing along the Bromley Road to Southend village, we walk over the former river-bed, at a present elevation of from 60ft. to 70ft., O.D. On the east is arable-and wood-land rising to about 200ft., O.D. Here angular water-worn flints may be found up to 160ft.

At Southend village the present stream widens out into a sheet of water, known locally as the " Duck Pond." This is apparently of artificial formation, for the water is dammed up to serve as a mill-pool for a flour mill. About half-a-mile further on is another mill, but the supply of water has so diminished in recent years, that it is no longer worked, and the enterprising occupier has apparently found it more profitable to devote his attention to the cultivation of watercress, for he has converted a former osier-bed, appurtenant to his mill, into watercress beds. I have had the run of these grounds for many years, and have found there—*Ranunculus heterophyllus, Caltha palustris, Lepidium draba, Stellaria aquatica, Impatient bijlora, A/thrum salicaria, Petasites officinalis, Kupatoria cannabinum, Symphytum officinale* var. b. *patois, Myosotis palustris, Veronica anagallis-aquatica, Polygonum amphibium* var. b. *terrestre, Salix fragilis, S. viminalis, Carex acuta, Glyceria plicata* var. b. *pedicellata,* and *G. aquatica*

At the foot of Bromley Hill the stream enters and passes through enclosed woods, but we may follow its former widely extended bed through roads recently opened up on the Bromley Park estate, towards Ravensbourne railway-station and Shortlands. Here the London Clay has been removed by subaerial erosion. The river has cut its way through the Blackheath pebblebeds into those of the Woohvich and Beading series, and, in places, the river-gravel rests upon estuarine clay containing shells of (*'yrena cuneiformis.* I have dug into the valley plain here at various points, and have always found river gravel beneath a few inches of black humus. Near Ravensbourne railway-station there is an old gravel-pit on part of a ladies' golf-ground, which shows a thickness of about 4ft. of gravel. The only approach to a worked flint I have found at this point is the oblong stone (no. 18 in the second photograph) (pi. ii., fig. 18). Had I not seen many other definitely worked stones of similar form in the collection of my friend, Mr. Benjamin Harrison, of Ightham, and other collections, I should have rejected it, but I think it exhibits traces of chipping. The elevations at this point are instructive. The valley-plain stands at about 50ft. O.D. Immediately to the east, at Bromley Hill, we find an elevation of 200ft., and on the west, or Beckenham, side, 180ft. The average width of the valley-plain, from the foot of Bromley Hill to a mile beyond that point, is about a quarter of a mile. By cubing the dimensions it will be seen that in this one mile only of its course, at least 27,878,400 cubic yards, representing about as many tons in weight, of earth has been excavated and carried down to the Thames by this now insignificant stream. The quantity of material would, of course, be vastly increased by including the slopes on either side. It will repay any of my audience, who may lie visiting the neighbourhood of Bromley, to proceed to the top of the recreation ground near Bromley church, and take a view of the valley from that point.

The most important plants which I have recorded in this portion of the valley are—*Nasturtium palustre, Alyssum incanum, Lepidium campestre, I,, hirtum, Impatiens noli-me-tangere, I. roylei, Aster laevis, Bidens tripartita, atitra stramonium, Stachys annua, Lamiiim amplexicaule, Iris pseudacorus,* and *Typha latifolia* var. b *media.* Some of these are no doubt garden-escapes.

Teeth And Flints Frimi The Ceuvei. Of The Ravexsbourse Valley. *The South EtiHlern Xaturcilmt,* 1'.HM!.

We now proceed to a point on the Beckenham side of the valley, somewhat over half a mile west from Bromley South railwaystation, where I had the fortune to alight upon a terrace-gravel not indicated on the one-inch Drift-Map. The gravel is disclosed in a small disused pit at the top of a meadow foot-path. I have marked it on my rough hand-drawn six-inch map, and invite your attention to a section of the valley at this point; the section is reproduced in the third photograph (pi. iii.). The original drawing is made to a horizontal scale of 108 inches to the mile, and a vertical scale of a quarter of an inch to the foot. The line over which it extends is slightly more than half a mile in length from east to west, and starts on the east at a point one mile to the south of Bromley South railwaystation, and a mile to the west of the main road to Farnborough. The terrace-gravel is 25ft. above the valley-plain, and nearly half a mile distant from the present stream. When the river was running here, it may be assumed that the gravel was the lowest point in the locality. Supposing that the slope upwards from the *then* river was at the same angle as it is from the existing stream, we are again impressed with the magnitude of the erosion which has occurred in this locality. We now take a leap of two miles to the large gravel-pit near Hayes railway-station where, as previously stated, two now extinct rivers met. The gravel here is about a quarter of a mile wide, and 15ft. thick. It is impossible to say for what length these dimensions are sustained. The pebbles, sand, and clays of the Woolwich and Beading beds, which rise to 315ft.O.D., on West Wickham Common, close by, have all been carried away, and the old rivers cut deeply into the Thanet beds. At the pit, the surface elevation is 210ft. Occasionally a clean perpendicular section is exposed, and exhibits 2ft. of vegetable mould, 15ft. of river-gravel, and 8ft. of white Thanet sand, resting upon unworn green-coated flints. Many of the water-worn flints are of great magnitude, and weigh as much as 18 lbs. This conveys some idea of the power of the stream which rent them from the chalk a few miles higher up the valley, and brought them hither. The gravel is not adapted for the preservation of bones. Even the molars of *Elepha primiyenius* become disintegrated, as indicated by the specimen (numbered 9) in the second photograph (pi. ii., fig. 9). The molars of *Rhinoceros antiquitatis* and *F.quus caballus* (numbered 10, 11, and 12 in the same photograph) (pi. ii., figs. 10-12) are better preserved. The bones (numbered 18 and 14) (pi. ii., figs. 18-14) from the same gravel are too fragmentary for determination. These remains, and the flint implements (numbered 15 and 17) (pi. ii., figs. 15, 17), were all found at the bottom of the gravel. It is somewhat remarkable that in the late Mr. S. Laing's book on " Human Origins," there is an illustration of an implement found in Somali Land, which is of similar form to the one numbered 16 (pi. ii., fig. 16), found a little further up the valley, and, at the Horniman Museum, Forest Hill, there is an implement from Egypt which is almost a replica of the one numbered 17 (pi. ii., fig. 17). A year ago the workmen informed me that they uncovered a bone as large as a man's body. It was resting upon the Thanet Sand, and when they attempted to move it, it crumbled to powder.

Silicified cretaceous fossils, and lumps of pudding-stone, the latter probably from the Woolwich Beds, are occasionally met with.

1 produce a specimen which is a much worn silicified shell of *Inoceramus mlcatus,* and also two valves of a bivalve mollusc, apparently of the genus *Mactra,* separated by a wedge of flint.

Of the indigenous plants found in this pit, the most rare are *Senecio riscosus* and *Jasione montana,* which are usually abundant in the season. As garden-rubbish is shot there, casual aliens frequently occur. The most interesting I have recorded are *Alyssum incanum, Camelina sylvestris, Anthemis tinctoria, Calendula officinalis,* and *Phalarix canariensis.*

I can hardly leave this portion of the river-valley without mentioning the earth-works on West Wickham common, just above the gravel-pit. The plan reproduced in the third photograph (pi. iv) was made from actual measurements, and is intended to represent the works as they were originally formed. Elevated points are represented by light shading; sunken portions by dark. The area enclosed is about two acres. The outer bank has in places been levelled and the trench filled in by the construction of a road and footpaths. On the west side, a segment has been bitten out by the quarrying of pebble-gravel. The remains, however, indicate that it was a pyriform work. Camden refers to it as having been "recast" in the 15th century for training the country people.

It will be seen that my plan shows, at the narrow or southern end of the enclosure, a raised tumulus surrounded by a shallow trench. The diameter of the tumulus is 80ft., and its height about 12ft. At the opposite end is a hollow-topped tumulus. The edges of this are raised by a gradual slope to a height of 4ft. above the general level. The cup-shaped hollow is 12ft. in diameter, and 2 ft. deep, but as there is a thick deposit of vegetable mould at the bottom, it was originally much deeper. Towards the north side are two small hollows and a small mound.

It cannot be supposed that a defensive work of this character was formed at any great distance from a water-supply. The only place we can look for it is the extinct river below. Probably this was not represented even by a brook within the historic period. I am therefore inclined to regard this work as having been originally a Neolithic hill-top camp.

The course of the extinct river which flowed to the Hayes pit from Chelsham and Farley was, as near as may be, along the road from Addington. Nine years ago, *Verbascuin nit/rum* was abundant along the roadside. It is the only indigenous species of the genus which has a perennial root-stock, but that fact, to far from protect

'. AS J J

I ing it, will, I fear, lead to its extirpation. People are attracted by its beautiful flowers, and, having discovered that it will live on year after year, they dig it up in order to plant it in their gardens. I have so frequently seen this plant growing over old river-gravels that whenever I meet with it I look about for evidence of the former existence of a river.

Turning to the more important branch which came from Tatsfield, this also is marked by a road running along its former bed for upwards of a mile. The river bank on the north side is now represented by a steep slope, with occasional terraces, leading up to West Wickham and Hayes Common. To the south is the arid implementiferous tract of chalk-land, referred to in the early portion of this paper. At about a mile from the Hayes pit is the large disused pit also mentioned above. Here large water-worn flints are abundant, but I have not heard of any implements or mammalian remains being found.

The pit is enclosed, and is under the protection of a gamekeeper. Some years ago the late Captain Torrens gave me permission to botanize there. Plants of *Verbascum thapsus, V. lychnitis,* and *T. niijrum* may generally be found in this pit. In 1908 I found there a single plant only of each of the hybrids between *Verbascum thapsus* and *V. lychnitis;* and *V. thapsus* and *V. nigrum.* The same crosses occurred freely in my own wildflower garden, but never a cross between *lychnitis* and *nigrum,* although that hybrid is listed in the *London Catalogue.*

The surface level at this pit is 250ft. O.D. The valley-plain is here narrowed, and the ground rises steeply on both sides. On the south, river-flints are strewn plentifully over the land for a distance of half a mile, and to a height of quite 50ft. above the valley plain. It is upon this southern slope, which is in West Wickham parish, that so many Paheoliths have been found, first by Mr. George Clinch, and subsequently by Mr. Santer Kennard and others. Let it not be thought, however, that one has only to run down to the place to gather implements by the score. I have for many winters devoted an occasional half-day to prowling over the freshly-ploughed land, with a pound of clay clinging to each foot, and the only implement I have found is the large tongue-shaped one which is numbered 16 (pi. ii., fig. 16). The farm-bailiffs and labourers have been so well instructed by Mr. Kennard and others, for whom they reserve their finds, that there is little chance for strangers.

At a short distance beyond the last pit, the orange colour which indicates valley-gravel on the Drift-Map terminates. Nevertheless, the gravel is continued below the "clay with flints " until it enters the deep valley which runs parallel with the Westerham Road, to Biggin Hill, Cudham, and thence towards Tatsfield. On the west side of the road, at Biggin Hill, Norhead's Farm rises like an island in the middle of the valley to a height of 500ft. O.D., the valley-plain at this point standing at about 400ft. I have gone over the high ground at the farm to seek for traces of Lower Eocene beds, which undoubtedly once covered the chalk there. The small black pebble (numbered 19 in the second photograph) (pi. ii., fig. 19) was my only find, and this, I think, bears traces of workmanship upon it. I showed it to a foreman at the Hayes pit, who, four months later, found there the pebble numbered 20 (pi. ii., fig. 20), which so closely matches the other.

In regard to the climatic and other conditions which prevailed when this and other deep, and now dry, valleys in the chalk of the North Downs were carved out, I think that no one who has examined the localities will dissent from the following propositions: 1. That the remains of the mammoth and woolly rhinoceros found in the gravels suggest that a severe climate prevailed in the region whence the water which brought the gravels flowed.

2. That the streams which cut out the valleys were of considerable volume and force. 8. That this leads to the inference that a wide and elevated catchment area once existed which has since disappeared. 4. That such catchment area was afforded by the North Downs having then extended southward over the Holmesdale valley to, and beyond, the Greensand ridge.

Even now, the chalk at Old Terry's Lodge, near Wrotham, stands at 770ft. O.D. At Ide Hill, and Crockham Hill, on the Greensand ridge, we get 700ft.

, although the chalk has disappeared. There are patches of Lower Eocene pebble-gravel still remaining at high elevations on the North Downs, suggesting that those beds once covered the whole area. Mr. Clement Reid's discover' of Pliocene fossils in the pipes in the chalk indicates that beds of that system also once overlay parts at least of the North Downs.

Now, starting with 700ft. as the height of the Lower Greensand ridge, and adding to that the Gault, Chalk, and Lower Eocene beds, we should have an elevation of at least 1,500ft., which we believe extended over about 100 square miles.

The character of the extinct mammals is suggestive of long and severe winters. Given that the whole of the winter's fall was snow, which accumulated on the elevated area, and was rapidly dissolved in a short, hot summer, all the factors for the excavation of the valleys are supplied. In conclusion, I hope it will not be regarded as presumptuous if I, as a naturalist, who is now in "the sere and yellow leaf," impress upon our younger friends the supreme importance of work in the field. We hear much of nature-study. That which is so called is too often nature without the study, or study without the nature. There are too many arm-chair naturalists; and far too much manufacturing of books out of books.

By all means read everything upon which you can lay your hand relating to your special branch, but take it for granted that much which you find in popular literature is incorrect. Such literature frequently contains a large proportion of fancy to a very small proportion of fact. By constant observation you will find that the writers of our serious text-books were by no means infallible. Authors of the highest repute, as Lyell, and the Hookers, and Bentham, sometimes made a slip. Locke's advice that " they who would advance knowledge should lay down this as a fundamental rule, ' not to take *words* for things,'" was never more needed than it is to-day when of " making many books there is no end." Therefore, whenever it may be possible to do so, test by your own observation the accuracy of all statements in regard to natural phenomena which you may hear or read.

The instruction I have derived in working out the Geology of the Upper Kavensbourne valley induces me to advise the young people present, first to read Huxley's "Physiography," and then to examine as thoroughly as possible the nearest river-valley to their place of residence.

Pevensey And Its Lords.
By Rkv. E. E. CRAKE, M.A., Rector of Jevington.

What a wonderful old ruin Pevensey Castle is—there is nothing like it in Sussex—scarcely in England! It is not in majestic, though fallen, buildings that the charm lies—Pevensey possesses them not—but in its hoary antiquity. Was there ever an era when Pevensey was not a great stronghold—we trow not. Its very position marked it out as a place of refuge, even in the time of the Ancient Britons. In the dim bygone ages, when history was unwritten, when the Romans bad not found us out, when Pevensey was an island, and the waters of the Channel encompassed it, and ran far up into the country to where Lewes now stands, Pevensey must surely have been a stronghold.

Two thousand years ago the Romans came and begirt the island with mighty walls. It was just opposite Port Itium on the continent, and they made it their seaport. That gate leading into Westham was the great Porta Decumana, the watch towers on the walls are of their building. You mark the solid stone blocks of which they are built, like the Porta Nigra at Treves, and the thin line of red brick which served to bind all together. Those Romans were mighty builders! The Saxons called it the Andredceaster, or the camp of the Andredsweald, for it stood in sight of the mighty forest which, for 140 miles, cut off our county from the rest of England. In the Notitia of Pancriollus it is called Anderida. I know that seven other sites are claimed for this lost city, but antiquaries have now generally come to the opinion that our Pevensey is the true one. Then the time came, it was about 404a.D., when the Romans departed and left the luckless Britons to themselves. Was it here that the Britons came down to the shore and bewailed the departure of the legions? Was it from Anderida that they wrote that pleading letter to the Roman consul, "the barbarians drive us into the sea, and the sea throws us back on the barbarians." What had become of their own young men—had Britons become so degenerate that there were no stout hearts and strong arms among them to beat back the fierce Picts and Scots?

They had been drawn off from the Island to fight Rome's battles abroad, and Constantine had no finer soldiers in his legions than the Britons. So they were left to themselves, and we find that they forthwith took possession of Pevensey (as we call it) and made it their great southern stronghold. Its strength was soon to be tried. The *Saxon Chronicle* tells us: "491 A.d.—This year Ella and Cissa besieged Andredceaster, and slew all bhat dwelt therein, so that not a single Briton was there left." In the "Chronicle of Fabius Ethelwerd" the date is put "492 A.d.," but the account he gives of the event is much like the other—perhaps he had it before him—it runs thus: "492 A.d.—After three years jElla and Cissa besieged a town called Andred-Cester, and slew all the inhabitants, both small and great, leaving not a single soul alive." It was in 477 A.d., that Ella and his three sons, Cissa, Cymen, and Wlencing, landed at Cymenesora—now Heynor—near Wittering. Cissa took Chichester, which he called Cissanceaster, and with fire and sword the fierce invaders drove the Britons eastward into the fastnesses of the mighty forest of the Andreads Weald. At Mercreadsburn—probably near Seaford, in the valley of the Cuckmere—the decisive battle took place in 485 A.d., and the vanquished Britons took shelter behind the Roman walls of Anderida, which they dreamed to be impregnable. There they were besieged, and Henry of Huntingdon gives us the following account of it. "The Britons then collected like bees and beat the besiegers in the day by stratagems, and in the night by attacks. No day, no night, occurred wherein unfavorable and fresh tidings would not exasperate the minds of the Saxons; but rendered thereby more ardent, they beset the city with continual assaults. Always, however, as they might assail, the Britons pressed them behind with archers, and with darts thrown by thongs, wherefore, quitting their walls, the pagans directed their steps and arms against them. Then the Britons, excelling them in fleetness, ran into the woods and again came upon them from behind as they approached the walls. By this artifice the Saxons were long annoyed, and an immense slaughter of them was made, until they divided their army into two parts, so that while one part should storm the walls they might have a line of warriors arrayed against the charges of the Britons. But then the citizens, worn down by long want of food, when they could no longer sustain the multitude of assailants, were all devoured by the sword, so that not an individual escaped."

The Saxons now occupied the fortress and strengthened it— repairing the breaches they had made in the outer walls. You may still see evidence of their work—especially on the southeastern side of the outer walls, where the "herring bone" masonry is conspicuous. According to Henry of Huntingdon, the great Romano-British stronghold was devoid of buildings within its mighty walls until the Normans came in 1066 A.D. But the "Portus Anderida" is frequently mentioned by the chroniclers. Earl Godwin and his warrior son, Harold, entered it, and captured the shipping that lay there—and here his son, Sweyn, landed with eight ship3 when he returned to England after temporary banishment. It was on September 26th, 1066, that Duke William landed at Pevensey. The Bayeux Tapestry tells us of this: "Hie Wilhelm Dux, in magno navigio mare transivit et venit ad Pevenesani," and the "Chronicle of Battle Abbey" records the same event; "Dux ergo.... navigationem aggressus, prospere tandem prope castrum Pevenesel dictum applicuit." What a wonderful scene the watchers from the lofty Roman walls must have witnessed on that fair September evening. They had watched the mighty fleet of over 1,000 vessels approach the shore—the sun gleaming on the white sails and flashing on the steel armour of the 60,000 men-atarms who crowded the vessels. They had sailed from S. Valery the day before, and a gentle wind had wafted them safely across the Channel. There were none to resist them—it was unlike the landing of Caesar at Deal on August 26th, 55 B.C, when the courageous Britons had rushed into the water to encounter their foes. It was the eve of S. Michael, and amid shouts of rejoicing, and the flourishing of trumpets, William had leapt ashore. As he leapt he fell — perhaps encumbered with his heavy armour—and the onlookers were dismayed at the omen. But William, with ready wit, grasped a handful of sand, and holding it aloft exclaimed, " Thus I take seizin of mine inheritance." The great army formed up on shore— encamped for the night, and, leaving a guard for the fleet, commenced the next day to march along the ridge of hills which lead to the field of Senlac. We are not told whether they occupied the stronghold, but there is little doubt of it, for William was a prudent commander, and would make the base of his operations secure. It needs not to tell the story of the great battle with Harold and his huscarles. There Harold and his brothers, Gurth and Leofwine, fell—you may see the exact spot in the grounds of Battle Abbey. There fought, by the side of William, Robert, Earl of Mortain—the future Lord of Pevensey—we are told that he "never went far from the Duke's side, and brought him great aid." There fought Robert, Earl of Eu, and for his services received the Rape of Hastings, which included Bodiam and Herstmonceux. There fought William de Warenne, who obtained the Lordship of Lewes, William de Braose, who became Lord of Bramber, Roger de Montgomerie, afterwards Earl of Chichester and Arundel, De Monceux and De Aquila. But we turn to our story of Pevensey.

From the time of the Norman conquest its port began to decline in importance, though, even in the time of Henry III. (1220), it is spoken of as " a considerable port" ("Charters of the Cinque Ports"). Later on, great changes in the coast line, eastward of Beachy Head, took place, and, eventually, Pevensey was left more than a mile from the sea, and "rich perennial pastures took the place of a watery plain" (Horsfield). There is little doubt but that Robert de Mortain commenced to build the present Norman Castle—and accomplished so much of it that, within a few years of his death, it endured and withstood the attack of a besieging army for six weeks. Odo, the turbulent Bishop of Bayeux, had retreated thither from Kent—to take refuge with his brother, the Lord of Pevensey. On

A the approach of famine, he surrendered to the forces of Rufus— with whom eventually he made his peace. In the reign of Henry I., the then Earl of Mortain rebelled against the king, and, being taken prisoner in Normandy, was committed to the Tower for life, and his eyes were torn out. It was a barbarous deed—but a very common one in those days. Is it not told us that the king put out his brother Robert's eyes after he had shut him up in Cardiff Castle? Henry now bestowed the Lordship of Pevensey upon the famous family of De Aquila—commonly known as the "Lords of the Eagle"—and, in this "lordly line," it remained for more than a century. Gilbert de Aquila was returning with his royal master from Normandy, when the "White Ship" was wrecked, carrying to a watery grave three children of the king and two of de Aquila's. These de Aquilas were great warriors and magnificent founders of religious houses. Gilbert de Aquila, the Third, founded Michelham Priory, a house of the Augustinian Canons, for the health of his soul, and the soul of Isabel, his wife. One is glad that he thought of his wife as well as of himself!

In 1235, Gilbert Marshall, Earl of Pembroke, became Lord of Pevensey— he lost his life at a tournament at Ware in 1241. Time fails to tell of the siege of the castle by Simon, Earl of Montford, and again by the forces of King Richard when Dame Pelham bravely defended it—her Lord being in the north with Henry of Lancaster. Queen Joan of Navarre—the last wife of Henry IV.—was his prisoner here. Another royal prisoner was Prince James of Scotland—afterwards James the First of that realm. We read that Sir John Pelham was allowed JE700 a year to provide his captive with food, clothing, and all necessaries—a huge sum in those days. In 1460, Sir William de Fiennes was made Constable of the Castle for life—a member of the family of the Lords of Herstmonceux. In 1620, " the honour of Pevensey was in the Crown," Horsfield tells us. The Castle passed through various hands, and is now the property of the Duke of Devonshire —a worthy successor to the " Lords of the Eagle."

Michelham—-A Sussex Priory.

By Rev. E. E. CRAKE, M.A., Rector of Jevington.

A walk to Michelham, when spring has well advanced, is a delight to one who loves rural scenery.

The ancient Priory is approached on the east and west sides by roads leading through woods (which have little varied in character since they formed a part of the impenetrable forest of the Andredsweald), sweet-scented with bluebells, wild violets, anemonies, and many another woodland charm. The primroses line the banks and gleam in the glades of the wood, and as you approach "The Hyde" —as the wide expanse of the common is called—masses of furze, just bursting into a yellow sea of blossom, add their charm to the scene.

The rabbits scamper away in droves, and you do not wonder that, in the dietary of the monastic houses around—and of more pretentious places, such as Hurstmonceux Castle—rabbits formed a leading feature. I wonder why it is that hares find no place in the dietaries we have left us of Hurstmonceux in the times of Sir John de Fiennes, the founder of the Castle. There is no mention of them, and yet they must have abounded then as now.

As we approach the Priory it strikes you as a place built for a secluded life—where men might live and die—
'' The world forgetting.
By the world forgot."

No public road leads to it, it is far from the towns and the " busy haunts of men." But this was not always the case. When the Priory was in its glory, the great main road from Lewes and the Valley of the Cuckmere passed close to it, leading to Battle, Hastings, Bexhill, etc. Thus we read that Prince Edward, on his way from Hurstmonceux to the fatal field of "The Mise of Lewes," halted here for the night, and he visited it again 20 years later by the same road, when he was King Edward the First. Again, we read that on June 16th, 1283, the Priory received a visit from the Archbishop of Canterbury (Peckham), who had spent the previous day at Battle, and left next day for Bexhill, returning on the 18th to the Priory, and leaving again the next day for Rochester. But that day is passed, and the world's mighty ones do not visit Michelham now-a-days.

The Priory is surrounded by a lovely moat, five and a half acres in extent, it is fed by the waters of the river Cuckmere, which passes through it. The banks of the moat are beautifully fringed with maple, elm, and willow, and the water is well-nigh covered with a variety of aquatic plants, water-lilies growing luxuriantly. It teems with fish—once the main supply of the refectory of the Priory, especially on fast days. Pardon a slight digression, a reminiscence occurs to me. I was once visiting the place, rod in hand, and I sought permission to fish from the then tenant of Michelham. "Do you come from Eastbourne," he asked. "No," I replied, " But why do you ask that?" "Oh," said he, laughing, "Because *they never catch anything!"* But "revenons a nos moutons!" We approach a noble embattled stone tower, which is called "The Gateway," the entrance into which is over a strong bridge, which spans the moat. It is of later date than the rest of the Priory, and was, perhaps, built by Prior John Leem, who ruled here from 1876 to 1416. Under the tower is a lofty arch, of the Tudor style, over which are four small gothic windows, with trefoil heads; below, level with the moat, is a dark apartment called the dungeon. You ascend by a stone staircase to two large rooms —now most carefully restored—possessing fine chimney places, and so up to the leads, whence a fine view of the surrounding country can be obtained. The Priory grounds, enclosed by the moat, are about eight acres in extent. The building was occupied for many years by a tenant farmer, and the part of it which was used as a farm house was kept in a fair condition, the rest was falling into utter ruin.

The present owner (J. E. A. Gwynne, Esq.) commenced its restoration some ten years ago, and the work which he has accomplished deserves the highest commendation. Slowly and carefully every part of the old building is being renovated, room after room has been lined with stout oak panelling, adorned with beautifully carved bosses here and there. Noble chimney pieces have been disclosed to view where they had been blocked up, fine stone arches reconstructed, all in the spirit of the ancient days, and the grand crypt, which had been used as cellars, and was subdivided, is once more as perfect as it was 500 years ago! Horsfield tells us of this crypt: "Springing arches support the groined roof, and concentrate on the capital of a massive round pillar in the centre. The intersections of these arches are ornamented with an arager's head, a rose, and other devices."

Mr. Salzmann, in his fine *History of Haihham,* devotes some 60 pages to Michelham Priory. He says of this crypt: "It is a square chamber with very massive walls; it was recently divided into four small rooms, but has now been opened by Mr. Gwynne. The handsome vaulted roof is supported by massive ribs springing from a circular central pillar, and eight corbels, two on each wall; at the intersection of the ribs are circular bosses, of which one is carved into the semblance of a clownish face, the second bears an heraldic rose—possibly added at a later period—the third appears to have the emblem of the Holy Trinity slightly cut upon its surface, and the fourth is plain."

The monks' cells have disappeared—we read that they were " in bad repair" at the Dissolution. The church was utterly destroyed when the end of the Priory came, its very foundation cannot be traced.

The refectory was unroofed, and the upper part of its walls thrown down, subsequently it was divided into two stories, and cut up into a number of small rooms. There are quaint passages, notably one called " The Slype," windows with fine tracery of the decorated period, but of the monastic building little remains. We now turn to its history.

It had a noble founder—Gilbert de Aquila, Lord of Pevensey, and third of his name. Dugdale gives the actual charter of endowment. It runs thus: "I, Gilbert, Lord Aquila, by the permission of King Henry III., for the welfare of my soul, and the souls of Isabella my wife, and of my children, of my brothers and sisters, of my ancestors and descendants, have given to God and the Church of the Holy Trinity of Michelham, and to the Prior and Convent of Canons serving God in that place," etc. It was richly endowed with manors, churches, and lands.

We learn that its income in 1291 amounted to £1500 of our money, and this was increased by subsequent benefaction, so that at the Dissolution it was little less than £2800 of our money. "The order of monk placed in this House," says Horsfield, "was that of Canons Regular of St. Augustine, sometimes, from the colour of *their hair,* denominated Black Canons." I wonder whether Mr. Horsfield supposes that all the monks of the Order had *black hair.* We should have supposed that the name " Black Canons" was given to them on account of the *coarae black robe* they wore.

The Canons Regular of St. Augustine were, perhaps, the least ascetic of the monastic orders. Enyot de Provins, writing in the thirteenth century, says of them: "Among them one is well shod, well clothed, and well fed. They go out when they like, mix with the world, and talk at table." They were little known in England till the tenth or eleventh century, and their first house in this country was at Colchester in the reign of Henry I. They increased so rapidly in number, that, according to the " Monasticon," 216 houses of their Order were established in England.

The Priory of Michelham was founded in 1229, and one Roger appears to have been the first Prior. It had accommodation for twelve or fourteen inmates, though at the time of the Dissolution we find that the number had fallen to eight. None of these Priors were distinguished men, the one of whom we know most was John Leeiu, who was probably a native of Willingdon. Through his efforts the churches of Alfriston and Fletching were obtained for the Priory in 1898, and other large donations in lands made to it. The last Prior was Thomas Holbeme, who ruled the house from 1518 to 1587. It fell with the "lesser monasteries," *viz.,* those whose early income amounted to less than £200 per annum (money of the period). It thus preceded by two years the fall of the great houses (such as Battle and Lewes), of which the preamble of the first Act of Suppression had spoken as being "great solemn monasteries wherein (thanks be to God) religion is right well kept and observed." Michelham Priory had not an unblemished reputation—probably its monks were no better and no worse than their neighbours—but we know enough of it to gather with certainty that the monks had fallen from the high ideal with which Gilbert de Aquila had founded it. The form and ceremony remained, but the soul had departed from the lifeless body, the monks had become careless and indifferent, they were despised by the poor, and despoiled by the rich—and so the end came!

The Priory was given by Henry VIII. to Thomas Cromwell, Earl of Essex.

Notes On The Flora Of Eastbourne As Observed During The
Congress.
By W. H. GRIFFIN.

It has been my privilege to attend our annual gatherings for several years in company with Mr. B. T. Lowne, ray co-Secretary of the Catford and District Natural History Society. Upon each occasion we have, by rising early, and making use of every opportunity between the meetings, acquired some knowledge of the flora of the localities visited. As I shall have occasion to refer to the comital numbers given in *The London Catalogue of British Plants* it is well to explain that the late Mr. H. C. Watson, in his *Topographical Botany,* divided the British Islands into 112 vice-counties, naming those in which each species had been recorded. Since Mr. Watson's death other botanists have added to his vice-county records, and the comital number stated in the London catalogue for every indigenous species and naturalised alien has been arrived at by coordinating the records.

On the first day of the Congress, Mr. E. M. Holmes and Dr. Treutler, two members of the Union, who are familiar with the flora of Beachy Head, very kindly assisted me in exploring the slopes below the hotel, and the piece of undercliff below Cow Gap. Here *Rosa pimpinelli folia, Orchis usttdata,* and *Habenaria viritiis* were observed. The gentlemen named had both seen *Bipleuriim aristatum* on the grassy slopes in former years, but we failed to find it on this occasion. The *London Catalogue* comital number for it is only 2, but we believe it has been found in two or three new stations since the last edition was published. We were pleased to note the plentifulness of *Raphanus maritimus* about the gap leading to Holywell. The comital number for it is 26. In the *Flora of Kent,* by Hanbury and Marshall, it is said to have been recorded at Broadstairs, and no other station for it is mentioned in the county.

Another rare and handsome plant seen about the chalkcliff was *Lavatera arborea.* The comital number for this is 15. In the *Flora of Kent* it is mentioned as recorded at Broadstairs, Walmer, near Dover, and under the Lees at Folkestone. Except for the station near Dover, the Kentish records are old and somewhat doubtful. Upon a subsequent occasion we explored the upper part of the beach and the banks above it, beyond the western end of the Parade. Many plants of *Crambe maritima,* not yet flowering, were seen growing on the shingle *below* high-water mark, as well as on the chalk above that point. The comital number for this species is 32. A matter which struck us as singular was that, whereas the great majority of the plants of *Diplota.ris muralis* observed here were typical, the majority of those seen in west Kent and east Surrey are of the variety b. *babinytonii.* Miss Milner, B.A., who is, we believe, since the recent lamented decease of Dr. Whitney, the most active field-botanist of the Eastbourne Natural History Society, very kindly became our mentor and guide in a most interesting botanical ramble over the shingly flats, and amongst the marsh-ditches between Eastbourne and Pevensey. *Silene maritima, Saponaria officinalis,* and *Trifolium subterraneum* (comital number 89) are all plentiful on the grass-clad shingle, and we also saw there *Trifolium striatum, Vicia yemella* (= *tetrasperma),* and *Lathyrus nissolia.* The very beautiful but fragile water violet *(Hottonia palustris)* was most abundant in the ditches on both sides of the Pevensey road. There we also obtained *Myriophyllum spicatum, Utricularia vulgaris,* and *Lemna trisulca.* The fronds of the latter were remarkably large, a single frond measuring in. x Jin., whereas the largest frond we can find in many specimens taken from the Thames marshes below Dartford is in. x TVn-On an average, the Eastbourne plants are about twice as large as those we have seen in the Thames marsh-ditches.

A most interesting problem is presented by comparing the Flora of Eastbourne with that of Dover. It may be stated in the following questions:— Why should *Raphanus maritimus* be plentiful on the chalk of Beachy Head and very scarce (even if it occurs there at all) on the chalk at Dover, while with *Brassica oleracea* the opposite is the case? They are both cruciferous plants. Why also should *Silene nutans* be plentiful below Shakespeare's Cliff at Dover and absent below the cliffs of Beachy Head? The difference in latitude is slight. Judging from the geology, the chemical constitution of the soil must be very similar, and we are led to suppose that some small difference in the meteorological conditions is the controlling factor. Exigencies of space must not prevent an expression of our thanks to Mr. Holmes, Dr. Treutler, and Miss Milner for their kindness in imparting information and pointing out localities, also to our kind host and hostess, Mr. and Mrs. Hollway, for accommodating their domestic arrangements to our unconscionable hours. Nothing could exceed the kindness we experienced at Eastbourne on every hand.

Nature Study (summary Of An Illustrated Address).

By WILFRED MARK WEBB, F.L.S.

Mr Webb said:—Two years ago, at the Maidstone Congress, I had the honour of presenting to this Union a formal paper on " The Teaching of Nature Study." Since then I have found no reason to change my opinions to any material extent, so that having been invited to deal with the question again at Eastbourne, I think it best, by way of a change, that I should offer some general remarks, and show, with the help of lantern slides, what has been or might be done.

Every child or creature inherits potentialities, and it requires a certain environment to develop them properly. Our life has become more and more artificial, and the education given to children in the 19th century was found for this reason to be open to several grave objections. Speaking generally it created no interest, it was unnatural, it repressed activity, and did not satisfy inherent longings and instincts such as those for exploration or travel, hunting and the like.

As a matter of fact it has been pointed out that many of the most celebrated of men have been those, like Thoreau, Robert Louis Stevenson, Walt Whitman, and Darwin, who broke away from the trammels of school, played truant and educated themselves in certain important directions. The education of the schools, of which the feature has been the imparting of second-hand information, has been based on the assumption that the mind of the child is like that of an adult, and the result has been the loss of power on the part of pupils of acquiring knowledge for themselves and of acting under newly arisen circumstances.

It may be mentioned that the folly of manufacturing machines in a similar way in the army has at last been recognised, and the whole system of military training has been changed.

The aims of Nature Study are to obviate the objections to ordinary education, which have been described. In its highest form it should be carried on out-of-doors and information given must be reduced to a minimum. Bearing in mind also that we are dealing with a part of the general education of young children who should "face existence whole," we should, while insisting on accurate observations and deductions, avoid the introduction of logical connections that would turn our work into what is called elementary science, which should come at a later stage. The brain of the child stores up this and that piece of experience that interests it, and later on each fragment is assigned to its proper position with regard to the others. I do not, however, suggest that observations on one particular object should not be continued, for the value and interest of them is thereby increased.

There are many difficulties in the way of the teachers, and I would not minimise them in the slightest. I cannot help thinking, however, that, in many cases, the chief one is the jumping of the chasm that separates "thinking that nothing can be done" and "seeing whether something could not be attempted."

To prove this Mr. Webb described, with the help of lantern slides, what has been done in town schools which, though most severely handicapped in many ways, often produce some of the best work. The various questions of observations on clouds and their forms, the apparent movement of the sun, pets, school gardens (as gratifying the desire to possess property and for observation work, rather than for manual or technical instruction),were next gone into. Outdoor school rambles and excursions were also discussed and photographic records dealt with.

The materials for Nature Study, it was pointed out, were endless, and the choice depended on local conditions. A teacher of Nature Study would be the better for a scientific training in biology at least, but it has been shown to be possible for the teacher to begin with the children, for the whole object of nature study is not to afford information, but, by the gratification of properly directed curiosity, to develop mental powers. What the specialist considers of interest, few others do, and it would be better even for the teacher with a scientific training to take up some branch of work that is new to him when he wishes to introduce nature study. At the same time the direction taken should be congenial to the master and pupils. To fulfil the latter condition it may be necessary to lead off from something that is already known to the children.

By way of offering further suggestions, Mr. Webb dwelt on the apparently unpromising subject of collections, and showed photographs of cases illustrating the various sides which should be found in a school museum. Having already mentioned the living side under the heading of pets, he dwelt on preserved specimens. On the nature study side there were records of work done and suggestions as to what might be pursued, and in a few cases special slides were shown carrying the matter further. Incidentally an effective but inexpensive way of making a school museum, without monopolising floor space, was described. The importance to children of collecting was characterised by the seeking and finding, and not in the killing and hoarding, of specimens. Allusion was made to the need for directing attention to groups of plants and animals which were not beautiful in the eyes of the public, and to the aid which might be given by schools to naturalists studying neglected classes.

In conclusion, Mr. Webb appealed for help in working out the distribution of Centipedes and Millipedes, offering in return a printed pamphlet of hints and a supply of collecting tubes.

LIFE MEMBERS. *(Instituted at the Canterbury Congress, 1902.)*

Ailkin, Robert, F.E.S. 4, Lingards Koad, Lewisbam, S.E. (Hon. Treas.)

Adkin, Mrs. B. ,,,,

Bennett, F. J., F.G.S. "The Acacias," West Mailing, Kent.

Brown, B. Weir, LL.B. Langley Tower, Eastbourne.

Bullen, Rev. B. Ashington, F.L.S., F.G.S. The Locks, Hurstpierpoint, Sussex. (Hon. Sec.)

Bullen, Mrs. Ashington. The Locks, Hurstpierpoint, Sussex.

Coomaraswamy, Ananda K., F.L.S., F.G.S. Walden, Worplesdon, Guildford.

Foran, C. Elm Grove, Southsea.

Gray, H. Norman, P.A.S.I. "Belvedere," Woodville Road, New Barnet, Herts. (Autumn Meetings' Sec.)

Howorth, Sir H. H., K.C.I.E., F.R.S., F.G.S. (V.P.) 30, Collingham Place, Earl's Court, S.W. (Ex-Pres.)

Meeson, F, 2, Marchmont Gardens, Richmond.

Merrifield, F., F.E.S. (V.-P.). 14, Clifton Terraoe, Brighton.

Neate, P. J., J.P. Watt's Avenue, Rochester.

Oke, Alfred, W. 82, Denmark Villas, Hove.

Reid, Miss Elizabeth. The Mount, Meads Road, Eastbourne.

Rudler, F. W., I.S.O., F.G.S., Ac. 18, St. George's Road, Kilburn, N.W. (Ex-Pres.).

Sargant, Miss E., F.L.S. Quarry Hill, Reigate.

Stebbing, Rev. T. R. R., F.R.S., F.L.S. (V.P.). Ephraim Lodge, The Common, Tunbridge Wells. (Ex-Pres.).

Stebbing, Mrs. T. R. R., F.L.S. Ephraim Lodge, The Common, Tunbridge Wells.

Stebbing, Miss Grace. Catton, Southborough.

Stirling, Sir J., Bart., F.R.S. Finchcocks, Goudhurst, Kent.

Turner, Miss E. L., F.L.S. Langton Green, Tunbridge Wells.

Vardon, Rev. S. A., M.A. Langton Green, Tunbridge Wells.

Walker, A. O., F.L.S. Ulcombe Place, Maidstone.

Walmsley, A. T., M.l.C.E. Atherstone, Castle Avenue, Dover.

Whitaker, W., F.R.S., F.G.S. (V.-P.). 2, Camden Road, Croydon. (Ex.-Pres.).

DELEGATES.

Adkin, Bobt., F.E.S. (South London). 4, Lingards Road, S.E.

Baldook, J. H., F.C.S., F.R.P.S. (Surrey Photo, and Croydon N.H.S.). 3, St. Leonard's Road, Croydon. Barton, A. (Maidstone and Mid-Kent). Sunny Croft, Holland Road, Maidstone. Bedford, E. J. (Eastbourne). Anderida, Oorringe Road, Eastbourne. Beadle, Clayton, (Sidcup). Halewood, Station Road, Sidoup. Blackie, A. (Tunbridge Wells). 50, Grove Hill Road, Tunbridge Wells. Brabrook, Sir Edward, C.B., F.S.A. (Balham). 178, Bedford Hill, S.W. Bullen, Mrs R. Ashington (Holmesdale). Hurstpierpoint.

Connold, Edward, F.E.S. (Hastings and St. Leonard's). 7. Magdalen Terrace, St. Leonard's.

Davey, Henry, M.A. (Brighton and Hove). 15, Victoria Road. Brighton.

Edwards, Stanley, F.L.S., F.E.S. (West Kent). 15, St. German's Place, Blackheath, S.E.

Findon, Hugh, F.L.S. (Hampstead). 58, Carlton Road, Tufnell Park, N.

Frisby, G. E. (Holmesdale). 9, Fengate's Road, Redhill.

Frost, R. C. (Woolwich Antiquarian). Sudbury House, St. John's Road, Plumstead.

Gilbert, E. G., M.D. (Tunbridge Wells N.H.S.). Cantley, Madeira Park, Tunbridge Wells.

Gray. H. Norman, P.A.S.I. (City of London College). New Barnet, Herts.

Griffin, W. H. (Catford). 6, Rutland Park, Perry Hill, S.E.

Gwinnell, W. F., F.G.S. (Polytechnic). 33, St. Peter's Square, Ravenscourt Park, W.

Hembry, F. W., F.R.M.S. (Sidcup). Langford, Sidcup.

Hepworth, J. (Rochester Nat.). Linden House, Rochester.

Kensett, Miss E. J. (Horsham). 100, New Street, Horsham.

Lowne, B. T. (Catford). Bromley Road, Catford, S.E.

McDakin, Capt. Gordon (East Kent). 12, Pencester Road, Dover.

Mathews, P., M.A. (Rochester Phil.). 3'2, South Avenue, Rochester.

Meeson, F. (Woking). 2, Marchmont Gardens, Richmond, S.W.

Moring, Percy (Dover). 23, Randolph Gardens, Dover.

Nicholson. C. S., F.L.S. (North London). 22, Crouch Hall Road. Crouch End, N.

Newington, F. (Lewes Photo. Soc). Lewes.

Otter, J. L. (Selborne). 20, Hanover Square, W.

Pankhurst, E. Alloway (Brighton and Hove). 3, Clifton Road, Brighton.

Parkin, T., M.A., F.Z.S. (Hastings and St. Leonard's).

Roberts, C. J., B.A. (Folkestone). 16, Cherilon Gardens, Folkestone.

Roods, A.. F.S.I. (Photo. Survey of Surrey and Croydon N.H.S.). 67, Thornhill Road, Croydon.

Rogers, J. T., (New Brompton). Athelney House, Gillingham.

Sanderson, Miss (Haslemere). Hindhead, Haslemere, R.S.O.

Sanderson, Miss E.,,,,,,,,,

Sparks, H. (Eastbourne). 2, Harttield Boad, Eastbourne.

Spitta, E. J., M.D. (Brighton and Hove). 41, Ventnor Villas, Hove.

Stenning, John C. (Sussex Photo.). Steel Cross House, near Tunbridge Wells.

Stedman, W. T. (Northfleet). Dover Lodge, Northfleet.

Trollope, W. T. (Tunbridge Wells N. H.S.). Hawthorndene, Tunbridge Wells.

Trollope, Mrs. W. T.

Vincent, W. T. (Woolwich). 89, Burrage Boad, Woolwich.

Webb, Douglas M. (Dover). 9, Waterloo Crescent, Dover.

Webb, W. Mark, F.L.S. (Selborne). Odstock, Hanwell, W.

Young, W. Plomer, F.E.M.S. (Battersea). 251, Lavender Hill, S.W.

Young, Mrs. W. Plomer (Morley Antiquarian)..,,,,,

Bursill, Philip C, Public Library, Woolwich (M).

Butler, Miss E. Hillside, Eastbourne (Ml.

Cawley, Rev. P. G., M.A. Youl Grange, Meads, Eastbourne (A).

Chapman, T. A., M.D., F.E.8. Betula, Reigate (M).

Chater, Miss Edith. 35, Jevington Gardens, Eastbourne (M).

Cbilcott, Miss M. 35, Hartfield Road, Eastbourne (M).

Connold, E., P.E.S. 7, Magdalene Terrace, St. Leonard's (Ref.).

Cooper, E. W. Greenhayes, Reigate (A).

Courtis, Miss R. Wannock Dene, Wannock (A).

Crafer, Mrs. 102, Beaconsfield Villas, Brighton (M).

Crowden, Rev. C, D.D. Derrailere, Carlisle Road, Eastbourne (A).

Cudworth, Mrs. Whitfield, Reigate (M).

Davis, Miss Clare, Hillside, Eastbourne (M).

Day, Walter E. 42, Earl Street, Maidstone (M).

Donisthorpe, H. St. J. K., F.Z.S., F.E.S. 58, Kensington Mansions, S.W. (Ref.).

Edmonds, F. Bernard. G, Clement's Inn, W.C. (M).

Farncombe, Alderman J. Saltwood, Eastbourne (M).

Fovargue, H. W. Cransley, Saffron Road, Eastbourne (M).

Foxley, G. 12, Grange Road, Eastbourne (M).

Fremlin, Ralph J. Heathfield. Maidstone (A).

Frisby, G. E. 9, Fengate's Road, Redhill (Ref.).

Frisby, Horace A. Hughenden, Upperton Road, Eastbourne (M).

Frost, R. C. Sudbury House, St. John's Road, Plumstead (M).

Frost, Mrs.,,,,,,,,,,; ,, ,, (M).

Gibb. Miss L. M. A. 21, Alwyne Mansions, Wimbledon (M).

Gibbs, S. 53, Terminus Road, Eastbourne (M).

Gilbert, H. V. Lynchmere School, Eastbourne (A).

Glover, H. J., F.C S. St. Catherine's, Westham, Hastings (A).

Green, Laurence, F.C.S. Oaklands, Maidstone (M).

Griffin, W. H. 6, Rutland Park, Perry Hill, S.E. (Ref.).

Griffiths, Inspector General R.S.P., R.N. Rosemead, Church Road, Horley (A).

Grinling, C. H, 17, Rectory Place, Woolwich (M).

Groves, James, F.L.S. 58, Jeffreys Road, Clapham Rise, S.W. (Ref.).

Gruner, Miss Joan B. Brackenhurst, Hindhead (M).

Habgood, H., M.D. Stafford House, Upperton Road, Eastbourne (M).

Habgood, Mrs.,,,,,,,,,,,,, (A).

Hannen, Hon. H. The Hall, West Farleigh (Ml.

Hardcastle, J. H. Public Library, Eastbourne (M).

Harris, Poulett, M.D. 98, Lower Addiscombe Road, Croydon (M).

Harper, A., M.D., 7, Chiswick Place, Eastbourne (A).

Hollams, Mrs. Dean Park, Tunbridge Wells (A).

Holmes, E. M., F.L.S. Ruthven, Sevenoaks (Ref.).

Hollway, J. J. Baylands, Orchard Road, Eastbourne (Local Sec, 1906).

Howse, Thos., F.L.S. Guy's Cliff, Richmond (Ref.).

Horrell, E. C, F.L.S. Copleston Road, Denmark Hill, S.E. (Ref.).

Hughes, F. Wallfield, Reigate (M).

Hunt, Gerald. Wannock Dene, Wannock (A).

Hutchinson, Prof. Jonathan J., F.R.S. , etc. Inval, Haslemere (V.-P.).

Jackson, G. H., M.D. Millgap Rond, Eastbourne (A).

Jay, Rev. W. P., M.A. St. Anne's Vicarage, Eastbourne (A).

Jay, Miss. St. Anne's Vicarage, Eastbourne (A).

Jenn«r, J. H. A., F.E.S. 209, School Hill, Lewes (M).

Jones, E. Dukinfield, F.Z.S. Castro, Reigate (M).

Kearton, R., F.Z.S. Ardingly, Caterham Valley (M).

Keeble, Fras. H. The Manor House,

Tatsfield (A).

Kelsey, Ellis. Malwood, Eastbourne (M).

Kensett, Miss E. J. Concord, 18, Barrington Road, Horsham (M).

Kerley, Mrs. G. Lismore Road, Eastbourne (A).

Kirwan, Miss M. B. E. 22, Bedford Grove, Eastbourne (A).

Klaassen, Miss A. C. Aberfeldy, Campden Road, Croydon (M).

Lemon, Mrs. F. Hillcrest, Redhill (A).

Littler, Frank M. Launceston, Tasmania (A).

Livett, Rev. G. M., B.A. Wateringbury Vicarage (M).

Lynton, Rev. W. R. Shirley Vicarage, Derby (Ref.).

McDakin, Capt. G. 12, Pencester Road, Dover (M).

McQueen, Thos. M. B. Bolton House, Eastbourne (M).

Marsh, Mrs. Woolwich (M).

Martin, E. A., F.G.S. 58, Whitworth Road, South Norwood (Photo. Seo.)

Masters, Mrs. Emma. 2, Bower Place, Maidstone (M).

Maude, A. H. Ivymount, Eastbourne (M).

McDonald, Miss Anne. 10, Blackwater Road, Eastbourne (A).

Miles, Miss Alice A. Fernside, Eastbourne (M).

Milner, Mrs. 6, Bedford Grove, Eastbourne (A.

Milner, Miss B. F., B.A (M).

Mitchiner, Philip H. Everton, Doods Road, Reigate (M).

Morgan, John. 10, Ambrose Place, Worthing (A).

Mooyaart, Rev. R., M.A. 33, Enys Road, Eastbourne (A).

Mooyaart, Mrs.,,,,,,,,, (A).

Munro, Wm. 138, Britton Street, Gillingham (M).

Newman, T. P. Hazelhurst, Haslemere (M).

Nicholson, C. S., F.L.S. 22, Crouch Hall Road, N. (M).

Nicholson E. A.,,,, (M).

Nicholson, W. E., F.E.S. Lewes (Ref.).

Nottidge, A. J. Yardley Lodge, Tonbridge (Ml.

Offley, Miss L. 28, Furness Road, Eastbourne (M)

Pankhurst, E. Alloway. 3, Clifton Road, Brighton (M).

Pettey, T., M.D. 93, South Street, Eastbourne (M).

Pierson, H. 57, Castle Hill, Hertford (A)..

Plomer, E, Rohais, 107, Enys Road, Eastbourne (M).

Plomer, Mrs.,, ,,,,,,,,,, (A).

Poole, H. H. 16, Heathcote Street, W.C. (M).

Pratt, Charles. Norfolk Villa, Eastbourne (M).

Probart, Miss M. E. Barrington,, (M).

Ragg, R. S., B.A. The School House, Reigate (M).

Rawling, W. The Nest, Eastbourne (A).

Reid, Capt. Savile G., R.E. The Elms, Yalding (M).

Roberts, Miss M. 57, Peartield Road, Forest Hill, S.E. (M).

Rodda, C. T. Wood Bank, Grove Road, Eastbourne (M).

Rodda, Miss Bessie.,,,,,,,,,,, (M).

Russell, Hon. F. G. Rollo (I'.R. Met. Soc). Dunrozel, Haslemere (M).

Salmon, C. E., F.L.S. Pilgrims' Way, Reigate (M).

Sunson, Miss L. B. 20, Bedford Grove, Eastbourne (M).

Soames, Rev. H. A., M.A., F.L.S., etc. Lyncroft, Otford (Ref.).

Sparkes, H. 1, Hartfield Road, Eastbourne (Loc. Sec, l'JOti).

Sparks, Miss M. ,,,,,,,, ,, (M)

Sparrow, Wm. 16, Pevensey Road, Eastbourne (M).

Starling. E. A., M.D. Chillingworth House, Tunbridge Wells (M).

Stebbing, W. P. D., F.G.S., F.S.A. 8, Playfair Mansions, S.W. (M).

St. George, Loftus. 2, Upperton Gardens, Eastbourne (M).

Storr, Raynor. Hindhead (M).

Swanton, E. W. (F. Myc. Soc). Rrockton, Hasleinere (Museum Sec.)

Tapstield, Miss C. 27, High Street, Maidstone (M).

Taylor, G. R. S. Clears Corner, Reigate (M).

Taylor, Rev. J., M.A., B.D. 81, Marine Parade, Dover (M).

Taylor, J. H. Angles. Royal Parade, Eastbourne (M).

Thomson, A. H. Sutton, Surrey (A).

Tonge, Alfred E., F.E.S. Aincroft, Reigate (M).

Treutler, W. J., M.D., F.E.S. 8, Goldstone Villas, Hove (V.P.).

Treutler, Mrs, (M).

Trustram, Miss Mary. 23, Bedford Grove, Eastbourne (A).

Trustram, Miss 35,,,,,,,,, (A).

Tutt, J. W., F.E.S. 11!), Westcombe Hill, S.K. (Editor).

Turner, H. J., F.E.S. 98, Drakefell Road, S.E. (M).

Turner, J. W., B.A., B.Sc. Lindfield Lodge, Folkestone (M).

Ward, H. Snowden. Hadlow (M).

Webb, Mrs. W. M. Odstock, Hanwell, W. (M).

Webb, Miss A. Sandringham, Eastbourne (A).

Webb-King, Mrs. Somersham, Woolwich (A).

Welsh, Miss. 23, Bedford Grove, Eastbourne (A).

West, W. 15, Horton Place, Bradford (Ref.).

Whelpton, Miss C. M. St. Saviour's House, Eastbonrne (A).

Whelpton, Miss G. L (A).

Williams, J. Aneurin. Wheelside, Hindhead (M).

Window, Miss. Crofts, Haslemere. (M).

Wright, J. C. Holmdene, Eastbourne (M).